£45.00

THE REGIMENTAL HISTORY OF 1ST THE QUEEN'S DRAGOON GUARDS

Both The King's Dragoon Guards and The Queen's Bays commenced their service to the Crown more than 300 years ago. Since 1959, when these two famous Cavalry Regiments joined to become 1st The Queen's Dragoon Guards, they have continued to add lustre to the many deeds of duty and valour inherited from their forbears.

As Colonel-in-Chief for the past 54 years, I am delighted to see the story of these centuries of brave and devoted endeavour set down in order that we can remember with thanksgiving the sacrifices of the past, and so that serving and future members of our regimental family may draw strength and inspiration to uphold these high standards.

Elizabeth R

Colonel-in-Chief

February 1992

THE REGIMENTAL HISTORY OF 1ST THE QUEEN'S DRAGOON GUARDS

MICHAEL MANN

1993

© Michael Mann 1993

Published by 1st The Queen's Dragoon Guards 1993
Produced for the regiment
by Michael Russell (Publishing) Ltd
Wilby Hall, Wilby, Norwich NR16 2JP
Typeset by The Typesetting Bureau
6 Church St, Wimborne, Dorset
Printed and bound in Great Britain
by Biddles Ltd, Guildford and King's Lynn

ISBN 0 85955 189 X

Contents

Acknowledgements

It would not have been possible to publish this history without the generosity of nine members of the regiment to whom all KDG, Bays and QDG, past and serving, owe a tremendous debt of gratitude: the late Brigadier A. W. A. Llewellen Palmer, DSO, MC, Colonel E. K. Savill, DSO, E. Vestey, Brigadier C. Armitage, DSO, MC, the late R. G Hollies Smith, the Hon. C. T. Law, General Sir Jack Harman, GCB, OBE, MC, D. Baird, and Brigadier G. N. Powell.

Then I must thank my fellow Regimental Trustees for all the assistance, encouragement and advice that they have so generously afforded over so many years. I must particularly acknowledge my debt for all the cheerful tolerance, patience and help given to me by Major K. D. McMillan at Home Headquarters, and G. S. Gill, our Museum Curator. Lieutenant Colonel T. J. D. Holmes has frequently perused the archives of the Public Record Office on my behalf, and has written and spoken to numerous members of the regiment, whom he knew to have knowledge of some event; his assistance has been outstanding. Most of the appendices are the result of his painstaking research. Mrs S. Carter, the mother of a member of the regiment, has drawn many of the maps, a time-consuming task that requires great accuracy, for which I am extremely grateful. Mr Burley, DCM, introduced me to Professor Montemaggi of Ravenna, who has most generously allowed me to draw on his knowledge of the Gothic Line battles. Mrs Evelyn Tiarks, now in her nineties, has opened a window on another side of regimental life in the past, for which I am most grateful.

The individual members of the regiment, past and present, who have contributed advice, documents, memories and photographs, have been too numerous for me to mention individually, but their willing help has been invaluable, and without it this history could not have been written. I have tried to acknowledge that help in the appropriate passage. The accuracy of each chapter up to the commencement of the Second World War was checked by that fount of regimental knowledge, and my lifelong friend, the late R. G. Hollies Smith. The later chapters have each been checked by at least four members of the regiment, whose knowledge of the particular period was outstanding. Nevertheless any errors, omissions or failures of emphasis and balance are entirely my responsibility. Inevitably, some with personal experience, on reading my account of an event, will wish that it had been described somewhat differently; but that is the lot of every author!

Ever since I joined the King's Dragoon Guards as a subaltern it has been my ambition to write this history of the family of which I am a proud member, as were my father-in-law, my two brothers-in-law, and in which my son was killed. Only those who have experienced the bonds of comradeship, affection and support that is given by such a close regimental family will understand my continuing debt of gratitude and loyalty.

MICHAEL MANN

1

Raising of the Regiments
1685-1688

On 6 February 1685 Charles II died and was succeeded by his brother, who became James II. James was a soldier of some experience and had served under the French Marshal Turenne in four campaigns, and with the Spaniards in two more. In later times the Duke of Wellington spoke of his military abilities with admiration. James wanted to increase the Standing Army, whose cavalry in Charles II's reign consisted only of two regiments of horse and one of dragoons. With memories of the Civil War and the Commonwealth, Parliament was very reluctant to sanction both the expense and the potential threat that it felt was posed by a Standing Army, and especially one that owed its allegiance to the monarch. When in June 1685 the Duke of Monmouth, the illegitimate son of Charles II and nephew of James, landed at Lyme Regis and raised the standard of rebellion, it gave the King the excuse that he needed, and at once he used the opportunity to raise further regiments of both cavalry and infantry. There were to be six new regiments of horse and two of dragoons.

The first and senior of the new regiments was given the title 'The Queen's Regiment of Horse' [KDG]. In those days regiments were the property of their colonel, and by a commission dated 6 June 1685 James gave Sir John Lanier the task of raising The Queen's Regiment of Horse. It was to consist of nine troops, each made up of a captain, lieutenant, cornet, quartermaster, three corporals, two trumpeters and sixty troopers. The men were recruited from 'the best yeomen or best serving men, having active and nimble bodies joyned with good spirit and ripe understandings'.[1] The regiment was soon recruited and was at first stationed near London, but was almost immediately sent to the West Country.

The colonelcy of the second of the new regiments was given to Henry Mordaunt, second Earl of Peterborough, by a commission dated 20 June 1685. It was formed by bringing together four troops of horse – one raised by Sir Michael Winkworth around Wakefield and Pontefract, a second by Sir John Talbot at Hounslow, a third by John Lloyd at Edgware, and a fourth by Lord Ailesbury in London – and was named the 3rd Regiment of Horse [Bays]. In both regiments each troop had a quartermaster, whose duties corresponded to those of the present-day squadron sergeant major and squadron quartermaster sergeant. They were appointed on a warrant from the colonel, and were therefore known as warrant officers. The quartermasters ranked with the sergeants of other branches of the Army, and the only other non-commissioned rank in the horse was the corporal, a distinction which is maintained to this day by the Household Cavalry.[7] The three troops of the 3rd Horse raised in and around London were despatched at once to the West Country to the scene of Monmouth's rebellion.

Sir John Lanier, the only commoner among the six new colonels of horse, was an experienced soldier. In 1672 he was a captain in Sir Henry Jones's regiment of light horse in France, and the following year was wounded in the face, losing an eye. From 1674 to 1678 he was a lieutenant colonel of Monmouth's Horse in the French service, when he came back to England and served as a brigadier in the expeditionary force sent to Flanders for a projected war with France. For six years from 1679 he was the Governor of Jersey, which he ruled with an iron hand until he took over the colonelcy of the new Queen's Regiment of Horse. The Earl of Peterborough, colonel of the 3rd Regiment of Horse, had taken an active part in the Civil War, and had had his horse killed under him at the Battle of Newbury. With the immediate alarm over Monmouth's rebellion subsiding, by the end of July the Earl of Peterborough had raised an additional troop and Sir John Egerton another troop to increase the strength of the 3rd Regiment of Horse, now to be generally known as Peterborough's Horse, to six troops. Peterborough's Horse was soon to be augmented to nine troops, and The Queen's Regiment of Horse and Peterborough's Horse were the only two newly raised regiments of horse to have an establishment of nine troops each.

We are fortunate in knowing the names of all the officers who formed The Queen's Regiment of Horse, many of whom, like their colonel, had previous military experience. The lieutenant colonel, William Legge, had served in the Blues, and Major Boade came from the Holland Regiment.[2] The nine troop captains and their officers were (to use the spellings of the original source): 1st Troop, Colonel Sir John Lanier, Lieutenant Samuel English; 2nd Troop, Lieutenant Colonel William Legge, Lieutenant Harrison; (Major John Boade, or Board, without a troop); 3rd Troop, Charles Nedby, Lieutenant Sir Thomas Bludworth; 4th Troop, James Fortrey, Lieutenant Henry Moore; 5th Troop, Lewis Billingsley, Lieutenant William Law; 6th Troop, John Staples, Lieutenant George Neares; 7th Troop, George Hastings, Lieutenant Edward St Lo; 8th Troop, Lord Charles Hamilton, Lieutenant Stedholme; 9th Troop, Henry Lumley, Lieutenant James Boucher. *adjutant* Thomas Freckleton; *chaplain* Thomas Morer (or Moore); *quartermaster* Thomas Watts; *chirugeon* Alexander Rubin.

	TROOP CORNETS	TROOP QUARTERMASTERS
1st Troop	La Case	Robert Stapleton
2nd Troop	Boyle	Thomas Smith
3rd Troop	Strother	Francis Nedby
4th Troop	Killigrew	Edward St Barbe
5th Troop	Thomas Milward	Christopher Billingsley
6th Troop	William Downing	Nathaniel Green
7th Troop	Philip Darcy (or Davy)	Thomas Beston
8th Troop	Edward Yarborough	Ralph Gardner
9th Troop	Thomas Beaumont	Fry Vicaridge

Captain Charles Nedby had served since 1676 in Monmouth's Horse in both France and England, and had just previously been a captain of the Royal Dragoons in Tangier.[3] The present Regimental Museum at Cardiff has the original commission of Captain James Fortrey, signed by James II on 8 June 1685. Fortrey had served in the Duke of York's Regiment since 1680. John Staples had for many years served in the Life Guards. Robert Stapleton, the quartermaster, raises an interesting point, for the records of the Royal Hospital at Chelsea show that he was appointed there as second major in 1707 on the recommendation of Prince George of Denmark, who put forward his claim on the grounds that he had been quartermaster to The Queen's Regiment of Horse at the Battle of Sedgemoor.[4] However, although both The Queen's Regiment of Horse and the 3rd Regiment of Horse had marched for the West Country, it appears that they did not arrive in time to participate in the Battle of Sedgemoor, and the rebellion was suppressed without using the newly raised regiments.

Both The Queen's Regiment of Horse and the 3rd Regiment of Horse were among the first of the new regiments to take to the field, and the first duty of the former was to escort the luckless Monmouth back to London as a prisoner. Peterborough's Horse was set to guarding the prisoners held at Winchester and, with the Royal Artillery, which was encamped at Devizes, it was used to patrol the roads and to bring in any rebels that were found.

Because the Earl of Peterborough's Regiment of Horse was not completed until the end of July 1685, there is no accurate record of the officers commissioned to form the regiment. On 12 July the Secretary-at-War, William Blathwayt, wrote from Whitehall to Peterborough; 'His Majesty commands me to signify his pleasure that you forthwith return an account, in what condition the Regiment under your command is at present. How armed and cloathed, and in what readiness for His Majesty's service.'[5]

The troop captains and lieutenants, so far as they are known, were (again to use the spellings of the original source): *troop captains* Colonel Henry, Earl of Peter-borough; Lieutenant Colonel Sir John Talbott; Major John Chitham (or Chetham); Sir John Edgerton; Lord Thomas Bruce. *Lieutenants* Hugh O'Conner (Captain Lieutenant); Robert Warner; John Chetham; John Lloyd. *Adjutant* William Poston. *Chaplain* Michael Poulton. *Chirurgeon* Thorowgood Meautys.   John Talbot had held a commission in the Foot Guards in 1661. John Chitham had served as a gentleman private in the Life Guards and in Lord Duras's Horse in France as a corporal in 1673. He was a lieutenant in Peterborough's Horse in 1678, and a brigadier in the Queen's Life Guard from 1678 to 1685. Lord Bruce became the Earl of Ailesbury in October 1685. Hugh O'Conner was an Irish Roman Catholic who had lost his inheritance on account of the Act of Settlement. He served as a captain in Sir George Hamilton's Foot in France, then in Thomas Dongan's Irish Foot and finally in the French Army from 1678 to 1685.[3]

In August 1686 Richard Cannon gives a list of the then serving officers: Colonel The Earl of Peterborough; Lieutenant Colonel Sir John Talbott; Major John Chitham; Captains John Lloyd, Henry O'Conner, Sir Michael Wentworth, Henry Lawson; Lieutenants William Barlow, William Scott, Ferdinand Kelly, Gilbert Talbot, Walters Mildmay, John Chitham; Cornets Samuel Yatches, Thomas Lloyd, Francis Lennard, Oratio Walpool, Francis Norris, William Bell; William Poston, adjutant; Michael Poulton, chaplain; Thoroughgood Meautys, chirurgeon.

The Queen's Regiment of Horse was soon brought back to London, and on 16 July 1685 each troop was ordered to be reduced by ten men. On 20 July the regiment furnished an escort for Princess Anne, who in later years was to become Queen Anne, when she moved to Sevenoaks, and it provided her guard for the period of her stay there. Later in July seven troops of the regiment were ordered to quarters in Hounslow, where on 22 August James II reviewed ten battalions of infantry and twenty squadrons of cavalry, including The Queen's Regiment of Horse on Hounslow Heath. Three days later, on 25 August, the seven troops were marched into quarters at Winchester, Ilchester and Blandford. By the late summer of 1685 Peterborough's Horse was quartered at Battersea and in the East End of London, where it was called upon to provide escorts for various members of the Royal Family and to guard convoys of specie on their way to Portsmouth as pay for the Navy.

At the end of July an order further reduced the establishment of each troop of horse to forty men. At this period regiments of horse wore crimson coats, in contrast to the red cloth of the dragoons. The coats were long and full-skirted, with deep cuffs and without collars, and they were decorated with a large number of buttons down the front, on the cuffs and on the pocket flaps as well as along slits at the back and side. The facings of The Queen's Regiment of Horse were the yellow of the Stuart livery, whereas those of Peterborough's Horse were of red, though this was changed in 1694 to a peculiarly deep shade of buff or drab. The troopers as well as the officers wore sashes of red, and their cloaks were made of red cloth. Both regiments wore cuirasses and back plates until 1687 or 1688, when these were returned to store, although the officers appear to have retained theirs. Skull caps of iron, known as 'potts', were also issued to be worn in action under a broad-brimmed black hat, turned up on one side and worn with a white feather. The officers' hats were additionally decorated with silver lace, and their gauntlet gloves had silver fringes. When not in action the iron 'potts' or 'skulls' were often carried on the saddle bows. The loose supple boot of the dragoon was stiffened into the jackboot of the horse, which had a wide and rigid top, designed to prevent the knees being crushed in close mounted action; these jackboots of the horse were known as 'gambados'. Other ranks may also have been issued with an undress, or fatigue coat, of plain grey cloth.

The *London Gazette* of 30 June and 4 July 1687 describes the coats of the quarter-masters of The Queen's Regiment of Horse as being decorated with large plated buttons, the sleeves of which were 'faced with silver tissue', and the corporals were to have coats of superior cloth to those of the troopers, although hats, cloaks, housings and holster caps were to be of one standard pattern. A clothing list of 1696 gives the hats as being edged with silver, and the men were issued with a cloth waistcoat, a pair of buff gloves, embroidered housings and holster caps, and a cartridge box in addition.[6]

The arms of the horse consisted of a sword carried in a broad shoulder belt, a pair of pistols with 14-inch barrels, and a carbine, also carried by a shoulder belt. The sword and carbine belts crossed on the chest were a distinctive mark of the horse.[6, 8]

In the Royal Library at Windsor Castle there is a manuscript book containing watercolour drawings of the standards and colours of all the regiments of 1685, which were drawn between October 1685 and May 1686. The Blues and The Queen's Regiment of Horse were the first regiments of horse to be granted the privilege of kettledrums and banners, and the livery of the kettledrummer and the eighteen trumpeters was to be the royal livery of crimson, blue and gold. The Queen herself paid for the livery of the trumpeters, and each suit cost her £36. 12. 2d. A Treasury Paper of 1692 states that the cost of the kettledrums and drum banners, together with the liveries of the kettledrummers and the trumpeters, amounted to £612. 8s. 2¼d. The nine standards, one for each troop, were provided for The Queen's Regiment of Horse by the Queen herself, each costing £40. 6. 8d. The cost of the standards, drum banners and the trumpeters' livery borne by the Queen came to a total of £1,058.12.5d.[9] The normal practice was for the cost to be borne by the colonel, who was granted an annual clothing allowance to cover the expense. The Windsor book gives a colour illustration of three of the standards of the 2nd [KDG] and 3rd [Bays] Regiments of Horse: the troops of the colonel and lieutenant colonel, and a representative drawing for the other seven troops. The three standards are roughly square, with a fringe, and are identical. Although no sizes are given, they were probably about 2 ft. 6 in. by 2 ft. 3 in. Those of The Queen's Regiment of Horse are of yellow silk damask with a fringe of gold, probably intermixed with yellow. In the centre there is an embroidered crown in natural colours over a monogram in gold thread of M B R (Maria Beatrix Regina, Queen Mary of Modena), interlaced and reversed. The Earl of Peterborough's Regiment had standards of white silk damask, with a fringe of silver and white, but with no monogram or other embroidery.

On 18 August 1685 James II wrote to William of Orange,

> I design to have a rendezvous of most of my new Horse and Foot on Saturday next at Hounslow before I send them to their several quarters; those I have seen already I am very satisfied with, and hope I shall be with the rest...Of Horse there were twenty squadrons and one of Grenadiers on horseback, and of Dragoons, and really the new troupes of both sorts were in very good order, and the Horse very well mounted. I was glad that the Mareschal d'Humières saw them for several reasons.

This camp on Hounslow Heath became a regular feature. The King used it as a training ground for his new troops, with the additional consideration that this concentration of soldiers would be near enough to London to act as an overawing influence on its inhabitants and on Parliament. In November 1685 James asked Parliament for £1,400,000 to enable the new Standing Army to take over the role formerly exercised by the militia. Parliament was opposed to this idea and, fearful that James was filling the Army with Roman Catholic officers, would only vote £700,000, whereupon the King prorogued Parliament and never recalled it. James had concentrated some 13,000 to 16,000 troops at the Hounslow camp, which now became a permanent feature of London life. Each troop erected its own 'baraques or hutts', and a military hospital was established with its own matron and female servants. Londoners, especially the women, flocked to watch the troops in their red coats, and the camp soon became a place of entertainment and amusement, with the lines crowded with a varied throng from all classes, watching the mock battles and sieges, the troops drilling – in regiments or in brigades – or rehearsing taking their positions in line of battle. It was the first time that a British Regular Army had been so trained, and it was not to happen again until Aldershot and the Curragh were established after the Crimea. The King often visited the camp, and had a tent and a chapel erected at the camp for his own personal use. The concentration of troops at Hounslow gave rise to a number of rumours: the *London Gazette* of 17 June 1686 said:

> Notwithstanding the many false and scandalous reports that have been raised and spread abroad by ill men concerning the great sickness and mortality of H. M. Forces encamped on Hounslow Heath, it appears upon enquiry, that the whole number of sick and lame since the time of Encamping hath been but 138, of which several came so hither, that only two have died, and that the rest are either cured and returned to their Colours, or in a way of recovery.

On 30 May 1686 The Queen's Regiment of Horse marched into quarters at Staines, Colnbrook and Egham, remaining there until 24 June before rejoining the camp on Hounslow Heath. On 30 June both The Queen's Regiment of Horse and Peterborough's Horse were reviewed on Hounslow Heath by the King. In the Royal Library at Windsor there is an engraved plan of the layout of the Hounslow camp on 19 July 1686, giving the position of the tents and huts of each regiment, and the order in which the individual regiments formed the line of battle for the King's inspection.

The Queen's Regiment of Horse left Hounslow on 5 August 1686 and was quartered at Canterbury and Maidstone. Peterborough's Horse had been moved in October the previous year to the Oxford area, to be quartered at Woodstock and Oxford. The move was occasioned by an undergraduates' demonstration against the Master and Fellows of University College, who had turned the place, according to Macaulay, into 'a Roman Catholic seminary'. A travelling company of actors had performed a play – under the windows of the Master's lodgings – in which one of the characters had the same name as the Master. They then altered the lines to propose that this character, an old hypocrite, should be hanged for changing his religion. The King was sufficiently provoked to order Peterborough's Horse into the city, in concern about the temper of the university. Whilst at Oxford the regiment was reviewed by General Sir John Lanier, the colonel of The Queen's Regiment of Horse, who with Sir John Fenwick had been appointed by the King to visit the quarters of the various regiments and to report on their efficiency. It returned briefly to Hounslow in February 1686, then went back to the Oxford area for two months before being brought down again to the Hounslow camp. In the autumn of 1686 the regiment went on a route march to Liverpool, collecting revenue monies on the way, and then returned to Oxford. It provided a guard of honour for James II at Salisbury, and moved between Oxford and Hounslow throughout 1687.

At this period troopers of horse were paid 2/6d a day, out of which they were expected to maintain their horses. Among the officers, a cornet received 14/-, which included allowances for two horses at 1/- each, and two servants at 2/6d each; a captain of horse got 10/- and 11/6d in allowances; colonels and lieutenant colonels received a captain's pay and allowances, plus 19/6d and 9/- respectively. Majors with no troop got 20/- and 7/6d in allowances.

The Queen's Regiment of Horse returned to Hounslow Heath in June 1687 for two months, but in August it was on the march again to Reading, from where it accompanied James II on a tour of the kingdom. It also provided a guard and escort to the Queen while she was staying at Bath. During this period, with a growing feeling throughout the Army in favour of the Protestant religion, James started to replace those whom he considered to be unreliable Protestants with Roman Catholics on whose loyalty he could depend. Lord Peterborough was persuaded to become a Catholic, and Catholics too superseded a number of Protestant officers of Peterborough's Horse. The lieutenant colonel, Sir John Talbot, was replaced by Major Chitham, whose place was taken by Captain O'Conner; Lieutenant Scott was removed in favour of Lieutenant Riley; and among the cornets, Wiltshire replaced Lennard, George Carpenter replaced Yatches, and Count Gavemberti relaced Norris. As a result of these changes, James looked on Peterborough's Horse as a trustworthy regiment and brought them back to Hounslow in the summer of 1688. In 1688, too, The Queen's Regiment of Horse also returned to the capital and was stationed at Colnbrook, Chertsey and Byfleet.

On 25 September 1688 Samuel English was promoted to captain from lieutenant in The Queen's Regiment of Horse. On 3 December an advertisement appeared in the *London Gazette*:

> Whereas Captain George Hastings, of H. M.'s The Queen's Regiment of Horse, commanded by the Hon Sir John Lanier, having barbarously murdered Captain Samuel English in his bed, at Henley upon Thames on 1 December, being of the same Regiment, and immediately made his escape with one, Samuel Reynor, Corporal in his own Troop, accessory to the said murder, the Captain being upon a black horse and the Corporal upon a large black gelding with one eye: whoever shall keep and secure the said Captain and Corporal, and give notice to Mr Freckleton in Tuttle Street, Westminster, shall have a considerable reward.

Freckleton was the regiment's agent, and his son, Thomas, was adjutant in 1685. Nothing more is known about the outcome of the crime or the reason behind it, or, indeed, of the fate of the culprits.

On 10 June 1688 William Blathwayt wrote to Peterborough from Whitehall:

My Lord,

It having pleased Almighty God, about 10 o'clock of this morning, to bless His Majesty and His Royal Consort the Queen with the birth of a son, and His Majesty's dominions with a Prince, His Majesty has commanded me to signify the same to you, that upon notice given of it by you to the several Troops of the Regiment under your command, they may join in the public thanksgiving to be observed in these parts on the 1st July next, and give such other demonstrations of their joy, for so great a blessing, as you shall judge fit on this occasion.

This baby was to become known as the Chevalier de St George, but the rejoicing was to be short-lived. Storm clouds were gathering on the horizon, as discontent grew among the populace and within the Army over James's attempts to strengthen Roman Catholic influence.

SOURCES

1 Earl of Ilchester's MSS.
2 C. T. Atkinson, '250 Years Ago', *Journal of the Society for Army Historical Research*, vol. 14.
3 John Childs, 'Nobles, Gentlemen and the Profession of Arms in Restoration Britain, 1660-1688, *SAHR Special Publication 13*.
4 G. T. Dunne, *The Royal Hospital*, 1950.
5 F. Whyte and A. H. Atteridge, *The Queen's Bays, 1685-1929*, 1930.
6 C. C. P. Lawson, *A History of the Uniforms of the British Army*, vol. 1, 1940.
7 Major N. P. Dawnay, 'The Badges of Warrant and Non-Commissioned Ranks in the British Army', *Journal of the Society for Army Historical Research*, vol. 27, supplement pp. 8-13.
8 Revd Percy Sumner, 'Uniforms and Equipment of Cavalry Regiments from 1685 to 1811', *Journal of the Society for Army Historical Research*, vol. 13, p. 82.
9 War Office Records.

2

The Glorious Revolution, Ireland
1689-1690

The position of the Army was ambiguous. Parliament was reluctant to make any move that might compromise its own position, and whilst neither common law nor statute actually forbade the existence of a Standing Army, it remained outside the control of Parliament and under the command of the King. James used the Army to enforce his rule, and because he depended upon its loyalty, he continued to replace officers of known Protestant belief with those who were Catholics, on whose loyalty he could depend. This stirred up fears of autocratic rule and Catholic supremacy. Quarrels broke out between Protestant and Catholic soldiers at the Hounslow camp, and these were aggravated by agitators distributing tracts calling on the Army to defend liberty and the Protestant religion.

When James imprisoned seven bishops in the Tower for refusing to read out a declaration suspending the penal laws against Catholics, their guards drank their health. On the news of their acquittal reaching the Hounslow camp, the soldiers cheered. Whereupon James broke up the camp and dispersed the regiments.

The actions of the King and his attempts to increase the power of the Catholics led a number of prominent officers secretly to side with William, Prince of Orange. When he landed at Torbay in November 1688, The Queen's Regiment of Horse [KDG] was ordered to march to Warminster via Newbury and Marlborough to join the advance guard of James's army. Peterborough's Horse [Bays] was moved to Salisbury, where the main body of the Army was being concentrated.

On 22 September the King, in anticipation of William's move, had raised the strength of each troop of horse and the colonels had been told that 'all the Troops of our Regiment of Horse under your command are to be forthwith filled up and recruited with ten troopers with horses fitt for service in each troop, more than the present complement'. Before the two regiments marched, the rank and file were ordered to leave their breastplates and 'potts' with the civil authorities in each place where the various troops were billeted. The officers were allowed to retain their breastplates, if they so wished. An officer of the Ordnance then took charge of the breastplates and 'potts', which were removed to the Tower.

James had intended to take command of the Army at Salisbury, but when he saw the extent of the daily desertions to William, he returned to London and ordered the troops to take up a position beyond the Thames. Both The Queen's Regiment of Horse, under Sir John Lanier, and Peterborough's Horse, under the Earl of Peterborough, remained loyal to James so long as he was king, with The Queen's Horse forming the rearguard of James's army and occupying the most advanced post up to the end.

Any thought of resistance to William, however, vanished with the wholesale accession of officers to his cause, and James fled the country. He seemed unable to realise that an army of Englishmen would share the general anti-papal sentiments of the nation, and would not be willing to follow him without question. 'Nothing is more pathetic in the career of this grave, conscientious, industrious, obstinate and sadly imperceptive King than his remark to Dartmouth in 1688, "This I may say, never any Prince took more care of his sea and land men as I have done, and been so very ill repayed by them." '[1]

Princess Anne (later Queen Anne) was on the way to be with her husband, Prince George of Denmark, who had joined William. On nearing Oxford on 15 December, she was met by Sir John Lanier at the head of The Queen's Regiment of Horse. The cavalry led the procession under the Earl of Northampton, and the princess was escorted by a troop of gentlemen headed by the Bishop of London, clad in a purple coat and military attire, armed with pistols and a drawn sword. William then ordered The Queen's Horse to Cambridge, and later extended their quarters to Royston and Newmarket.

Peterborough's Horse was not so easily accepted by the new King. The Earl of Peterborough was removed from command by an order dated 31 December 1688, and this was followed by his impeachment — an action which was subsequently dropped — for having become 'reconciled to the Church of Rome'. The new colonel was to be a Protestant, the Hon. Edward Villiers, Lord Grandison's heir. As lieutenant colonel of the Life Guards he had commanded the cavalry brigade at Sedgemoor and had won Wiiliam's confidence by his cool handling of the royal guard at Whitehall as James fled from London. The regiment was immediately strengthened by posting to it two troops of Lord Brandon's Horse and fifty troopers from Colonel Slingsby's regiment.

On 15 March 1689 The Queen's Regiment of Horse was moved to Huntingdon, and then on to Lincolnshire, where it became involved in suppressing a mutiny. The Parliament at Westminster, without consulting the Scottish Assembly, transferred the allegiance of the Scottish troops from James to William. On 8 March the Royal Scots were suddenly ordered to proceed to Holland, an order that was received with much discontent. They marched as far as Ipswich and there, under the influence of some of their Jacobite officers, they mutinied, seized four cannon and marched north. William at once sent General De Ginkel with Sir John Lanier and The Queen's Horse, together with Colonel Langston's Horse [Princess Anne of Denmark's Horse, disbanded 1692] and some Dutch troops, after the mutineers, having declared them to be rebels under a proclamation. De Ginkel overtook the Scots at Sleaford in Lincolnshire, where both sides drew up for battle.

The Scots formed a semicircular line with a strong hedge on one flank and a stretch of water on the other, and their four cannon in front. The Queen's Horse outflanked the Scots line and, wheeling, made to charge their rear. At this the Scots asked for a parley, which only persuaded De Ginkel to promise that he would use his influence to try to secure a general pardon, whereupon the Scots 'yielded themselves to the King's mercy'.

The Scots, who numbered about 500 men, including some twenty officers, were taken under guard to Folkingham, where they were subsequently pardoned but nevertheless sent off to Holland. However, this was not the end of the affair. About forty Scots had gone ahead of the main body to Sleaford, where The Queen's Horse found them at the Angel Inn. 'They had been drinking so hard on free cost, that, more resolute than their main body, though with less prudence, they provoked the Dragoons to fire upon them, killing a sergeant and a private soldier, mortally wounding two others.'[2]

The Queen's Horse returned from this affair to Newmarket and Royston. On 9 April they and Villiers' Horse were ordered to march to Scotland. The Duke of Gordon had refused to accept the authority of William and Mary, and was holding Edinburgh Castle for King James. Moreover Graham of Claverhouse had raised some of the Highland clans, notably Macdonald of Keppoch, who marched south against William's forces and defeated General Mackay at Killiecrankie. Villiers' Horse never reached Scotland, but was halted at Ripon, where it was sent back to the south. The Queen's Horse remained under Sir John Lainier investing Edinburgh Castle with a body of foot, and the castle surrendered on 13 June. After the defeat at Killiecrankie, The Queen's Horse joined what was left of General Mackay's forces. In the middle of August, it and Berkeley's Dragoons [4th Hussars] attacked a party of the Scottish rebels near Forfar and drove them back towards Atholl; but the cavalry was not able to follow up this success owing to the nature of the ground.

The clans, dispirited, started to disperse to their homes. This enabled the King to transfer his forces to meet the threat posed by James's landing in Ireland, where he had raised a considerable army. Villiers' Horse was the first to make ready for Ireland, embarking at Hoylake in Cheshire on 2 September. It was delayed by bad weather, and Colonel Trelawney wrote to the Earl of Shrewsbury that as 'it has blown a storm these two days it will be difficult to ship the horses'. Some 100 troop horses were indeed lost on the voyage across the Irish Sea. In October The Queen's Horse embarked in Scotland and landed at Belfast on the 13th. Soon after the arrival of Villiers' Horse, its colonel was promoted to the rank of brigadier general, and placed in command of all the cavalry.

The winter of 1689/90 was a miserable one for both regiments. The men suffered from the hard conditions and resulting sickness, as well as from the disturbed state of the country, with the Irish peasantry carrying on an intermittent guerrilla war-fare against the troops. Because the country had been laid waste, the cavalry had to ride for some distance in order to forage, and the Roman Catholic peasants formed themselves into bands of reparees, numbering up to two or three hundred men, who would lie in wait for isolated or small bodies of troops and ambush them, stealing horses and supplies and killing stragglers. Whenever they were discovered and attacked, they immediately dispersed, hiding in the bogs and woods, which made it difficult for the horse to come to grips with them.

Villiers' Horse, stationed at Carlingford, was given the responsibility of keeping open the Pass of Newry. William's army was commanded by Field Marshal the Duke of Schomberg, a vigorous eighty-year-old, who on Christmas Day asked Villiers and Sir John Lanier whether it was possible to force the River Shannon in the vicinity of Athlone, where they would be opposed by a large force of James's troops under the command of Patrick Sarsfield. Inefficiency and delay, however, lost them the opportunity, and Villiers informed Schomberg that 'it was not to the King's service to take a post on the other side of the Shannon'.[3]

In February 1690 Sir John Lanier took The Queen's Horse from Newry as part of a reconnaissance towards Dundalk, which was held by the enemy. They entered and burnt the west side of Dundalk, capturing Bedloe's Castle, its garrison and 1,500 head of cattle before returning to Newry. On 3 March Schomberg wrote to the King to say that he considered Villiers to be 'careless of his Regiment. The cavalry must be put in better condition, and they have not the men fitted to do it. Brigadier Villiers is but little qualified to command, though very active in looking over his own interests.' Villiers' Horse was engaged in the early part of May in the siege of Claremont, which surrendered on the 12th.

William arrived in Ireland during June and proceeded to the camp at Dundalk. At the same time Villiers' Horse was reinforced by seventy men and horses, bringing up its strength to six troops of fifty men each, and the regiment moved into the Dundalk camp. The Queen's Horse had a strength of nine troops of fifty men each. On 17 June William reviewed his horse, which consisted of the Life Guards, the Blues, Lanier's, Villiers', Coy's [5 DG], Byerley's [6 DG], Schomberg's [7 DG], Russell's, Langston's and Wolseley's Horse [all disbanded later], together with an

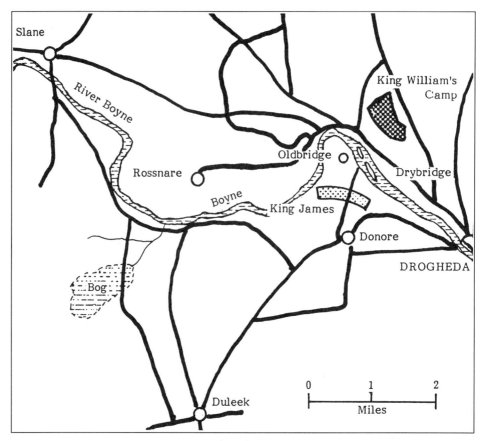

Battle of the Boyne

ad hoc troop commanded by Harbord. William's army now considerably outnumbered that of James, being estimated at somewhere between 36,000 and 40,000 men, whereas although James had probably 45,000 throughout Ireland, he could only concentrate some 20,000 to 26,000, and these he gathered in a strong position overlooking the River Boyne. James knew that he ought to retire into the mountains and let William wear out his troops in chasing him, but he also knew that his Irish soldiers would not hold together under such defensive tactics, and that they would desert in large numbers. Indeed William counted on the impulsive and undisciplined nature of the Irish soldiery as much as he did upon the steadiness and coolness of his British, Dutch and Huguenot veterans.

William reconnoitred the enemy's position on the day before the battle, and was rash enough to breakfast on the banks of the Boyne opposite the spot where the Irish were also reconnoitring. The Irish swiftly brought up two cannon, and the first shot killed Prince George of Hesse's horse. The second 6 lb shot grazed William on the shoulder, while three of his escort were killed. It proved to be only a flesh wound, although it bled profusely; however, William remained in the saddle for nineteen hours, and determined to attack the following day. Against Schomberg's advice, he decided on a frontal assault by the main body of his infantry in order to force a passage across the Boyne, crossing by a ford near the Old Bridge, while the bulk of the cavalry, with some infantry, crossed the river by the Slane Bridge on the right flank and attacked the enemy's left flank. On his left flank the remainder of the horse were to force a crossing of the river and place themselves between James's camp and Drogheda. The Boyne was tidal from its mouth at Drogheda as far as Old Bridge, and in July was only fordable at Old Bridge by ten o'clock in the morning.

The Queen's Horse and Villiers' Horse advanced with the rest of the army from Ardee to the banks of the River Boyne, reaching there on 30 June and encamping for the night. The rank and file of The Queen's Horse, under command of Sir John Lanier, numbered 360 men, Villiers' Horse 248 men. The village of Old Bridge lay in a bend of the river on high ground dominating the ford below, and it was here on 1 July 1690 that James concentrated his main defensive position, with his Irish infantry behind earthworks down by the ford. Halfway up the slope between the ford and Old Bridge were the French infantry and cavalry under Lauzun. Because a frontal assault could prove so costly, William, at Schomberg's behest, despatched a strong force of cavalry to reconnoitre towards his right flank in order to try to cross the Boyne at the Slane Bridge, some eight miles to the west, and so turn James's flank. It was feared, however, that the Irish had destroyed the Slane Bridge. On their way William's troops discovered a nearer ford at Rossnare, only three miles from Old Bridge. It was found to be guarded only by O'Neill's dragoons, who resisted stoutly but were forced to retire, leaving behind about seventy casualties, including O'Neill himself. William sent a reinforcement of two more brigades, bringing up the force on his right to about 10,000 men. Lauzun moved his French contingent to the support of O'Neill's dragoons and to counter the threat to his left flank. James then made the error of further reinforcing this flank with some of his Irish troops, thereby weakening his centre. But neither side was able to come to grips as they were separated by extensive and previously unknown bogs, which proved to be impassable.

It was now ten o'clock in the morning, and only nine battalions of Jacobite infantry and James's right-wing horse and dragoons were left to face William's main attack, some 5,500 men against 15,000. As soon as William learnt at 10.15 a.m. that his right wing was across the Boyne, and seeing that the tide was on the ebb, he ordered the main attack to go in. The assault was led by William's

Dutch Blue Guards, followed by some Huguenot regiments and Hanmer's English brigade. The Dutch, after steadily clearing the Irish from protecting hedge and wall, were then charged twice by the Irish cavalry. They repelled both charges, and Schomberg, crossing the river himself, hurried the Huguenots and Hanmer's brigade, who were still in mid-stream, to their assistance. Tyrconnel, in command of the Jacobite centre, launched cavalry charge after cavalry charge in support of his retreating infantry. He managed to halt Schomberg's advance, but that intrepid old soldier rode forward to encourage his troops and in doing so was shot dead. However, the Jacobites had been forced between half a mile and a mile up the hill away from the ford.

At this point William made his decisive move. At the head of 1,500 horse and dragoons, including The Queen's Horse and Villiers', William led them to his left and managed to force a crossing of the Boyne near Drybridge, despite having to cross a morass. The Inniskilling Dragoons charged and broke a Jacobite dragoon regiment, but were then themselves routed by a counter-charge of Irish horse. William led two regiments of Danish dragoons to their assistance, but they were also caught up in the rout.

The Irish then charged Cutts's regiment of English foot [disbanded], who stood firm and repulsed the attack, thus giving William time to re-form his cavalry. William himself behaved with conspicuous bravery: always in the thick of the action, he had the left heel of his boot shot away and then the cap of his pistol struck by another shot. He now led his left wing to threaten James's right flank.

In the centre Tyrconnel was rallying his infantry to attack down the hill towards the ford. On the order to charge, only the Irish Foot Guards obeyed, the rest of the Irish infantry wavering at the sight of William's cavalry coming in towards their right flank. Again the Irish cavalry charged William, winning enough time for the infantry to re-form on the heights of Donore, taking up positions around the church and hedges. As the infantry of William's centre and the cavalry of his left flank attack pressed their advance, the Irish infantry broke and fled, but once again their complete destruction was prevented by the bravery of the Irish cavalry. A general cavalry skirmish ensued around Platin House, halfway between Donore and Duleek, which continued for about thirty minutes before the Irish cavalry gave way.

The Irish cavalry's brave action enabled the main body of the Jacobite army opposite Rossnare, where both James and Lauzun had spent the day, to retire towards Duleek and Dublin. This retreat turned into a rout, and James forsook his army, leaving Lauzun to do the best he could. The French fell back in good order, but the Irish vanished into the surrounding hills. At Duleek the French covered the panic-stricken Irish as they tried to cross the River Nanny, and William's cavalry were unable to dislodge them. As the French retired from Duleek, William's cavalry pursued them for three miles, but then fell back on the main army.

The casualties on both sides were not severe, the Jacobites losing about 1,500 killed, wounded and taken prisoner, whilst William's losses were only 500 killed and wounded – among them the redoutable old Duke of Schomberg. The main effect of the Battle of the Boyne was to shatter the morale of the Jacobites. A few days after reaching Dublin, James left for France and never returned. The battle is also said to have brought fame to some half a dozen regiments, of which one was specifically named as The Queen's Regiment of Horse, led by the gallant Sir John Lanier, and Villiers' Horse who crossed the Boyne immediately under the King's eye.

Following the battle, both The Queen's Horse and Villiers' bivouacked near Duleek. On 2 July the King, having heard that James had evacuated Dublin, sent the Duke of Ormonde in command of 1,000 cavalry, of which Villers' Horse formed a part, to occupy the Irish capital. On the same day Drogheda surrendered and the rest of the army advanced on Dublin. On the 7th William reviewed The Queen's Horse at Finglass, and then marched on Limerick with a detachment of The Queen's Horse and four troops of Villiers' Horse, leaving two troops of Villiers' and the bulk of The Queen's Horse in Dublin. The Irish had formed a strong camp in front of Limerick, and on 3 August a force of 1,000 men, including a detachment of The Queen's Horse, advanced on the Irish camp and drove three regiments of cavalry and two of Irish infantry out of their entrenchments. A week later, on 9 August, the Irish were forced to retire into the town. That evening Limerick was formally summoned to surrender.

The two troops of Villiers' Horse left in Dublin were ordered to escort the siege guns and ammunition on their way to the King at Limerick. On the 11th they had reached the ruined castle of Ballyneety, seven miles from Limerick. It was a fine summer's day; the guns were parked while the men pitched their tents, let out the horses to graze and rested. In the meantime William had learnt that a body of Irish cavalry, under Major General Partick Sarsfield, had crossed the River Shannon to intercept and destroy the guns. Sir John Lanier was ordered to take his regiment, The Queen's Horse, to meet the guns at Cullen, so as to reinforce the escort. But Patrick Sarsfield was able to surprise the resting gunners and troopers, sabring Lieutenant Ball and the men of Villiers' Horse, even though they put up such resistance as they were able, and massacring several Irish peasants who were bringing food to sell to the troops. Sarsfield then collected the artillery and all the ammunition waggons, set fire to them, and made off. By this time Sir John Lanier was within three miles of Cullen. Hearing the noise of the ammunition exploding, he took The Queen's Horse across country in an attempt to cut off the Irish at

Kilcullen Bridge. Sarsfield, however, managed to make his escape by way of Ath-lone; for this feat James created him Earl of Lucan.

William was determined to take Limerick and on 30 August he attempted to carry the town by storm, only to be repulsed by the strength of the Irish resistance. Conditions for the besiegers were appalling: rain fell continuously, making the boggy and swampy countryside impassable and filling the trenches and encampment with water. With the loss of his siege guns and with the increasing danger of fever decimating his army, William decided to raise the siege, and on 30 August the army departed and the King returned to England.

In September Major General Kirke commanded a force, which included The Queen's Regiment of Horse, sent to the relief of Birr Castle, which was being besieged by some 5,000 Irish. As Kirke drew near, the Irish withdrew and the siege was raised. Brigadier Villiers took a force, which included his own regiment, to lay siege to Kinsale. Villiers prosecuted affairs vigorously, and 'played upon the fort for three days with 8 pieces of cannon of 24 lb, and 2 mortars. One of our English ships of sixty guns unfortunately blew up and not a man escaped. The waters are so far out that the enemy dare not venture to attempt the relief of Kinsale, but have retired towards Limerick.'4 Villiers' efforts were rewarded by the fall of Kinsale on 15 October.

SOURCES

1 R. H. Whitworth, 'James II, the Army and the Huguenots', *Journal of the Society for Army Historical Research*, vol. 63, p. 134.
2 *London Gazette*, no. 2438.
3 State Papers, 1689. Lord Lisburn's Dispatch.
4 State Papers, 1690. Dispatch.
5 *London Gazette*, 1691.

3

Ireland, Aughrim, Flanders, Landen
1691-1694

Both regiments spent the winter of 1690-91 chasing bands of reparees and in foraging. In March 1691 Captain Carpenter of Villiers' Horse [Bays] led a very successful reconnaissance. (Carpenter later became Lord Carpenter and colonel of the 3rd Dragoons.) Then on 15 March Lieutenant Spicer and Cornet Collins of Villiers' heard that the Irish were planning to capture a supply of excellent remount horses newly arrived from England. These two officers were only able to gather about a dozen troopers, as the remainder of the regiment was away on duty, but this small party rode out to meet an enemy numbering at least one hundred. They sighted the Irish at Cappoquin in County Waterford, and charged with such spirit that they drove them into a wood. When another eighteen troopers who had galloped to their assistance arrived, they dismounted, entered the wood and sabred or shot forty of the enemy, capturing a captain and seven men, all of which was effected 'without the loss of one man'.[1]

The English cavalry were not always so successful against the reparees. A letter written by Patrick Sarsfield to Lord Mountcashel in February 1691 stated,

> Our mountaineers take horses from them every day, and if we had a little money to reward them, our cavalry would be very well mounted at the enemy's expense, which would be a double advantage for us, for while accommodating ourselves, there would be as many dismounted men for them. We have already had more than a thousand this winter, and they have brought me thirty seven from Lanier's quarters, of which twenty two were out of his stable.[2]

In February 1691 The Queen's Regiment of Horse [KDG] formed part of a force under General De Ginkel and Sir John Lanier assembled to deal with a concentration of the Irish at Ballymore. They left Streamstown on 26 February, and had advanced only four miles when they came to a pass, barred by palisades and defended by 2,000 Irish. As the English infantry advanced, the Irish broke and fled, pursued by The Queen's Horse towards a hill where the main body of the Irish was formed, near the Moat of Greenonge. The Irish withdrew into the town, which was defended by a trench, but then continued to retreat. The Queen's Horse charged and got amongst the enemy foot while they were still in the town, slaughtering a large number and scattering the remainder into the surrounding terrain. They then chased the Irish cavalry along the causeway towards Athlone with such energy that many of the Irish troopers left their horses, took off their boots, and sought refuge among the bogs and woods alongside the causeway.

On 19 April two troops of Villiers' 3rd Horse, under command of Major James Kirk of the regiment, took part in a raid on Macroom in County Cork, together with 140 dragoons and militia. They killed some twenty Irish and took prisoners, horses and cattle. Kirk was promoted to lieutenant colonel for his drive and initiative in this operation.

In early May a troop of The Queen's Regiment of Horse, together with a detachment of the Queen Dowager's Foot, scattered a party of the enemy near Wyandstown. On 6 June the regiment advanced from their quarters at Mullingar to the siege of Ballymore, which surrendered on 8 June. On 16 June both The Queen's Regiment of Horse and Villiers' Horse took part in the siege of Athlone. Faced with the River Shannon, the cavalry were employed as dismounted infantry in the trenches, where Lieutenant Colonel Kirk, who was beloved by all ranks, was killed by a cannon ball. On the 30th Athlone was taken by storm: the infantry and dismounted troopers forded the Shannon, breast high, overcame every obstacle and captured the town within thirty minutes. The Irish left 1,000 dead on the field, and the spoils included a number of cannon and Major General Maxfield. For this service De Ginkel was created Earl of Athlone.

Twenty of The Queen's Horse, together with ten mounted grenadiers of the Queen Dowager's Regiment of Foot, reconnoitred the Irish camp at Ballinasloe on 4 July. In Clanoult woods they were surrounded by 400 Irish horse, but defended a bridge with great bravery until half their number had been killed. The survivors managed to withdraw. General St Ruth, sent by Louis to replace Lauzun, who had been recalled to France, had believed Athlone to be impregnable. In order to avenge its loss, he decided to give battle in the open, against the advice of the Irish generals, who wanted to concentrate their forces at Limerick. St Ruth chose a strong position along a ridge, some three and a half miles down the road from Ballinasloe to Loughtrea in the county of Galway. His right rested on the hill of Kilcommodon, his left was secured by the Castle of Aughrim; around the castle and along the whole of the front of his position there was an area of marsh and red bog following the course of the River Meldham. Facing this on the opposite side of the river was another, lower, ridge, which St Ruth overlooked. There were only two places where cavalry could cross, the first on St Ruth's right by the Tristaun bridge, the second where the road from Ballinasloe crossed the Meldham by means of a causeway and bridge just below Aughrim Castle.

William's troops crossed the Shannon on 10 July, and on the 11th they were encamped near Ballinasloe. At 6 a.m. on Sunday, 12 July the army left Ballinasloe and crossed the River Suck, closing onto St Ruth's position at Aughrim just before noon. The Queen's Regiment of Horse was in Major General Ruvigny's brigade and started the day on the right wing.

The 3rd Horse on the left wing was in a brigade commanded by its colonel, Brigadier Villiers. St Ruth placed his Irish cavalry on his flanks, leaving only one regiment of horse in reserve. His infantry lined the ridge behind hedges, walls and ditches, which had been reinforced and made into a more or less continuous breastwork.

De Ginkel, with 19,000 English, slightly outnumbered the Jacobite army, but the latter had the advantage of height and a strong prepared position. De Ginkel decided to make diversionary attacks on the Irish right and centre, but to deliver the main stroke on their left against Aughrim. The battle opened at about 3 p.m. with a strong attack against the Tristaun bridge, which was weakly held. The 3rd Horse in Villiers' brigade fiercely engaged the enemy, and soon the ferocity of the battle in this sector caused both sides to bring up reinforcements. The Queen's Horse in Ruvigny's brigade was moved across from the English right, and helped the 3rd Horse to push the enemy back up the hill.

At about 4.30 p.m. the English infantry in the centre crossed the Meldham river and the adjoining bogs to assault St Ruth's centre. At considerable cost they managed to drive back the defending Irish musketeers, but then the English infantry on the left centre pushed too far forward and were charged by the Irish cavalry, who broke them and forced them back into the bogs at the bottom of the hill.

About 6.30 p.m. De Ginkel ordered the English infantry on his right forward against the Irish left at Aughrim. Advancing steadily, they got to within twenty yards of the Irish lines without a shot being fired, then marched through a murderous fusillade to push back the Irish defenders. Once clear of the defences, however, they were also charged by the Irish cavalry who swept them down the hill again. De Ginkel then ordered his right-wing horse forward, and led by the Blues the English cavalry scrambled along the narrow causeway of the Ballinasloe road by Aughrim Castle, only wide enough for them to ride two abreast. They then formed ranks and charged the Jacobite left wing.

At this point the Irish infantry of the centre had rallied and were about to advance, when they were charged by Ruvigny's brigade on the English left, with The Queen's Horse among them. De Ginkel later claimed that this gallant charge, which 'bore down all before it with astonishing impetuousity', decided the fate of the day.[3]

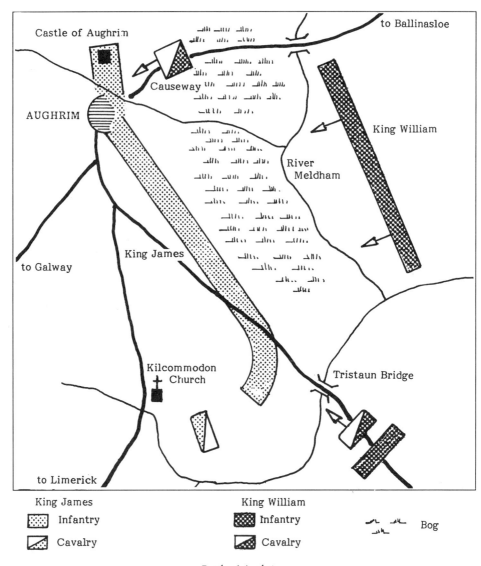

Battle of Aughrim

It was now 8 p.m. As St Ruth was trying to rally his men for a last counter-attack, he was decapitated, either by a chain shot or a cannon ball. His French Guards removed his body and retreated from the field; at the same time the Irish musketeers ran out of ammunition. These two events took the heart out of the Irish resistance, whose cavalry had been superb and whose infantry had fought doggedly. The Jacobite centre gave way and fled. The right continued to resist for another half hour before retreating. The troops holding Aughrim Castle, now completely isolated, surrendered.

The Queen's Horse and the 3rd Horse took up the pursuit with the rest of the English cavalry, who did great execution among the fleeing Irish before a heavy misty rain and oncoming darkness put an end to the slaughter.

The whole of St Ruth's artillery was captured and about 2,000 prisoners were taken, among them two generals, Hamilton and Dorrington. Some 4,000 of the Irish were killed, against English losses of 2-3,000 killed and wounded. The Queen's Horse had twenty-three men killed and five wounded, and lost eleven horses killed and twenty-four wounded. The 3rd Horse had a captain, two lieutenants, a cornet and thirty men killed, and a captain, a lieutenant and twenty-four men wounded. Forty- one of their horses were killed or wounded.

The remnants of the Irish army that escaped from Aughrim took refuge with the garrison at Limerick, where Tyrconnel still hoped for help from France. De Ginkel resumed the advance and on 21 July took Galway. He then secured Portumna, Banaghan and the other crossings of the Shannon before closing up and investing Limerick. In the meantime The Queen's Horse had been detached to Charleville where they had some brushes with the Irish. From within Limerick Tyrconnel attempted to seduce the loyalty of De Ginkel's men:

> We do hereby publish, declare and engage that every trooper, dragoon or soldier who at any time hereafter quits the service of the Prince of Orange and repairs to Limerick, Athlone, Banagher, Sligo or Galway, shall receive the following sums: every such trooper or dragoon, having his horse with him, shall have two pistoles in gold or silver, and every common soldier a pistole in like coin.

No troopers of The Queen's Horse or the 3rd Horse betrayed their allegiance. [4]

During August Tyrconnel died, allegedly of poison, and command of the Irish troops fell on Patrick Sarsfield, now Earl of Lucan. The 3rd Horse had little chance of distinguishing themselves during the siege, being engaged in the trenches and in support of the infantry. The siege ended on 25 September with the signing of the Treaty of Limerick. This contained two articles, one military and the other civil. The military article allowed for the exchange of prisoners, and free passage for any Jacobite troops to go abroad if they so wished. Some 11,000 Irish left the country for France and this led to the establishment of the Irish Brigade in French service, the celebrated 'Wild Geese'. The civil article attempted to secure some degree of religious toleration for the Roman Catholic Irish, but the Protestant Irish Parliament refused to ratify the treaty for six years; and even after ratification many Catholics were victimised and much of their land was confiscated.

In spite of sectarianism and the resulting ruthlessness of the war in Ireland, there were then, as now, redeeming incidents. In 1690 The Queen's Horse was quartered in the village of Glenavy in County Antrim, and in consideration of the kindness shown them by the local people, the officers presented the church with a silver chalice, which is still used. On the lid, which serves for a paten, is inscribed, 'This plate was given to ye Church of Glenavy by the Officers of ye Queen's Regmt. of Horse, commanded by ye Honble. Major Sir John Lanier, in the year 1690. In honorem Ecclesiae Anglicanae.'[5]

Both The Queen's Horse and the 3rd Horse went into winter quarters after Limerick. At Christmas The Queen's Horse received orders to embark for England, and the last division of the regiment landed at Whitehaven on 17 January 1692, marching from there to Northampton. The 3rd Horse remained on the Irish establishment until April 1692. On the 8th it landed at Barnstable and marched for Salisbury, where it was billeted until June, when it moved to Devizes, Chippenham and Shaftesbury. The regiments of horse received the following order intended to fill up the gaps caused by the Irish campaigning:

> Whereas we have thought fitt to add nine Troopers to each troop of our Regiment of Horse under your command. These are to authorise you to raise the said additional number of men with horses fitt for service, as you shall raise any of them you are after the 31st day of this instant January to give notice thereof to our Commissary-General of the Musters, or to our Chief-Deputy Commissary, that they may be mustered accordingly, from which Muster they are to commence and be in our Pay and entertainment. And all Magistrates, Justices of the Peace, Constables and other our officers whom it may concern, are required to be assisting you herein and in providing Quarters and otherwise as there shall be occasion. Given, etc. this 17th January, 1692/3.[6]

The Glorious Revolution had had an unsettling effect upon English society. Men had been called from their civilian employment to become soldiers in a brutal war in Ireland, and when demobilised did not easily return to peaceful pursuits. The 3rd Horse was moved to the vicinity of London, and was busily engaged for the whole of 1692 in acting as a mobile police force, in suppressing robbery and civil unrest. Its marching orders directed it to patrol the roads, paying particular attention to the areas of Blackheath and Hounslow Heath, notorious for highwaymen.

In the spring of 1692 The Queen's Horse was ordered to hold itself in readiness for foreign service, and in the middle of August it landed at Williamstadt in North Brabant, under command of Lieutenant Colonel Lumley. It then joined the

Confederate Army under King William at Lambeque in the camp at Deynse, on the banks of the River Lys, some three weeks after the Battle of Steenkirk.

The Queen's Horse had lost their colonel at the Battle of Steenkirk, when Sir John Lanier was killed while acting as a lieutenant general on the staff. He was succeeded by the lieutenant colonel, now promoted to be brigadier general, Henry Lumley. Sir John Lanier was one of the old professional soldiers, with a long experience of foreign service, untroubled by changes of allegiance at the close of a campaigning season, and not over-concerned with religious or political loyalty. The shift of master from James to William was seen merely as a change of employer, with the promise under William of Continental war and so a better chance of fame and fortune. But it was the Laniers, Churchills, Kirks and Trelawneys that made up the core of officers that held together the newly raised and inexperienced regiments.

The Queen's Horse, not actively engaged during the summer of 1692, marched in September with the army to Genappe and Waterloo, and went into winter quarters. Each regiment of horse was a self-contained unit, clothed and equipped by its colonel, who was also responsible for feeding and quartering it. Pay was drawn from Richard Hill in Antwerp, and on campaign William Blathwayt, the Secretary of State, issued orders and acted as a go-between for correspondence with the regimental lieutenant colonels and for general administration. Jacobite influence was still strong, and many soldiers deserted to the French over the winter of 1692/93. During that winter The Queen's Horse, along with the First Foot Guards and several infantry regiments, came under suspicion of disloyalty.

William wanted to curb the power of the French, and sought to achieve his ends by forming a confederacy of Holland, Austria, Spain and the independent states of Germany to block King Louis's military ambitions. In the spring of 1693 William established a strong camp at Parck, on the River Senne near Louvain, and The Queen's Horse marched there via Brussels, forming part of the left wing of the Confederate army. The French commander, the Duke of Luxembourg, sat in front of Parck until 6 July, when he moved to besiege Huy in the hope of enticing William out of his position. William marched to relieve Huy, but on reaching Tongres heard that the town had fallen. He detached ten infantry battalions, totalling 8,000 men, to reinforce Liège and Maastricht, and then withdrew to Neer-Hespen, where he set up a new camp.

Luxembourg, having succeeded in making William reduce his force, made a swift march, and on 28 July confronted the King's army, numbering some 50,000 British, Dutch, Hanoverians, Brandenburgers and Danish, with 80,000 of the finest French troops. William was determined to fight, and took up a position in front of his camp. He had the open ground between the villages of Neerlanden on his left flank

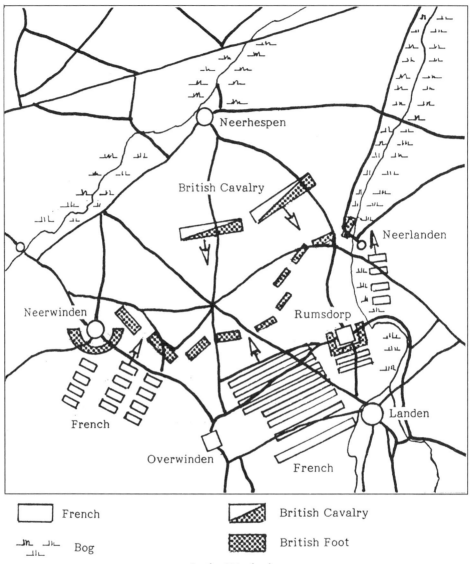

Battle of Neerlanden

and Neerwinden on his centre right entrenched. The key villages of Neerwinden and Laer, on the right, were fortified, and there he placed the cream of the British infantry; on his left flank he stationed another British infantry brigade in the village of Rumsdorp. The cavalry were posted to the rear of the Confederate line, with The Queen's Horse on the left wing together with the Life Guards and the 3rd, 4th and 6th Dragoon Guards.

At dawn on 29 July the Confederate artillery opened up on the massed cavalry and infantry of the French centre, and it was not until eight o'clock that Luxembourg attacked Neerwinden and Laer with four columns, outnumbering the defenders by three to one. After heavy hand-to-hand fighting, the British, led by William in person, managed to beat back the assault, inflicting heavy losses on the enemy. Another attack on Rumsdorp and Neerlanden on the left was also beaten off, although here the French outnumbered the defenders by four to one.

Luxembourg reinforced his left wing with a fresh 7,000 troops and attacked Laer and Neerwinden a second time. Again, the British in Neerwinden, taking advantage of its slightly forward position, brought flanking fire to bear on the French columns and beat them back. But it had now become a battle of attrition, and the superiority of the French numbers was beginning to tell. The defending Confederate infantry had suffered heavy casualties, losing 4,000 out of their original total of 14,000 men, whilst the six hours of close quarter fighting had exhausted the survivors.

Luxembourg now brought up 12,000 fresh infantry from his reserves and launched a third assault on Neerwinden, led by the Gardes Françaises. The Grenadier and Scots Guards, together with the Dutch Guards, resisted this attack until their ammunition ran out, and then they were forced to yield Neerwinden to the French. At the same time the Gardes Suisses advanced in the centre on the Confederate line, where the Coldstream Guards and the Royal Fusiliers beat them off, only to be charged in flank by the Maison du Roi, the French Household Cavalry. The Coldstream not only beat off this attack but captured a standard, and then the Bavarian cuirassiers in a brilliant charge drove back the French cavalry.

The overwhelming French superiority in numbers was now beginning to tell, and as Luxembourg brought up yet more fresh troops for a final assault, William ordered nine battalions to move from his left flank to support his hard-pressed right. Sadly, they did not have time to reach their destination, for as they vacated their position, the Marquis de Fequières led forward the French cavalry against the now weakened entrenchment. The nine battalions hesitated, halted and turned about, but too late to prevent their old position being carried.

The six regiments of English horse had spent the whole day dismounted and disgruntled in the rear. William now galloped back and put himself at the head of the Life Guards, The Queen's Horse, the 3rd, 4th and 6th Dragoon Guards and Galway's Horse, and led a series of desperate charges to cover the retreat of the Confederate infantry. The Queen's Horse was the only regiment that was able to form line before charging, and the Duke of Ormonde, who commanded the second troop of Life Guards, too impatient to wait for his own men, charged at the head of one of the squadrons of The Queen's Horse. The other regiments, having brought up their horses at the gallop, charged as they arrived. The disciplined impact of The Queen's Horse had great effect, although Ormonde had his horse killed under him and he himself was taken prisoner.

The Queen's Horse re-formed, and charged a second time with such impetus that the French cavalry retired and the infantry were able to withdraw. But the French pressed on again, and William personally led the English horse in another five or six charges to relieve the retreating infantry.

As the Confederate left wing drew back in reasonable order, covered by the English horse, the right collapsed and turned into a rout. William ordered The Queen's Horse, which was brigaded with the 3rd Dragoon Guards and Galway's Horse, to cover his retreat over the bridge of Neer Hespen. The pursuing French were only held back with the utmost difficulty, and the King himself, being one of the last to leave the field, was nearly captured.

William was a brave soldier, but an unlucky general. The Confederate losses in killed, wounded and prisoners numbered 12,000, whereas the French casualties were 8,000. The French captured eighty guns and numerous colours, including two troop standards of The Queen's Horse, one being crimson with M.R and crowns, the other yellow with M.R.[7]

On 5 August 1693 King William reviewed The Queen's Horse, along with other regiments, at Wemmel, and in September the army marched to Ninove, before moving into winter quarters early in October. The Queen's Horse was in cantonments around Brussels until early November, when it marched to Ghent, where it remained until the following spring.

At the end of May 1694 The Queen's Horse was brigaded with Schomberg's [4 DG] and Wyndham's [6 DG], and moved from Ghent to billets in villages between Brussels and Dendermond. In early June the regiment was reviewed by William at Tirelmont, and was reported as having made 'a very gallant showing; the horses being in very good order, and the men very well cloathed and armed'.[8]

SOURCES

1 *London Gazette*, 1691.
2 French State Archives.
3 State Papers, 1691; 'De Ginckell's Despatch'; Records of The King's Dragoon Guards.
4 Irish State Papers, 1691.
5 *The K.D.G.*, vol. 1, no. 6, December 1932.
6 Ministry of Defence Records, War Office 26.

7 *Journal of the Society for Army Historical Research*, vol. 16, p. 88.
8 J. M. Brereton, *History of the 4/7th Royal Dragoon Guards*, p. 43.

4

Flanders, Portugal
1694-1703

The 3rd Horse [Bays] was relieved of its policing duties around London by a regiment of Dutch horse, and was ordered to provide a relay of travelling escorts for the King, each of one officer and twenty-seven men, as William made for Harwich to embark for the war on the Continent. The regiment was then ordered to hold itself in readiness for foreign service, but this was countermanded owing to the threat of a French invasion. When this danger was averted as a result of the naval victory over the French fleet at La Hogue, the 3rd Horse spent the winter of 1693/4 in cantonments around Birmingham, Coventry and Northampton.

On 19 January 1694 General Villiers died and the vacant colonelcy was filled by the appointment of Brigadier Richard Leveson from the 3rd Dragoons [3rd King's Own Hussars]. In the middle of March the 3rd Horse, together with the 3rd Dragoons, marched to London, and on the 26th King William reviewed the two regiments in Hyde Park, complimenting them on their bearing and appearance.

On 28 March the 3rd Horse embarked in transports on the River Thames and sailed for Flanders, landing at Williamstadt in North Brabant in early April. The regiment then marched to join the army at Mont St André, where it was brigaded with Wood's [3rd Dragoon Guards], Wyndham's [6th Dragoon Guards] and Galway's [later disbanded], the brigade being under the command of Brigadier Leveson, their new colonel. In early June The Queen's Horse [KDG] joined the Confederate camp at Mont St André, where with Schomberg's Horse [4th Dragoon Guards] it was posted in the first line of cavalry.

In the autumn the 3rd Horse took part in the covering operations for the siege of Huy. When the town had been taken without too much difficulty, it moved with the rest of the cavalry, including The Queen's Horse, into winter quarters at Ghent. In April 1695 both The Queen's Horse and the 3rd Horse were ordered to provide thirty men from each troop, to muster between Deynse and Ghent, as the French had started to form an extensive series of field works on the Flemish frontier between the Lys and the Scheldt. However, after a reconnaissance in force, the troops returned to their regiments around Ghent.

On 29 May both regiments moved from Ghent to join the army then encamped at Arseele. There they were again reviewed by William and 'made a very gallant show; the horses being in good order, and the men very well clothed and armed'. Brigadier Leveson, colonel of the 3rd Horse, was promoted to the rank of major general.

On 2 June 1695 William marched from Arseele, having by skilful manoeuvres drawn the French army to the Flanders side of their entrenchments, and so opened the way for the investment of Namur. The Governor of Namur was the able Count de Guiscard, reinforced by the brilliant Marshal de Boufflers, who managed to enter the town with seven regiments of dragoons, though even then only raising the French garrison to 15,000 men. William opened the siege on 6 July. In the meantime the main French army under Marshal Villeroi moved to attack the Confederate lines, and The Queen's Horse was detached from the left wing, to-gether with three other regiments of cavalry, and moved to the right in order to cover the Confederate camp. The 3rd Horse kept open the army's lines of com-munication, occupying the open country between Charleroi and Mons, eventually taking post in August in the village of Waterloo. Marshal Villeroi found the Confederate position too strong to attack, but on 20 August he marched towards Perwys, trying to force a passage at the springs of the Mehaigne. To counter this move William detached twenty squadrons of horse, including The Queen's Horse, to Taviers and Boneffe. Villeroi moved with forty French squadrons to attack Boneffe, whereupon the Confederate cavalry retired, covered by The Queen's Horse, which was attacked by the French horse as it withdrew. In the ensuing fracas Lieutenant Alexander of the regiment was captured, and several men and horses were killed. The French endeavoured to follow up their success by forcing a passage between the right of the Confederate army and the River Maese, but The Queen's Horse and the rest of the cavalry were too strongly posted, and Villeroi retired.

After a heavy bombardment and two assaults on 18 and 27 July, William gained a partial lodgment in Namur, but at a heavy cost of life. Boufflers then made a gallant sortie, which was repulsed, and after a further bombardment de Guiscard and Boufflers surrendered, having seen that there was no hope of relief by Villeroi. On 26 August the remnants of the French garrison marched into captivity with full honours of war. Namur having fallen, both The Queen's Horse and the 3rd Horse returned to their former winter quarters at Ghent.

On 30 May 1696 William reviewed The Queen's Horse near Ghent, and the regiment 'made a very noble appearance; both men and horses being in good order'. Both The Queen's Horse and the 3rd Horse spent 1696 serving under the command of the Prince de Vaudemont in defence of the maritime provinces of Flanders, and were encamped around Bruges and Ghent.

The campaign of 1696 was confined to desultory skirmishing and partisan war-fare for the sake of plunder. The partisans of both sides were dressed as peasants, with their weapons hidden, and would try to surprise isolated detachments, seizing horses and baggage, and, if possible, an officer for whose return a ransom would be extracted. In October both The Queen's Horse and the 3rd Horse returned to their winter quarters at Ghent. By now the long years of war were forcing economies, and in 1696 the cost of clothing drummers for The Queen's Horse was reduced from £2. 10s to £1. 15s, and for hautboys from £3. 10s to £2. 10s.[1]

In mid-May 1697 both The Queen's Horse and the 3rd Horse left their winter quarters around Ghent, and on the 14th The Queen's Horse crossed the Scheldt and on the 16th, together with the 3rd Horse, joined the army, then encamped at St Quentin Linnick. The French were besieging Aeth, and William sent Brigadier Lumley in command of The Queen's Horse and a detachment of dragoons and infantry to make a reconnaissance towards Enghien. Lumley soon came across a strong party of 200 French carbineers and 150 hussars near Brussels, which he at once attacked. The Queen's Horse put the French to flight, capturing the colonel of hussars in command of the French party, together with two lieutenants, a cornet and forty men with their horses. Following this action on 3 June, Brigadier Lumley was promoted to the rank of major general in recognition of his leadership.

On 8 June a squadron of the 3rd Horse set out from the camp at Promelles to cover a foraging party and encountered sixty French carbineers and forty of their dragoons. The 3rd Horse killed twenty, and took a captain, two lieutenants and forty troopers prisoner.

William did not attempt to raise the siege of Aeth, but took possession of a strong camp before Brussels in order to protect the city. Both The Queen's Horse and the 3rd Horse were encamped at Wavre before Brussels, until the Treaty of Ryswick terminated hostilities on 20 September 1697. Shortly afterwards The Queen's Horse and the 3rd Horse returned to England via Ghent, the former landing at Hull in December, and being quartered at Salisbury and Dorchester; the latter arriving in England in November, and occupying quarters in Yorkshire. After only two months in England, however, the 3rd Horse was ordered to Ireland, sailing from Liverpool for Dublin, where it arrived for garrison duties in March 1698.

With peace came the inevitable reductions in establishment. The Queen's Horse retained all their officers, but lost 78 men and 38 servants, leaving the regiment with 9 troops, and a strength of 37 officers and 353 non-commissioned officers and men. The 3rd Horse, being on the Irish establishment, was reduced to 6 troops, with 24 officers and 244 non-commissioned officers and men. At this time the rates of pay were: colonel 12s; lieutenant colonel 10s; major 6s. 8d; captain 7s;

lieutenant 5s; cornet 4s. 6d; adjutant 2s. 6d; surgeon 3s; chaplain 3s. 4d; quarter-master 3s; corporal 1s. 6d; trumpeter 1s. 6d; drummer 1s. 6d; private 1s.

During the autumn of 1699 The Queen's Horse moved from Salisbury to London, where it provided the King with a relay of escorts on his return from Flanders. Leveson, colonel of the 3rd Horse, had been promoted to lieutenant general, but died early in 1699, and was succeeded by Daniel Harvey, who had been the lieutenant colonel of the 2nd Troop of Life Guards (later to become the 2nd Life Guards) by a commission dated 25 March 1699. In June 1701 The Queen's Horse moved to the vicinity of Windsor, where a daily guard had to be found for Princess Anne, and travelling escorts for her and King William.

Peace was shortlived. Charles II of Spain died without an heir, and Louis XIV supported the claims of his grandson, the Duke of Anjou, to succeed as Philip V of Spain. The Emperor of Austria at once protested, whereupon the French seized Milan and the Spanish Low Countries, while a French fleet occupied Cadiz and the Spanish West Indies. William joined in a confederacy against France with Sweden, Prussia, Austria and the States General of Holland. In June 1701 John, Earl of Marlborough, was sent to the Continent to negotiate the alliance and to command the Confederate army. Even then Parliament was lukewarm, but while these preparations were in progress James II died and Louis and Philip of Spain immediately declared his son to be James III of England. Louis XIV's designs against England were apparent, and the smouldering antagonism against the French burst into a flame of British patriotism.

With more preparation for war, orders came on 18 February 1702 for the augmentation of The Queen's Horse, 'Major General Lumley's Regiment consisting of Nine Troops of 30 Private Troopers in each, ordered to be recruited to 54 private men with Ye addition of a Corporall and a Trumpeter in each Troop. Effective Private Men 243; Non Commission Officers 18; Totall 261.'[2] In the same month The Queen's Horse was moved to Romford, and embarked at Woolwich for Flanders on the 27th. After a rough passage, when several horses were lost, the regiment landed at Helvoetsluys on 17 March and was quartered forty miles inland at Breda with three other regiments of horse and two of infantry. Foreign service was not to everyone's liking. The *London Gazette* of the 5-9 March 1702 recorded a deserter of The Queen's Regiment of Horse who absconded with a bright bay gelding.

On 8 March 1702 King William died. Queen Anne, coming to the throne, declared war against France, and so the long War of the Spanish Succession began. Accordingly, on 21 June 1702 The Queen's Horse, together with the other regiments at Breda, marched, under command of Lieutenant General Lumley, to join the main army. The French, aware of this movement, attempted to intercept the small force, but, by forced marches, the detachment evaded the enemy and joined the Earl of Marlborough's camp at Duckenburg, near Nijmegen on the Rhine, on 26 June. The French, who had amassed a large army, commanded by the Duke of Burgundy and Marshal Boufflers, were threatening the Dutch frontier. Marlborough, with a concentration of 60,000 men, of whom some 12,000 were British, recrossed the Waal and camped at Ober Hasselt within six miles of the French. There he was frustrated by the caution and delays imposed by the Dutch generals. On 26 July he crossed to the left bank of the Meuse, and by five forced marches arrived at Hamont on the frontier of the Spanish Netherlands.

Boufflers in alarm broke camp, called Marshal Tallard to his aid from the Rhine, and on 2 August an exhausted French army lay in an ill-chosen site, between Peer and Bray. Marlborough, with a fresh army, concentrated two or three miles to the north and made immediate preparations to attack, but was prevented again by the Dutch deputies, imposed upon him. Boufflers was able to slip away, and a rare opportunity was missed.

Boufflers again advanced north, hoping to capture a convoy of stores on its way to Marlborough, who quickly returned to Hamont, collected his convoy, and then placed his army across the line of the French retreat. On 11 August the French, in disorder, bumped into the Allied army at Helchteren, and Marlborough at once gave the order to advance. The artillery of both sides were engaged by three o'clock, and at five Marlborough ordered his right wing to destroy the disordered French left. At this point the Dutch general in command of the right wing refused to move, and yet another opportunity went begging. The following day Marlborough prepared for a general assault, but the Dutch again insisted on a further day's delay. With the French making off during the night, a third chance was lost.

Marlborough then decided to lay siege to the Meuse fortresses, detaching a part of the army, under his personal command, to cover these operations. The Queen's Horse was with this covering force. Venloo was invested on 18 August, and fell twenty-seven days later. Stevenswaert, Ruremonde and Maseyk were quickly captured. The French, now thoroughly alarmed lest Marlborough advance on the Rhine, detached Marshall Tallard and sent him back to Cologne. Boufflers, with a weakened army, decided to reinforce Liège, and arrived there on 12 October, to find himself confronted by Marlborough.

For the fourth time the Dutch deputies prevented Marlborough from attacking, and Boufflers left Liège to its fate. The Queen's Horse was among the first to enter the city, and on the 23rd the citadel was stormed and taken. With the campaign of 1702 at an end, The Queen's Horse returned to winter quarters in Holland. In spite of the lost opportunities caused by the intransigence of the Dutch, Marlborough had secured the safety of the Dutch frontier. For his services he was created Duke of Marlborough.

His plans for the campaign of 1703 were to invade French Flanders and Brabant, but the Dutch were not prepared for such an ambitious undertaking, so at the end of April Marlborough besieged and captured Bonn. The Queen's Horse formed part of the first line of the right wing of the army, being with the 3rd, 5th, 6th and 7th Horse in Wood's brigade. Regimental headquarters consisted of a lieutenant colonel, major, adjutant, chaplain, surgeon and kettledrummer; there were nine troops, which were grouped into three squadrons, each troop having a captain, lieutenant, cornet, quartermaster, two to three corporals of horse, between forty and sixty troopers and a trumpeter. Marlborough reverted to his original plan and moved to attack Huy, and on 10 June a detachment of The Queen's Horse was involved in a skirmish with the French.

The Dutch, however, again failed to back these moves, and eventually Marlborough was forced to confine his operations to the capture of Huy and Limburg in August and September. The Queen's Horse helped to cover these sieges. So a second year was wasted, leaving Marlborough so irritated at the 'great caution of the Dutch in not hazarding a battle' that, for a time, he seriously considered resigning his command.

In the summer of 1703 Portugal had joined the Confederacy, and Queen Anne decided to send a British force to support the Archduke Charles of Austria in his attempt to secure the Spanish throne. The 3rd Horse and the Royal Dragoons, together with six regiments of foot, made up the contingent. The King of Portugal had promised to supply horses 'of a superior description', and so the regiment handed over all its horses to other corps before embarking in Ireland in September 1703. The force was commanded by the Duke of Schomberg, and the cavalry by Major General Harvey, the colonel of the 3rd Horse. However, the fleet was not ready to sail, and the regiment was landed on the Isle of Wight, where it went into quarters until it re-embarked in November.

Further delays meant that the troops did not land in the Tagus until March 1704. On arrival it was found that the French had bought up all the best horses for the use of their own troops in Spain, and General Harvey was compelled to purchase horses of a very inferior quality. As a result only some twenty men per troop were mounted and fit for service. The 3rd Horse, having arrived at Abrantes in April in a largely dismounted state, was only able to send about 120 mounted men to the Alentejo on the Spanish frontier, where they encamped on the banks of the Tarra, near Estremos. Because of a misunderstanding between the Duke of Schomberg and the Portugese, there were delays in opening the campaign. The French were able to invade Portugal, and the 3rd Horse was engaged in defensive operations

only. In the meantime Major General Harvey, 'having received orders to take by
force all the horses that he should have occasion for to reinstate his Regiment upon
paying a reasonable price for the same, has been so diligent therein that he has
procured a sufficient number and has marched for the army'.[3] By the summer the
3rd Horse was fit for action, and advanced to invade Castile. On reaching the
Agueda, near Ciudad Rodrigo, they found the French in force opposite them, and
after some inconsequential manoeuvring the regiment returned to Portugal and to
winter quarters in villages on the Alentejo.

Marlborough, having spent the winter in England, returned to Holland in April,
having decided that he could only achieve his aims by keeping the utmost secrecy.
His plan was too bold and ambitious for the cautious Dutch ever to agree to it. He
intended to bring to battle and defeat the main French army, if possible on the
Danube, by drawing the French after him, and leaving the defence of the Netherlands to the Dutch. On 29 April he wrote to Godolphin in England,

> By the next post I shall be able to let you know what resolutions I shall
> bring these people to; for I have told them I will leave this place on
> Saturday. My intentions are to march all the English to Coblenz, and to
> declare here that I intend to command on the Moselle. But when I come
> there to write to the [Dutch] States that I think it absolutely neccessary, for
> the saving of the Empire, to march with the troops under my command
> and to join those in Germany that are in Her Majesty's and the Dutch pay,
> in order to take measures with Prince Lewis [Grand Duke of Baden] for the
> speedy reducing of the Elector of Bavaria. What I now write I beg be
> known to nobody but Her Majesty and the Prince [Consort].[4]

The army, consisting of forty battalions and eighty squadrons, left its winter
quarters early in May. The Queen's Horse camped at Bois-le Duc on 4 May with
the rest of the cavalry, leaving 'Boilduck' as the troops called it, for Bedburg,
seventy miles away near Cologne. It crossed the Maas by a bridge of boats
near Ruremonde, and concentrated there with the rest of the army on 12 May.
Villeroi was preparing to attack the reduced Dutch forces, but on hearing that
Marlborough had set out, he was ordered to follow him to the Moselle.

On 20 May Marlborough started out, riding ahead with the cavalry, including
The Queen's Horse, leaving his brother General Charles Churchill to bring on the
infantry and artillery after him. Because of the poor state of the roads, made worse
by heavy rain and the passage of so many cannon and waggons, the cavalry were
ordered to cut fascines each morning, which were then dropped into the worst of
the ruts and potholes. They then rode on ahead to prepare the night's camp for the
infantry, and to ensure that there were ample supplies of bread and meat, all of

which were paid for in cash on the spot. As a result the army was never short of rations; and as the strictest march discipline was enforced and any 'marauders to be hang'd without mercy', the soldiers retained the goodwill of the local people as they passed by. In order to deceive any enemy vedettes observing the march, the men were aroused at 1 a.m. and marched until noon, when they were halted at the previously prepared camps and left to rest for the remainder of the day. This routine kept them in good heart, and obviated the risk of their progress being betrayed by the clouds of dust that usually attended such marches.

On 26 May the cavalry reached Coblenz, where the Moselle flows into the Rhine. The general expectation was that Marlborough would turn up the Moselle valley, but he crossed the river and went on to cross the Rhine the following day. Deane records that at Coblenz 'Here my Lord Duke left us and tooke the horse with him before us into Germany.'[5] Even then the French thought that he was probably heading for Alsace, and Marlborough gave credence to this belief by having a bridge of boats built across the Rhine at Philippsburg. Tallard and Villeroi moved to cover such an advance, while Marlborough crossed the Main and then by more forced marches the Neckar, and headed for Bavaria and the Danube.

At Mainz the amy was reviewed by the Elector, a 'Golley handsome man', accompanied by a 'bundance of coachis with fine Ladeys in ym'.[6] There were fourteen battalions of foot, and seven regiments of horse,'The Horse as followeth: Lt Genll Lumley's regiment [KDG]; Majr. Genll Woods regiment [3 DG]; Brigadeir Windhams regiment of Carbineers [6 DG]; Duke Schomberghs regiment [7 DG]; and Col Coddugans squadron [5 DG]; and Col Rosses Draggons [5 L]; and Col Cunninghams regiment of Draggons [8 H].'[5] 'The Army appeared that Day fully as Clean and Compleate, as if they had marched out of their Quarters into the Field. Of which His Electoral Highness took notice, and Exprest himselfe in the following manner, "I have Lived to have a hoary head, but never Saw such a Body of Men so Cleand and Compleat in my Life." '[6]

On 10 June Prince Eugène of Savoy rode into Marlborough's camp at Mundelsheim, and on the next day both commanders rode to Gross Heppach to inspect the cavalry. 'His Highness was very surprised to find them in so good a condition after so long a march, and told His Grace that he had heard much of the English Cavalry and found it to be the best appointed and finest he had ever seen. But, says he, money, which you don't want in England, will buy clothes and fine horses, but it cannot buy that lively air I see in every one of these troopers' faces.'[7] At Gross Heppach Marlborough and Eugène were joined by Prince Louis of Baden, and their various actions were planned: Eugène was to march to the Rhine and keep Tallard and Villeroi engaged, while Marlborough and Prince Louis would lay waste to Bavaria in order to force that Elector to change sides. The unsatisfactory part of this agreement was that Marlborough and Baden were to share the command, each being in charge on alternate days.

SOURCES

1 British Library Harl. MSS 7018.
2 Public Record Office, WO 55/342 of 18 February 1702.
3 R. Cannon, *Present State of Europe*, p. 24.
4 G. M. Trevelyan, *Blenheim*, p. 354.
5 J. M. Deane, *A Journal of Marlborough's Campaigns during the War of the Spanish Succession, 1704-1711*, Society for Army Historical Research, Special Publication, no. 12, 1984.
6 'Diary of Sergeant Wilson', MS.
7 G. M. Trevelyan, *Blenheim*, p. 364.

5

Schellenberg, Blenheim
1704

Marlborough laid down his requirements for the discipline of the horse:

> It is sufficient for them to ride well, to have their horses well managed and
> trained up to stand fire; that they take particular notice what part of the
> squadron they are in, their right and left-hand men and file-leaders, that
> they may, when they happen to break, readily know how to form. That
> they march and wheel with a grace and handle their swords well, which is
> the only weapon our British Horse make use of when they charge the
> enemy; more than this is superfluous.

The Duke would allow each man only three charges of powder and ball for a
campaign, and that only for guarding their horses when at grass and not to be
made use of in action. The British contingent totalled some 14,500 men; the Dutch
and Hesse troops brought Marlborough's command up to 40,000. After meeting
with Eugène and Baden, Marlborough pushed on towards Bavaria, but the weather
had broken, and the roads became almost impassable. Riding ahead with the
cavalry, he constantly urged the squadron leaders forward, as, soaked to the skin
and bespattered with mud, The Queen's Horse [KDG] rode through Ebersbach and
Goppingen and on 11 June 1704 cleared the heavily wooded and easily defendable
Geislingen Pass, which the French had neglected to guard. Marlborough was
forced to wait with the cavalry at Gingen for two days, while the infantry and
artillery cleared the two miles of the pass. On 20 June the Duke wrote, 'I was never
more sensible of heat in my life than I was a fortnight ago, we have now the other
extremity of cold; for as I am writing I am forced to have fire in the stove of my
chamber. But the poor men, that have not such conveniences, I am afraid will suffer
from these terrible conditions.'[1]

The Army now descended into the valley of the Danube, and by the end of June
had joined forces with Baden, while the Duke of Württemberg came with more
German infantry and the Danish cavalry. The army with all its baggage and
artillery had marched from the Netherlands, a distance of 300 miles, in five weeks.

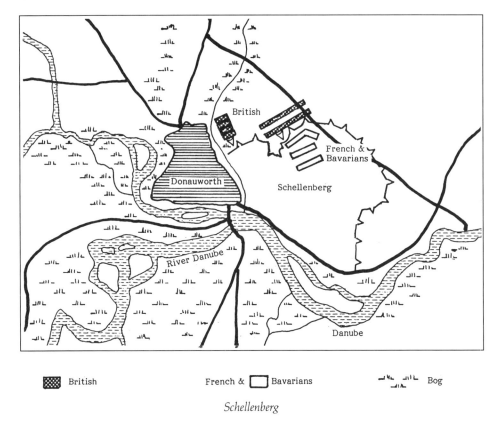

British   French & ☐ Bavarians   Bog

Schellenberg

Marlborough decided to capture the strategic town of Donauworth, and 'there settle a magazine for the army'.[1] The bad weather had given Marsin and the Elector of Bavaria time to concentrate their forces in a fortified camp near Dillingen, whence they reinforced Donauworth with 10,000 infantry and 2,500 cavalry, under command of the Comte D'Arco, a Piedmontese. Donauworth is dominated by a steep bell-shaped hill, the Schellenberg, and D'Arco, on arrival, immediately started to improve its neglected trenches and fortifications. On 2 July, being Marlborough's day of command,

> He marched by three in the morning at the head of thirty squadrons, three Imperial regiments of grenadiers, and a detachment of 7,000 foot; the whole army marching close after them. It was four o'clock before he reached Donauworth. He saw the Count D'Arco's men hard at work throwing up entrenchments on the hill at Schellenberg. As soon as the British troops were all come up, he formed a disposition for attacking them. The hill was very steep and rough, and difficult to ascend.[2]

As time was at a premium, Marlborough launched a frontal assault as soon as his guns were ready.

> About six o'clock in the afternoon the English Guards began the attack, the whole line going on at the same time. The thirty squadrons [including The Queen's Horse] kept in the rear of the foot, as close as the nature of the ground would permit. The enemy maintained their ground with great resolution for an hour and ten minutes; by which time the whole army being come up, and supporting the attack, at length they gave way, and a terrible slaughter ensued, no quarter being given for a long time. Count D'Arco, with the greater part of them made down the back of the hill to the Danube, where they had a bridge of boats; but this breaking under them, great numbers were drowned.[2]

Private Deane of the Grenadier Guards wrote, 'They being strongly intrenched they killed and mortyfyed abundance of our men both officers and souldiers. Gods assistance we driving them out of there works and possessing ourselves of them. Our horse likewise persued and killed abundance of them driving severall hundreds of them into the river Danube, who there made there exit.'[3]

The assault had been made in six lines, the first four being of infantry, and the last two of cavalry, who followed closely behind with Marlborough at their head. As soon as the French broke, Marlborough ordered Lumley to take the cavalry forward. The Queen's Horse charged through the gaps in the infantry, and with the rest of the cavalry fell upon the fleeing French and Bavarians. The troopers, enraged by the heavy casualties which the infantry had suffered, gave no quarter, and their swords rose and fell, cutting and driving the broken enemy into the Danube, and capturing thirteen colours. Darkness, accompanied by sheeting rain, finally closed over the scene of slaughter.

The action lasted for one and a half hours, and although the Allied losses were severe, of the 12,000 brave French and Bavarian defenders, less than 4,000 survived to rejoin their army. The five regiments of British cavalry lost two officers killed, and eight wounded; with thirteen troopers killed and fifty-three wounded. In The Queen's Horse Lieutenant Colonel Palmer and Cornet Law were wounded, but only two horses were recorded as having been lost.[4, 5] Chaplain Samuel Noyes of the Royal Scots wrote, 'Collonel Palmer is shot thro the Body but whether mortal or not I can't say. I have not seen him yet, but by the account given me of him I fear very much for him.' A later letter said:

> I went and waited on Collonel Palmer in Donawert. I found him up and his cloaths on, to my great Surprise. I askt him how he did. He told me he

was ful of pain, for the Ball was still in his Reins. I went to Lt General
Lumly's and enquired how t'was with Him, and they told me they had
great hopes of him, that he was very easy and without much pain and in
no fear of a Feavour, but the Bullet not found when the Surgeon was last
with him.

A fortnight later Noyes reported they had found the ball and taken it out and
that he was 'recovering very fast and past all manner of Danger.' But a later letter
reported, 'I was mighty willing to send such good news, but now I wish I had not
been so eager, for that very day about 4 in the afternoon the Collonel dyed. He
swooned away as his wounds were dressing and never came to himself again.'[6]
Marlborough in his despatch wrote, 'They [the horse and dragoons] supported our
foot in the attack, and stood within musket shot of the enemy during all the action.
Our horse was commanded by General Lumley.'[7]

The capture of Donauworth opened the way into Bavaria. The Elector retreated
to Augsburg, where he entrenched himself. At first he seemed inclined to come to
terms, but when he heard that Marshal Tallard and the French army were on the
way to reinforce him, he opted to hold out. In order to put pressure on the Elector
before the French could reach him, Marlborough decided to lay waste Bavaria.
Thirty squadrons of horse were despatched to burn and plunder as far as Munich.
On 'the 23rd July the burning party was ordered out and burnt all the villages
round our camps'.[3] Parker of the 18th Foot records, 'A great number of parties
were sent out far and near, who burned and destroyed all before them; insomuch,
that it was said there were 372 towns, villages and farmhouses, laid in ashes; and it
was a shocking sight to see the fine country of Bavaria all in a flame.'[2] In a letter to
Sarah, his wife, Marlborough wrote, 'This is so contrary to my nature, that nothing
but absolute neccessity could have obliged me to consent to it. For these poor
people suffer for their master's ambition.'[1]

In the meantime Tallard was marching to the assistance of the Elector. He wasted
five days besieging Dillingen on the way, but encamped before Augsburg on 23
July, and on the 26 joined forces with the Elector and Marsin. Tallard wanted to
remain at Dillingen, but the Elector persuaded him to camp at Hochstadt in order
to attack Eugène. So a compromise was reached whereby the French and Bavarians
left their strong position at Hochstadt and moved forward to ground between the
villages of Lutzingen and Blenheim. On 12 August the Franco-Bavarians set up
their tents on the flat open stubble fields that extended without a hedge from the
village of Blenheim on the banks of the Danube to the pine-covered hills around
Lutzingen, four miles away, protected, as they thought, by the Nebel stream with
its marshes to their front and by the Danube on their right flank.

Eugène had marched north of the Danube and parallel with Tallard, reaching Münster on 26 July, while Marlborough was encamped at Schobenhausen some twelve miles to the south of the river. Eugène rode over to concert plans with Marlborough and Prince Louis of Baden. To the relief of both Marlborough and Eugène, the touchy Prince Louis agreed to besiege Ingoldstadt, taking 15,000 men with him. Marlborough and Eugène then joined forces on 11 August, the British having crossed the Danube, and the united army camped between the villages of Kessel and Munster, only about four miles from Tallard.

Early on 12 August Marlborough and Eugène rode forward with a strong cavalry escort to reconnoitre the ground. Observing the enemy's cavalry some distance off, they climbed the church tower at Tapfheim, where they could clearly see the French and Bavarian forces marking out their camping ground between Blenheim and Lutzingen. The French were on their right, using the village of Blenheim as a heavily fortified post by the Danube – twenty-six battalions of the best French infantry and twelve squadrons of dismounted dragoons formed its garrison. The French cavalry were on the flat stubble in the centre, adjoining the Elector of Bavaria's cavalry, who then continued their line up to the village of Lutzingen. So the two armies camped side by side, but with no thought of engaging in battle. Indeed the position was so strong that Tallard was convinced that the Allies would be forced to retreat, probably to Nordlingen. The French and Bavarians fielded some 56,000 men and 90 guns, to the Allies' strength of 52,000 with 60 guns. Marlborough and Eugène agreed, from their eyrie in the church tower, to attack the next morning.

At 2 a.m. on 13 August Marlborough had the general call sounded, followed an hour later by the assembly, when the army marched, crossing over the River Kessel by pontoon bridges in a dense mist. They moved off in eight columns, the infantry in the four centre columns, the cavalry in the two outer on each flank. On reaching Tapfheim, where Marborough had posted two outpost brigades, the army halted, and a ninth column was formed from these two brigades, reinforced by eleven of the best British battalions. The Queen's Horse, comprising three squadrons under command of Lieutenant Colonel Crowther, marched with this extreme left flank column. The army then resumed the advance towards the Nebel stream, with the British taking post from the Danube on their left to the village of Oberglau in the centre of the French and Bavarian position. Eugène was to be responsible from Oberglau to Lutzingen, where he was to keep the Bavarians and the French under Marsin fully occupied, whilst Marlborough made the main effort against Tallard. By now the mist had cleared and Marlborough with Eugène pushed on ahead with a strong escort, who drove in the enemy's outposts, the main army spreading out like a fan behind them, 'advancing cheerfully and showing a glad countenance',[8] as

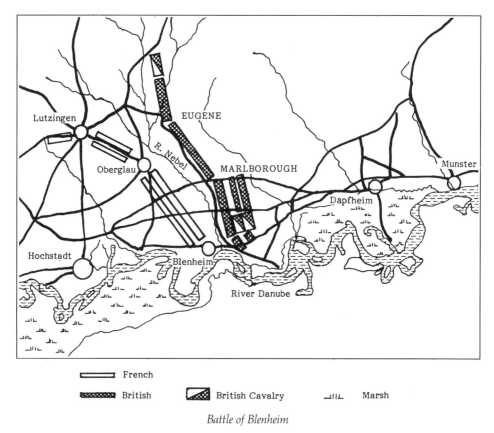

French

British   British Cavalry   Marsh

Battle of Blenheim

they marched to their allotted positions. By 7 a.m. the British troops were in place within a mile of the Nebel and in full view of the enemy.

The French were taken completely by surprise, Tallard being convinced that Marlborough was retiring to Nordlingen.

The Comte de Merode-Westerloo wrote,

> I slept deeply until six in the morning, when I was abruptly awoken by one of my old retainers. This fellow shook me awake and blurted out that the enemy were there. I asked, 'Where?' and he at once replied 'There' — flinging open the door of the barn. The door opened straight on to the fine, sunlit plain beyond — and the whole area seemed to be covered by enemy squadrons. I rubbed my eyes in disbelief.[9]

At last Tallard realised that the Allies intended battle, and in a great hurry the French and Bavarians took up their positions, but in their haste as they were camped, as two separate armies, rather than strategically as a combined force.

The result left their centre very weak in infantry. Tallard's cavalry formed in two extended lines in the centre, as they had been camped, supported by only nine battalions of young and inexperienced infantry. Marsin and the Elector on the left wing compounded this error by leaving most of their infantry in defence of Lutzingen, with the cavalry in the centre where they also had been camped.

From 7 a.m. until noon the British and the French sat and watched each other, while Eugène, who had more difficult ground to cover, marched to his allotted position. The artillery of both sides opened fire, Marlborough ordering the infantry to lie down, while a stone bridge across the Nebel was repaired and five more pontoon bridges were laid. At last at 12.30 p.m. word arrived from Eugène that he was ready, and Marlborough gave the order to cross the Nebel and advance. The Queen's Horse, together with Wood's [3 DG], Cadogan's [5 DG], Wyndham's [6 DG], Schomberg's [7 DG], Hay's [Greys], and Ross's [5 L], were on the extreme left in support of the infantry assaulting the strongly defended village of Blenheim. The British infantry were charged in flank by the elite French Gens d'Armes, but the situation was restored by the brave action of some Hessian infantry in the second line.

Brigadier Cutts, seeing further cavalry attacks being prepared, asked General Lumley for support for his flank, and five squadrons from the 3rd, 6th and 7th Dragoon Guards crossed the Nebel, forming in line. Tallard ordered eight squadrons of the Gens d'Armes to eliminate this threat, and the French attacked using their usual tactics of coming to a halt to fire their horse pistols. The five British squadrons immediately charged home with the sword and routed the French, much to the dismay of Tallard, who was 'strangely confounded' and affected by this early defeat. Cutts meanwhile continued to attack the village of Blenheim, but it was too strongly held to be taken; however, the ferocity and bravery of the British infantry in these assaults not only kept the huge French garrison locked in Blenheim, but also drew in some of the reserve French infantry brigades.

Marlborough had now ordered Lumley to bring up the rest of the horse, and the troopers of The Queen's Horse dismounted in order to lead their horses by the bridle over the fords and marshy ground of the Nebel. As soon as they were on firm ground they remounted, but because of the bad going their horses were blown and overheated. Then, while they were forming, they were charged by fresh French squadrons, who succeeded in scattering The Queen's Horse, pushing them back to the edge of the Nebel.

The British infantry, however, through whose intervals the horse had been able to retire, poured in a heavy flank fire in support of their cavalry comrades, and drove back the French. Then the second line of horse charged the French.

Marlborough, writing later to Robert Harley, said, 'The bravery of all our troops on this occasion cannot be expressed; the generals as well as the officers and soldiers behaving themselves with the greatest courage and resolution, the horse and dragoons having been obliged to charge four or five several times.'

Lumley, having rallied his disordered squadrons, moved the horse, including The Queen's Horse, to the centre. Here the British infantry had been formed in line, again with intervals to allow the cavalry to pass and retire through them.

> Our left wing cavalry passed the rivulet pell-mell in the centre, as did our right, having made several passages with diverse pieces of wood which they found at hand. So all passed and drew up in order of battle as well as the ground would permit on the other side of the rivulet. The enemy gave us all the time we wanted for that purpose and kept very quiet on the hill they were possessed of, without descending to the meadow towards the rivulet, insomuch that even our second line of Horse had time to form themselves; and to this capital fault we ought principally to ascribe the victory.[10]

The horse now advanced up the hill at a measured trot, and soon drove back the French cavalry. Tallard sought to correct his earlier errors by bringing up his nine centre battalions of infantry, but he was unable to reinforce them as all his reserves had been drawn into Blenheim. These young soldiers, advancing in square, poured their fire into the British horse, emptying many a saddle and checking their charge. Lord Orkney related, 'By this time I had gott over with what foot was left with me, and marched straight to sustean our horse whom I found to some confusion and calling out for foot; I went to severall Esquadrons and got them to rally and maike a front till I enterlined them with foot.' Marlborough then attacked these nine battalions with a battery of artillery and three battalions of Hessian infantry. The young French infantry stood firm and closed ranks for a time, but their losses became so severe that they started to disintegrate. Tallard ordered his cavalry to charge to their relief, but the cavalry refused to obey his order. The Queen's Horse and the rest of the first line then charged the hapless French infantry . 'The poor lads never moved, but were mowed down in their ranks and lay in straight lines of white coated corpses.'[11]

'The battle went on from right to left very brave, the horse charging most furiously on both sides and I must say our confederate forces behaved themselves to a miracle being led by brave and prudent generals and commanding officers, cutting and hewing them to a degree.'[3] The cavalry then returned and re-formed in front of the line of infantry.

It was now about 4.30 p.m. Marlborough, clearly visible to all ranks on his grey charger, rode along the lines of horse, drew his sword and ordered the trumpeters to sound the charge. The Queen's Horse, with the rest of the first line, again advanced steadily uphill at their controlled trot, to be met by the ranks of the still considerable numbers of the French cavalry, who again resorted to their outdated tactic of meeting the British charge with fire from their pistols and carbines sitting at the halt. The horse then charged the French, who wheeled on their supporting second line and broke. Thirty French squadrons fled to the rear of Blenheim, pursued by the Allied horse, 'driving severall squadrons of them in to the river Danube where they all perrished men and horse, and likewise abundance of them taken prisoner.'[3]

> At the same time our squadrons pursued at their heels, cutting down all before them; for in all such close pursuits, 'tis very rare that any quarter is given. In short, they were almost all of them killed or drowned; and the few that reached the far side of the river, were killed by the boors of the villages they had burnt. Tallard fled that way, but finding the bridge broken, he turned up the river by Hochstadt, and was taken.[2]

The Comte de Merode, on the French side, related:

> We were borne back on top of one another. So tight was the press that my horse was carried along some three hundred paces without putting hoof to ground, right to the edge of a deep ravine; down we plunged a good twenty feet into a swampy meadow; my horse stumbled and fell. A moment later several men and horses fell on top of me, as the remains of my cavalry swept by, all intermingled with the hotly pursuing foe.[9]

Marsin and the Elector, seeing the collapse of Tallard's army, set fire to the villages of Oberglau and Lutzingen, and withdrew in some semblance of order, with Eugène following them up.

Merode, himself in the centre, continued:

> Meantime the enemy sealed all the exits from Blenheim at their leisure, and the general commanding the village rode straight into the Danube and was drowned, instead of organising the twenty seven battalions and fourteen dragoon regiments into a square to fight their way out. As it turned out, however, these forces fought on until after seven o'clock, bereft of an experienced commander, and then surrendered as prisoners of war when they saw the enemy bringing up all his artillery.[9]

Chaplain Noyes wrote:

> God be praised we have got a most prodigious great victory. We have taken 41 pieces of cannon, and 25 General Officers, Killd one Lt General

and drowned another, and tis generally allowed we have of private men 12,000, some say many more, and about 1,500 Officers. The number of the slain is very great, abundance drownd in the Danube, which is ful of Jack Boots likewise that others threw off the better to swim over, and even of these great numbers are returned to Us chusing rather to be in our hands than venture the mercy of the countrey Boors.[6]

The Queen's Horse had two officers wounded, Lieutenant Barton and Cornet Charles Law, and lost fifty-eight horses. [4, 12] In later years Alexander Inglis, the regiment's surgeon at Blenheim, was appointed to the Royal Hospital, Chelsea, and in 1710 became Surgeon General to the Forces.[13]

THE QUEEN'S REGIMENT OF HORSE: THE BLENHEIM ROLL

[Each man's bounty is given after his name] *Lieutenant Colonel Commanding* Thomas Crowther, £88; *Major* John Deane, £81; *Captains* Christopher Billingsley, Thomas Panton (on staff), John Morey, Patrick Lisle, James Bringfield (on staff), William Goodwin, *all* £64. 10s; *Lieutenants* Charles Wiseman, Peter Law (on staff), Charles Alexander, Benjamin Bishop, James Stalker, Robert Wilson (on staff), Thomas Stirrop, *all* £45, Roger Barton (wounded) £90; *Cornets* William Benbow, John Usher, Thomas Jackson, Tristram Dillington, Nathaniel Law, *all* £42, Charles Law (wounded) £84; *Quartermasters* William Shaw, Anthony Dodsworth, Nicholas Hallman, George Hartwell, Henry Whitaker, John Dodsworth, John Scott, Francis Kingston, *all* £25.10s; *Chaplain* John Gaile, £20; *Adjutant* Francis Kingston, £15; *Surgeon* Alexander Inglis, £18; *Volunteer* Matthew Pitt; *other ranks*: 27 corporals, *each* £2. 10s; 17 trumpeters, *each* £2. 10s; 1 kettle drummer, £2. 10s; 412 troopers, *each* £2.

SOURCES

1 Archdeacon W. C. Coxe, *Memoirs of John, Duke of Marlborough*, 1820.
2 *The Military Memoirs of Robert Parker*, Longman, 1968.
3 J. M. Deane, *A Journal of Marlborough's Campaigns during the War of the Spanish Succession, 1704-1711*, Society for Army Historical Research, Special Publication no. 12, 1984.
4 'The Blenheim Roll', *English Army Lists*, vol. 5.
5 *War Office IV*, vol. 3, p. 163.
6 Letters of Samuel Noyes, Society for Army Historical Research, vol. 37, p.130: letter 9 of 22 June 1704; letter 10 of 2 July 1704; letter 12 of 16 July 1704; letter 13 of 19 July 1704; letter 14 of 10 August 1704.

7 Records of The King's Dragoon Guards, Regimental Museum, 1st The Queen's Dragoon Guards, Cardiff.

8 British Museum, Additional MSS 9114, Dr Hare's Journal.

9 *Comte de Merode-Westerloo*, Longman, 1968, p. 166.

10 Sergeant John Milner, *A Compendious Journal, 1701-1712*, 1735.

11 G. M. Trevelyan, *Blenheim*, p. 401.

12 *War Office IV*, vol. 3, p. 163.

13 C. G. T. Deane, *The Royal Hospital, Chelsea*, 1950, p. 173.

6

Neer Hespen, Ramillies
1704-1706

Marlborough arranged for the disposal of the prisoners and the booty and then followed the French retreat, marching on 19 August 1704 in three columns, crossing the Rhine near Philippsburg on 8 September and leaving Louis of Baden to besiege Landau. He himself moved back into the valley of the Moselle and occupied Treves. He then laid siege to Trarbach, whose fall was soon followed by that of Landau. On the march the cavalry had difficulty in obtaining sufficient forage in the devasted areas of Bavaria, and The Queen's Horse [KDG] spent much time in sending out small detachments to scour the countryside.[1] In addition the dreaded scourge of glanders had spread from the French camp to the British troop horses, and Richard Pope of Schomberg's Horse commented on 'the pestilential distemper that carries off 40 or 50 a day'. By the middle of November the weary troopers were able to move into their winter quarters at Breda in Holland, having marched a thousand miles in less than five months.

The 3rd Horse [Bays], having spent the winter on the Alentejo in Portugal, rejoined the assembling army on 24 April 1705 at Estremos, for a projected advance through Estramadura to Madrid. The regiment formed part of the force that besieged Valencia de Alcantara, which was taken by storm on the 8 May. It then assisted in the covering operations for the siege of Albuquerque, which surrendered on 22 May. Following these operations the 3rd Horse went into camp on the banks of the Chevora river, where on 27 May a party of fifty French cavalry attacked a foraging party, and captured several mules. Thirty-nine of the 3rd Horse galloped out of camp and charged the French, recapturing the mules and taking one prisoner.[2]

The regiment was not engaged again until the autumn, when it took part in the unsuccessful siege of Badajos. There was not sufficient strength available to invest the town, and on 14 October a relieving force managed to enter the fortress; but on that day, too, the 3rd Horse crossed the Guadiana and, led by General Wyndham, charged some squadrons of Spanish cavalry near the Chevora, cutting them up and driving them back across the river. The siege of Badajos had to be abandoned, and the regiment returned to Portugal, where it spent the winter in villages along the frontier.

Meanwhile Marlborough had returned home after the campaign of 1704. In April 1705, 'Lt. Genn'll Lumley commanding in His Grace's absence, having recd orders from His Grace for the forces to march out of all garrisons and to joyne on there march, the Lt. Genn'll sent his commands accordingly and on the 20th Aprill we marcht out of Breda.'³. Marlborough ordered 'the troops that had acted under him in the preceding campaign, to assemble at Maastricht, whither he soon came, and marched towards the Moselle. By easy marches we arrived on 24 [May] on the banks of that river.'⁴ Parker relates, 'Here it was that the Prince of Baden had promised to join the Duke. The Duke waited above a month, and no appearance of the Prince of Baden. At length he sent the Duke word he was so ill of the gout that he could not possibly join him.'⁴ The other Rhineland Princes and the Austrians had also failed him, and the Dutch were again being less than helpful.

Villeroi, in Marlborough's absence, had moved forward.

At this time the Duke received an express that the enemy had taken Huy, and were in full march to Liège. This made him hasten his march; in-somuch, that we were but half the time in returning, that we took going up. When we had advanced as far as Aix-la-Chapelle, another express arrived. Upon which the Duke hastened away with all the Horse and Dragoons, each of them taking a Grenadier behind him. But as soon as the enemy heard of the Duke's approach, they drew off from Liège, and retired within their lines.⁴ Villeroi had retreated to the security of the fortified Lines of Brabant. These consisted of sixty miles of defences stretching from Namur to Antwerp, which Marlborough decided to force.

Villeroi, with the Elector of Bavaria, mustered a superior army of 70,000 men to oppose Marlborough, but one which had to be spread along the fortified lines. Marlborough

at this time became acquainted with a gentleman of that country, through whose estate the enemy's lines ran; and as he wanted to get clear of such troublesome neighbours, he acquainted the Duke with the situation of two barriers, where the enemy kept but slight guards, and he supplied him with two trusty peasants to guide him thither. On the 16th July, [the Duke] ordered the army to strike camp. The Veldt-Marshal marched away to the left with all the left wing [on the]17th; at the same time a detachment of 10,000 foot was made from the right wing, which drew up on the right of all. The enemy soon had an account of all this. Whereupon Villeroi edged

with their right wing away to the right to observe the motions of the
Veldt-Marshal, and the Elector drew all the troops that were on their left,
close to him, to oppose the Duke.

Thus matters stood until night came on, at which time our right wing stood to
their arms, and the horse mounted.[4]

At the last moment the horse were ordered to carry a truss of forage on their
saddle, and it was these trusses, dropped to make the muddy way easier, that
guided the infantry.

> The Duke at the head of the right wing of horse, kept close to them, and
> our right wing of foot close to the horse. Thus we marched all night.
> About daybreak [on]18 July the detachment came up to the barriers, where
> they found but a lieutenant and forty men in each of them, who being
> much surprised, gave us only one fire and made off.[4]

The place was between the villages of Elixhem and Neer Hespen on the river
called the Little Geete, very near William III's battlefield of Landen. Churchill
describes the horse on that summer morning: 'On the right, the Scots Greys; next
the Royal Irish Dragoons; next the King's Dragoon Guards; the 5th Dragoon
Guards; the 7th Dragoon Guards; then the Carbiniers; and finally the 3rd
Dragoon Guards. Such is the array.'[5] The Queen's Horse had three squadrons
present, being one of the first regiments to cross the Little Geete and pass the
lines. Parker relates:

> The Elector and Villeroi were strangely confounded; when, to their
> surprise, they saw our right wing of Horse, and a body of foot behind
> them, drawn up in great order, and also the remaining part of our foot
> crowding over the lines, as fast as they could. However the Marquis and
> the Count, according to their orders, drew up [their forces].[4]

Marlborough formed his thirty-eight squadrons into two lines, the first being
entirely British. Taking post on the right flank at the head of the Greys, he drew his
sword and led the red-coated troopers forward at the collected trot. The French
and Bavarian cavalry again halted to receive this onslaught, firing a feeble volley
from the saddle, while The Queen's Horse, quickening into the charge, crashed into
the enemy line. There were a few moments of strenuous sword work, and then The
Queen's Horse broke several squadrons of the Bavarian cuirassiers, riding into
their second line and throwing it into confusion. So the whole mass of horsemen
surged through the French infantry, who turned and ran. At this point the disordered British horse were charged in turn by five fresh French squadrons and were

themselves driven back, giving some of the Bavarian infantry the chance to re-form and retire steadily in square.

Marlborough was himself nearly captured, but rallied his squadrons. A second charge broke the French and Bavarians completely, capturing their guns, and driving their cavalry from the field, while The Queen's Horse cut down a battalion of infantry.

The French lost between 3 and 4,000 men, mostly prisoners, eighteen guns, a pair of kettle drums, and 'many standards and colours'.[1] The total Allied loss was only some fifty men. Marlborough wanted to press the pursuit, but again the Dutch refused, enabling Villeroi to retreat towards Louvain and take up a position along the River Dyle, so saving Brussels and Flanders.

On 20 September the *London Gazette* reported:

> Yesterday the Army made a forage to the right and a guard being posted for the security of the foragers at the village of Wickstadt, a party of the enemy came to the village, and being challenged by the sentinel, were retiring to the wood some distance from them, but Lieut Alexander of Lumley's Horse [KDG] posted himself with his guard of 30 Horse and as many Hussars, between the wood and the village to cut off their retreat.
>
> He completely succeeded in his attack, killed 35 of the enemy and made 40 prisoners.[6]

The regiment helped to cover the investment of Sandvliet. Deane tells us:

> Wch siege was carried on wth. great vigour & expedition; Our men never leaving of cannonading and bombarding them until the governour beet a parley and surrendered themselves upp prisoners of war. The place surrendered the 19th of October. While we lay here my Lord Duke took his leave of the army and set forwards on the 17th towards the city of Vienna to conferr there.[3]

The Queen's Horse marched into winter quarters at Breda, where it rested and was joined by some remounts from England.

Early in 1706 the 3rd Horse was sending detachments to raid into Spain, and on 31 March an Allied army, composed mainly of Portugese troops and numbering some 17,000 men, advanced once more into Spain, under command of General das Minas. The British element, commanded by Galway, amounted to less than 3,000, comprising the 3rd Horse and five infantry battalions. On 7 April the regiment crossed the Selor and advanced on Brocas, where the Duke of Berwick commanded a small enemy force.

As the Allied advance guard debouched from the winding mountain road into the plain in front of the town, Berwick retired down the woody defiles covering the road to Carcares. The Portugese squadrons in front skirmished with the retreating enemy, but were repulsed. The 3rd Horse, together with some Dutch dragoons and Portugese cavalry from Biera, emerged at this moment from the thickest part of the woods and charged the enemy, capturing Major General Don Diego Monroy and the Conde de Vanilleros, together with eighty other prisoners. They continued the pursuit through the forest, returning finally to Brocas, where they camped for the night.[7]

The 3rd Horse marched from Brocas to take part in the siege of Alcantara, situated on a rock near the Tagus in Spanish Estramadura. Alcantara soon fell, yielding 4,000 prisoners, and the regiment marched to the valley of the Xerte, near Plasencia. On 1 May Galway drove the enemy from the banks of the River Tudar, and then sent forward a detachment of the 3rd Horse who successfully blew up the bridge at Almaraz. The advance was beset by trouble, however, as the Portuguese plundered the villages and progress was slow. Then the Portuguese generals refused to advance any further, and the army changed direction, investing Ciudad Rodrigo, which was occupied after token resistance on 26 May. On 1 June Salamanca was reached. The Portuguese then agreed to advance on Madrid and, with the 3rd Horse forming the advance guard, the Guadarrama Pass was secured without opposition and the regiment occupied the northern outskirts of Madrid on 24 June. Three days later the main body arrived and camped near the city, while 'the Spaniards came in such crowds to the camp with music and dancing, that the scene resembled more an assemblage of troops for a spectacle, and review on a festive or gala day, than the camp of a hostile army of foreigners.'[8]

The Archduke Charles was proclaimed King of Spain in his absence. By the time he did eventually arrive, the Duke of Berwick with a French army had resumed the offensive, and was assisted by the partisans of Philip of Anjou, whose numbers had increased because of the brutal Portuguese behaviour. Communications with Portugal were severed and the army left Madrid in a north-easterly direction, hoping to join up with Peterborough and the Austrian forces. Galway reached Guadalajara, where at last Peterborough arrived, but with only 3,000 Austrians, and the advance was continued to Ginjon in the province of Toledo, where the army halted for a month.

As Berwick and the French advanced on Madrid, and more and more Spaniards rallied to his side, the position of the Allied army became more and more dangerous. In early August it was decided that the only course was to retire on Valencia, where the Royal Navy could guarantee a safe base. So the 3rd Horse retired across the Tagus, through the region of La Mancha, across the Xucar and into winter quarters at Valencia, arriving at the end of September. Thus the regiment had marched some 450 miles around Spain and through a thoroughly unsatisfactory campaign.

Back in the Low Countries, the winter quarters were at Breda, 'Lt Genll. Lumley commanding in the Duke of Marlborough's absence.'[3] During this winter of 1706 the Revd Thorold was collecting funds in order to build St Mary's English Church at Rotterdam, and the church records show a list of officers of The Queen's Horse who generously subscribed: Lieutenant General Lumley gave £30, Colonel Palmer £10, Lieutenant Colonel Crowd (Crowther?) £5, the captains and subalterns £2 or £3 each, the quartermaster, John Dodworth, with five others, £4.10. On 12 March a draft of 80 men and 172 horses set out for Lumley's Horse in Flanders.[9] Marlborough arrived back at The Hague in the middle of April, 'and on the 27th of Aprill the whole garrison of Breda marcht out'[3] and the army started to assemble. Early in May the British and Dutch joined forces at Bilsen, soon to be reinforced by the Danes, bringing Marlborough's strength up to about 60,000 men. Then, to his astonishment, he learnt that Villeroi had left the security of his lines behind the River Dyle and was moving forward on Judoigne.

Early on the morning of Sunday, 23 May (Whitsunday), after a night of rain, Marlborough marched towards Ramillies, the plateau between the Geete and the Mehaigne that Villeroi was also hastening to occupy. Soon after eight o'clock, as the early morning mist started to clear after the night's rain, the British horse met a detachment of the enemy crossing the plateau, and a little later the united armies of the French, Spanish and Bavarians under Villeroi and the Elector of Bavaria were to be seen, with their right bounded by the Mehaigne and the village of Taviers, and their left by the village of Autre Eglise. In the centre of the plateau were the villages of Offuz and Ramillies, and a little distance in the rear was a prominent mound known as the tomb of Ottomond. The whole line extended for some four miles.

Marlborough at once saw the weakness of Villeroi's position, for he had thrown his wings so far forward that the French line was drawn up in the shape of a crescent. Moreover the French guns, concentrated in the villages of Taviers and Ramillies, were too far apart to allow effective cross-fire, and the boggy nature of the ground in front of Offuz and Autre Eglise would lock the French troops into their positions and stop them from launching any counter-attack.

At the beginning of the action The Queen's Horse, with the rest of the British horse, was posted in reserve on the heights of Foulz on the extreme right of the Allied position, in support of the British infantry, whose main role was to tie down the considerable enemy forces stationed in Offuz and Autre Eglise. Marlborough opened his attack with a strong feint by the British infantry against the French left; ' Villeroi perceiving this, immediately ordered off from the plain an entire line, both

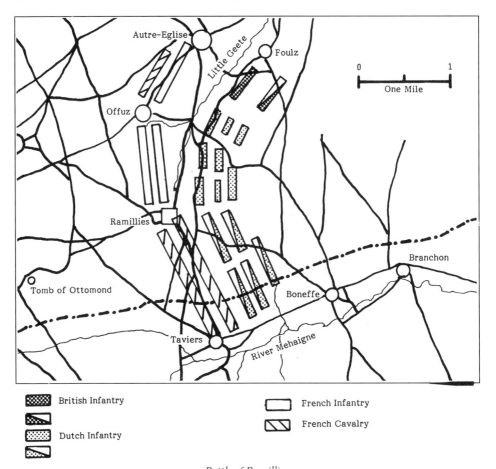

▨ British Infantry   ▭ French Infantry

▨ Dutch Infantry   ◰ French Cavalry

Battle of Ramillies

of horse and foot, to reinforce those on the Geet.'⁴ In the meantime the Dutch on the left of the Allied position had assaulted and taken Taviers, and twelve of their battalions had attacked Ramillies. On the right the British infantry, wading through the marshes of the Geete, were pressing back the French defenders of Autre Eglise and Offuz when they were ordered to withdraw. The men, not understanding the reason for this order, fell back sullenly.

> When the Duke observed that these (the French reinforcements) had arrived there, he sent orders to our right wing to retire easily up the hill, without altering their aspect. This we did, until our rear line had got on the back of the rising ground, out of sight of the enemy. But the front line halted on the summit of the hill in full view of them, and there stood,

ready to march down and attack them. As soon as our rear line had retired
out of sight of the enemy, they immediately faced to the left, and both
horse and foot marched down to the plain as fast as they could.[4]

As these troops were following a convex line as opposed to the French crescent-
shaped position, they rapidly reinforced the main attack on Ramillies. Villeroi was
caught completely off balance, with his reserves reinforcing Autre Eglise on his left
wing and unable to make the much longer march back to Ramillies in time, which
his crescent- shaped position demanded.

The Dutch cavalry now charged over the plain between Ramillies and Taviers
and routed the first line of the French cavalry, but were driven back in confusion by
the second line. During the mêlée Marlborough was nearly captured, and only
escaped because to the bravery of his aide-de-camp and equerry, the latter having
his head carried away by a round shot.

The Dutch cavalry re-formed and were reinforced by the Danish cavalry on their
right. At the same time the assault on Ramillies was developing, with the Dutch
infantry managing to advance from Taviers around the right flank of the French.
As the Allied cavalry charged again, the Dutch and Danish horsemen cut the
French Maison du Roi to pieces, and then drove the rest of the French cavalry from
the field. They then turned on the enemy infantry and rode them down, sabring
and capturing great numbers.

'The Horse on our left fell upon there Foote, on there right, of which they slew
great numbers, cutting about 20 battalions to pieces whose coulers we took and
likewise there cannon.'[3] Villeroi tried to bring forward his remaining cavalry from
the left, but they were hampered by the careless manner in which the French had
left their baggage encumbering the ground. The enemy infantry in Ramillies now
started to retreat in good order, as did the garrisons of Offuz and Autre Eglise,
covered by the cavalry of the French left wing.

At this point Marlborough ordered the British infantry left crowning the heights
on his right, and supported by the British horse, who had remained on the heights
of Foulz all day, to advance. It was now 4.30 p.m. and at last the horse were able to
seize their opportunity, with the 5th Dragoons and Greys on the right and The
Queen's Horse next in line to them, the seven regiments of British horse (also the
3rd, 5th, 6th and 7th Dragoon Guards) charged in pursuit of the retreating enemy.

The Horse on there left wing seemed to make a stand to gayne time for
there Foot to retire, but our folks charged them soe quick and with soe
much bravery yt. the enemyes Horse clearly abandoned there Foott, and
our Dragoons pushing into the village of Auterglisse made a terrible
slaughter of the enemy. The French Kings own regiment of Foott called

the Regiment du Roy begged for a quarter & delivered up there arms and
delivered up there coulers to my Lord John Hayes Draggoones.[3]

Certainly Lord John Hayes accepted the surrender of the Régiment du Roi, but
his dragoons, the Greys, were not alone in achieving their surrender. As David
Chandler points out, 'The Scots Greys and the King's Dragoon Guards thundered
forward at the gallop (a rarity at this period) to round up the Régiment du Roi as
the French fled for Judoigne.'[10] The Regimental Records of the King's Dragoon
Guards also relate that 'they compelled the Regt du Roi to lay down their arms &
surrender themselves prisoners of war'.[1]

The Queen's Horse continued the pursuit until 2 a.m. the following morning.
'Near the cross roads of St Pierre-Geest, the Elector of Bavaria and Marshal Villeroi
were almost captured by General Wood when, at the head of the English Dragoon
Guards, he crashed at a gallop into the Bavarian Horse, capturing their kettle
drums and two Lieutenant Generals.'[11] A painting of The Queen's Horse seizing
these kettledrums now hangs in the Officers' Mess.

The spoils were enormous, it was literally the 'Annus Mirabilis': most of the
enemy's baggage was captured, fifty-six guns, eighty standards and colours, and
several more sets of kettledrums; 600 officers and 4,000 men had been taken
prisoner, and the total enemy losses came to more than 15,000. But at Ramillies the
brunt of the hard fighting was borne by the Dutch horse and foot, with the British
horse only being fully engaged towards the end of the day, and in the pursuit.

Most of the fugitives headed blindly in the direction of Judoigne,

> but the ways were blocked by broken-down baggage waggons and
> abandoned guns, and the crush and confusion was appalling. The British
> Cavalry, being quite fresh, quickly took up the pursuit over the tableland.
> The guns and baggage fell an easy prey, but these were left to others,
> while the red-coated troopers pressed on, like hounds running for blood,
> after the beaten enemy. Not until two o'clock in the morning did the
> Cavalry pause, having by that time reached Meldert, fifteen miles from the
> battlefield; nay, even then Lord Orkney with some few squadrons spurred
> on to Louvain itself, rekindling the panic which set the unhappy French
> once more in flight across the Dyle.[12]

For a couple of hours the weary troopers off-saddled and snatched a few moments' rest with their horses' reins looped over their arms, but then the pursuit was
resumed. The Dyle was crossed on 14 May and for six more days Marlborough
pressed the demoralised enemy. Finally he gave his exhausted troops a halt, as the
shattered French retired towards Ghent. The whole of the Spanish Netherlands had

been freed at a single blow, for Louvain, Lierre, Ghent, Bruges, Brussels, Malines and Alost fell into Allied hands, and when the garrisons of Oudenarde, Antwerp and Ostend were summoned, they also surrendered. The Queen's Horse was engaged in the siege of Ostend, which lasted from 6 to 27 June, along with the Scots Greys and the 5th Dragoons.[13] The garrison was allowed to march out with its baggage, but without military honours.

As soon as Ostend had fallen the army was assembled for the siege of Menin, a fortress which was one of Vauban's masterpieces. It fell in early August. Dendermond and Ath were then besieged and taken before the victorious troops were allowed to return to their winter quarters.

SOURCES

1 Records of The King's Dragoon Guards, Regimental Museum, 1st The Queen's Dragoon Guards, Cardiff Castle.

2 *London Gazette*, 1705

3 J. M. Deane, *A Journal of Marlborough's Campaigns during the War of the Spanish Succession, 1704-1711*, Society for Army Historical Research, Special Publication, no.12, 1984.

4 *The Military Memoirs of Robert Parker*, Longman, 1969.

5 W. S. Churchill, *Marlborough, His Life and Times*.

6 *London Gazette*, 1 October 1705.

7 *London Gazette*, 1706.

8 R. Cannon, *Historical Records of the British Army, 2nd Dragoon Guards*, 1837.

9 *War Office IV*, vol. 4, entry of 12.3.1706.

10 D. Chandler, *Marlborough as a Military Commander*, 1973.

11 G. M. Trevelyan, *Ramillies*, 1932.

12 Sir John Fortescue, *History of the British Army*, vol. 1, 1910.

13 *Journal of the Society for Army Historical Research*, vol. 13, p. 202.

7

Almanza, Almenara, Oudenarde
1707-1709

In the Peninsula reinforcements had reached Galway at Valencia during January 1707. Galway and Das Minas decided to destroy the French and Spanish magazines in Murcia, and then move up the Guadalquivir and on to Madrid. The 3rd Horse [Bays] took to the field with the rest of the army early in April, and on the 10th of that month crossed into Murcia. After driving some enemy detachments back into Castile, they camped on the plains at the foot of the mountains near Villena, to which Galway laid siege. Information was then received that the French and Spanish, under the Duke of Berwick, who were expecting reinforcements led by the Duke of Orleans, were advancing on Almanza, some twenty-five miles distant. Galway and Das Minas determined to engage the French, but apparently without finding out their strength. Berwick had a force of about 25,000 French and Spanish, and a strong train of artillery, whereas Galway could only muster some 15,000, of whom only a third were British, half were Portuguese, and the remainder were Dutch, German and Huguenot.

Berwick drew up his army in a strong position in two lines on the plain to the south of Almanza, his infantry in the centre and the cavalry on each flank, the French on the left and the Spanish on the right. On 25 April Galway marched his army fifteen miles through rough country so that the men, arriving at Almanza at three o'clock in the afternoon, were tired and parched. Galway and Das Minas nevertheless determined to attack at once, in spite of the enemy's numbers and the fact that in the event of a reverse they would have to retreat across an open plain. Das Minas claimed the post of honour for his Portuguese on the right wing, while the British and Dutch formed up on the left, but with some Portuguese cavalry squadrons in their second line. Being short of cavalry, Galway interspersed his squadrons with battalions of foot. On the extreme left were the 1st Royal Dragoons with the 4th and 7th Hussars, next came Wade's Brigade which had four regiments of foot together with the 8th Hussars and Peterborough's Dragoons; then came the 3rd Horse, in the midst of two regiments of Dutch horse on their left and four more on their right. Berwick, thinking that the British contingent would be on the right, posted his French troops to oppose them. As soon as he saw that the British were, in fact, on the left, he transferred his best French troops back to his right. The shouts of the French infantry, 'Tout à l'heure, messieurs, tout à l'heure', could be heard as they marched to their new positions.

Galway immediately opened his assault, without any preliminary artillery preparation, leading an advance of the 3rd Horse and the rest of the cavalry of the left wing against the Spanish. They were driven back by sheer weight of numbers, but were assisted in their withdrawal by two of the infantry battalions, the 6th [Warwicks] and 33rd [Duke of Wellington's] Foot, whom Galway had interlined with them, and who now opened a galling fire on the flank of the Spanish cavalry. This gave the 3rd Horse time to re-form, and then, with a second spirited charge, they drove back the Spanish cavalry in confusion.

The rest of the British foot had advanced directly on the enemy infantry, and had beaten them back, in spite of their superior numbers, onto their second line. The Guards and 2nd Foot [Queen's] even managed to break their way through the second line, pursuing the fleeing Spanish infantry up to the walls of Almanza.

Whilst the battle was going well on the left, the Portuguese on the right had done nothing. Berwick, seeing their inaction, directed the French cavalry of his left wing to attack the Portugese cavalry, whereupon the Portuguese first line fled from the field, to be followed at once by their second line. The Portugese infantry, left alone to face the full force of the French attack, resisted courageously for a while, but were then overwhelmed by sheer numbers and driven from the field.

Berwick was now able to bring his French troops against the British right flank, and at this point 'two French regiments being far advanced, the Lord Tyrawley ordered Colonel Roper, who commanded Major-General Harvey's Horse [Bays], to attack them, which was done with so much vigour that they broke through them and made them beg for quarter, before the enemy's cavalry could come to their assistance.'[1] The troopers, losing their formation in the aftermath of the charge, became dispersed, and were in turn badly mauled by the arrival of fresh French squadrons. Colonel Roper was killed, as were the commanding officers of the other four British cavalry regiments. The 3rd Horse also lost Captain Nicholson and Lieutenant Bridger killed, and Lieutenant Gee, Cornet Broughton and Quarter-master Sarden wounded and taken prisoner. There is no record of the rank and file casualties.

The Irish regiments in French pay had many men who recognised their op-ponents in the 3rd Horse from their years of campaigning in Ireland, and they shouted to various officers and NCOs of the 3rd Horse to stop their attacks.

The British and Dutch infantry fought on doggedly, but were soon surrounded, and cut down or captured. The 3rd Horse, having re-formed, saw the 11th Foot

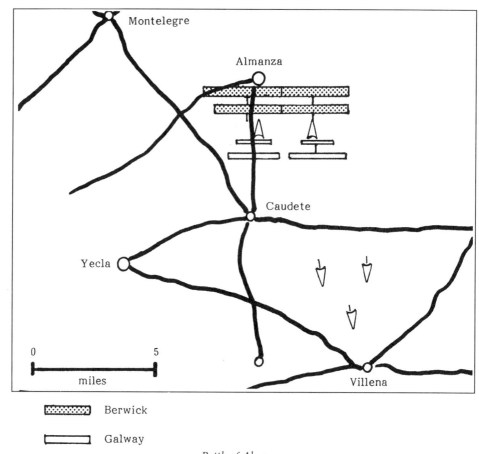

Berwick

Galway

Battle of Almanza

[Devons] still fighting bravely, and with the shout 'Go and save those brave men' charged once more to give what help they could.

General Shrimpton managed to bring some 2,000 foot off the field and retreat to the Caudet hills, but there they were surrounded and forced to surrender the following day. The survivors of the 3rd Horse retreated from the field in good order. Galway was not a great general, but he was a veteran soldier and managed to extricate 1,000 horse and 1,500 infantry, together with the six British guns. This sad remnant retired some twenty miles to Alcira.

The action had lasted for two hours, costing Galway 4,000 killed and wounded, and 3,000 prisoners of war. Fortescue says:

> The British alone lost eighty eight officers killed, and two hundred and eighty six captured, of whom ninety two were wounded. The simple

fact was that, as the bulk of the Portuguese would not fight, the action resolved itself into an attack of 8,000 British, Dutch and Germans upon thrice their number of French and Spaniards, in an open plain. The most singular circumstance in this fatal day was that the French were commanded by an Englishman, Berwick (the natural son of James II), and the English by a Frenchman, the gallant but luckless Ruvigny (a Huguenot).[2]

The only Portugese cavalry to behave well were those led personally by Das Minas, who was followed into action by his mistress, dressed as a man. She was killed at his side.

The Allied losses were so severe that Galway was forced back onto the defensive for the rest of the year, retiring to Catalonia, where he used the 3rd Horse and the remainder of his cavalry with the utmost skill. The French besieged Valencia, Xatvia, Lerida and Denia, and then retired to winter quarters.

During the winter of 1707-08 the losses were made good from England, and the 3rd Horse rapidly regained its form:

> The Regiment of Horse of General Harvey is certainly one of the finest regiments that ever was seen, and the worst horse they have is worth fifty pistoles. The Count de Noyelles [Portuguese commander in chief in Catalonia], who some days before his death [1708] reviewed that regiment, was so well pleased with it that he returned thanks to Colonel Goring, who commands it, and to the rest of the officers, and told them that in his letter to the Duke of Marlborough he would take a particular notice of the good condition wherein he had found it.[3]

In the spring of 1708 the 3rd Horse camped in a valley near the town of Monblanco, then moved to the plain of Cervera, thirty miles from Tarragona, where they remained on the defensive, confining their activities to a few skirmishes amidst the mountain passes of Catalonia. That autumn they went into village cantonments in Catalonia.

On 14 January 1707 the first set of 'Clothing Regulations' was issued, 'The sole responsibility for the pay and equipment of a regiment rests with the Colonel, who is held responsible in his fortune, and in his character, for the supplies of his regiment.' By a warrant of Queen Anne in 1707, a Board of General Officers was formed, whose duty it was to select, seal and issue patterns for the clothing of each regiment by the colonel. The clothing was supplied by the colonel, then inspected by a regimental board, who compared it with the sealed patterns of the General Board. Any complaints were heard by the General Board, whose decision was final.[4]

In February 1707 the War Office arranged for 180 horses and 120 men to be sent out as reinforcements for Lumley's Horse [KDG] in Flanders,[5] who were in their winter quarters at Ghent. [6] At this time the horse were again issued with cuirasses, an item of equipment which they thought they had seen the last of in 1698. Marlborough wrote to Eugène that 'the Queene has decided to give cuirasses to the English cavalry', but he only allowed the breastplate to be worn under the scarlet coat, presumably because the British horse never showed their backs to an enemy. On 10 June,'This day his Grace the Duke of Marlborough reviewed the horse and dragoons of the right wing of the first line, those of Great Britain appeared for the first time in their cuirasses; and the whole was very complete and in good order.'[7] The reference to Great Britain was a reminder that the Act of Union between England and Scotland had just been passed in January.

> The city of Ghant [sic] being the winter quarters for the major part of the English forces; & Lt. Genll Ingoldsby commanding the garrison in his Grace and Duke of Marlborough's absence. Lt. Genll Lumley being come over from England, on 2nd May, 1707 he revewed the whole garrison; and he having recd. orders from his Grace the Duke of Marlborough, our Capt. Genll, to march out the forces, he according to his Grace's orders marched out from Ghant on the 5th of May, being Munday, both foote and horse together with the trayne of artillery.[8]

Throughout 1707 Marlborough 'was back in the same tutelage that had ruined all his plans of campaign except those of Blenheim and Ramillies',[9] with his allies crabbing every proposed move. Villeroi had been replaced by Vendôme, with an army of 100,000 men against Marlborough's 80,000, and operating from the strong bases of Mons, Lille and Tournai, whereas Marlborough had to cover the weakly fortified towns of Brussels, Bruges and Louvain. Furthermore the successes at Almanza and against the Germans at Stolhofen had raised French morale.

The Queen's Horse were in Palmer's brigade on the right wing together with the 3rd, 5th, 6th and 7th Dragoon Guards.

> The Duke of Vendôme drew the French army behind their lines near Mons; and though he outnumbered us considerably, yet he had positive orders not to hazard a battle, unless it was in defence of his lines. We lay at Meldert near six weeks, in hope of drawing him from behind them; at length he ventured out, and encamped at Genappes, six leagues from us. The Duke suffered him to lie quiet about eight or ten days; then on our beating Tattoo, we decamped on a sudden. We marched all night, and by the time it was day, the Duke with the right wing of the horse had advanced within less than half a league of the enemy: but they had just

struck their tents, and were marching off in some confusion. The Duke immediately ordered Count Tilly to advance with the horse and fall on their rear. But Tilly could do nothing in the enclosures, as he had none but horse, so Vendôme marched off at his leisure ... The van of our horse had advanced very near Vendôme's camp. He was not expecting this second visit, struck his camp in a great hurry, marched off as fast as he could, and never stopped till he found himself safe within his lines ... We lay near three weeks weather bound. The French had thrown up new lines; and the Duke wanted to take a view of these lines. With this view he ordered the whole army to make a grand forage that way, he ordered also sixty squadrons and twenty battalions to cover the foragers, and to guard him while he was viewing the ground. But the Frenchman took care to keep within his lines. As the country was very plentiful, our camp was provided with forage till the middle of October, when we went into quarters.[10]

Foraging caused ill feeling and led to partisan activity, Deane relates how 'A partyzan party of about 30 men, lying in a wood upon the roade side, fell upon and plundered severall wagons and stript some.' His comment on 1707 was, 'I am sorry I cannot give you an account of more action in this campaigne but it was not our ffault, ffor still as we advanced they retracted from one place to another, still threatening what they would do but all proved but wind and a French gascoig-nade; and this wonderfull never beeten Duke Vandosme this year did nothing.'[8] The Queen's Horse returned once more to Ghent for the winter.

During the spring of 1708 more than a hundred men and horses joined The Queen's Horse, bringing the regiment up to strength, and 'on Tuesday May the 11th marcht out the Horse and Foott'[8] from Ghent to Brussels, where it was posted on the right of the first line of cavalry. The French now numbered nearly 100,000 men, with the son of the Dauphin, the Duke of Burgundy, in supreme command, Marshal Vendôme acting as both tutor and guide; in addition, both the Duke of Burgundy's brother, the Duc de Berry, and the Pretender, or as he titled himself, the Chevalier de St George, were present. Marlborough was joined by the Prince of Hanover, later to become George II, and then by Eugène.

On 26 June 'his Grace was informed of the treacherousness of the city of Ghant, and that the enemy had full possession of the same, both toun and castle, and the city of Brugge likewise'.[8] The citizens of these two towns of the Catholic provinces of the Netherlands loathed the high-handed administration of the Dutch, and had been bribed by the French to open their gates. Vendôme left strong garrisons in both places and decided to besiege Oudenarde. It was outside Oudenarde that Marlborough, after a series of forced marches, confronted the French on 11 July.

The Queen's Horse were now brigaded with Cadogan's [5 DG] and Palmes's [Carabiniers] under Brigadier-General Kellum. Even though the regiment carries Oudenarde as one of its battle honours, the British cavalry took no active part in the battle. Having crossed the Scheldt by the town bridge in Oudenarde in the late afternoon, they took post on the heights of Bevere, where they covered the Allied right wing against the threatened attack of a large body of French troops, which never materialised. There they remained without drawing sword for the whole of the engagement. 'Out on the slopes beyond the Ghent road long lines of scarlet horsemen sat motionless upon their horses, as if at a review.'[11]

Oudenarde was a great victory, and Deane commented: 'The brave and wise conduct of the Duke of Marlborough, and Prince Eujeane likewise, wch. is never wanting, cannott but be worthily commended for theire courage in this wonderful undertaking.'[8] Marlborough, in writing to his wife the following day, said, 'I thank God the English have suffered less than any of the other troops; none of our English horse having been engaged.' As night fell over the battlefield the troopers dismounted, picketed their horses, and made the best they could on 'the cold and wett ground, for it was a verry wett morning'.[8]

At daybreak Lieutenant General Lumley gathered forty squadrons, including The Queen's Horse, and set off in pursuit of the French. The French rearguard was soon overtaken and attacked. By the end of the day the cavalry were within two miles of Ghent, and had forced a series of lines which the French had constructed from Ypres to the River Lys. By the evening of 13 July the whole of Marlborough's army was camped on French territory. Columns were sent out to forage and levy contributions, while the outskirts of Arras were put to the torch, bringing home to the French that the war was now on their own soil.

Marlborough next decided to besiege Lille, the capital of French Flanders. It was invested on 12 August, The Queen's Horse covering the operations and providing escorts for convoys of ammunition and stores en route to the besiegers from Ostend via Menin. But Lille proved a difficult place to capture, with the French making every effort to succour and relieve it. Vendôme, with Burgundy and Berwick, entered the plain of Lille to find themselves confronted by Marlborough and Eugène. They hesitated, and then retreated. During September the Allies made several abortive assaults on the city, while Vendôme took up positions along the Scheldt and Scarpe. From here the French attempted to interfere with the convoys coming to the besieging army, but were driven off. By September 'all manner of provisions began to grow very scarce, and the ammunition of the besiegers began to fall short'.[10] Major General Webb was ordered to escort a convoy of 600 waggons for Lille with 6,000 foot and 100 horse. 'The Duke of Burgundy had an account of all this, and ordered La Motte to march with a body of 24,000 horse and foot and twelve pieces of cannon to intercept this convoy. La Motte came up with Webb as he was passing on the back of the wood of Wynendael.'[10] Deane, who was engaged, writes: 'But on the 17th we had a regular battle with the enemy, in wch engagement the enemy lost on the spot above 2,000 men, and as many or more wounded; and we kept the feilde and pass and brought our convoy safe. The rest of our horse could not come up by reason of a morross.'[8] But the presence of The Queen's Horse was felt, for Parker commented, 'And after all, had not Cadogan come up with his squadrons, it would be hard to say what might have been the consequence.'[10]

On the night of 26 November Marlborough, with The Queen's Horse, forced the line of the Scheldt, taking 1,000 prisoners. He then sent back the main part of the force to Lille and pushed on with the cavalry and two battalions of the Guards to Alost. On 9 December Boufflers, who had conducted the French resistance at Lille with distinction, eventually surrendered, the French garrison marching out with the honours of war. Marlborough and Eugène at once invested Ghent, which soon surrendered, to be followed by Bruges. Having now re-established his position, Marlborough sent the troops into winter quarters.

Deane, the private soldier, commented:

> Thus after a very long, tiresome, troublesome, mischeivous and strange, yet verry successfull campaigne, we are safe arrived in garrison; for wch we ought to returne thanks to God for preserving us in the many dangers we have from time to time been exposed unto; and endeaver to live as we ought to doe, like men who carry our lives in our hands; not knowing how soone it may be our turns to be cutt off, as we have been eyewittnesses that many brave fellows have been before us; that soe we may expect still greater success the Summer ensueing.[8]

'As this spring was very wet, it was the beginning of June before we could take the field.'[10] The Queen's Horse marched from Ghent on 17 June 1709, joining the army near Menin on the 21st, where Marlborough confronted the French on the plain of Lille. Villars, now in command of the French,

> drained all the garrisons near him to give them a warm reception; particularly from Tournai he drew 3,000 of their best troops. This is what our generals wanted, for the very evening that these troops had joined him, on beating Tattoo, our army decamped, and marched away to the left. When day appeared, they found themselves before Tournai, which was invested immediately. [10]

The Queen's Horse, with the 3rd and 5th Dragoon Guards and the Greys, formed part of a force of thirty squadrons and seventeen battalions under General Lumley, whose task it was to invest the town from the opposite side of the River Scheldt to the main army. Here, 'from the constant attempts of the enemy to throw in a body of troops to the relief of the garrison, the duty of this detachment was uncommonly severe'.[6] Tournai, fortified by Vauban, was extremely strong but, with its garrison diminished, it surrendered on 3 September.

Back in the Peninsula in the spring of 1709, the Allied army was too weak for any major move. During the summer months the 3rd Horse encamped on the banks of the River Segre. On 26 August the regiment forded the river and advanced through fertile country to the siege of Balaguer, which was soon captured. The regiment then moved twelve miles to Ager, to learn that King Philip was advancing at the head of a French and Spanish force to relieve the besieged towns. The enemy army proved to be more menacing than committed, and there was only some slight skirmishing. The Allies, having taken Ager, placed garrisons in the two captured towns, recrossed the Segre, and the 3rd Horse went into winter quarters.

On 12 June 1710 Philip advanced, camping at Belcayre, and the following day came within cannon shot of the Allies, but soon retreated. The 3rd Horse, led by Lieutenant General Stanhope, charged his retiring right wing with great courage and threw several enemy squadrons into confusion. On 26 July Stanhope was sent forward to secure the strategic pass of Alfaras, taking with him four regiments of cavalry, including the 3rd Horse. The next day Stanhope, well in advance of the rest of the army, found nineteen enemy squadrons supported by some infantry on the march to take possession of a plateau overlooking Almenara. Stanhope at once occupied some rising ground to his front, and as soon as the main body of the Allied army had come up, sought permission to charge the enemy. By now the French and Spanish force had increased to a first line of twenty-two squadrons. Their right flank was guarded by a body of infantry in a church and they were backed by a second line of another twenty squadrons and nine battalions of infantry. Stanhope's sixteen squadrons were reinforced by a further six squadrons and, with Stanhope himself at the head of the right-hand squadron of the 3rd Horse, they advanced to the attack in the light of the evening sun. The enemy's first line moved forward at the same time, and the two opposing ranks met in the shock of a cavalry charge. King Philip's Life Guards fought the 3rd Horse in a fierce hand-to-hand contest, but were soon routed. Stanhope killed General Amenzega in personal combat, and the enemy's first line broke and fled, pursued by the 3rd Horse, who then charged and broke the second enemy line. As the enemy cavalry fled, losing the standard and kettledrums of their Life Guards and the standard of the Regiment of Granada, their infantry threw away their weapons and joined the mad rush from the field. The onset of darkness added to the terror of the flight, but it prevented the 3rd Horse and British cavalry from taking large numbers of prisoners amongst the passes and defiles of the mountains.

The British lost 73 troopers killed and 113 wounded, testifying to the fierceness of the combat. The loss among the officers was proportionately more severe: Brigadier Generals Earl Rochfort and Count Nassau, Colonel Travers, Captain La Porte, Cornets Garson and Webb and the quartermaster were killed; while Generals Stanhope and Carpenter, Lieutenant Colonel Bland, Captains Ravenel, Willis, Moor and Naizon, Lieutenants Mills, Patterson, Jobber, Heron and Wood, and Cornets Wildgoose, DuCase and a quartermaster were wounded.

Almenara, entirely a cavalry affair, brought the greatest credit to the 3rd Horse and the other British cavalry engaged. After the battle the troopers camped in front of Almenara before marching towards the Cinca, to take the war into Aragon.

SOURCES

1 R. Cannon, *Historical Records of the British Army, 2nd Dragoon Guards*, 1837. 'The Annals of Queen Anne'.

2 Sir John Fortescue, *History of the British Army*, vol. 1, 1910.

3 R. Cannon, *Present State of Europe*, vol. 19, p. 225.

4 Lieutenant Colonel E. A. H. Webb, *A History of the Services of the 17th Regiment*, 1911.

5 *War Office IV*, vol. 5, February 1707.

6 Records of the King's Dragoon Guards, Regimental Museum, 1st The Queen's Dragoon Guards, Cardiff Castle.

7 *London Gazette*, 10 June 1707.

8 J. M. Deane, *A Journal of Marlborough's Campaigns during the War of the Spanish Succession, 1704-1711*, Society for Army Historical Research, Special Publication no. 12, 1984.

9 G. M. Trevelyan, *Ramillies*, 1932.

10 *The Military Memoirs of Robert Parker*, Longman, 1968.

11 W. S. Churchill, *Marlborough, His Life and Times*.

12 C. T. Atkinson, 'Gleanings from the Cathcart MSS', *Journal of the Society for Army Historical Research*, vol. 29, 1951, p. 98.

8

Malplaquet, Brihuega
1709-1711

Back in Flanders in 1709, Robert Parker writes,

> Villars crossed the Scheldt near Valenciennes, with a design of cutting off
> the detachment [which included The Queen's Horse, and had been sent
> forward] under the Prince [of Hesse]: but perceiving that our army was
> very near him, he stopped short at Malplaquet; and as the place was well
> situated, either for disturbing the siege, or standing a battle, he set his men
> to work in cutting down the trees of those woods, within which he had
> drawn up his army; and in throwing up several strong entrenchments, one
> behind another, in the intervals of the woods.[1]

Cathcart comments that the Allied detachment under the Prince of Hesse was left
alone because 'The French did not stir and by 1 p.m. on the 7th September the
"Grand Army" was arriving to sustain Hesse's advanced-guard, whose position
was thus secured. . .'[2]

Marlborough, having been reinforced by the troops besieging Mons, decided on
a frontal assault on the French position at Malplaquet. 'So on the 11th September, about eight in the morning the battle began. Prince Eugène commanded on
the right, the Prince of Orange and Count Tilly on the left, and the Duke of
Marlborough in the centre.'[1] The Queen's Horse [KDG] was stationed on the right
wing, together with the 3rd, 5th, 7th Dragoon Guards, the Greys and the 5th
Dragoons, under command of Brigadier General Kellum. After a preliminary cannonade by the artillery of both sides,

> General Schulenberg, at the head of 36 German battalions, attacked the
> enemy on their left, in the wood of Sart: the Earl of Orkney, with the other
> British Generals, attacked the village of Tasnières, and the opens between
> the woods of Sart and Janfart: the Prince of Hesse attacked the village of
> Blaregnies, and the opens between the woods of Lamert and Janfart; and

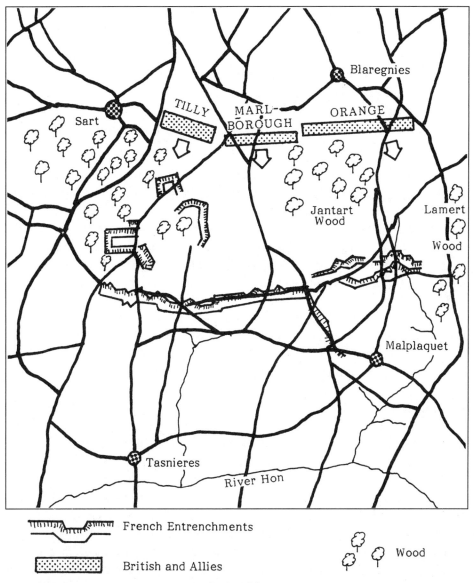

French Entrenchments

British and Allies

Wood

Battle of Malplaquet

the Prince of Orange attacked the village of Malplaquet, and their right in the wood of Lamert.

Thus were the infantry on both sides engaged with a most terrible fire, and prodigious slaughter. As our infantry advanced, the cavalry kept close in the rear of them.[1]

Deane describes the action: 'The dispute was verry hott and continued soe for 3 long hours.' At length Orkney's British Brigade in the centre,

> drove them cleare out of the wood, and there, some of our forces, being drawne up, fell upon them and broke them confusedly. And they runn, all that could runn, behind their horse for shelter; for the French doth really think that they have no business to stand against us in the feild except they have eyther a ligne or breastwork, a wood or a wall, before them to cover them.[2]

The battle swayed to and fro, until Boufflers at the head of the French Gens d'Armes drove back the infantry from their hard-won gains, but was eventually checked at a parapet lined by the British foot. Cathcart tells how Marlborough personally brought up the thirty squadrons of the right wing, which included The Queen's Horse. Then the British cavalry charged the enemy cavalry 'that stood in three lines upon the hill' and broke them. The supporting Hanoverian and Prussian squadrons, however, 'did not show a proper forwardness' and turned about, failing to support the British first line and so allowing the French to rally. This lack of 'forwardness' was only partially corrected, because 'our foot could not persuade them to halt but by firing upon them'.[3]

At this point the French Household Cavalry, the Maison du Roi, charged The Queen's Horse and the British cavalry, penetrating their first two lines and throwing the third into disorder. Eugène personally brought up more of the Allied cavalry, and the battle developed into a series of charges and counter-charges — 'smashing work there was on both sides for a good while, and sometimes they gave ground and sometimes wee. Soe that for a great while one could not tell wch. way it would goe. Our horse gave them genll. onsett and broke them at a great rate, and begun to drive them Jehue-like, and forced them to turn tayle and runn as lustyly as they used to.'[2] Finally, at three o'clock in the afternoon, the French were forced to retreat. The Allied foot were 'too exhausted' to pursue and make the victory 'more compleat'.[3] However, 'there being 40 squadrons of our horse ordered to pursue them – the wch. they did, giving but verry few quarters, and made a very great slaughter.'[2] Another account tells how the Allied cavalry, including The Queen's Horse 'rushed upon the enemy with overwhelming fury, and threw part of the French army into disorder, and a most fearful slaughter followed. The pursuit was continued as far as the village of Quiverain.'[4] Lord Orkney, in command of the British infantry, wrote: 'We broke through them, particularly four squadrons of English. Jemmy Campbell, at the head of the Grey Dragoons [Greys], behaved like an angel, broke through the lines. So did Panton, with little Lord Lumley at the head of one of Lumley's [KDG] and one of Wood's [3 DG].[5]

Malplaquet was the bloodiest of Marlborough's battles, neither side giving quarter, the fighting savage. The French suffered 12,000 casualties, most of them among their cavalry in the centre, and the Allies lost not less than 20,000, a quarter of their entire strength (of whom 11,000 were Dutch). The Queen's Horse was fortunate for it had only Lieutenant Stormont and ten troopers killed, with five officers and thirty-nine men wounded. It also lost fifty-two horses.[3,4] (Dalton's *English Army Lists*, volume 6, gives Stormont's name as Lieutenant Stirrup!) The Allies captured sixteen pieces of cannon, twenty-six standards and twenty colours, with some 3,000 prisoners. The news of this costly victory shook Europe with its tale of slaughter and gave Marlborough's enemies grounds for the discontent that eventually undermined him.

Marlborough himself was deeply affected by the scale of his losses, and it was three days before the army returned to besiege Mons. The siege progressed slowly because of heavy rain and the marshy nature of the ground.

> The enemy were thought likely to attempt some interference with the siege. The cavalry [which included The Queen's Horse], in addition to having to find large escorts for the siege train which came from Brussels, had to keep a vigilant watch on Boufflers's movements and send out large patrols and covering parties. Just before Mons fell a reported French move to Bavai seemed to herald an attack and the covering troops were kept in readiness, though foraging parties went out as usual. No attack developed.[3]

Mons surrendered on 9 October, 'and on Sunday the 17th there was a generall thanksgiving throughout the camp for the taking of Mons, and our glorious victorye this campaigne although with greate effusion of blood. On the 18th of October the army broke up camp and sent forwards ther marches all Regimts. towards ther appointed quarters.'[2]

The campaign of 1710 was one of sieges rather than battles. The Dutch were 'as unlikely to agree to our attempting anything risky; when there's a battle our all's at stake.'[3] The French, too, were on the defensive; both sides were suffering from the losses at Malplaquet. The Queen's Horse was moved from Brigadier Kellum's brigade to that of Brigadier of Horse Sybourg, joining up with Stair's [Greys] and Ross's Dragoons [5th Lancers], which formed, along with the rest of the British cavalry, part of of the first line of the right wing of the Allied army.[6]

Operations opened with the capture of the castle of Mortagne, and on 20 April the army

> in the evening marched up to the enemy's lines, and passed them without

opposition at Pont-à-Vendin, and advanced into the plains of Douay. Villars could not bring his army into the field time enough to oppose our passing their lines. Upon this our generals invested Douai, in which there was a garrison of 10,000 regular troops. The works were strong, and a great many of them, and well stored with all manner of necessaries. Soon after Douai was invested Villars advanced and drew up within cannon-shot of the covering force [which included The Queen's Horse]. He seemed determined to give the Duke battle, and began to cannonade us with great fury. The cannonading held till night; at which time he retired out of reach of gun-shot, and there stood looking at us all the next day. Upon this we fell to throwing up a sort of entrenchment, to cover our men from his cannon: and Villars retired before the plains of Arras.[1]

While Douai was under siege, on 12 June, 'His Grace vewed the English Horse in particular, who appeared verry noble to the true satisfaction of the generalls and others that attended my Lord Duke'.[2] Douai, 'after a siege of eight weeks, capitulated, and surrendered both the town and fort of Scharpe. After the surrender of Douai, our generals advanced up to Villars, to try if he would stand a battle: but he thought fit to retire behind the Sensee, where he secured his army, and at the same time prevented us from laying siege to Arras. Upon this we invested Bethune.'[1]

There was a minor incident while The Queen's Horse was covering the operations at Béthune. 'Severall squadrons of the enemy advanced within a mile of my Ld. Dukes quarters – but was boldly intercepted by severall squadrons of ours who ware commanded by my Lord Duke to give them the meeting; and a small skirmish theire was and about 60 of the enemy was kild and disabled and some few of ours. But the enemyes squadrons was glad to retire to theire grand camp.'[2] On a later occasion, 'August the 9th our Genlls. ordered 22 esquadrons of Horse and 12 regiments of Foot to cover our seiging armies who incamped between Aire and Bethune – the enemy having lately attempted to disturbe our seiging army wth. about 30 squadrons of theire cavalry, designing to have secured a pass; but now our folkes have spoyled theire designes & hindered them.'[2]

Béthune surrendered on 28 August. 'We afterward laid siege to Aire and St Venant at the same time. The latter surrendered in a fortnight: but Aire held out till toward the end of October.'[1] The Queen's Horse was mainly employed in patrol work and in escorting convoys. Deane describes this work: 'The bread wagons and convoys and likewise paymasters of the army had likely to have been intercepted by the enemy coming from Lisle; but Genll. Cadduggan went wth. a strong detachment – and the enemy hearing of it they sheered off, and our folkes came home verry safe; and care was taken to line the road wth. squadrons &

detachments.'[2] On foraging he wrote: 'June the 7th it was a genll. forrage day, and our folkes forraged towards Bethune and wide of it; and our horse and huzzars fell upon a strong party of the enemy and took about 200 of them and severall good horses.' And again, 'On the 17th the enemy sent parties of hussars to fall upon our horses as they were grazeing, and they took some sure enough; but the next day our hussars went and lay perdue [hidden] for them & fell upon ym. and kild many of ym. and brought [back] 200 horses.'[2] Cannon comments on this period: 'During the time these sieges were in progress, there was much skirmishing, and some hard fighting between detached parties, in which the Queen's Regiment of Horse had a considerable share. The French, fearful of hazarding a general engagement, carried on a desultory warfare against convoys, detached posts, and foraging parties, which proved extremely harassing to both armies.'[4] The siege of Aire, in which the regiment was closely involved, 'was prolonged by our blundering engineers, and though more tedious than bloody was over early in October, though St Venant held out until November.'[3] Both places having been captured, The Queen's Horse went into winter quarters with the rest of the covering force.

Marlborough had now lost the Queen's favour and was being subjected to insult and slander. He wrote, 'I suppose that I must every summer venture my life in battle, and be found fault with in the winter for not bringing home peace, though I wish for it with all my heart and soul.' These are words which many of his soldiers would have echoed.

In the Peninsula, after the brilliant affair at Almenara, King Philip retired, both armies crossing the Ebro on 19 August 1710. The French and Spanish took up a strong position in front of Saragossa, with their left wing resting on the Ebro and their right covered by a steep hill. The action started at midday with the Allies attacking the French and Spanish position with courage and determination. For some time the battle swayed fiercely to and fro without either side gaining the advantage. During this mêlée the 3rd Horse was held in reserve, watching the contest. After two hours of fighting, the Walloon regiments in the service of King Philip started to show signs of giving way, and at this point the 3rd Horse was 'let loose'. As its advance broke into the charge, it hit and broke the partly disordered ranks of the Walloon infantry and shattered them. The regiment then rallied and was engaged in repeated charges and counter-charges until, in spite of their greater numbers, the Spanish fled. Mr Burton was able to write from the battlefield: 'We have beat them entirely out of the field, leaving behind them 5,000 prisoners and near 4,000 killed, with all their cannon and ammunition and about 200 coulers and standards. This victory hast not cost us 600 men.'[7] The booty included twenty-two guns with their ammunition, standards and colours, and the baggage and plate of King Philip. The 3rd Horse pursued the fleeing enemy, killing numbers on the way, until the Spanish managed to get clear of their pursuers – though they saved only about 8,000 men.

The Anglo-Portugese Allies now marched upon Madrid, entering it on 21 September and garrisoning the city with British troops. This was the moment for bold and resolute action, and General Stanhope encouraged the hesitant King Charles to unite his forces with some Portugese and British foot who were operating on the borders of Portugal under Count Staremberg. However, the Portugese generals retreated back into winter quarters in their own country, leaving the main army isolated in Madrid. Following Saragossa the Duke de Vendôme had been ordered from France to take command of the remnants of the French and Spanish army, which he joined by a series of forced marches bringing large reinforcements. He then brought in their troops in Estremadura, joining them with his own, and also organised the support of the armed peasantry and auxiliaries. So by November he was able to advance upon Madrid. King Charles, hearing that the enemy had been so substantially augmented, forsook Staremberg, now in command, and Stanhope, and retreated to Barcelona with a strong escort. By 10 November the French and Spanish were in touch with Staremberg's sadly depleted force in Madrid.

Staremberg and Stanhope, confronted with such overwhelming numbers, were forced to abandon Madrid. For convenience of forage and supplies the army marched in five national columns, Portugese, British, Spanish, Dutch and German. The British troops, comprising the 3rd Horse, with a squadron of the Royal Dragoons, Pepper's and Stanhope's Dragoons, a battalion of the Foot Guards, and the remnants of seven regiments of foot, formed the rearguard on the left under Lieutenant General Stanhope. They retreated along the River Tarjuna, and reached the village of Brihuega on 6 December, fourteen miles from the main body. Their retreat had not been easy. The Spanish peasantry, enraged by the depredations of the Portugese, refused all supplies and forage, and seized every opportunity to harass and plunder the retreating troops, who were in any case withdrawing in the worst of wintry weather, without any tents. The local people also refused to pass on any information about the French movements, although they kept Vendôme fully informed of the state and progress of the British and Portugese.

Stanhope, noticing that he was being followed by a large body of horse, reported the matter to Staremberg. He was ordered to halt for a day at Brihuega in order to collect supplies. Stanhope, denied information by the Spanish peasantry, had no idea that the French were so close to his position, and he was astonished on the morning of 8 December to find a body of 2,000 French horse, supported by infantry, on top of the hill which overlooked the town. In fact Vendôme's men had covered one hundred and seventy miles in seven days by a series of forced marches. Brihuega had little in the way of defences, with only a broken down and narrow old Moorish wall. It was also a town of considerable size, with narrow streets, all of which were commanded on every side by the surrounding hills and within range of the enemy's artillery. By five o'clock in the evening the town was fully invested. Stanhope rejected a summons to surrender, in the hope that if he held out, Staremberg might come to his relief. He sent his aide de camp through the enemy's lines to inform Staremberg of his plight. The evening and night were used to barricade the gates and to dig such entrenchments as time allowed.

At midnight King Philip and Vendôme arrived with the main body of the French and Spanish troops, bringing up the besieging army to over 20,000 men. By morning the French had erected two batteries, which opened fire at nine o'clock on opponents who had no artillery with which to reply. A mine, too, was being dug. Two breaches were soon made: these were kept under such fire that it became impossible to repair them; in addition the meagre British force of 2,500 men was desperately short of ammunition.

At three o'clock in the afternoon the French and Spanish made their first assault on the breaches, and at the same time the mine was exploded, making a third breach through which the enemy poured into the town. With the men of the 3rd Horse fighting dismounted alongside their comrades of the foot, the British managed to drive the enemy out of the town and repulse this first storming of the breaches. The French then entrenched themselves around the breaches and waited for reinforcements. In due course they made a second assault, which was again driven back by the accuracy and intensity of the British fire. However, as their ammunition began to give out, 'the little garrison standing firm with the bayonet, contested every inch of ground, horse and dragoon fighting by the side of the foot, and every man doing his utmost.'[8]

The British were eventually forced back from the breaches into the town, setting fire to the houses as they retreated, and throwing stones and rocks at the enemy when their ammunition had run out. After four hours' fighting there was still no sign of relief by the main body of the army, and at seven o'clock Stanhope surrendered, being loath to sacrifice any more of the lives of his gallant band. The British had lost 600 in killed or wounded, but had inflicted three times those casualties on the enemy.

It was not until the next morning that Staremberg appeared with the main army, and by that time the 2,228 British prisoners had been escorted to the rear. Among those who surrendered were Lieutenant Generals Stanhope, Wills and Carpenter, Major General Pepper and Brigadier General Gore. The 3rd Horse remained in captivity until they were exchanged the following year, returning in October 1711 to Kingston-upon-Thames, where they went about recruiting and re-mounting the regiment.

In Flanders meanwhile, the campaign of 1711 started with the army reassembling in May, and on the 29th 'His Grace ordered a genll. review of the English Horse wch. was accordingly done'.[2] The uniform at this time consisted of a 'red coat faced with yellow, broad silver lace on the sleeves, sleeves and pockets bound with narrow lace, yellow waistcoat and shag breeches, silver laced hat, brown wig with a black bag'.[9]

The French had constructed a formidable series of lines from the coast to Bouchain. These were so strong that Villars wrote to King Louis that he had brought Marlborough to his 'ne plus ultra'. As the Duke's army had not the numbers to force the lines by assault, he was compelled to resort to stratagem.

> From Arras the Sensee branches out in several riverlets, which form a great morass from thence on to Bouchain; over this morass two causeways had been made for the convenience of the country people. The uppermost of these was called Arleux, and as this was the best and largest, Villars had caused a strong fort to be built on the higher end of it. When the Duke found that he could not draw Villars from behind his almost inaccessible lines, he set that great genius of his to work.[1]

There then occurred an incident which reflected little credit on the cavalry and which involved all the British regiments, including The Queen's Horse. On 6 July General Hompesch had arrived at Douai with the cavalry to watch the approaches to Bouchain, at the causeway at Arleux.

> Hompesch and most of the general officers lay in the town, and those of them who lay in the camp were not under any apprehension of danger, as they were immediately under the cannon of Douai; insomuch that they neglected to keep the common outguards of their camp. Villars had an account of this, and immediately ordered a good body of horse and dragoons to pass the morass at Arleux and Bouchain. Villars putting himself at the head of them, they marched with all the silence imaginable; till about twelve at night, without being so much as challenged by any one sentry, they fell upon the right flank of our horse, trampling and cutting down all before them; and had they not fallen to plunder too soon (a bewitching thing to soldiers of all nations), they might have driven through our whole detachment; but while they were rifling the horse, the quarter-guards of the foot firing on them gave the alarm. [The French] suffered little or nothing, but killed and wounded many of our troopers, and carried off a considerable number of our horses.[1]

Deane commented that they 'killd the centinells & then cut our troops horses loose from the picquett, & alarumed our men who were most of ym asleep: and so, as ya came out of theire tents the enemy cut them down & wounded ym at a sad rate before they could gett to theire arms, & plundered the sutlers, killing men, women and children.'[2]

Marlborough managed to deceive Villars of his intentions, first capturing the fort at Arleux, then leaving it poorly garrisoned so that Villars was able to recapture it and level it to the ground. He then pretended to confront Villars's main army, and while the French were distracted, on the night of 4 August he crossed the French lines with a swift countermarch across the very causeway at Arleux, with its now demolished fort. The Queen's Horse were one of the first regiments across the causeway, and the infantry on their night march were suddenly informed, 'Generals Cadogan and Hompesch crossed the causeway at Arleux at three o'clock this morning, and are in possession of the enemy's lines. The Duke desires that the infantry will step out.'[8]

Step out they did, and soon Marlborough was laying siege to Bouchain, The Queen's Horse forming part of the covering force; 'and so multifarious were the services performed, and so great were the obstacles to be overcome, that every corps was fully employed'.[4] For a time Lumley was pushed forward with thirty squadrons, including The Queen's Horse, to contain the garrison of Valenciennes, who made repeated sorties, all of them blocked. Bouchain surrendered on 13 September. 'After repairing the fortifications, the troops once more separated into winter quarters.'[4]

SOURCES

1　*The Military Memoirs of Robert Parker*, Longman, 1968.
2　J. M. Deane, *A Journal of Marlborough's Campaigns during the War of the Spanish Succession, 1704-1711*, Society for Army Historical Research, Special Publication, no.12, 1984.
3　C. T. Atkinson, 'Gleanings from the Cathcart MSS', *Journal of the Society of Army Historical Research*, vol. 29, 1951, p. 95.
4　R. Cannon, *Historical Records of the British Army, 1st Dragoon Guards*, 1837.
5　H. H. Craster, 'Letters of the First Lord Orkney', *English Historical Review*, vol. 19, April 1904.
6　*The Post Boy*, 25 April 1710.
7　*Strafford Papers*, vol. 2.
8　Sir John Fortescue, *History of the British Army*, vol. 1. 1910.

9 Captain R. G. Hollies Smith, 'A Brief History of the Uniform of the King's Dragoon Guards', *The K. D. G.*, vol. 5, no. 5, 1958.

9

Preston, Peacetime Duties
1711-1733

When Marlborough arrived in The Hague in November 1711, he found himself accused of fraud, extortion and embezzlement. Parker relates: 'These resolutions being laid before the Queen, she was pleased to dismiss the Duke of Marlborough from all his employments; and at the same time the Attorney General was ordered to prosecute him. The Duke of Ormonde was declared Captain-General in his room.' So one of this country's most distinguished generals was rewarded by a series of trumped up accusations, which were all proved to be false.[1]

Ormonde took command with a 'restraining order' that he was not to fight a battle nor besiege a town, and in July 1712 he was ordered to suspend hostilities for two months and to withdraw the British troops from Prince Eugène.

> Then the troubles began. The auxiliary troops in the pay of England flatly refused to obey the order to leave Eugène, and Ormonde was compelled to march away with the British troops only. Even so the feelings of anger ran so high that a dangerous riot was only with difficulty averted. The British and the auxiliaries were not permitted to speak to each other, lest recrimination should lead either to a refusal of the British to quit their old comrades, or to a free fight on both sides. The parting was one of the most remarkable scenes ever witnessed. The British fell in, silent, shamefaced, and miserable; the auxiliaries gathered in knots opposite to them, and both parties gazed at each other mournfully without saying a word. Then the drums beat and regiment after regiment tramped away with full hearts and downcast eyes, till at length the whole column was under way, and the mass of scarlet grew slowly less and less till it vanished out of sight. At the end of the first day's march Ormonde announced the end of hostilities with France at the head of each regiment. He had expected the news to be greeted with cheers: to his infinite disgust it was greeted with one con-tinuous storm of hisses and groans.[2]

As the British retired to the Netherlands, deserting their allies, the Dutch, in disgust, refused the troops entry into their towns, so that the army had to march across country and build pontoon bridges to cross each river. The bread contracts, too, were dishonoured, and the men received only bad bread. The troops grew near to mutiny. The Queen's Horse [KDG], having camped for a short time at Drongen, was then quartered in Ghent and Bruges, with Major General Sabine commanding in the former and Brigadier General Sutton in the latter. There the regiment remained when the Treaty of Utrecht was signed in April 1713. Not until the spring of 1714 was it ordered back to England, landing at the Red House, Battersea at the beginning of April. It then marched into quarters at Northampton, Daventry and Wellingborough. Here the cuirasses which had been reissued by Marlborough in 1707 were finally withdrawn and returned to store.

With the coming of peace each troop was reduced to thirty men, giving the regiment an overall strength of 337 officers and men. There were other changes too. In August 1714 Queen Anne died and the Elector of Hanover became king as George I. As there was no Queen Consort, George I conferred on The Queen's Horse the new title of 'The King's Own Regiment of Horse' in recognition of its distinguished services during the recent campaigns under Marlborough, who was now reinstated as commander in chief. At the same time the regimental facings of yellow were changed to blue, as a 'King's' regiment. Hollies Smith points out, however, that there is some doubt about the date of the change in facings:

> An oil painting by Peter Tillemans is in existence which shows a mounted portrait of William Pritchard Ashurst, who served as a captain in the regiment for a short time in 1722. He is portrayed wearing a low cocked hat edged with silver lace and buttons and very long cuffs reaching above the elbow, a yellow waistcoat and breeches, buff gauntlets and no sash. The saddle cannot be seen, but he has no saddle cloth or holster caps, and a detachment of the regiment is shown in the background riding four abreast with trumpeters in yellow coats mounted on greys. This would seem to show that the regiment was still using yellow as a facing colour in 1722, but it may be that the coat shown is an informal one and not the full-dress one.[3]

The 3rd Horse [Bays], having re-established themselves after their captivity, had a new colonel appointed on 1 January 1712: John Bland succeeded General Daniel Harvey. In November the regiment took up garrison duties in Ireland. It was partly based in Dublin, where it mounted the guard, but it was also employed in marching from town to town, and participating in reviews, inspections and various parades. On 7 February 1715 the colonelcy changed again, when it was conferred on Colonel Thomas Pitt (later to become Lord Londonderry).

George I was concerned by the upsurge in Catholic intrigue and the preparations of the Pretender for invasion and rebellion. In order to curb the growing lawlessness, several regiments, including the 3rd Horse, were recalled from Ireland. The regiment landed at Hoylake in Cheshire in July 1715 and was quartered in Staffordshire. At the same time the prospect of a rising in Scotland led to thirteen new regiments of dragoons and eight of foot being raised.

The Jacobite Earl of Mar raised the standard of rebellion in Scotland in September, and by the late autumn had some 10,000 Highlanders assembled under his command. He had, however, very little energy or military capacity, and this gave the Government in London time to prepare. Because there was a fear that Jacobite sympathy might extend throughout the kingdom, troops were widely deployed. The strength of The King's Horse was increased by an additional ten men per troop, and the regiment was moved to the West Country to contain the strong Jacobite sentiments in that area. Information was then received that the late Secretary-at-War, Sir William Wyndham, had arranged for a large quantity of arms and ammunition to be stored in Bristol, ready for distribution for a planned local uprising. Two squadrons of the King's Horse immediately marched on the city, seized the arms and arrested Wyndham. The rest of the regiment marched to Hertford, where it was used to pacify the surrounding districts.

The Earl of Mar, encouraged by the lack of resistance to him in Scotland, sent Brigadier General Macintosh with a small force into East Lothian and Midlothian, which they plundered. There they were joined by English Jacobites from Cumberland and more Lowland Scots, bringing their strength up to 2,300. At this point General Fraser took over command and marched on Lancaster, brushing aside militia forces of 10,000 men at Penrith and an equal body at Kendal, neither of which had any stomach for a fight. In the face of this threat, Major General Wills, commanding in Cheshire, assembled what few regular troops were available to him, consisting of the 3rd Horse and the 26th Foot [Cameronians], together with five of the newly formed dragoon regiments [later to become the 9th Lancers, 11th, 13th and 14th Hussars, and Stanhope's, which was disbanded]. The total force amounted to about 2,000 men.

In the meantime the rebels had been augmented by 1,200 Lancashire Jacobites, bringing their numbers up to nearly 5,000, though most had little military training and were poorly equipped. Coming down the Ribble Valley they entrenched themselves around Preston, barricading the entrances to the town and loopholing the houses. They neglected, however, to secure the bridge over the River Ribble.

Major General Wills set out for Preston, arriving in front of the town on the afternoon of 12 November. He immediately prepared to attack the rebels, with the 3rd Horse advancing down the main road from Manchester and the 26th Foot,

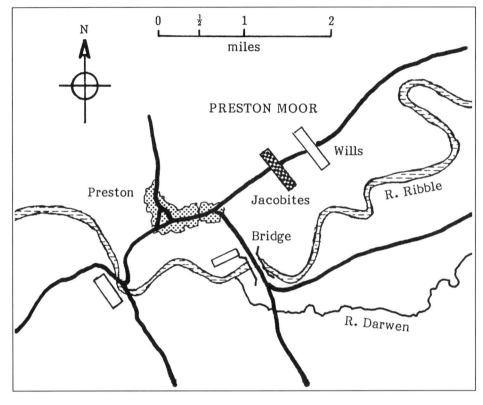

Battle of Preston

with the newly formed regiments of dragoons dismounted, assaulting the town from either side. They forced the barricades and fought their way into the town. The rebels, stationed in large numbers in the loopholed houses, were able to pour down fire on the King's troops, forcing them to retreat back to the captured barricades. While this attack was in progress, 1,000 fresh troops under General Carpenter arrived to reinforce Major General Wills.

Wills now adopted new tactics. He had the mainly wooden houses set alight from both ends of the town, then his troops advanced up the streets as the flames progressed, firing their volleys into the houses. Nightfall brought a halt to the firing, amid scenes of horror as the Jacobites were burnt alive. The 3rd Horse was posted to close all the roads out of Preston and to guard the crossings over the Ribble. During the night the regiment cut down numbers of Jacobites who were attempting to escape, until it was said that the Ribble flowed red with Jacobite blood.

At dawn on 14 November the work of firing the houses was started again, but those Jacobites who were left began to surrender and soon all resistance ceased. Desertion and slaughter had reduced the number of prisoners taken to 1,489, of whom 118 were people of distinction, including the Earl of Derwentwater. Many of these captives were later tried and executed for high treason, and many more were transported. The casualties of the 3rd Horse amounted to only one trooper wounded and two horses hurt, while Major General Wills's total casualties were three officers and fifty-three men killed, and thirteen officers and eighty-one men wounded. The 3rd Horse received the thanks of George I for their services in this action, and the honour of a new title as 'The Princess of Wales' Own Royal Regiment of Horse'. They remained at Preston as garrison until 1716. Meanwhile the squadrons of the King's Horse who were not at Bristol came up from Hertford to escort the Jacobite prisoners from Preston to London. With the suppression of the rebellion, the regiments of horse were again reduced in strength, each troop being cut down to twenty-five men only.

With so many regiments being formed and disbanded as need arose, the matter of precedence became an issue of some importance. On 23 April 1713 a warrant was issued which laid down the following order of seniority: The Life Guards, The Horse Guards, The Queen's Horse and the 3rd Horse.[4] A further warrant issued by George I on 19 February 1715 confirmed this order of seniority.[5] A Report of General Officers laid down the principle that

> they ought to take post from the time of their coming upon the Establish-
> ment of England and not from the time they were raised. The Rules settled
> by King William which had governed during the continuance of two wars
> should remain a Standing Regulation, thereby revoking the said Order of
> the Late Queen which Order was the only instance wherein this Rule had
> been set aside.[5]

Both regiments now had a period of moving around England. In 1716 the King's Horse was reviewed near Barnet by the Prince of Wales, and in October of that year The Princess of Wales' Horse marched to Windsor, Egham, Staines and Colnbrook with a strength of 28 officers, 6 quartermasters, 12 corporals, 6 trumpeters and 240 troopers. Whilst around Windsor the regiment assisted the Life Guards in providing travelling escorts for the Royal Family, and in November 1716 it was also reviewed by the Prince of Wales. In May 1717 The Princess of Wales' Horse was relieved at Windsor and proceeded to the Kentish coast, while The King's Horse that year was employed in protecting the royal parks and forests against the activities of a band of poachers known as 'The Blacks', who were believed to have carried off nearly 2,000 deer over the preceding months.

In 1717 General Lumley decided to resign as colonel of The King's Horse, a matter which was to give rise to considerable official stirrings. George I was keen to put an end to the practice of the buying and selling of colonelcies, but this was so prevalent and the vested interests so strong that it was not until the reign of George III that it was finally stopped. In November 1717 the Secretary-at War wrote to Lord Irwin that 'Mr Lumley has desired the King's leave to sell his regiment. He will expect 9,000 guineas for it, and indeed no regiment can be in better condition. I have the King's order to make you the proffer for it.' The money was partly to be found by Lord Irwin selling his regiment of foot, the 16th [Beds & Herts], to its lieutenant colonel, John Cholmley. This would provide 3,000 guineas. Lord Irwin was to put up another 3,000 guineas from his own money, but £1,500 of this was to be borrowed on the next assignment of money from the Government for the clothing of his regiment, and £1,500 by sale of stock and a loan. This initial £6,000 had to be paid by a particular date. The balance of £3,000 was to be found by Lieutenant Colonel Cholmley selling, in his turn, the 16th Foot to one of his majors for £2,100, and raising a further £441 by the sale of his own lieutenant colonel's company in the 16th Foot as a separate transaction. But there was to be an additional complication, for Lord Irwin disputed the state of the 'off reckonings' of The King's Horse. (The pay of a regiment was in two parts: subsistence, which was paid to the men, and 'off reckonings' which were paid to the colonel to clothe and equip the regiment, and which produced a handsome profit for him.) Irwin was taking over what was then seen as a business concern, the administration of a regiment of horse, and he wanted to ensure that he was not going to be the loser. By 21 December Lumley had received all his money except for £700, which was to come within a week. Feeling generous, he wrote to Lord Irwin: 'Mr Lumley again set forth the fineness of Your Lordship's own Troop, and that he would present yr Lordship with the 2 horses used by the Kettle Drummer and Trumpeter, And also the silver Trumpet that was to his late Troop, And how that at Your Lordship's coming to Town he wou'd let him into the True Character of Everie Officer of the Regiment.'[6]. So on 13 December 1717 The King's Horse had a new colonel appointed in the person of Lord Irwin. George I's concern was understandable.

In April 1718 The Princess of Wales' Horse moved to Bedfordshire, with some detached troops being posted as far away as Cornwall. During this year, too, they were made to reduce their establishment by ninety men. In 1719 they went on a route march into Warwickshire, where they occupied quarters, and then in 1720 moved back to Bedfordshire and Buckinghamshire, where during the autumn they provided a series of escorts for George I on his return from the Continent. The next move was in May 1722, when the regiment camped near Andover, and on 30 August it was reviewed on Salisbury Plain by the King, along with the 3rd Dragoon Guards and the 4th and 10th Dragoons, and seven regiments of foot. After the review the regiment escorted the King to Winchester, and on the next day to Portsmouth, where George 1 reviewed the garrison and then the shipping in the harbour. That autumn The Princess of Wales' Horse struck their tents on Salisbury Plain, and marched to Newbury, where they were quartered in the town and its neighbourhood. Throughout this period The King's Horse and The Princess of Wales' Horse regularly furnished travelling escorts to the Royal Family, and assisted the Life Guards with their household duties at the court and in London.

In 1719 Philip of Spain decided to fit out an expedition under the command of Ormonde (who had supplanted Marlborough in Flanders, and who had now changed sides), in order to invade England in the cause of the Pretender. The King's Horse was ordered to the West Country to meet this threat, but the Spanish fleet was dispersed and wrecked by a storm, and so the projected invasion came to nothing.

Stirred by the furore over the sale of colonelcies, George I determined to regulate the practice of buying and selling commissions generally, to obviate any temptations to corruption and malpractice on the part of officers. In February 1720 a Board of General Officers issued a regulation fixing the amounts of money that were to be paid for commissions in each regiment. The amounts laid down for The King's Regiment of Horse and for The Princess of Wales' Horse were: colonel and captain, £9,000; lieutenant colonel and captain, £4,000; major and captain, £3,000; captain, £2,500; captain-lieutenant, £1,500; lieutenant, £1,200; cornet, £1,000; adjutant, £200. This was fortuitous, as Viscount Irwin, after all his trouble to purchase the colonelcy of The King's Horse, died early in 1721, and Richard, Viscount Cobham came from the 1st Royal Dragoons to be colonel of the regiment.

The King's Horse was encamped on Hounslow Heath during the summer of 1722, where it was reviewed by the King on 5 July, who 'was pleased to express very great satisfaction at the good appearance the Regiment made, and with the performance of its exercise'.[7] The regiment was again reviewed on Hampton Court Green by General Earl Cadogan, who was acting as the General Commanding in Chief. During 1723 The Princess of Wales' Horse was stationed around Northampton, Daventry and Stony Stratford, and then during the winter of 1723-24 it was back around Windsor at Colnbrook, Uxbridge and Chertsey. In the spring of 1724 the regiment moved back to the Midlands around Warwick and Coventry, and in 1725 it marched to quarters in Devonshire; for the winter of 1726 it was back in Northamptonshire.

The colonel of The Princess of Wales' Horse, Colonel Pitt, had become the Earl of Londonderry, and on succeeding to the title he retired from command of the regiment, to be replaced by one of the regiment's most distinguished colonels, John, Duke of Argyle, by a commission dated 26 August 1726. Argyle, born in 1678, had fought with Marlborough at Ramillies, Oudenarde and Malplaquet, although he later helped in the intrigues which led to Marlborough's dismissal. He was, however, a kindly man, who was known as 'The good Duke of Argyle'.

On 11 June 1727 George I died at Osnabruck, to be succeeded by his son, who became George II. The Princess of Wales now became the Queen, and so in June 1727 the title of The Princess of Wales' Own Regiment of Horse was changed to 'The Queen's Own Royal Regiment of Horse'. The regiment was at once ordered to London, and on 26 July it furnished all the mounted guards for the Royal Family. On 27 July (the anniversary of Almenara) the regiment provided a squadron of forty rank and file as a guard of honour, when the King reviewed four troops of the Life Guards and two troops of the Horse Grenadier Guards in Hyde Park. During the review the regiment also provided the King's Guard at Whitehall, consisting of four officers, one trumpeter and forty privates; the Queen's Guard consisting of two officers and eighteen privates; and the Princess of Wales' Guard at Kensington Palace made up of one officer and sixteen privates; a total of seven officers, one trumpeter and sixty-four corporals and privates, apart from the guard of honour.[8] Following the review, The Queen's Horse provided the travelling escorts for the Royal Family, and on 2 August the regiment was reviewed at Kew Green by the King, who was accompanied by the Queen and by many generals and other persons of distinction. The smart appearance of the men and their horses, and the skill of their movements were 'highly approved'.[9]

The King's Horse was also involved in royal duties, providing a detachment on 11 October 1727 at the Coronation of George II in London. In April 1728 the regiment provided a relay of travelling escorts for the King on his way from London to Newmarket, and throughout 1729 a detachment of the King's Horse was in constant attendance on the Prince of Wales at his home at Richmond Palace.

In February 1728 The Queen's Horse, having been relieved of their royal duties, marched to new quarters at Nottingham and Stamford, where they remained for a year, proceeding in March 1729 to Warwick and Coventry.

During 1719 the size of a trooper's horse for all regiments of horse was laid down as 'a strong well-bodied horse from fifteen hands and an inch, to two inches, and not exceeding'.[10]

On 20 May 1730 the King reviewed the Life Guards and Horse Grenadier Guards in Hyde Park, and The Queen's Horse was brought down to London that month, where it provided a squadron to keep the ground for the review. At the same time the King's Horse provided the guard at Whitehall, and a personal escort of a captain, a lieutenant, a trumpeter and twenty-four troopers to George II; a second escort of a lieutenant, a trumpeter and twelve troopers for the Queen; a third escort of a quartermaster and eight troopers for the Prince of Wales; and a fourth escort of eight troopers for Prince William. After the review The Queen's Horse took over the duty of providing the travelling escort for the Royal Family.

On 24 June 1730 both The King's Horse and The Queen's Horse were reviewed together by the King on Datchet Common. 'The splendid appearance and high state of discipline and efficiency of these distinguished Regiments excited general admiration, and they received the approbation and thanks of their Sovereign.'[11] The King 'was pleased to express his entire satisfaction at the good appearance and with the performance of the exercise'.[7] On 1 January 1731 Wade's Horse [3 DG] relieved The Queen's Horse of their royal escort duties, and the regiment moved to Kent and Sussex, where it spent the summer, moving for the winter back to Nottinghamshire, Derbyshire and Bedfordshire, where it stayed until the winter of 1733. After the 1730 review, the King's Horse was stationed at Canterbury, until in 1732 it was once again called upon to provide the royal travelling escort duties.

On 9 February 1731 William Pitt, later 1st Earl of Chatham, was gazetted as a cornet in the King's Horse. Four years later Pitt was to enter Parliament as the Member for Old Sarum, and was dismissed from the Army for his maiden speech. Walpole had managed to persuade the King, still inexperienced in British ways, that every Member of Parliament who was a serving officer and who had opposed his Excise Bill, should be dismissed the service.

The King's sentence of dismissal fell not only upon the young cornet William Pitt, described by Walpole as 'the terrible cornet of horse', but also upon the colonel of the regiment, Lord Cobham, so that on 22 June 1733 the King's Horse received a new colonel in the person of Henry, Earl of Pembroke, and in August of that year The Queen's Horse also had a new colonel as a result of the Duke of Argyle being made colonel of the Royal Horse Guards. Lieutenant General William Evans from the 4th Dragoons was appointed.

SOURCES

1 *The Military Memoirs of Robert Parker*, Longman, 1968.
2 Sir John Fortescue, *History of the British Army*, vol. 1, 1910.
3 Captain R. G. Hollies Smith, 'A Brief History of the Uniform of the King's Dragoon Guards', *The K. D. G.*, vol. 5, no. 5.
4 Public Record Office, W. O. 26/14.

5 Public Record Office, W. O. 71/3.

6 James Hayes, 'The Purchase of Colonelcies in the Army, 1714- 63', *Journal of the Society for Army Historical Research*, vol. 39, 1968, p. 3.

7 Records of The King's Dragoon Guards, Regimental Museum, 1st The Queen's Dragoon Guards, Cardiff Castle.

8 *War Office Route Book.*

9 *London Gazette*, 1727.

10 Lieutenant Colonel E. A. H. Webb, *A History of the Services of the 17th Regiment*, 1911.

11 R. Cannon, *Historical Records of the British Army, 1st Dragoon Guards*, 1837.

10

Flanders, Aschaffenburg, Dettingen
1733-1743

The story of a cavalry regiment on home service at this period can sound like a recitation of place names, and to a certain extent this is inevitable, given the system of billeting in those days. There were very few barracks for the soldiers, whether horse or foot, and the men had to be scattered over the countryside, because by the terms of the Mutiny Act soldiers could only be accommodated in ale-houses, and there were not enough inns to allow for a concentration of troops. There were indeed small barracks in the garrison towns, but these were reserved for the cannon and those few men needed to look after them. The system was extremely unpopular, and particularly because it laid upon the innkeeper the obligation to provide food, fire and candle, and upon the cavalry the duty of quelling any civil disturbance. The King's Horse [KDG] in 1742 had, for instance, three troops in Newbury, two in Farnham, one in Alton, one in Henley, one at Wokingham, and one at Maidenhead.[1] When a whole regiment of dragoons was sent to Southampton, the civil authorities objected bitterly, even though the colonel protested that, owing to the regiment's being scattered, he had been unable to give it any training. The friction worked both ways: a cavalry cornet, having been cheated by an innkeeper, ordered eight of his troopers to march up and down outside the inn, knowing that the sight of redcoats would drive away all custom. Echoes of Kipling –

'I went into a public house to get a pint of beer,

'The publican he ups and says "We want no redcoats 'ere".'

The King's Horse was also known at this time as the 2nd Horse (the 1st Horse being the Royal Horse Guards), and was moved around the southern counties of England between 1733 and 1742. In 1736 it provided 'the usual detachments to attend upon the Court', being stationed around Windsor, until it was relieved from this duty in January 1737 by The Queen's Horse [Bays], which had itself seen many different quarters. At the start of 1735 it was spread as far afield as Hampshire, Shropshire and Chester; in April 1735 it went to Newcastle, Burton and Ashburn, only to be moved in June to Coventry, Warwick and Stratford, where it spent the winter. March 1736 saw the regiment on the road to new billets in Northampton, Towcester and Stony Stratford, until it relieved The King's Horse at Windsor. At Windsor it provided guards of honour for the Royal Family in June when the Life Guards were reviewed, and the guard at the Castle in July when the Foot Guards were reviewed. It was relieved by Wade's Horse [3 DG] in September 1737 and went back to Coventry and Warwick.

On 17 July 1737 Philip Browne received a commission as a cornet in The King's Own Regiment of Horse, with which he served until 1745. During this period he wrote a number of letters which have survived and which tell us much about the life of the regiment. His first letter was written from Market Harborough, dated 28 September 1737: 'Monday morning I march the Troop to Lutterworth, about seven miles from hence, there being a horse fair to be kept here, and shall return here again, and then if nothing unforeseen happens go to Northampton for the winter.'[2] There were three short letters from Northampton in December, and the next was from Newport Pagnell in August 1738, saying that during the previous week the regiment had been reviewed there by their colonel, Lord Pembroke. In May 1739 he was back in Northampton, writing to say 'Shall be in Putney the first of June, that being the day we relieve the Duke of Argyle's Regiment [Royal Horse Guards] and shall continue there for some time.' But on 8 July 1739 he wrote: 'Uxbridge. Our notice was so short to come here, that had not time to acquaint you of it.

Next Wednesday [11 July] is fixed for our Review, by His Majesty, at Hounslow Heath, where if seeing as fine a body of men and horses, as can be, will give you any pleasure, I believe you will not be disappointed.' Letters arrived from High Wycombe, 29 June 1740; from Newbury on 5 August and 15 September; and from Hereford and Worcester in October. In camp at Newbury, 'The tents are never dry, the men are continually wet, and the horses stand fetlock deep in water frequently.' At Hereford, 'We leave this place on account of the scarcity of forage.'[2]

The Queen's Horse was also being moved from pillar to post: during 1737 and 1738 the regiment occupied various quarters in Essex, Kent and Northampton-shire, and in June 1740 was back in camp in Windsor Forest doing royal duties again, until relieved in January 1741 by Wade's Horse [3 DG]. On 6 May 1740 a new colonel was appointed on the death of General Evans, in the person of John, Duke of Montague. Then, after a spell in Essex and Kent during the winter of 1741, the regiment returned to London in May 1742 to resume royal duties and to provide a relay of escorts for the King and the Duke of Cumberland on their return to England after the Battle of Dettingen in the autumn of 1743.

When the Emperor Charles VI of Austria died, the French backed the claims of their old friend the Elector of Bavaria against those of the rightful heiress Maria

Theresa. This led to the War of the Austrian Succession, with Bavaria, Saxony, Prussia and Spain siding with France in the hope that Austria would be divided among them, and Hanover and Holland joining Britain in supporting Maria Theresa. Preparations for an expeditionary force were made throughout the summer of 1741, by concentrating a number of regiments at Colchester.

Philip Browne wrote home from Worcester on 3 June 1741:

> It is certaine that The King did our Regiment the Honour, before he went to Hanover, to appoint it for one of them, that are to be sent abroad. We are in daily expectation of orders to march. I can't but say that I wait for the Order with chearful expectation and pleasure, being ready and willing to risque my life, when The King's Service and the Publick Good requires it. And as both now demand it, it is most agreeable to me to Court dangers and hardships abroad, rather than to continue in Safety and Ease at Home. And my only Wish and desire is, that I may behave gallantly and die like a soldier, or return with Honour and in Higher Rank.[2]

The King's Horse was first ordered to Hounslow, where on 26 June 1742 'We was reviewed by His Majesty at Hounslow Heath, when it was said the transports would be ready for us in a fortnight or sooner. The reason for our being ordered here is that there is not Quarters sufficient for us nearer London, owing to the numbers that are all ready thereabouts.'[2] An officer of Ligonier's [7 DG] wrote of this review: 'Upon their march for embarkation was reviewed, without respite or preparation, at Hounslow, by the King, in the centre between the Oxford Blues [RHG] and Pembroke's Horse [KDG], of nine Troops each, newly and completely appointed, and which had only marched from the neighbouring cantonments for that purpose.'[3] The regiment was increased in strength to 535 officers and men.

On 20 July Browne wrote, 'We are to embark with the Horse Guards, on the return of the transports, which are now taking aboard Hawley [1 RD] and Campbell's [Greys] Dragoons.'[2] The King's Horse marched on 7 August, embarking at Deptford in *The Liberty and Property* on the 8th, and on the 9th 'We weighed anchor at five o'clock from Gravesend and dropped anchor that day off North Foreland and have continued hereabout ever since for want of a fair wind.'[2] On 12 September 1742,

> I writ you from Ostend 30th August, where we landed, after a very disagreeable and fatiguing passage of twelve days. That morning we disembarked the horses upon the sands, the tide being halfway up their legs, where we remained till eight at night, and then marched to Bruges, where we arrived at one o'clock in the morning. Men, nor horses, having had any

refreshment that whole day. We was obliged to threaten to break open the doors of my Inn, before we could get admittance, after which enquiring for a bed they said we must lay in the kitchen, but we took the liberty to take possession of the Landlords. Next morning we marched for Ghent, the roads very sandy. With regard to living it is much the same as in England, only we fair better for the money. Please to direct to me of The Earl of Pembroke's regiment of Horse, it being known here better than by the King's own regiment.[2]

The passage across the Channel had cost the regiment three horses lost in the bad weather.[4]

The King's Horse was brigaded with the Life Guards and Ligonier's Horse, and with them at Ghent were the Blues [RHG] and six regiments of dragoons [1 RD, Greys, 3 H, 4 H, 6 D, 7 H]. The army was commanded by the Earl of Stair, who reviewed The King's Horse on 30 October 1742,[4] and beforehand issued an order which must have kept the men busy:

> The Horse and Dragoons to take all their small accoutrements to pieces and see that they be very well cleaned and blacked, and then put them together again. The bosses, bits and curbs to be as bright as hands can make them. The boots to be as black as possible, and their knee-pieces not to appear above three inches above the boot-top. All their arms to be as bright as silver. The whole buff accoutrements to be of one light buff colour, the swords to be all brightened. The hats new cocked. Three straps to each cloak. Care to be taken that the men do not ride too long. Horses to be all trimmed and made as clean as possible. No Trooper or Dragoon to appear in the streets with his cloak on.[5]

The knee-pieces were the white worsted hose worn over the knee to prevent the boots from chaffing the breeches. In 1742 the Duke of Cumberland had had produced *A Representation of the Cloathing of His Majesty's Household and All the Forces upon the Establishments of Great Britain and Ireland*. This shows a trooper of The King's Horse wearing a red coat, with blue facings but no lapels, brass buttons, blue waistcoat and breeches, buff crossbelts with blue flask cord, yellow laced cocked hat, red horse furniture embroidered with the Royal Arms, and beneath the saddle a rolled red cloak lined with blue.[6]

Once the regiment was in winter quarters there was little for it to do. Many of the officers applied for leave back to England, and the Earl of Stair wrote to the Secretary-at-War, 'I thought it hard to refuse them leave, when they said their preferment depended on the interest of their friends at Court. They had no notion that it depended on their exertions here.'[7] The years of dispersal in England and the boredom of the winter routine, in quarters which the men detested, had damaged discipline, and in November Stair hoped that 'a little patience and some time ' might help him to establish some reasonable standards. A week later he was pleased to note that the execution of a trooper in Lord Pembroke's Horse had 'produced a wonderful effect'.[8]

On 18 March 1743 Philip Browne was writing from Brussels, 'Upon our arrival here, we drew up on the Grand Parade, near the Palace, the Windows & Balcony's of the Houses was filled with Ladies, & Count Harrach came out to review us, & saluted all the Officers, was greatly pleased at our appearance, we being after two days march in as much order, as if to be reviewed by the King in Hyde Park.'[2] On 18 April, whilst the regiment was still in Brussels, the Earl of Pembroke having retired as colonel of the regiment, General Sir Philip Honeywood, colonel of the 3rd Dragoons, was appointed in his place.[4,9] On 20 April Browne was writing, 'Bruxelles, we marched from yesterday morning to Louvain, and today here (Tirlemont)', and on 20 May from Coblenz,

> We passed the Maese, and yesterday morning the Rhine, over a flying bridge of boats. The weather during our whole march hath favoured us greatly, but it hath not been a little fatiguing, owing to the force marches we have made. We have frequently had nothing but straw to lye upon, and nothing to eat but bacon and eggs. Chearfullness runs through all the Troop, notwithstanding the bad accommodation we have had.[2]

On 4 June the whole army, British and Hanoverian, was united at Hochst on the River Main, and there it was joined by a body of Austrians under Marshals Aremberg and Nieuberg.

Louis XIV, hoping to prevent the junction of the British with the Austrians, had despatched a force of 60,000 men under Marshal Noailles to move against the Earl of Stair, and another army to oppose the Austrians under Marshal Coigny. Stair marched to Aschaffenburg, where, on 19 June, he was joined by George II and the Duke of Cumberland, the King taking over command of the army.

Noailles managed to cut the British communications with Frankfort and to block the passage of the Main. Stair wrote to the Secretary-at-War, 'It will be impossible for us now to find forage. The French being masters of one side of the river, forage cannot be brought down to us by water, so we must move upward.' King George, unwilling to retreat, lingered in camp until he found himself in command of a starving army, in which discipline was slipping and looting increasing. When news came that some Hanoverian and Hessian reinforcements had reached Hanau, it was decided to retire there. Noailles, however, had not been idle, and as soon as he heard that the British were on the move, he sent Count Grammont with 28,000

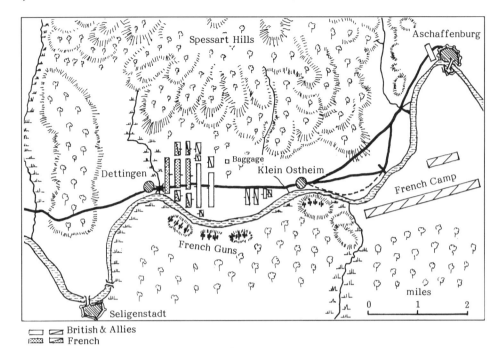

☐ ▭ British & Allies
▨ ▨ French

Battle of Dettingen

men across the Main to take up a blocking position by the village of Dettingen. As the British line of retreat was confined to the banks of the Main because of the dense woods covering the hills running down to the river, Noailles had placed a number of batteries on his side of the Main to bring down fire on their march.

At 4 a.m. on 26 June The King's Horse and the Life Guards led the rest of the British cavalry, followed by the Austrian cavalry, in advance of the British and Austrian infantry, with the Guards and Hanoverian cavalry acting as rearguard. As the army marched they could see the French crossing the Main behind them. Noailles was closing the trap. As The King's Horse reached the village of Klein Ostheim at 8 a.m., it was ordered to wheel and face the river to cover the passage of the army through the single road that wound through the village. The French batteries were positioned across the Main for this very purpose, and for four hours The King's Horse had to endure their galling fire. Browne wrote home after the battle,

> For several hours we stood the cannonading of the enemy, from several batteries they had erected, which commanded the line of march, so as not only to annoy us, but frequently went beyond us. Our men and horses stood it, without the ranks being the least disordered, and as soon as men

or horses was killed they closed again, and at the same time we could see that as our cannon played upon them, they sett up a gallop in great disorder. They begun by cannonading us on our march.[2]

As the veteran Spanish warrior the Marquis da Santa Cruz had commented in 1735, 'It is no way to inspire courage among your cavalrymen to force them to stand cold-bloodedly under prolonged cannonades.' When shells exploded in a closely ranged regiment, they were capable of killing up to ten horses at a time and stampeding many more. The ability of the British horse to stand and take this punishment before Dettingen, and then to move and manoeuvre straight into action, was the cause of much admiration and wonder at the time. A trooper of the 3rd Dragoons described the ordeal:

> The sarvants of the rigement went into the rear of the rigement with their led horses, I had a led horse so I was there. We stayed there till the balls came flying all around us. We see first a horse with baggage fall close to us. Then seven horses fall apeace, then I began to star about me, the balls came whistling about my ears. Then I saw the Oysterenns [Austrians] dip and look about them for they dodge the balls as a cock does a stick, they are so used to them.[11]

At about midday the cavalry were moved forward between the villages of Klein Ostheim and Dettingen to form the right wing alongside the Austrian infantry. Browne remembered: 'About one o'clock the Brigade of Horse Guards and Horse was ordered in the front; about a mile before we came to the ground, where we engaged, our eyes was presented with numbers of dead bodies, and some that was shot and slain and not expired, which we could not help riding over and passing through; I saw numbers that the Foot put an end to by firing their pieces in their ears.'[2]

The King's Horse was brigaded in the second line of cavalry, together with Ligonier's [7 DG], Campbell's [Greys], Bland's [3 H] and Rich's [4 H]. The French Gens d'Armes and Maison du Roi started to parade in front of the armies 'without a thought except for the fine figure it was cutting in the sight of both armies',[10] and they then advanced to attack in great numbers before the British left flank had been formed. Bland's Dragoons were moved across to fill a gap in the line, and the French assault was gallantly met by these two weak squadrons, who charged and charged again in support of the hard-pressed infantry, but with heavy loss.

The King's Horse and Ligonier's, having been summoned from the right, came galloping up and fell on the Gens d'Armes, but were repulsed, partly because the charge was delivered with more dash than order, and partly because of the weight of the opposing numbers. The French, too, were protected by helmets and breastplates capable of deflecting pistol shot and sword thrust.

Browne continues:

> Before the left of the Brigade of Horse was formed, the Gens-Arms, the best troops of France, advanced to attack us, and a battery of their cannon flanked us; upon their advancing to attack Genl Honeywoods [KDG] and Genl Ligoniers [7 DG] regiments we marched forward and meet them sword in hand; at the same time their cannon ceased, and they flanked us on the left with their foot; then we engaged and not only received but returned their fire; the balls flew about like hail, and then we cut into their ranks and they into ours. The Major was on my right; his scull cap turned two musket balls, but he received two deep cuts by their sabres; Cornet Allcroft, who was near me, was killed, and the standard which he bore was hacked, but we saved it; a Captain on the right of the Major was killed, another on my left was shot in both shoulders, three men and eight horses of the Major's Troop, to whom I am Lieut was killed, and eight men shot and cut, but not yet dead. Our Squadron suffered most, we being on the left. In the right Squadron Lieut Draper was killed, and his Captain wounded. Capt Watts and Cornet Lightfoot, who both charged in the center Squadron, are well, but one of their Captains is badly bruised by being trampled under the Horses feet, and his Lieut was shot through the thigh. I did not receive the least hurt but my left hand and shirt sleeve was covered with blood, which must fly from the wounded upon me. Had not the English Foot come to our relieve we had been all cut to pieces, the Gens Arms being nine deep and we but three, after which we rallied again and marched up to attack them again.[2]

George II had an unfortunate day, for his horse ran away with him to the rear, indifferent to a flow of Germanic oaths; but then the King returned on foot to encourage his troops. The French cavalry drove back The King's Horse and Ligonier's, and as the King's Horse wheeled away to the right it crashed into, and disordered, the Blues coming up to its assistance. Both regiments had to retire and re-form. The French cavalry, meanwhile, succeeded in breaking through the British foot, but the 21st [Royal Scots Fusiliers] and the 23rd [Royal Welsh Fusiliers], recovering themselves, turned inward and shot down the French horsemen. At the same time Rich's [4 H], Stair's [6 D] and two regiments of Austrian dragoons arrived, and together with the now rallied remnants of the Blues, King's Horse and Ligonier's, charged the French horse and managed to drive them back.

The French attacks on the British centre and right had not been pressed and were easily repulsed, so that once the French cavalry had been defeated, George II gave the order for a general advance. The British and Hanoverian infantry gave a loud cheer and swept forward, pausing from time to time to deliver their volleys of fire into the disintegrating French ranks. The French made headlong for the fords and bridges across the Main, many soldiers in their haste plunging into the river and drowning. In spite of Stair's strong advice the pursuit was not pressed, for King George was too thankful to have escaped from Noailles's trap, conscious that it was 'a victory won not by the generals but by regimental discipline, valour and a half-perfected musketry skill'.[8]

> The English, Hanoverians and Austrians remained masters of the field. We then proceeded on our march, and came to our ground at eight o'clock, it pouring then with rain and continued so all night, and not an Officer had a tent, the baggage not being come up, and we had nothing to eat nor drink, and we quenched our thirst by the rain that fell upon our hats, and we had nothing at all for our horses. We laid upon our arms that night, and got here [Hanau] last night at ten o'clock.[2]

The wounded were left lying on the field. One survivor wrote home: 'Tis a very shocking thing to see the poor Souls that are lying upon the Ground, and to hear their cries.'[7] Parties moved about the battlefield collecting those they could and getting them to the surgeons. Others, less merciful, came only to plunder the dead and dying of both sides. These marauders, of whom the sutlers were the worst, were an intrinsic part of every eighteenth-century battlefield.

The King's Horse lost Captain Meriden, Lieutenant Draper and Cornet Aldcroft, with eight men and twenty horses killed. Major Carr, Captains Saurin and Smith, Lieutenant Wallis, two quartermasters and twenty-eight troopers were wounded, as well as twenty-four horses. The despatch after the battle said, 'The King's Horse supported for eight or nine hours the most severe cannonade that was ever known and then attacked the [French] Household Troops, who, to do them justice, supported the ancient reputation of their corps with great bravery.'[4, 9]

The British remained in the camp at Hanau until the middle of August, when 'the whole Army repassed the Rhine, and encamped at Bebrick opposite to the City of Mentz ... The Army marched from Mentz [on 27 August] and arrived at Worms [30th] whence they marched for Franckenthal and arrived here [Spire] [1 September] ... The Army had suffered very much in their march, having had no Forage at all and provisions very scarce tho' the orders were very strict and the Officers very careful yet in spite of all their Diligence the Soldiers had riffled and plundered some villages as they passed along and that several of them had been hanged for Maroding.'[2] At Spire the army was joined by some Dutch auxiliaries, but no further action took place and 'The King left the Army the 5th [October] and the 6th we marched to our old camp at Mentz. To Morrow morning [15 October] the Brigade of Horse to Bruxelles which is to be the winter Quarters of us and the foot and horse Guards.[2]

SOURCES

1 *Miscellaneous Orders (Guards and Garrisons)*, vol. 325, p. 147.

2 'Letters of Captain Philip Browne, 1737-1746', *Journal of the Society for Army Historical Research*, vol. 5, 1926.

3 *United Services Journal and Naval and Military Magazine*, 1833.

4 Records of The King's Dragoon Guards, Regimental Museum, 1st The Queen's Dragoon Guards, Cardiff Castle.

5 J. M. Brereton, *A History of the 4/7th Royal Dragoon Guards*, published by the regiment, 1982.

6 Captain R. G. Hollies Smith, 'A Brief History of the Uniform of the King's Dragoon Guards', *The K. D. G.*, vol. 5, no. 5, 1958.

7 Michael Orr, *Dettingen, 1743*, 1972.

8 R. E. R. Robinson, *The Bloody Eleventh*, vol. 1, Devon & Dorset Regiment, 1988.

9 R. Cannon, *Historical Records of the British Army, 1st Dragoon Guards*, 1837.

10 Sir John Fortescue, *History of the British Army*, vol. 1, 1910.

11 Colonel H. C. B. Rogers, *The Mounted Troops of the British Army, 1066-1945*, 1959.

11

Clifton Moor, Fontenoy
1744-1745

During 1744 The Queen's Horse [Bays] remained in the south of England, the regiment being required to find reinforcement drafts for the cavalry regiments serving in Germany. Colonel the Duke of Montagu received a royal warrant in January 1744:

> GEORGE R. We having thought fit to Augment and Recruit Our several Regiments of Horse serving in Flanders with all Expedition, Our Will and Pleasure therefore is that you do forthwith make a Draught of Ten private men mounted out of every Troop in the Queen's Royal Regiment of Horse under your Command towards Augmenting and Recruiting the said Forces, which said Draughted men are to march to such place as shall be appointed, where they are to embark on Board such Transport ships as shall be ordered to receive them. Wherein the Civil Magistrates and all others concerned are to be assisted in providing Quarters, impressing Carriages and otherwise as there shall be occasion. Given at our Court of St James's This 20th Day of January 1744 in the Seventeenth Year of Our Reign.
> By His Majesty's Command,
>
> WILLM. YONGE.[1]

In 1745 Prince Charles Edward Stuart, the 'Young Pretender', landed in Scotland and raised the Highland clans in rebellion. Most of the British regular troops were in Flanders and Germany, so The Queen's Horse was immediately ordered to march to Nottingham and Derby, where it came under the command of aged veteran Field Marshal Wade, who was gathering what troops he could to meet the Jacobite threat. A number of new regiments were hastily raised, the Duke of Montagu raising a regiment of horse, known as 'Montagu's Horse' or 'The King's Carbineers', and a battalion of foot, called 'The Ordnance Regiment.'

Prince Charles Edward, having defeated General Cope at Prestonpans, entered Edinburgh and then marched south. Wade having established his headquarters at

Doncaster, The Queen's Horse with the rest of the cavalry pressed on from there, first to York, and then to Newcastle, where they arrived in October, to be joined by the foot on the 21st. However, the Jacobite army managed to elude Wade by entering England through Cumberland, besieging Carlisle and taking it, and then marching south through Manchester and on to Derby, which the rebels entered on 5 December. In the meantime Wade had marched his force through Durham and Darlington. On 8 December The Queen's Horse was at Doncaster. Wade then heard that the rebels were retreating back to the north, and he sent The Queen's Horse with the rest of the cavalry in pursuit under General Oglethorpe. The regiment left Doncaster on 10 December and rode so hard that they covered a hundred miles in three days through ice, snow, sleet and mud. They arrived at Preston on 13 December, having captured a number of rebel prisoners on the way, and there linked up with some of the cavalry from the Duke of Cumberland's main army, which was coming up with all haste from the south.[2, 3, 4]

On 29 December the cavalry caught up with the Jacobite rearguard, who were resting near Penrith at Clifton Moor in order to allow time for their transport, which was making poor progress over the bad roads. The rebel rearguard under Glengarry were in position at the edge of the moor so as to cover Clifton bridge. The cavalry attacked in the evening after dark: 'The rearguard of the fugitive force was overtaken and very roughly handled by The Queen's Horse.'[4] The foot engaged at Clifton Moor were not so fortunate. In the dark the claymore proved to be a most formidable weapon, and Glengarry drove off some of the horse, who were fighting dismounted in the dark, inflicting on them the loss of a hundred men, though this was counterbalanced by the capture of seventy prisoners.[5, 6] The Jacobites continued their retreat back into Scotland, leaving a small garrison in Carlisle Castle. The Queen's Horse was left with the force besieging the castle, while the rest of the army pushed north in pursuit. Carlisle Castle soon surrendered, and The Queen's Horse was then ordered to return to York, where it remained for the whole of 1746.[3, 4]

In May 1744 The King's Horse [KDG] moved out of their winter quarters, with Field Marshal Wade in command of the army.

> On the 8th we marched from Bruxelles & encamped at Assche. On the 22nd we came to our present camp [Beirlegem, near Oudenarde], and upon coming to our ground, we could distinctly hear the cannon from Menin. Part of there army [French] hath invested it, & the other part covers the Besiegers. And I am afraid for want of numbers it is not in our power to raise the siege. Yesterday we heard that our augmentation of ten men a Troop for the Horse was gott into Ghent, but that a detachment of french

had prevented the recruits. But a detachment is marched from Ghent to open the communication sufficient, I doubt not, to oblidge the french to retire. We want nothing but numbers; the army is greatly taken care of, having plenty of provisions and the forage and wood brought in carts to the head of the line, so that the men are not harassed. I doubt not if occasions should neccessarily cause it, you will hear as good an account as formerly of their behaviour, there appearing in the men an implacable hatred against the french, who are the enemies of all mankind.[7]

For the whole of May, June and July the army remained encamped, until at the end of July an order came to Wade from King George 'to commence hostilities of all kinds'.

On 31 July the Allies crossed the Scheldt, and on 3 August Philip Browne wrote from the plain of Lille: 'For a week past have undergone great fatigues. On Saturday we marched and it is sure a mortification to the french that we are living in their country at free quarter.' And in September, 'We have had no movements lately, nothing material hath occurred. We have had no skirmishes of consequence as yet, but have frequently a few men & horses both killed and taken owing to the imprudence of those that venture beyond the Guard.' At the end of October he was writing from Brussels:

We came into garrison here the 7th inst after a week of the greatest hardship we have undergone during the campaign owing to the excessive stormy winds & great rains, and our being destitute of tents the last two nights; the wind was so high that the soldiers tents tore like paper, & the officers had none at all, and our great coats and cloaks was so wet that they continued so after we came here.[7]

Many of the troops were being recalled to England to meet the Jacobite threat, but an order of 18 October from Cumberland specifically ordered The King's Horse to remain in Flanders.[8] The regiment stayed in winter quarters around Brussels until the spring of 1745.

The French army was commanded by Marshal Saxe, who took to the field at the end of April 1745. Feinting towards Mons, he marched on Tournai, to which he laid siege. On 3 May The King's Horse, now in the brigade of Sir James Campbell, marched from Brussels with the army under the command of the Duke of Cumberland, to attempt the relief of Tournai. An order of 9 May issued from Braffoel near the French position, of which the village of Fontenoy was the centre, laid down that 'The Army to be under arms at ten o'clock, leaving the tents standing, the Foot drawn up, and the Horses saddled. None of the Cavalry to march in their

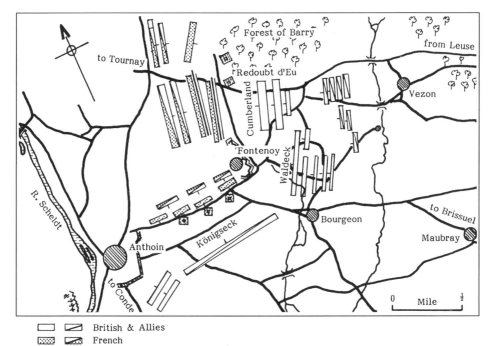

British & Allies

French

Battle of Fontenoy

Frocks, all in red Clothes.'[7] On the following day a squadron of the regiment helped to drive in the French picquets posted in front of the French army.

Saxe had taken up a strong position along a crest, with his right secured by the village of Anthoin on the banks of the River Scheldt, which ran through Fontenoy before bending back to the forest of Barry. In front of this forest lay the village of Vezon, which was occupied as an advanced post. These villages had been fortified and redoubts built in between. The Dutch and the Austrians were to attack the French right and centre, while the British would assault the French left.

At two o'clock on the morning of 11 May 1745, The King's Horse with the rest of the cavalry led the advance towards the French position. Taking their place on the right of the British attack, under Sir James Campbell, they cleared the enemy from their advanced post at Vezon and deployed in the plain beyond the village, in order to cover the infantry forming up for the main assault. The battle then developed into an infantry slogging match, which Saxe described as 'une affaire de mousqueterie', with the cavalry sitting on the flank watching the slaughter.

Philip Browne, writing after the battle, described the scene:

> Old Officers that served the late war say they never saw so continual and
> so long a fire as the French played upon us from there batteries, during

which time life was so uncertain to each one for a moment as we was during the whole time, upwards of nine hours, within the command of them; four men was killed and an officer wounded by the same ball in the Major's Squadron to whom I am Lieut. and three horses of his Troop was killed just behind by one ball likewise and they continually grounded and deadned before us, dropped among us and flew over our heads – which never put any one Squadron in the field into disorder, for so soon as the dead men and horses could be taken out of the ranks the whole was formed immediately again as at a review. There was no Cavalry charged at all, but the troops expressed great inclination to engage sword in hand; but as the enemy continued in there lines, that could not be unless they had been first carried, but they showed a spirit of what they would do if they had oppurtunity by the undauntiness and resolution they behaved with during the long cannonading.[7]

While the infantry attacks were being fruitlessly pressed with exemplary bravery, General Sir James Campbell, commanding the British cavalry, had his leg taken off by a cannon ball. The seventy-four-year-old veteran was carried dying from the field, bemoaning the fact that he could no longer lead his men. Eventually, after hours of repeated attacks, the British infantry were ordered to retire, and Browne commented,

Success is not allways to the valiant and brave; would intrepid calm courage and resolution have carried our point we had now not been a retreating – but our brave & not to be excelled forces are retiring in as much order as they advanced & we wish for no more then that the enemy would advance from behind there batteries, & if they should, my Life upon it, we should destroy them all. There was batteries continually playing upon our front & both flanks at the same time, during the whole attack which was made by the infantry and they supported by the cavalry.[7]

As the infantry reluctantly retired in perfect order, halting and turning every now and then to deliver controlled volleys at the French, the British cavalry 'pushed forward through the cross fire to lend what help they could'.[5] Cannon describes how 'after scrambling through a hollow way abounding with difficulties, the squadrons charged the enemy with great gallantry. In this attack The King's Own Regiment of Horse displayed its accustomed bravery.'[9] Philip Browne summed up the regiment's feelings:

I writ this from a pass where our Squadron with others are posted to protect & secure the regular and exact order in which our forces are retreating; it is about a mile from the field of blood & slaughter where true English courage & bravery hath been exercised & displayed in as high a degree as is possible for mankind to act and by the behaviour of the Hanoverians they may henceforward justly be stiled of the same nation.[7]

The French tribute to the courage and discipline of the British infantry was no less sincere, for Voltaire wrote, 'Les anglais ralièrent mais ils cédèrent: ils quittèrent le champ de bataille sans tumulte, sans confusion, et furent vaincu avec honneur.'

The Allied losses were severe, nearly 6,000 out of the 15,000 engaged were killed or wounded, and of these 3,660 came from the twenty British infantry battalions. The British cavalry lost more than 300 men and 600 horses, mainly owing to the cannon fire, but the King's Horse escaped this pounding fairly lightly, losing only eleven men and twenty-six horses killed, with Lieutenant Brace, four men and six horses wounded. The Cumberland Papers at Windsor Castle give the losses of clothing and equipment for The King's Horse as '8 suits of Privates clothes at £3. 15s; 12 cloaks at £2. 10s; 12 saddles complete with housings and holster caps at £4; 9 pairs of boots at £1. 5s; 16 swords at £1. 7s; 10 pairs of buff belts at £2. 15s.'[9]

The retreat continued to the fortress of Ath, where the army encamped under its walls, and from thence to Lessines, and on to Grammont and Brussels. Meanwhile the fortress of Tournai fell to the French, who also captured Ghent.

The Jacobite invasion of England caused a great many of the British regiments in Flanders to be called home, and a Distribution List dated 5 January 1746 shows Honeywood's Horse [KDG] as 'expected from Flanders'.[8] The regiment arrived on 1 March 1746, where it formed part of a force assembled to meet the threat of a French invasion on the south coast, which never materialised.[3, 10]

With the re-establishment of peace after the defeat of the Jacobites at Culloden, the Government again felt the need to cut expenditure. On 14 December 1746 the Adjutant General issued an order 'as a measure of economy' that the three senior regiments of horse on the English Establishment be reduced to the status of dragoons. As this meant not only a loss of pay but also a reduction in seniority, the blow was softened by allowing the three regiments concerned to call them-selves Dragoon Guards, and to take precedence over other regiments of dragoons on the English Establishment. A warrant was issued which established the new arrangement:

GEORGE R.

Whereas, We have thought fit to order OUR OWN REGIMENT OF HORSE, commanded by Our trusty and well-beloved General Sir Philip Honeywood; THE QUEEN'S ROYAL REGIMENT OF HORSE, commanded by Our right trusty and right entirely beloved Cousin and Counsellor, Lieutenant General John Duke of Montague; and OUR REGIMENT OF HORSE, commanded by Our right trusty and well-beloved Counsellor, Field Marshal George Wade, to be respectively formed into regiments of DRAGOONS, and their establishment and pay, as Dragoons, to commence the 25th December, 1746. And, whereas, it is become neccessary, by the said regiments being formed into Dragoons, that their former titles as regiments of HORSE should be altered; We are hereby graciously pleased to declare OUR ROYAL WILL AND PLEASURE, that Our regiment of Dragoons, now under command of General Sir Philip Honeywood shall bear the title of Our FIRST REGIMENT OF DRAGOON GUARDS; Our regiment of Dragoons, now commanded by the Duke of Montague, the title of Our SECOND REGIMENT OF DRAGOON GUARDS; and Our regiment of Dragoons, now commanded by Field Marshal Wade, the title of OUR THIRD REGIMENT OF DRAGOON GUARDS, and have rank and precedency of all other regiments of Dragoons in Our service. Nevertheless, Our further will and pleasure is, that the said three regiments of DRAGOON GUARDS shall roll and do duty in Our army, or upon detachments, with Our other forces, as Dragoons, in the same manner as if the word GUARDS was not inserted in their respective titles.

Whereof, the Colonels above mentioned, and the Colonels of Our said regiments for the time being, and all others whom it may or shall concern, are to take notice and govern themselves accordingly.

Given at Our Court of St James's, this 9th day of January, 1747, in the twentieth year of Our reign.

By His Majesty's Command,

H. FOX

By this measure the Government saved the sum of £70,000 a year, but as the men were placed on a lower scale of pay than formerly, every man was given the choice of taking his discharge with a gratuity of £3, and fourteen days' pay to enable him to reach his home. At the same time each man who elected to continue serving also received a gratuity of £3. Trumpeters were abolished and were replaced by drummers and hautboy-players; carbines were returned to store and muskets with bayonets were issued in their stead; the flask-string was removed from the pouch-belts; and the remainder of the equipment was brought into line with that of other dragoons. The officers whose pay had not been altered were to wear gold lace and gold embroidery on their regimentals, with a crimson silk sash to be worn over the left shoulder. The quartermasters were also to have gold lace, and to wear their sashes around their waists; while the sergeants were to be distinguished by narrow lace on their lapels, sleeves and pockets, and a worsted sash around their waist.

The regiments were given an establishment of six staff officers, and pay, consisting of:

The colonel, as colonel 15s, for servants 4s. 6d, 19s. 6d

Lieutenant colonel, as lieutenant colonel, 9s. 0d

Major, as major, 5s. 0d

Chaplain, 6s. 8d

Surgeon, 6s. 0d

Adjutant, 5s. 0d

For the KDG there were to be nine troops, each consisting of:

1 captain; 3 horses 3s, for servants 4s. 6d, 15s. 6d

1 lieutenant; 2 horses 2s, for servants 3s, 9s. 0d

1 cornet; 2 horses 2s, for servants 3s, 8s. 0d

1 quartermaster, for self & horse 4s, servants 1s. 6d, 5s. 6d

3 sergeants at 2s. 9d each, 8s. 3d

3 corporals at 2s. 3d each, 6s. 9d

2 drummers at 2s. 3d each, 4s. 6d

1 hautboy at 2s, 2s. 0d

59 dragoons at 1s. 9d each for man and horse, £5. 3s. 3d

Allowance for widows, 2s. 0d

Clothing lost by deserters, 2s. 6d

Recruiting expenses, 2s. 4d

Agency, 1s. 2d

8 troops more as above for the KDG, £68. 6s. 0d

TOTAL COST PER DAY £79. 7s. 11d

TOTAL COST PER YEAR £28,979. 9s. 7d

The Second, or Queen's Horse, had a strength of 6 Troops:

TOTAL COST PER DAY £53. 15s. 8d

TOTAL COST PER YEAR £19,630. 18s. 4d

In December 1746 the King's Dragoon Guards moved, with one troop at Dunstable, one at St Albans, one each at Ware and Hoddesdon, two at Hertford, and one each at Barnet, Hitchin and Hampstead, where they changed from horse to dragoon guards. Lieutenant Colonel Martin Maden retired, and was succeeded by Lieutenant Colonel Timothy Carr. Having formed the new establishment, the regiment moved to Herefordshire until the autumn of 1748, when it marched to Norfolk, where it was employed on coastal duties, which included 'Detachments or

Out Parties employed to prevent Owling and Smuggling, and in preserving the Public Peace'.[11] The following year, 1747, the regiment's responsibilities were extended into Essex.

During 1746 there occurred the death of a long-serving quartermaster of the Queen's Dragoon Guards who had served in the regiment for nearly fifty years. He was buried at Coventry, with the following inscription on his tombstone:

> Here lieth the body of Arthur Manley, late Quartermaster in the Queen's Royal Regiment of Horse, who served the Crown of Great Britain upwards of 56 years, from the 15th July, 1685 to the 24th August, 1741. He died June 7th 1746, aged 78 years.
>
> > The Israelites in Desert wandered but two score
> > But I have wandered two score sixteen and more,
> > In dusty campaigns; restless days and nights
> > In bloody battles oftimes did I fight,
> > In Ireland, Flanders, France and Spain.
> > At last here lies my poor mortal remains.
>
> I served in the Foot ten years, and in the above Regiment of Horse upwards of forty-six years.

Another old soldier from The King's Dragoon Guards was Captain Nathaniel Smith, who left The King's Own Regiment of Horse in 1748 to become the captain of Southsea Castle, and then the comptroller of the Royal Hospital at Chelsea, where his natural talents led him to become, first, the major of the Royal Hospital, and eventually its lieutenant governor.[13]

The Queen's Dragoon Guards re-formed onto the new establishment in York, and then moved from there to Derby and Nottingham in 1747, moving on to Bristol and the surrounding area of Somerset in 1748, then on again in January 1749 to Gloucester, Tewkesbury, Worcester and Pershore. In 1749 the Duke of Montague died, and was succeeded in the colonelcy on 24 July by a very famous soldier, Sir John Ligonier, from the 4th Irish Horse [7 DG], who had won great distinction in the preceding wars. Ligonier was made colonel of the Blues in January 1753, and wrote, 'I am so pleased with my Bay Guards, that if left to myself, I would rather keep them than have the three Squadrons [of the Blues].' In the autumn of 1749 The Queen's Dragoon Guards marched into the seaside towns of Sussex to perform the same duties as the KDG in Norfolk, assisting the revenue officers in preventing widespread smuggling.

The Peace of Aix-la-Chapelle was signed in 1748, and both The King's Dragoon Guards and The Queen's Dragoon Guards suffered a reduction in strength of one sergeant, one corporal and twenty-three men per troop. On 3 November 1748 the

War Office wrote to the colonels, "Your Grace may please to order proper Officers to repair to the said places to collect and receive the Arms of the Non Commission Officers and Private Dragoons belonging to the said Regiments so to be reduced, which are ordered to be delivered into His Majesty's Stores of Ordnance and give the neccessary Acquittances for the same. 1st D. G. Norwich. 2nd D. G. Bristol.'[11]

During 1750 The King's Dragoon Guards were involved in keeping the peace in Birmingham in aid of the civil magistrates, who called upon the services of the regiment in July: 'Three Troops of the Regiment obtained great credit for the temper, patience, and forbearance which they displayed when suppressing some disturbances among the populace.'[3] This duty in aid of the civil power was to become increasingly frequent over the next hundred years, earning the regiment the nickname of 'The Trade Union', not because their ranks contained an undue proportion of 'barrack room lawyers', but because of their reliability in times of civil unrest.

In 1751 the officers and men were directed to wear aiguillettes on the left shoulder, the former being of gold cord and the men's of worsted.[9] A warrant of 1 July 1751 gives details of the dress of the drummers and hautboys of The King's Dragoon Guards: they were to wear a red coat with blue facings, long hanging sleeves, blue waistcoat and breeches, blue cloth caps embroidered with the King's cypher and crown with the White Horse of Hanover and the motto 'Nec Aspera Terrent', and a blue 'turn-up' with a drum and '1. D.G.' However an inspection report of 1750 noted that The King's Dragoon Guards had no hautboys and that the drummers had taken the place of the trumpeters. When the regiment paraded on foot, small kettledrums, known as 'nakers', were carried on the back of a drummer boy. It is of interest that some years later, in 1776 and again in 1782, an inspection report of The King's Dragoon Guards mentions fifers, and that 'the Trumpeters are mostly fifers', for which there seems to be no easy explanation in a cavalry regiment.[13]

Sir Philip Honeywood died in July 1752, and Lieutenant General Sir Humphrey Bland, of the 3rd Dragoons, was appointed as colonel of The King's Dragoon Guards in his place. In the autumn of 1752 the regiment was ordered to Scotland, where it remained until September 1754, when it was stationed at York, Leeds and Wakefield. In the spring of 1755 the King's Dragoon Guards marched to quarters in the south of England.

SOURCES

1 *War Office Records*, p. 92.

2 *London Gazette*, 'History of the Rebellion, 1745'.

3 R. Cannon, *Historical Records of the British Army, 2nd Dragoon Guards*, 1837.

4 F. Whyte and A. H. Atteridge, *A History of the Queen's Bays*, vol. 1, 1930.

5 Sir John Fortescue, *History of the British Army*, vol. 2, 1910.

6 J. M. Brereton, *History of the 4/7th Royal Dragoon Guards*, 1982.

7 'Letters of Captain Philip Browne, 1737-1746', *Journal of the Society for Army Historical Research*, vol. 5.

8 *Journal of the Society for Army Historical Research*, vol. 22, 1944, pp.292, 295.

9 *Journal of the Society for Army Historical Research*, vol.25, 1947, p.9.

10 Records of The King's Dragoon Guards, Regimental Museum, 1st The Queen's Dragoon Guards, Cardiff Castle.

11 Public Record Office, W. O. 4/45, pp. 207, 209 of 3 November 1748.

12 C. G. T. Deane, *The Royal Hospital, Chelsea*, 1950.

13 C. ffoulkes, 'Notes on Early Military Bands', *Journal of the Society for Army Historical Research*, vol. 17, 1938.

12

Light Troop, Cherbourg, St Malo, Minden
1756-1759

Trouble broke out with the French in North America and in the West Indies, and as a precautionary measure the strength of both The King's Dragoon Guards and The Queen's Dragoon Guards [Bays] was augmented, the former by having one sergeant, one corporal and fifteeen men added to each troop, and the latter by the addition of a corporal and fifteen men. At the same time both regiments were given an additional light troop, consisting of one captain, one lieutenant, one cornet, one quartermaster, two sergeants, three corporals, two drummers and sixty troopers, including one farrier, but this was soon increased by another twenty-nine troopers, bringing the overall strength of the light troop up to a hundred men. The role of the light troop was reconnaissance, skirmishing and outpost duty, for which it was felt the heavier chargers of the regiments of Dragoon Guards were not so well suited.[1,2,3]

A warrant of 14 April 1756 stated that the men of the light troop were to be

> Height 5 ft 6½ to 5 ft 8, and to be light, active young men. Horses 14 hands 2 in. and not under; to be well-turned nimble road horses, as nigh to the colour of the regiment as can be got. Arms: carbine with ring and bar, 4 ft 3 in long, with a bayonet of 17 in; one pistol, 10 in in the barrel, and of carbine bore; straight cutting sword, 34 in in the blade, with a light hilt without a basket. Accoutrements: tanned leather shoulder belt, 3¼ in broad, with spring and swivel; tanned leather belt for sword and bayonet; tanned leather cartouche box, with a double row of holes to contain 24 cartridges, with a tanned leather strap, 1½ in broad. The saddle to be with small cantles behind, as the jockey saddles are, and to be 22 in long in the seat; on the right side of the saddle is to be the holster for the pistol, and on the left a churn, in which a spade and felling axe, or a spade and woodman's bill is to be carried. There is to be a bucket for receiving the butt of the carbine and a pipe to receive the end of the horse-picket. The bridle is to be a plain light bit with single reins; there is likewise to be a tanned leather headstall with a hempen collar.

The clothing and cloaks of these Light Dragoons to be the same as that of the rest of the regiment, only instead of hats the men are to have jockey caps ornamented in the front with H. M.'s cypher and crown in brass, and the number or rank of the regiment. The crest is likewise to be covered with brass, out of which is to be a tuft of stiff horsehair, coloured half-red, and the other half of the facing. Light jockey-boots with small stiff tops.[3] An inspection report of 1759 reported that the light troop of The King's Dragoon Guards was wearing 'brass caps'.[4]

The inspection report of The King's Dragoon Guards for October 1754 state that the officers' uniforms were old, and that the men were 'of large size and have white gaiters'. The horses were of a heavier type than most dragoon horses, with the drummers mounted on greys, with drums that were received in 1752. On 27 May 1755 the officers' hats and uniforms were not according to regulation, although by 24 October the 'Officers properly armed and accoutred; they report new uniforms are making conformable to regulation.' In 1756 the inspecting officer was able to report, 'Officers' uniforms new.'[4]

The inspection reports for The Queen's Dragoon Guards showed that in 1753 the 'Officers' uniforms new and according to regulation. Horse furniture old but good. Men have white gaiters. Horses very fine, have great spirit; all bays, except the Drums, who are mounted on greys.' In 1754 '12 Drummers swords received in 1750, Standards 1754.' In 1759 'Drummers and Farriers mounted on greys. Drums of brass, embossed, not painted, being made before the regulation.' The light troop 'had brass caps'.[4].

Snippets of information survive. On 15 December 1756 one of the older officers, Captain John Richardson, became an agent for the Out-Pensioners of the Royal Hospital, Chelsea.[5] In March 1757 The King's Dragoon Guards marched into quarters around London. Later that year the regiment was in camp at Dorchester; there was an artillery camp a mile and a quarter from the town, then a line of camps of six infantry battalions, with the KDG camp at the western end. The regiment had a standard guard in place of the more usual quarter-guard, and this consisted of a sergeant, corporal and eighteen troopers. They also provided pickets of a captain, two lieutenants, two sergeants, two corporals, two drummers, and five men from each troop. In addition the regiment provided what was known as the grand guard, which on active service acted as vedettes outside the line of camps. This was made up of a captain, two lieutenants, two sergeants, three corporals, two drummers and fifty men. The number of men required for guard, combined with frequent parades, fatigues and roll calls, made camp life fairly demanding.[6]

In 1757 Lieutenant Colonel Carr, commanding the regiment, died, and was succeeded by Lieutenant Colonel William Thomson. This was the year of the outbreak of the Seven Years War and the King asked Parliament to send an expeditionary force to assist Prince Ferdinand of Brunswick, who was engaged in fighting the French between the Rhine and the Main. On 23 June 1758 Pitt announced in Parliament that he was ordering 2,000 cavalry to reinforce Ferdinand, and The King's Dragoon Guards, together with the Blues, the 3rd Dragoon Guards, the Greys, Inniskillings, and 10th Dragoons were warned to prepare for foreign service. At the same time Pitt decided upon a raid on the French coast to keep French troops in France, and so prevent the reinforcement of their troops in either Canada or Germany. The light troops of nine cavalry regiments, including the KDG light troop, had been ordered to concentrate at Petersfield, where they were placed under the command of Colonel Elliott of the Horse Grenadier Guards.

On 1 June 1758 the force, consisting of the nine light troops (KDG, 3rd DG, 1st, 2nd, 3rd, 6th, 7th, 10th and 11th Dragoons) together with nine infantry battalions, under the command of Charles, Duke of Marlborough, embarked on transports, with twenty-four ships of the line as escort, and sailed for St Malo on the French coast. On 5 June the force arrived at Cancalle Bay, some eight miles from St Malo, and landed there the following day. Leaving a brigade to guard the landing place, the rest marched towards St Malo and on the 7th encamped at Parame. That night the light troops moved forward, entered St Malo, reached the harbour, and set fire to more than a hundred privateers and merchant vessels, as well as burning all the naval stores and magazines around the harbour. The next day they moved out beyond St Malo towards Dol, where they continually skirmished with various French detachments. The Duke of Marlborough made preparations to lay siege to the town of St Malo, but when he heard that considerable French forces were on the march to cut off his retreat, he retired to Cancalle Bay and re-embarked the troops. The fleet and transports then sailed twenty miles north east up the coast with the intention of raiding Granville in Normandy, and then repeating the process at Cherbourg, but continuing foul weather frustrated each attempt. The force therefore returned to Portsmouth on 1 July, where they encamped.

Pitt was determined to make another raid on the French coast, and preparations went ahead for a second expedition, which sailed from Portsmouth on 1 August 1758, with the same nine light troops of cavalry, and twelve battalions of infantry. A landing was made on 6 August in the bay of St Marais, six miles from Cherbourg, against opposition from a force of 3,000 French, who were driven off. The light troops disembarked the following day and advanced on Cherbourg, which surrendered at once. While the docks and fortifications were being destroyed and the shipping in the harbour burnt, the cavalry light troops 'scoured the surrounding country and levied contributions'.[1, 7] After ten days ashore the troops were re-embarked, together with a quantity of brass cannon and mortars taken from Cherbourg.

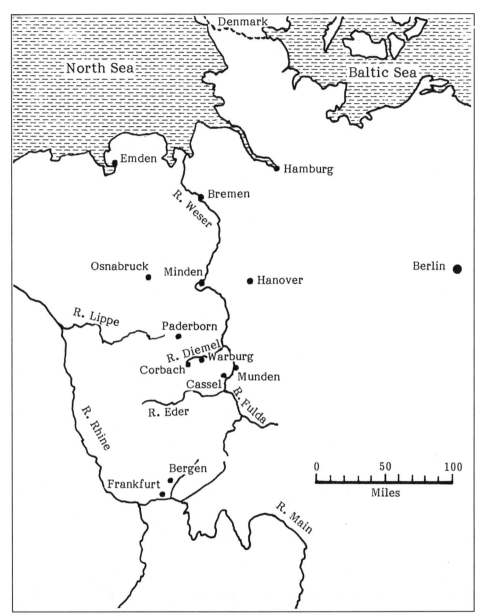

The Seven Years War
(1758-63)

Foul weather imposed a delay before the troops were again landed on 3 September at St Lunaire, twelve miles east of St Malo on the coast of Brittany. On the 4th and 5th there was some desultory skirmishing near the village of Matignon, but the weather was deteriorating and the Navy were forced to leave St Lunaire for the safer anchorage of St Cast, further to the west of St Malo.By the time the troops had reached St Cast, the French had brought up a considerable force which attacked the rearguard. The embarkation proceeded under the cover of the guns of the fleet, but then the rearguard began to run out of ammunition and had to make a dash for the boats. The French cannon opened fire, sinking many boats, and most of the rearguard were killed, drowned or taken prisoner. Altogether the expedition suffered 750 killed and wounded apart from the prisoners, and the troops returned to England in low spirits. The light troops were quartered in villages on the South Coast on landing.[1, 7]

In the autumn of 1758 The Queen's Dragoon Guards left Scotland for Yorkshire and Lincolnshire, with two troops in garrison at Hull, and the light troop detached at Northampton. General Herbert had died in 1757 and was replaced on 5 April by Lord George Sackville, who came from being the colonel of the 3rd Irish Horse [6 DG] and was shortly to be removed in disgrace. The regiment moved to the South of England in 1759.

The King's Dragoon Guards, preparing for foreign service, were reviewed by the King in Hyde Park on 10 July 1758 and on 26 July the regiment landed at Emden, marching on 5 August under the command of the Duke of Marlborough to join the army of Prince Ferdinand of Brunswick. Before the regiment left England several corporals and troopers were promoted to the rank of sergeant on transfer to the 11th Foot [Devons], there being no vacancies at that time in the King's Dragoon Guards.[1, 8]

On 17 August the regiment camped at Coesfeld, where it was joined by Prince Ferdinand and a detachment of his army. The march of the British troops from Emden through Rhede had been marked by a lack of discipline. This was largely Catholic countryside and there were complaints of wayside crucifixes being destroyed and of looting. The cavalry patrols of the KDG were ordered not to arrest offenders but to shoot them, and the grand provost was ordered to hang without mercy anyone caught looting. A surviving KDG order book of this period shows that of twenty KDG who came before a regimental court martial during 1759, half were accused of drunkenness or disorderly conduct in camp, two of theft, one of losing his sword, one of neglect of duty. An NCO was reduced to the ranks for condoning drunkenness, but there were no charges of 'maroding'.[9, 10, 11]

On 14 August Marlborough reported: 'Four days and nights rain en route, The Foot have marched near their middles in water; the horses in bad condition.' On the march the order book allows two waggons for the captains and one for all the subalterns, with another two per squadron, the latter carrying the tents for the men on the march, and their blankets at other times. Each squadron had a forage cart, and there are references to farriers' carts. Soldiers are not to be employed as servants, and officers' servants are to behave more quietly; they are warned, too, that they are subject to military discipline. There are regulations for the exact pitching of tents, and that there be no obstacles to free movement, nor to delay the men in turning out promptly. Butcher's offal is to be buried and the 'neccessary houses' must be at a distance from the lines. They are to be renewed frequently and must be used for disposing of the horses' dung. Horses are to be properly shod, and care is to be taken in the putting on of saddlery and accoutrements, especially so that a trooper may get at his cloak without dismounting. Horses are to be kept within their troops, so that any 'epidemical disorder' can be checked from spreading. In this connection especial care is to be taken lest any horses brought in by French deserters are glandered. Troop leaders are to make regular returns of all the horses in their troop, giving age, size, marks and name. [9]

On 20 August 1758 Prince Ferdinand inspected the British troops. He wrote afterwards, 'I inspected them yesterday, and it is impossible for me to express the satisfaction I felt at the good order in which I found these fine troops.' On the 27th the army moved to Lette and two days later to Dulmen, then back to Munster, and by 19 October had reached Soest, during which time there was some desultory skirmishing before going into winter quarters at the end of November. Seven squadrons of the British cavalry were cantoned under General Mostyn at Paderborn, and nine along the Ems around Osnabruck under General Elliott. The King's Dragoon Guards were at Rheine, with the 3rd Dragoon Guards at Meppen and the 10th Dragoons at Haselunen. On 20 October the Duke of Marlborough died, to be succeeded in the command by the colonel of The Queen's Dragoon Guards, Lord George Sackville. Lieutenant Colonel Thomson handed over command of The King's Dragoon Guards to Lieutenant Colonel Robert Sloper on 13 February 1759; Sloper was later to become the colonel of the 14th Light Dragoons and Commander-in-Chief in India. The new colonel was not satisfied with the speed at which the KDG turned out, and issued an angry order to the officers that they were to go round to inspect the men's tents to ensure that all their accoutrements were properly placed and to hand. [1, 9, 11, 12]

Early in the spring of 1759 the army reassembled, The King's Dragoon Guards being brigaded with the Blues and the Inniskilling Dragoons. The British infantry was left around Munster observing the French army under Contades, and protecting the approaches to Hanover and Hesse. Prince Ferdinand, taking The King's Dragoon Guards and the rest of the British cavalry with him, marched against a

second French army, under command of Marshal Broglie, and attacked him in a
stongly entrenched position at Bergen on 13 April. Broglie repulsed all of Ferdinand's assaults, forcing him to retire with the loss of 2,600 officers and men killed
and wounded, as well as five guns. This was the first action in which the British
cavalry played a part – albeit a small one covering the retreat - in the Seven Years
War.[7, 10, 11]

The whole of the early summer was spent in intricate manoeuvring between the
French and Ferdinand. Contades managed to capture Cassel, the capital of Hesse,
and laid siege to Munster, Hameln and Lippstadt. On 10 July, when Broglie took
Minden, the French seemed to be poised to invade Hanover, having control of the
line of the Weser. Contades was positioned to the south of Minden, but in communication, by three bridges across the Weser, with Broglie, who was encamped to
the east of the river. The French position seemed impregnable: Broglie's right
rested on the town of Minden and the Weser, his left on a range of wooded hills,
while the whole of his front was covered by a wide morass. Ferdinand, with a feint,
managed to draw off some of Contades's troops and decided to attack the French
at Minden.

The French also planned to attack. On the night of 31 July Contades marched his
force across the bridges in eight columns and took up his position under the walls
of Minden, with his left flank covered by the morass, and with Broglie's troops on
the right flank, resting on the Weser. So the French infantry formed the two wings
of their position, with a mass of cavalry in the centre, consisting of fifty-five
squadrons with eighteen more in reserve.

A separate corps of the Allies under Wangenheim had meanwhile closed up to
the French position, ahead of Ferdinand's main army, opposite Broglie, where it
was strongly entrenched with several guns covering the only outlet through which
the French could debouch from their entrenchments. Ferdinand hoped that Contades could be lured out of his position to attack this corps, so that he could fall on
the French flank with the main army. On the evening of 31 July the cavalry was
ordered to be ready to move by one o'clock the next morning, and to have the
horses saddled and ready. At five in the evening the horses were unsaddled and
the men returned to their tents to sleep in their boots and clothes. Then, early on
the morning of 1 August, they turned out again to the picket lines and saddled up.
About three hours previously two French deserters had come in to say that the
French were advancing, but no one took much notice of this information and it was
not until 3 a.m. that Ferdinand realised what was happening. He at once gave
orders for the army to advance in eight columns, with the infantry in the centre
and the cavalry on both wings.

The leading right-hand column consisted of the British and Hanoverian cavalry

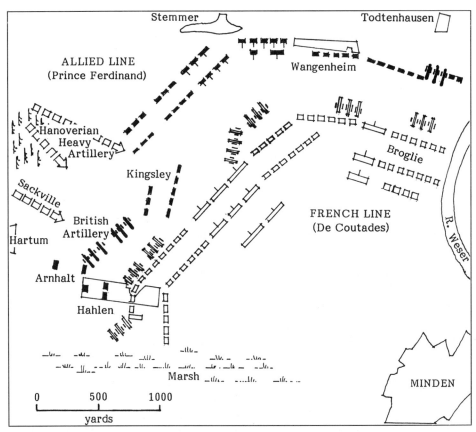

Battle of Minden

commanded by Lord George Sackville, with The King's Dragoon Guards, the Blues, the 3rd Dragoon Guards, the Greys, and the 10th Dragoons. Speed was of the essence if Wangenheim was to be helped in time, but Sackville, who was sleeping some distance from his troops, did not reach his command until it had been formed and waiting for half an hour. He ordered the line to form column by wheeling the files to the right in quarter ranks, then, advancing at the walk-march, he led his cavalry onto the field, breaking into a slow trot but frequently halting to adjust the dressing of his eighteen squadrons on their frontage of 1,950 paces. Lieutenant Colonel Sloper of The King's Dragoon Guards was later to describe the pace of the advance as 'not above half stepping out'. Sackville's twenty-four squadrons eventually drew up on the right wing in front of the village of Hartum, with fourteen squadrons in the first line and ten in the second, a total of 3,300 cavalrymen, of which 1,900 were British and the rest Hanoverians. The three squadrons of Bland's [KDG] were on the right of the line, next to two squadrons of the 6th Dragoons.[1, 7, 10, 13]

In the meantime Ferdinand sent an order to the first line of British infantry that when its time came to advance, this should be done with drums beating; but the order was misunderstood as a direct command to attack the French centre. The front rank of the 12th [Suffolk], 37th [Hampshire], and 23rd [Royal Welsh Fusiliers] under Brigadier Waldegrave, followed by the second, made up of the 20th [Lancashire Fusiliers], 51st [King's Own Yorkshire Light Infantry], and the 25th [King's Own Scottish Borderers] under Brigadier Kingsley, supported by three Hanoverian battalions, advanced, to everyone's amazement, straight on the massed French horse. For the first 200 yards the French guns tore great gaps in their ranks, and then eleven squadrons of the French horse charged them. The British halted and, when the cavalry were within ten yards, gave them a volley which covered the ground with men and horses. Then the infantry again advanced steadily.[7, 10, 11]

Ferdinand at once sent Winzingerode to tell Sackville to advance and sustain the infantry. Winzingerode found Sackville sitting in front of The King's Dragoon Guards. He gave him the order twice in French, but Sackville queried it. Later, at Sackville's court martial, Colonel Sloper of the KDG gave evidence that Sackville said, 'I do not comprehend how the movement is to be made.'

Meanwhile the French had rallied their cavalry and, together with four infantry brigades and thirty-two guns enfilading the British foot from the French left flank, they attacked the nine isolated battalions a second time. The British and Hanoverian foot again blasted the second line of French cavalry and then turned on the French infantry and dealt with them. Ferdinand sent a second messenger, one of his aides de camp, to Sackville, a British cavalryman, Captain Ligonier, to repeat his order. Sackville drew his sword and ordered the cavalry to advance to their front, rather than to their left in support of the hard-pressed infantry. At this point a third messenger arrived from Ferdinand ordering the cavalry to advance to the left. Sackville, thoroughly confused, halted his men.

In the centre the French third line of cavalry of gendarmerie and carbineers charged the nine indomitable battalions, and succeeded in breaking through the first line, only to be shattered by the volley fire of the second line. A fourth messenger now came from Ferdinand, but Sackville professed not to understand the order, and this time he trotted off to Ferdinand to get personal confirmation. On riding up he saluted Ferdinand, who said coldly, 'My Lord, the opportunity is now passed.' While this was happening a fifth messenger had been despatched, who in Sackville's absence rode up to Lord Granby, commanding the second line of the cavalry. Granby immediately gave the order to advance and The King's Dragoon Guards and their comrades had come forward three or four hundred yards at the full trot, when they were met by Sackville on his way back from Ferdinand. Sackville at once halted them. There the troopers sat,'gloomy and shame-faced and cursed Mylord Sackville'.

The courage of the British and Hanoverian foot had enabled Ferdinand to join up with Wangenheim's corps on his left, while the French were forced to retreat behind the walls of Minden, which surrendered on the following day.[1, 7, 10, 11, 13, 14, 15]

George II, on receiving Ferdinand's report of the battle, at once ordered Sackville to return to England, his command being taken over by Lord Granby. In April 1760 a court martial found Sackville guilty of disobeying orders, and a prominent witness against him was Lieutenant Colonel Sloper of The King's Dragoon Guards. The sentence of the court found him 'unfit to serve His Majesty in any military capacity whatever'. As a result, his colonelcy of The Queen's Dragoon Guards lapsed automatically, and by a commission of 10 September 1759 John, (later Lord) Waldegrave was appointed in his stead. The Queen's Dragoon Guards had now had four colonels in ten years. In later years Sackville was to reappear again as Secretary of State for the Colonies in Lord North's Administration and there his actions were said to have lost the American colonies.[2, 7, 10, 15, 16]

After Minden Ferdinand followed up the retreating French, and on 28 August The King's Dragoon Guards, together with the Blues and the 6th Inniskilling Dragoons with whom they were brigaded, came upon Colonel Fischer's corps of 2,000 men at Wetter. The order was given to draw swords and the KDG charged, killing some sixty of the enemy, wounding many more and taking 400 prisoners, as well as many horses and all the French camp equipment. The pursuit was continued for 200 miles, over very difficult country, including precipices, bogs and areas of flooding, but always pressing upon and harrying the rear of the French army, and taking many prisoners. Eventually, in November 1759, The King's Dragoon Guards went into winter quarters near the River Lahn. Even so, a desultory warfare was carried on throughout the winter months, and the troops got little rest.[1,12,17]

SOURCES

1 R. Cannon, *Historical Records of the British Army, 1st Dragoon Guards,* 1837.
2 R. Cannon, *Historical Records of the British Army, 2nd Dragoon Guards,* 1837.
3 Revd Percy Sumner, 'Uniforms and Equipment of Cavalry Regiments from

1684 to 1811', *Journal of the Society for Army Historical Research*, vol. 14, 1935.

4 *Journal of the Society for Army Historical Research*, vol. 3, 1924, p. 231.

5 C. G. T. Deane, *The Royal Hospital, Chelsea*, 1950.

6 E. J. Priestley, 'Army Life in 1757', *Journal of the Society for Army Historical Research*, vol. 52.

7 Sir John Fortescue, *History of the British Army*, vol. 2, 1910.

8 R. E. R. Robinson, *The Bloody Eleventh*, Devon & Dorset Regiment, 1988.

9 'An Order Book for the Seven Years War', *Journal of the Society for Army Historical Research*, vol. 30, p. 2.

10 Piers Macksey, *The Coward of Minden*, 1979.

11 H. C. Wylly, *History of The King's Own Yorkshire Light Infantry*, vol. 1, 1924.

12 W. Clowes, *King's Dragoon Guards 1685-1920*, 1920.

13 C. Duffy, *The Military Experience in the Age of Reason*, 1987.

14 Sir George Arthur, *The Story of the Household Cavalry*, vol. 2, 1909.

15 Rupert Furneaux, *The Seven Years War*, 1973.

16 F. Whyte and A. H. Atteridge, *The Queen's Bays*, vol. 1, 1930.

17 *The K. D. G.*, vol. 1, no. 7, p. 237.

13

Corbach, Warburg, Vellinghausen
1760-1762

In April 1760 The Queen's Dragoon Guards [Bays] were ordered to prepare for foreign service, embarking in transports on the River Thames and sailing at the beginning of May. The regiment arrived in the River Weser, just below Bremen, on 17 May, landed immediately, and then had a long march to join Prince Ferdinand's army at Fritzlar on 14 June. Here it was at once brigaded with The King's Dragoon Guards and the 3rd Dragoon Guards, under the command of Brigadier General Webb. During the ensuing campaign the KDG and the Bays fought side by side.[1, 2, 3]

The King's Dragoon Guards, having spent an uneasy winter at Osnabruck, marched on 5 May for Fritzlar, arriving on the 20th. The French had been considerably reinforced during the winter, and Broglie could now rely on a total strength of 150,000 men, part commanded by St Germain and part by himself, against the 80,000 available to Ferdinand. Webb's brigade set out from Fritzlar on 24 June and was soon providing advanced guards and pickets against the overwhelming strength of the French, which forced Ferdinand to remain on the defensive and involved the army in some extensive manoeuvring, when both The King's Dragoon Guards and The Queen's Dragoon Guards were engaged in a number of skirmishes. The brigade formed part of a force commanded by Major General Griffin, with the task of maintaining communications between the various parts of the Allied army. It was initially stationed at Walthersbruck.[1, 2, 4, 5]

Lord Pembroke, one of Granby's cavalry brigadiers, wrote home:

> Broglie is with an incredible mob full double ours, whatever political fal-
> sifiers may say in England. We have too many rouses for correspondence,
> and very little rest or belly-provender in return, for never poor devils lived
> harder, or earned their pay more than we all do, lying on one's arm night
> after night in damned bad weather, sleeping and starving au bivouac, or on
> a stone under a hedge. Notwithstanding all this, as poor beggars generally
> are, we are vastly jolly and happy.[6]

Broglie ordered St Germain to march from Dortmund on 4 July and meet him at

Corbach not later than 10 July. Ferdinand, hearing of these moves on the 5th, pushed forward a strong advanced corps, including Webb's brigade, to occupy the defile through which Broglie's army must pass in order to join up with St Germain. The KDG and The Queen's Dragoon Guards marched through Sachsenhausen, but found on arrival that St Germain, by forced marches, had traversed the defile and was forming up with Broglie on the heights of Corbach. As the cavalry arrived, it appeared that only about 10,000 of the French were deployed, and Ferdinand at once attacked. However, more and more French regiments came up and the Allies were forced to retire, the retreat being covered by the 5th [Northumberland Fusiliers], 50th [Royal West Kent] and the 51st [King's Own Yorkshire Light Infantry]. The French managed to rout a number of the German regiments and were pressing the rearguard, whose situation was becoming desperate, with cavalry and artillery. Ferdinand put himself at the head of a squadron of The King's Dragoon Guards and with a squadron of the 3rd Dragoon Guards, backed by The Queen's Dragoon Guards, led Webb's brigade in a charge against the advancing French.[1, 4, 5]

The KDG squadron, commanded by Major Sandys Mill,

> dashed forward with that intrepidity which characterises the charge of British cavalry; the troops of the enemy were thrown into disorder, and numbers fell beneath the conquering sabres of the English Dragoon Guards. This repulse damped the ardour of the foe, and the allied infantry, rescued from impending destruction by the distinguished bravery of the 1st and 3rd Dragoon Guards, was enabled to make an undisturbed retreat. In this charge Ferdinand was himself wounded in the shoulder.[1, 4, 5, 6, 7, 8, 9, 10]

Corbach was a defeat for Ferdinand, but one which would have been very much worse had it not been for the steadiness and bravery of the three British regiments of foot, and of the 1st and 3rd Dragoon Guards. The French casualties amounted to 819 killed and wounded, but Ferdinand lost 15 guns, 7 of them British, and 27 officers and 797 men. The King's Dragoon Guards and the 3rd Dragoon Guards had 3 killed, 8 wounded and 79 missing; of these Major Mill's squadron of the KDG went into action 90 strong, and only 24 returned. The regiment's final casualties in this action were 47 men and 58 horses killed or missing, with Lieutenant Jacob, 7 men and 2 horses wounded. The despatch after the battle read, 'Our Battalions would have suffered considerably had it not been for the bravery of the Hereditary Prince [Ferdinand], who, putting himself at the head of one of Bland's squadrons [KDG] and of Howard's Regiment of Dragoons [3 DG], charged the enemy so furiously as to enable our infantry to make a safe retreat.'[1, 4, 5, 7, 8, 10]

On 16 July the newly raised regiment of the 15th Light Dragoons, commonly known as 'Elliott's Tailors' because of their having been recruited among the journeymen and tailors of London, fought a brilliant and successful action at Emsdorff, capturing 2,600 prisoners, 9 colours and 5 guns. This dashing engagement considerably restored the Allies' morale after their defeat at Corbach.[5, 6]

Broglie now divided his army into three, threatening Ferdinand on both flanks as well as to his front. Ferdinand accordingly retired to the north west. Broglie next despatched the Chevalier de Muy, who had replaced St Germain, to Warburg with 30,000 men, to cut off the Allies from Westphalia. Ferdinand at once decided to attack de Muy, who was in a strong position along a ridge running across a bend on the north bank of the River Diemel, with his right resting on the village of Warburg and his left on Ochsendorf. Ferdinand ordered his nephew, the Duke of Brunswick, to take two columns, totalling 14,000 men, including the Royal and 7th Dragoons, in a flanking movement to attack the French left wing at Ochsendorf. Ferdinand, with the main body and the rest of the cavalry, would cross the Diemel at Liebenau and simultaneously attack de Muy from the front.[5, 6]

Brunswick's columns marched from their base at Corbeke at 6 a.m. on the morning of 31 July; Ferdinand, who had further to go, set out at 9 p.m. on the 30th. The KDG and The Queen's Dragoon Guards with the 3rd Dragoon Guards were on the right of the line, with the KDG on the right of Webb's brigade. The regiments in Honeywood's, Elliott's and the Earl of Pembroke's brigades had only two squadrons in action; only the Blues and the KDG mounted three squadrons. The British cavalry of the right wing was commanded by the Marquess of Granby, who also took charge of the first line, with the second line under Lieutenant General Mostyn. [1, 2, 5]

The main body of the army was delayed in crossing the Diemel at Liebenau, and it was 5 a.m. before it formed up on the heights of Corbeke, while the cavalry pushed forward to a wood near the Desenberg hill, about five miles from Warburg. The Duke of Brunswick, who had only ten miles to cover, managed to slip past the French outposts in the early morning mist and, with only a few grenadiers, seized a high hill surmounted by a tower in the rear of the French left wing. De Muy immediately tried to eliminate this threat to his position and launched a heavy attack on the isolated post. As each side brought up more men, the battle raged for four hours with increasing intensity, but Brunswick managed to hold his own.[5, 6]

Ferdinand, hearing the sound of battle, urged on his infantry to relieve his nephew, in spite of the marshy ground. The troops struggled in the heat, and at 12.30 p.m. he ordered Granby with his twenty-two squadrons of British cavalry and the horse artillery, 'to come up with the enemy with the utmost expedition'.

The cavalry, eager to avenge the unmerited disgrace of Minden, at once started

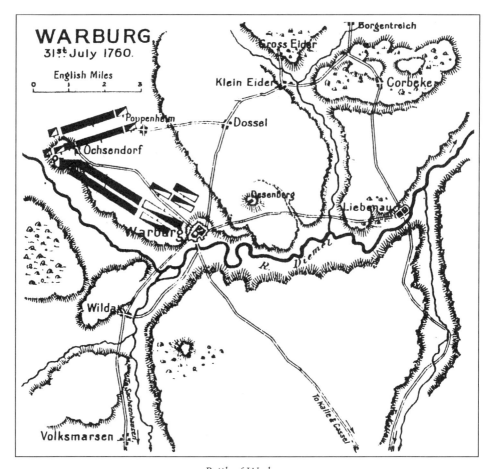

Battle of Warburg

off at a fast trot in columns of squadrons, followed by the three brigades of horse artillery. They covered the five miles to Warburg in less than an hour, with the guns bumping and careering along over very rough ground behind the 2,000 horsemen 'at a speed which amazed all beholders'. Fortescue describes the scene:

> The pace was checked for a brief moment as the squadrons formed in two lines for the attack. In the first line from right to left were the First [KDG], Third and Second Dragoon Guards [Bays] in one brigade, the Blues, Seventh, and Sixth Dragoon Guards in another; in the second line were the Greys, Tenth, Sixth, and Eleventh Dragoons.Then the advance was resumed, Granby riding at the head of the Blues, his own regiment, and well in front of all.[1, 2, 3, 5, 6, 9, 11, 12, 13]

The trumpet sounded. Granby turned, waving his sword to encourage the troopers pounding along behind; as he did so, his hat, and then his wig, flew off – 'a big bald circle in his head rendering the loss more conspicuous'. A delighted roar of laughter came from troopers. 'But he never minded; stormed still on, bare bald head amongst the helmets and sabres; and made it very evident that had he instead of Sackville led at Minden there had been a different story to relate.' So Granby unconsciously gave the English language the phrase of 'going at it bald-headed'.[15]

> As the trot grew into gallop and the lines came thundering on, the French squadrons wavered for a moment, and then, with the exception of three only, turned and fled without awaiting the shock. The scarlet ranks promptly wheeled round upon the flank and rear of the French infantry, whereupon the three French squadrons that had stood firm plunged gallantly down on the flank of the King's Dragoon Guards and overthrew them.[5]

Granby ordered two squadrons of the Blues, under Lieutenant Colonel Johnston, to wheel and come to the aid of the KDG. As the Blues galloped down, they rode over and annihilated the three gallant squadrons of French cuirassiers, enabling the KDG to re-form. Fortescue continues: 'The French infantry, finding itself now attacked on both flanks, broke and fled; and the whole of de Muy's men, horse and foot, rushed down to the Diemel, throwing down their arms and splashing frantically through the fords.'[5, 6, 12]

The three brigades of horse artillery, following the cavalry charge, came down to the river's edge, unlimbered and opened fire on the disorganised French to such good effect that they were unable to re-form. Granby then led ten squadrons, including the KDG and The Queen's Dragoon Guards, across the Diemel to complete the destruction of de Muy's corps, halting only at nightfall on the heights at Wilda, some four miles in front of Ferdinand's main body. Only fragments of the shattered French regiments managed to escape to Volksmarsen.

The French lost 6-7,000 men, killed, wounded and captured, as well as 12 guns, 10 colours, and 28 ammunition waggons. The French prisoners totalled 78 officers and 2,100 men. The total British cavalry casualties were 48 killed and 90 wounded, with about 215 horses killed and wounded; of these the KDG lost 7 men and 17 horses killed, with Cornet Earl, 28 men and 4 horses wounded. The Queen's Dragoon Guards had 3 non-commissioned officers and 9 men killed, 8 men missing, and Captain Arnot, Lieutenant Mattack, Cornet Callender, one non-commissioned officer and 10 men wounded. Their loss in horses was 10 killed, 2 wounded and 8 missing.[1, 2, 3, 4, 5, 6, 8]

Colonel Pierson of the Life Guards wrote:

I may speak of Lord Granby in a way he can't do of himself. There never could be a day more for the honour of the English cavalry, of which Lord Granby put himself at the head and charged in the manner that was always expected of him. Neither horse nor foot could stand against it, and a general confusion ensued as soon as they began to act.[6]

The day after the battle Prince Ferdinand issued a general order:

His Serene Highness again renews his compliments of thanks to the generals, officers, regiments, and corps who were engaged, and who, by their valour and excellent conduct, gained so complete a victory over the enemy: and orders that his thanks be publicly given to Lord Granby, under whose orders all the British cavalry performed prodigies of valour, which they could not fail of doing with his Lordship at their head.[1, 6, 11]

But the last word lay with Granby himself, writing to the Secretary of State: 'I should do injustice to the General Officers and to every Officer and Private Man of the Cavalry if I did not beg your Lordship would assure His Majesty that nothing could exceed their gallant behaviour on this occasion.' And later, writing to Lord Newcastle, he said, 'Finer troops I believe never were, and at the head of them I should be very happy to receive a visit from the enemy.'[16, 17]

After the Battle of Warburg both the KDG and The Queen's Dragoon Guards remained on outpost duty on the heights of Wilda until 3 August, when Webb's brigade returned to the main body of the army at Warburg. The brigade was stationed along the River Diemel, with the KDG at Borcholz, although from 14 to 29 September the regiment occupied an advance post at Geissmar. Broglie's troops occupied the opposite side of the Diemel, and his superior numbers enabled him to engage in various probing operations, which involved isolated parties of Webb's brigade in a number of skirmishes and contacts with the French during the months of October and November. In December there were heavy falls of snow, and both the KDG and The Queen's Dragoon Guards were withdrawn into winter cantonments among the villages of the Bishopric of Paderborn.[1, 2, 3, 4, 9]

The results of years of warfare had left the district bereft of food and forage, so much so that on 13 April 1761 a subscription was raised among the troopers of The Queen's Dragoon Guards for the relief of the local peasantry.

As the miserable and distress'd situation of the Poor inhabitants of the Villages upon the Dimel is an object of the greatest Compassion, and as a Collection is making for them by the German troops, Lieut-General Howard, being persuaded that no people can feel for such distress more than Ourselves, he thought it incumbent upon him to give the Troops notice of it, not doubting but the Commanding Officers of Regiments and all others will most cheerfully contribute to so charitable and humane a purpose. The Collection is desired to be made Regimentally, and given in to Lieut-Colonel Faucitt at Paderborn, as many Officers in each Regiment are absent.[3]

Both the British troopers and their horses also suffered from a scarcity of rations and forage. This resulted in much sickness among the men, and Cannon remarks that many 'of our gallant veterans, after fighting the battles of their country with heroic bravery, were, at the end of a long and toilsome campaign, hurried to an untimely grave by diseases produced by a want of food'. The horses fared a little better: a letter from an unknown officer commented that 'The consumption of horseflesh has been very great, which is not surprising, considering the marches they have made, the scrapes they have been in, and that since the 1st June and this time they have wanted about 50 days corn. The oats are mixed with chopped straw, and if care is taken and they are fed often, little more is requisite.' In the spring of 1761 remounts arrived from England to replace the severe losses of the previous year.[1, 11]

On 25 October 1760 George II died, and on 18 November Lord Granby issued an order of mourning: 'black crepe on the Colours, drums and banners; Officers to cover their swords, knots and sashes with black crepe, to wear black crepe around their arms, and plain hats with crepe hat bands.'[4]

In February 1761 the brigade of Dragoon Guards, including both the KDG and The Queen's Dragoon Guards, was ordered to march into Hesse-Cassel to attack the winter quarters of the French army. The advance was made in snowy conditions and over icy roads, but the French retreated before the cavalry and several magazines of stores and forage were captured. It was not the only successful winter operation: on 13 February Sporcken defeated the French near Gotha, on the 14th Granby captured the castle at Weissenstein, and on the 15th Ferdinand took Fritzlar. In March the regiments returned to Paderborn.[1, 2, 3, 4]

After the severity of this short winter campaign the troops required two months' rest to re-equip and train the recently arrived remounts. Both the KDG and The Queen's Dragoon Guards were brought into Lieutenant General Conway's corps and were employed in a number of harassing marches and counter-marches throughout Westphalia. On 15 July, when the Allied army was encamped in front of Hamm, its left lying behind Werle and its right on the River Lippe, it was attacked by the combined armies of Marshals Broglie and Soubise. The ground from Vellinghausen to Kirch-Denkern was held by the British under Lord Granby, and the French were driven back after a long and bloody contest. It was principally the infantry's day, as the ground was not suitable for cavalry. A farrier described the area as 'so bad and full of hedges and ditches'.[1, 2, 5, 18, 19]

At 3 a.m. on the following morning, 16 July, the French renewed their attack. The brigade of Dragoon Guards was posted in the rear of the Guards on the heights between the villages of Illingen and Wambeln, but took little part in the day's severe contest, apart from guarding prisoners handed over to them by the infantry. The French were eventually driven off, losing 6,000 killed, wounded and prisoners, but the pursuit could not be pressed because of the difficulties of the ground.[1, 2, 3, 5, 18, 19]

After the battle the Dragoon Guards manoeuvred around Paderborn and on the Westphalian plain, being involved in some skirmishes, until on 24 August Ferdinand led the brigade back across the Diemel, pushing back the French outposts. At Dringenberg the KDG captured 300 prisoners. The French then reinforced their outposts, but Ferdinand pushed them back again. On 26 August the brigade encamped within twenty miles of Cassel for four days, before retiring back across the Diemel to Corbeke and Buhne. On 17 September the KDG crossed the Diemel a second time and attacked a strong French post at Immenhausen, and again drove back the French outpost line almost to Cassel. King George III and Queen Charlotte were crowned in Westminster Abbey on 22 September, and the troops of the Dragoon Guard brigade paraded and fired a *feu de joie* in their honour.[1, 2, 3]

Ferdinand issued an order from the camp at Ohr on 17 October: 'His Serene Highness intends to make a Gratification to the Troops, and has ordered the Commissariat to deliver to each Battalion and to every Four Squadrons one Barrell of Brandy and six cwt. of Rice or Peas.'[3]

On 5 November The Queen's Dragoon Guards forced out a French regiment from its position at Capelnhagen, and then marched to Eimbeck near Hanover, where with the KDG they were involved in a smart skirmish with the French. On the night of 7 November the Dragoon Guard brigade marched through a heavy snowstorm to Furwohle in Hanover. On arrival late at night the tired troopers were erecting their tents when suddenly the trumpeters sounded 'To Horse'. (Even though at this stage Dragoon Guards had drummers and not trumpeters on their establishment, all the accounts of the Seven Years War speak of the regiments being summoned by trumpet.) The KDG and The Queen's Dragoon Guards hastily saddled up, remounted, and charged the advancing French, driving them off and inflicting considerable losses. Ferdinand, on hearing the trumpets sound, galloped up to see what was happening and congratulated the troopers personally on the bravery they had shown, in spite of their weariness after the day's march.

The Queen's Dragoon Guards had another encounter with the French at

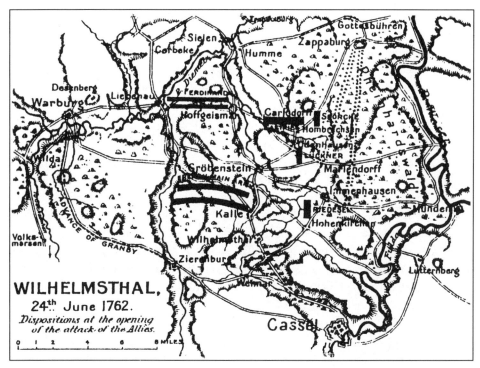

Battle of Wilhelmsthal

Furwohle on 9 November, and then with the KDG they marched and took up a position on the heights between Lithorst and Mackensen. Detached parties of both regiments continued with sporadic skirmishing, at a heavy cost in sickness to the men. The horses suffered too, from fatigue, poor rations and the severe weather, until in early December both regiments moved into winter quarters in East Friesland.[1, 2, 3]

Both the French and the Allied armies remained in their winter quarters until May, the countryside having been so devastated that neither army was able to move until the corn was sufficiently grown to provide forage. The KDG were with the 3rd Dragoon Guards in the Earl of Pembroke's brigade in General Mostyn's division. Both the KDG and The Queen's Dragoon Guards left East Friesland in May and joined the army at Brakel, to the east of Paderborn, on 18 June 1762. The Queen's Dragoon Guards were encamped on the heights of Tissel, and the KDG posted to the rear of the centre of the army.[1, 2, 3, 10]

The French, under the command of Marshals Soubise and d'Estrées, were encamped at Groebenstein with their headquarters at Wilhelmsthal. Ferdinand decided to attack, and on the morning of 24 June the KDG and The Queen's

Dragoon Guards were both part of the centre column of the Allied army, leaving their camp at daybreak and crossing the Diemel at Liebenau by 4 a.m. After a march of nine miles they seized the entrances to Langenberg, and then advanced on the French camp. At the same time two other Allied columns converged on the French left and right wings. The three columns, arriving at much the same time, caught the French by surprise. Forsaking their tents and baggage, they fled towards their headquarters at Wihelmstahl. The main body managed to make good their escape, but several of the finest of the French regiments under Stainville took up a position in the woods around Wihelmstahl, to cover the retreat of the main body. The KDG and The Queen's Dragoon Guards, pressing forward in pursuit, were part of the force which then surrounded these gallant French, killing many and compelling the remainder, including whole regiments, to surrender. The French corps was annihilated, 1,500 being killed and over 3,000 taken prisoner. The cavalry, including both the KDG and The Queen's Dragoon Guards, were then sent forward in pursuit of the main French army, capturing many more prisoners, guns and colours before bivouacking for the night on the heights between Holtzhausen and Weimar.[1, 2, 3, 5, 12]

The Queen's Dragoon Guards marched to Hoff, where they were later joined by the KDG and the Foot Guards. On 15 July, the main body of the army having arrived, the KDG with the Grenadiers and the Highland brigade forded the River Eder, reconnoitring a French position at Feltzberg, recrossing the Eder and camping on its banks that afternoon. The next day the KDG marched to Gundersberg, and on 22 July crossed the Eder a second time, camping near Kraetzenhausen before advancing on Homberg the following day. On the 24th they drove a strong French detachment from the heights around the town, and then occupied an old French position near Hilgenberg. Both the KDG and The Queen's Dragoon Guards were subsequently engaged in various further manoeuvres around Melsungen which resulted in the taking of Cassel and the start of peace negotiations.[1, 2]

SOURCES

1 R. Cannon, *Historical Records of the British Army, 1st Dragoon Guards*, 1837.

2 R. Cannon, *Historical Records of the British Army, 2nd Dragoon Guards*, 1837.

3 F. Whyte and A. H. Atteridge, *The Queen's Bays, 1685-1929*, vol. 1, 1930.

4 H. C. Wylly, *History of The King's Own Yorkshire Light Infantry*, vol. 1, 1924.

5 Sir John Fortescue, *History of the British Army*, vol. 2, 1910.

6 H. Peirse, 'The Battle of Warburg', *Cavalry Journal*, vol. 4, 1909, p. 146.

7 *The K. D. G.*, vol. 1, no. 7, 1930, p. 237.

8 William Clowes, *King's Dragoon Guards, 1685-1920*, 1920.

9 F. Grose, *History of the English Army*, vol. 2, 1801.

10 Records of The King's Dragoon Guards, Regimental Museum, 1st The Queen's Dragoon Guards, Cardiff Castle.

11 J. M. Brereton, *History of the 4/7th Royal Dragoon Guards*, 1982.

12 L. B. Oatts, *I Serve*, 1966.

13 E. W. Sheppard, 'The British Cavalry in Germany 1759-62', *Royal Armoured Corps Journal*.

14 R. Hargreaves, 'The Marquis', *Royal Armoured Corps Journal*.

15 De Mauvilon, *Geschichte Ferdinand's Herzogs von Braunschweg*.

16 *London Gazette*, 1760.

17 Newcastle Papers.

18 R. E. R. Robinson, *The Bloody Eleventh*, Devon & Dorset Regiment, 1988.

19 R. Savory, *His Britannic Majesty's Army in Germany during the Seven Years War*, 1966.

14

Peacetime Duties
1763-1792

The Seven Years War was drawing to its close. On 19 November 1762 The King's Dragoon Guards and The Queen's Dragoon Guards [Bays] withdrew into winter quarters in the Munster area, while the peace treaty was being signed. On 13 January 1763 the two regiments received the thanks of Parliament for their 'eminent and meritorious services'. On 2 January The King's Dragoon Guards started their march through Holland, via Guelderland, Nijmegen and Breda, to the port of Wihelmstadt, where they embarked for England. The Queen's Dragoon Guards followed at the beginning of February, with a strength of 1 officers, 32 men and 40 horses; in the regiment's train followed 31 officers' servants and 31 women.[1, 2, 3]

On arrival in England the KDG were scattered in small numbers throughout, chiefly, the county of Sussex – at Chichester, Lewes, Rye, Hastings and Havant – and were charged with 'coast duty'; The Queen's Dragoon Guards were comfortably quartered in Worcester, Pershore and Hereford. With the end of the war, both regiments were reduced in strength, the KDG to forty men per troop, and The Queen's Dragoon Guards to three officers, one quartermaster, two sergeants, two corporals, one drummer and one hautbois, and twenty-eight men per troop. In March the light troops of the Dragoon Guards, who had in any case served away from their parent regiments, were disbanded; eight men per troop, however, were equipped as light dragoons on smaller horses, to act for each regiment on much the same principle as the light companies in the infantry. These light troops were also given the task of providing travelling escorts for members of the Royal Family, taking the duty in turns – previously a duty which had been performed by the regiments of horse and Dragoon Guards, who now ceased to find such escorts.[1, 2, 3, 4]

Lieutenant General Bland died early in 1763, and his place as colonel of The King's Dragoon Guards was taken by Lieutenant General Mostyn, by a commission dated 13 May 1763.[1, 7]

In July 1764 changes were made in the uniforms of both the KDG and The

Queen's Dragoon Guards. Officers and men were no longer to wear knots or aiguillettes on the left shoulder. These were to be replaced with epaulettes. The embroidered edging on the officers' coats was to be discontinued, and the heavy jackboots abandoned in favour of a lighter boot. An inspection return of the KDG dated 31 October 1764 states that 'Officers' uniforms to be lapelled with blue, laced with gold, blue waistcoats and breeches. Accoutrements coloured white'.[1, 2, 3, 5, 6, 7]

An order from the War Office dated 27 July decreed that

> His Majesty having been pleased to order that all His Regiments of Horse and Dragoons, except the Light Dragoons, shall be mounted only on such horses as shall have their full tails, without the least part taken from them; all breeders and dealers in horses for the service of the Army, are desired to take notice, that, for the future, no horses but such as shall have their full tails, without the least part taken from them, will be bought for any Regiments of Horse and Dragoons. Both regiments were ordered to be remounted on long-tailed horses. In November the size of men and horses was fixed by regulation: the men were not to be above five foot ten inches in height nor less than five foot eight inches; the horses were not to be above fifteen hands two inches, and not less than fifteen hands.[1, 2, 3, 4, 7]

In 1766 trumpeters were officially brought in to replace drummers, and in 1767 The King's Dragoon Guards were ordered to change their waistcoats and breeches from blue to buff. An inspection return of 16 May notes, '9 Trumpets received in February, 1767, no drums. This Regiment has black half-gaiters. horses with long tails.' The following year not much progress had been made, for on 12 May the report stated:

> Standards 1761. Housings, etc 1763. Officers horses, 6 long tails. Officers uniform, red, faced & lapelled with blue, gold-laced buttonholes, buff waistcoats and breeches. Half-gaiters bad, new whole gaiters to be provided shortly. Regiment has no music. Men had all watering-frocks, 2 suits of clothes & 2 hats. The new bits differed from the old in having no bosses.

But by 1769 the reporting officer was able to say that there were now ninety-four long-tailed horses.[6]

From 1763 for the next thirty years both regiments moved, on an almost annual basis, from town to town throughout England and Scotland. Over the period of 1763 to 1773 their stations were:

	KDG	Queen's Bays
1763	Sussex	Worcestershire
1764	North of England	Worcestershire
1765	Scotland	London
1766	Scotland	London
1767	Scotland	South of England
1768	England	Edinburgh
1769	England	Manchester
1770	Musselburgh	South of England
1771	Warwick & Coventry	London
1772	Newbury	Colchester
1773	Newbury	York

On 7 December 1765 the Adjutant General wrote to the colonels of The King's Dragoon Guards and The Queen's Dragoon Guards asking, 'What utensils and accoutrements the Farriers have when the Regiment passes in review? Whether they have swords? pistols? one or two?' On 11 March 1766 an order was issued, 'For Farriers neither firelock, pistols, or swords.' Finally a warrant of the 19 December 1768 ordained:

> The Farriers of the Dragoon Guards to have blue coats with blue lining and blue waistcoats and breeches. The lapels of the Dragoon Guards to be blue. The capes and cuffs of the sleeves to be of the colour of the facing of the Regiment, except those of the Royal Regiments, which are faced with blue, whose capes and cuffs are to be red. To wear a small black bearskin cap, with a horseshoe on the forepart, of silver- plated metal on a black ground, and to have churns and an apron.[8]

On 3 June 1765 The Queen's Dragoon Guards were reviewed by George III on Wimbledon Common, together with the 10th and 11th Dragoons. In October 1766 a detachment of the regiment escorted Princess Caroline Matilda, recently married to the King of Denmark, to Harwich, where she embarked to join her husband. At the same time a poor harvest involved the regiment in helping the customs officers to enforce an Order in Council prohibiting the export of wheat and flour. In 1767 the regiment, having been mounted on bay horses (the rest of the heavy cavalry, with the exception of the Scots Greys, were mounted on blacks), came to be called The Queen's Bays, a title it retained up to amalgamation in 1959.[2, 3]

During 1766 and 1767 there were a number of administrative changes made for both regiments. The price of commissions was fixed by a board of general officers at: lieutenant colonel, £,205; major, £4,250; captain, £3,150; captain-lieutenant, £2,100; lieutenant, £1,365; cornet, £1,102. On 11 February 1767 a warrant was issued,

> For the more effectual maintenance of good order and discipline in our Regiments of Dragoon Guards that one Field Officer be always present with the regiment, one Captain with each Squadron, and one Subaltern with each Troop, and that a monthly return of their attendance be made to the Secretary at War and to the Adjutant General. Every Officer on appointment is to join his regiment within four months, and every Officer who has not served in any other cavalry regiment is to remain in Quarters till he shall be perfected in riding and all regimental duty. Moreover all Officers, while present with their Corps, are constantly to wear uniform.

On 15 September the War Office ruled that regiments of Dragoon Guards 'shall for the future have black gaiters in which they are to do all duties on foot. Those regiments that have half-gaiters may wear them out. No regiments are to have white gaiters.' On 21 September a Royal Warrant was issued ordering that the regimental number must appear upon the buttons of both Officers and men, but a clothing warrant of 1768 allowed The King's Dragoon Guards to have K. D. G. on their buttons rather than the numeral '1'. The same warrant allowed the coats to have turn-down collars which could button up to the neck, the epaulette was to be blue for the KDG with yellow tape and fringe, and the waist sash was to be of crimson with a stripe of the facing colour. In 1767 the establishment of The King's Dragoon Guards was fixed at nine troops, each having one captain, one lieutenant, one cornet, one quartermaster, three sergeants, three corporals, one hautbois, one trumpeter, forty troopers and forty-eight horses.[7, 9, 10, 11, 12]

On 27 November 1767 a warrant was issued regulating the price to be paid for horses:

> No more than twenty guineas should be given for each Recruit Horse; and whereas it hath been humbly represented unto Us that that sum hath not been sufficient for providing horses for remounting our said Regiments, for the future a sum not exceeding twenty two guineas may be given for each Recruit Horse, including all Expences until the delivery of each Horse at any place which may be appointed by the Commanding Officer if the Regiment be quartered in England; but if the Regiment be quartered in Scotland, the seller is then to deliver each Horse for the said price of twenty two guineas, all expences included, at such place as shall be

appointed by the Commanding Officer, provided such place be not farther northward than York.[3]

When The Queen's Bays were ordered to Scotland in February 1768, they marched to Doncaster and Pontefract, where they halted for a month. On resuming their march, several troops were called to the assistance of the civil authorities of Newcastle-upon-Tyne. The seamen in London were demanding higher wages, and ships were being forcibly prevented from entering or sailing. These disorders had extended to other ports, including Tyneside, where the seamen were conducting a reign of terror, blackmailing citizens to contribute to their cause and banding together to roam the district committing 'several outrages'. The Queen's Bays soon 'reduced the riotous seamen to obedience', and then continued on their way to Edinburgh. But early in 1769 the regiment was called south again to restore order to the coalfields of Manchester, Blackburn and Warrington. Even with the help of other troops, this took several months to accomplish. [2, 3]

During the summer of 1771 The Queen's Bays were again reviewed by George III on Wimbledon Common, and the following May the regiment provided the guard at Whitehall while the King reviewed the Life Guards and the Horse Grenadier Guards. In 1773 Earl Waldegrave took up the colonelcy of the Coldstream Guards, and on the 15th July, Field Marshal the Marquess Townshend succeeded him as colonel of The Queen's Bays.[2, 3]

Inspection reports of The King's Dragoon Guards for 1768 and 1769 spoke of 'Officers' uniforms embroidered with gold; arms and furniture handsome. Trumpeters finely mounted. 52 longtails . . . Officers' uniforms very good without any new alteration; buff facings; buff waistcoats & breeches; gold embroidery; epaulettes; laced hats. Horses bay of a large size.' On 6 June 1771 the regiment was inspected and reviewed by the Duke of Argyll at Musselburgh, just before marching south; on 16 May 1772 it was reviewed by Major General Pitt at Coventry; by Major General Pitt again at Newbury on 26 April 1773; by Major General Preston at Wimbledon on 13 May 1774; again by General Preston at Blackheath on 5 May 1775; and then by George III on 8 May 1775 at Blackheath.[1, 6, 11]

Purchasing rank in times of peace was not the only method for an officer to obtain promotion. Captain Humberston of The King's Dragoon Guards sought his majority by transferring to a new regiment that was being raised by his cousin, Kenneth Mackenzie, Earl of Seaforth. Humberston had joined the KDG as a cornet in 1771, and became a captain the same year. During 1778 and 1779 he wandered around the Highlands searching for recruits for the Seaforth Highlanders, the regiment which his cousin was raising. He wrote to Lord Greville on 25 January 1779: 'Every Officer who obtains a commission in these Regts. has it on condition that he on a certain day brings in such a number of men to the Regt.' He also wrote to the Secretary-at-War requesting permission to sell his commission in the KDG, but this was turned down by Lord Amherst and a year later Humberston was still a captain. In 1780, however, his efforts in helping to raise the Seaforth Highlanders were rewarded with a lieutenant colonelcy in the 100th Foot.[13]

Over the following ten years the KDG and The Queen's Bays served in a number of areas:

	KDG	Queen's Bays
1774	Wimbledon	Scotland
1775	Wimbledon, Norwich	Warwick, Lichfield, Coventry
1776	Norwich	Worcester
1777	Ipswich	Sussex
1778	Salisbury, Dorchester	Salisbury Plain
1779	Dorchester, Salisbury	Salisbury
1780	Exeter	Norwich
1781	Bath	York, Leeds, Bradford
1782	Devizes	Scotland
1783	Coventry	Manchester, Dorchester, Weymouth

On 20 April 1775 Banastre Tarleton purchased a cornetcy in The King's Dragoon Guards from John Trotter, who had in turn bought a vacant lieutenancy, joining the regiment at Norwich in late July. He joined a troop commanded by Major Peete, and for the next five months trained and learnt the role of a cavalry officer. A muster roll of 24 December 1775 marks Tarleton as 'Absent by King's Leave', for the young cornet had applied to his commanding officer, Lieutenant Colonel Robert Sloper, to volunteer for duty with Lord Cornwallis in the Americas, where rebellion had broken out. In a letter Tarleton wrote home from Brunswick on 25 May 1777, he said, 'I have just received a Letter from Genl. Mostyn's Regiment indulging me with 12 months more Leave, for which they apply'd to the King. The Favor was granted. They congratulate me on my good Fortune in America, & wish me success as to Promotion.' Tarleton remained on the strength of the regiment until 1778, while he was winning fame as a dashing cavalry commander of a force of Loyalists which he raised and called the Legion, but which soon became known as Tarleton's Green Dragoons. His ruthlessness and repeated successes against the Americans earned him the name of 'Bloody Tarleton'. Finally he met defeat at the Battle of Cowpens, and later capture at Yorktown. In 1818 his services were recognised by his appointment as colonel of the 8th King's Royal Irish Hussars.[14, 15, 16]

Over the period The King's Dragoon Guards were regularly reviewed – in 1776 at Norwich by Major General Johnson, the following year by Lieutenant General Oughton, again in 1778, 1779 and 1782 by Johnson, now a lieutenant general; in 1780, 1781 and 1783 by Major General Ward. In 1779 General Sir John Mostyn died, and on 21 April Lieutenant General Sir George Howard, from the 7th Dragoons, was appointed colonel. Sir George Howard was also Governor of the Royal Hospital, Chelsea from 1768 until the year before his death in 1796. In April 1779 the men who had been formed into light troops were removed from both The King's Dragoon Guards and The Queen's Bays, as well as from the 4th and 10th Dragoons, to form a new regiment, which was numbered as the 19th Light Dragoons. In a list of 'Detached and Unnumbered Corps' on the English establishment 'Waller's Corps' of infantry is mentioned as being raised in 1781 and disbanded in 1783. Captain Henry Waller, the major-commandant of this regiment, who enjoyed only this short time in command, was a captain in The King's Dragoon Guards.[1, 2, 3, 5, 11, 17]

The Queen's Bays were reviewed by George III on Wimbledon Common in May 1777. The following year, owing to the American War, the strength of the regiment was increased by 100 men and horses, and by a further 50 men and horses later in the same year. But with the end of the American War, both The Queen's Bays and The King's Dragoon Guards were again reduced in strength, the former by ninety men and horses, the latter by one sergeant, one corporal and twelve men per troop, a total of 126 men, giving the KDG a troop establishment of one captain, one lieutenant, one cornet, one quartermaster, two sergeants, two corporals, one hautbois, one trumpeter and twenty-eight troopers.[1, 2, 3, 11]

The King's Dragoon Guards were involved in 1782 in a romantic affair, which attracted some attention at the time. John Fane, 10th Earl of Westmorland, as a young man of twenty-two, fell in love with Sarah Anne, aged only eighteen and the only child of the banker Robert Child. Earning his nickname of 'Rapid Westmoreland', and knowing that Mr Child had other ideas for his daughter, at dinner one evening he asked the father, 'Child, supposing you were in love with a girl, and her father refused his consent, what should you do?' 'Why! Run away with her, to be sure!' the banker rashly replied. Fane, taking this advice, stole away with Sarah Anne from her home at 38 Berkeley Square early on the morning of 17 May 1782 and the two headed for Gretna Green. The girl's departure was soon discovered and parents set out in pursuit. However, Fane, passing a detachment of The King's Dragoon Guards en route and recognising the officer in command as an old friend, begged him to delay the angry parents. When the Childs arrived, they found the road obstructed by the troops and were forced to wait while a lengthy manoeuvre was completed. This gave the eloping couple such a head start that the parents gave up the chase at Baldock, and the pair were married at Gretna Green. All was soon forgiven, and a more regular marriage in church then took place.[18]

From 1784 to 1793 the two regiments were stationed at:

	KDG	Queen's Bays
1784	Sussex, Kent, East Anglia	Dorchester, Weymouth
1785	Lincoln, Boston, Stamford	Dorchester, Weymouth
1786	York, Durham, Newcastle	Ashford
1787	Scotland	South of England
1788	Manchester, Stockport	South of England
1789	Coventry, Devon	Scotland
1790	Dorchester	Manchester
1791	Winchester, Birmingham	Exeter, Taunton
1792	Ashford	Dorchester

The King's Dragoon Guards were assembled at Canterbury on 31 March 1784 and then marched to Greenwich and Deptford to be reviewed, first, on 21 May at Blackheath, by Lieutenant General Ward, and then on the 24th by George III. Before moving to Lincolnshire in 1785, the regiment was reviewed by General Philipson on 9 May, and in 1786 by Major General Wynyard at York. In 1787 the regiment was stationed at Dumfries, Prestonpans and Musselburgh, where it was reviewed by Major General Leslie on 21 May before being scattered by troops to Haddington, Dumfries, Dalkeith, Linlithgow and Dunbar. Further annual reviews were by Lieutenant General Sir William Erskine at Manchester in 1788, at Coventry by Major General Harcourt in 1789, by Major General Scott at Exeter in 1790, at Salisbury by Major General Lascelles in 1791, and again at Ashford in 1792, when the King also reviewed The King's Dragoon Guards on 29 April.

In 1784 the colonel of The Queen's Bays, Lord Townshend, applied to the King for the facings of their uniform coats to be altered from buff to black, a request which George III granted. Two years later the King personally reviewed the regiment on Ashford Common. Cornet Le Marchant purchased a lieutenancy in the regiment in November 1789, coming from the 6th Dragoons, and was promoted to captain in 1791, leaving The Queen's Bays for a majority in the 16th Dragoons in 1794. He was later to become the lieutenant governor of the Royal Military College, and then a major general commanding a brigade of heavy cavalry in the Peninsula. He was killed at the head of his men, charging the French at Salamanca on 22 July 1812. On 28 April 1784 Richard Vyse became commanding officer of The King's Dragoon Guards. Born in 1747, the son of an archdeacon, he was to become colonel of the regiment and a lieutenant general of some renown. 'Vyse affected much the pithy style and spirit of Frederick of Prussia; but though studiously laconic, he was somewhat partial to pompous language, and not without a turn for, dry caustic humour.' Vyse issued recruiting instructions in December 1787 which included among twenty-one orders such advice as 'You are to enlist no man who is not a Protestant, and a native of Great Britain; you are to enlist no man who has ever been in the sea-service or in the Marines.' No one under seventeen years or above twenty-three years of age was to be enlisted 'except any very fine Boy, who is likely to grow, who has been well educated, and whose parents are respectable people'. The Queen's Bays enticed recruits by telling them 'You will be mounted on the finest horses in the world with superb clothing and the richest accoutrements; you are everywhere respected; your society is coveted; you are admired by the fair — followed by buxom widows and rich heiresses; your privileges are equal to two guineas a week.'[19, 20]

A recruiting poster of the King's Dragoon Guards of a few years later still exists, and is on display in the Regimental Museum:

FIRST, or

KING'S

Dragoon Guards

Commanded by

General Sir Wm. Augustus Pitt, K.B.

A Few

DASHING LADS

ARE NOW WANTED

To complete the above well-known

Regiment to a new Establishment

Any YOUNG MAN who is desirous to make a Figure in Life, and wishes to quit a dull laborious retirement in the Country, has now an Opportunity of entering at once into that glorious State of Ease and Independence, which he cannot fail to enjoy in the

KING'S DRAGOON GUARDS

The Superior Comforts and Advantages of a Dragoon in the Regiment, need only to be made known to be generally covetted. All Young Men who have their own Interests at Heart, and are fortunate enough to make this distinguished Regiment their Choice are requested to apply immediately to

Serj. TIBBLES, at the Angel Inn, Honiton,

where they will receive

THE HIGHEST BOUNTY

And all the advantages of a Dragoon

As Recruits are now flocking in from all Quarters, no Time is to be lost;

and it is hoped that no young Man will so far neglect his own Interest as not embrace the glorious Opportunity without Delay.

N.B. This Regiment is supposed to be mounted on the most beautiful, fine, active black Geldings this Country ever produced.

The Bringer of a good Recruit will receive a Reward of THREE GUINEAS. [11,19]

As a result of the unrest caused by the French Revolution, rioting broke out in Birmingham. A mob, believing that British institutions were under threat, attacked the houses and meeting places of anyone suspected of republican principles. The King's Dragoon Guards were called in to restore order, which they did in such a spirited manner that it gained them the thanks of the King, which was conveyed to the regiment through a letter dated 26 July 1791 from the Secretary-at-War. [1, 7]

The period of thirty years of peacetime soldiering was coming to an end for both regiments as the excesses of the French revolutionaries made their impact.

SOURCES

1 R. Cannon, *Historical Records of the British Army, 1st Dragoon Guards*, 1837.

2 R. Cannon, *Historical Records of the British Army, 2nd Dragoon Guards*, 1837.

3 F. Whyte and A. H. Atteridge, *The Queen's Bays, 1685-1929*, vol. 1, 1930.

4 *Journal of the Society for Army Historical Research*, vol. 31, 1953, p. 138.

5 *Journal of the Society for Army Historical Research*, vol. 4, 1925, pp. 16, 228.

6 *Journal of the Society for Army Historical Research*, vol. 3, 1924, p. 231.

7 William Clowes, *King's Dragoon Guards, 1685-1920*, 1920.

8 *Journal of the Society for Army Historical Research*, vol. 13, 1934, p. 212.

9 *Journal of the Society for Army Historical Research*, vol. 14, 1935, p. 129.

10 Sir George Arthur, *Story of the Household Cavalry*, vol. 2, 1909.

11 Records of The King's Dragoon Guards, Regimental Museum, 1st The Queen's Dragoon Guards, Cardiff Castle.

12 Captain R. G. Hollies Smith, 'A Brief History of the Uniform of The King's Dragoon Guards', *The K. D. G.*, vol. 5, no. 5, 1958.

13 *Journal of the Society for Army Historical Research*, vol. 57, 1979, p. 40.

14 R. D. Bass, *The Green Dragoon*, 1957.

15 *Journal of the Society for Army Historical Research*, vol. 62, 1984, p. 127.

16 J. M. Strawson, *Irish Hussar*, Queen's Own Irish Hussars, 1986.

17 C. G. T. Deane, *The Royal Hospital, Chelsea*, 1950.

18 *Guide to Osterley Park*, National Trust and Victoria & Albert Museum.

19 *The K. D. G.*, vol. 5, no. 1, 1954, p. 59.

20 *Regimental Journal of 1st The Queen's Dragoon Guards*, vol. 3, no. 4, 1976, p. 359.

15

Campaign in the Netherlands
1792-1794

On 14 July 1789 the Bastille in Paris was stormed by the mob. In 1791 King Louis fled, only to be recaptured and guillotined in 1793. The French revolutionaries declared war on all the monarchies of Europe, and invaded the Austrian Nether-lands, declaring war on Britain on 1 February 1793. So started twenty-two years of conflict.

After twenty years of peace the British Army was not in a good state. A board of general officers in 1788 had examined the many varieties of swords used by each regiment, and for the Dragoon Guards recommended the adoption of the pattern used by the 6th Inniskilling Dragoons – the grip to have half-basket protection and the blade to measure 3 ft 3 ins from the guard to the point. In addition officers were told that their swords should be the same pattern as those of the troopers.[1,2]

In 1792 directions were issued 'For Manoeuvring' but these were for a 'Review of the King's Dragoon Guards before His Majesty'.

> Lieutenant Colonel Vyse has drawn up the following short Abstract of Directions for Manoeuvring, which he requests the Officers and Men of the Regiment will carefully consider, and execute with that precision, activity and alertness of attention, that can alone secure to them a continuance of that approbation, the confirmation of which he knows they are all so equally and justly emulous of receiving from their Sovereign, when they have the honour of appearing before him.

The directions continue, 'From the Men, silence, and attention to their Officers only is required', but sixteen more pages are devoted to pleasing His Majesty rather than preparing a regiment for war.

On 25 December 1792 The King's Dragoon Guards were augmented by ninety men, giving an extra ten men and ten horses per troop; this was followed on 25 January by another increase of an additional sergeant, corporal, nine troopers and nine horses to every troop. The regiment was to be increased from nine to twelve troops, six of which were to hold themselves in readiness for foreign service. Each foreign service troop was to consist of one captain, one lieutenant or cornet, three sergeants, three corporals, one hautbois, one trumpeter, forty-seven troopers and fifty-three troop horses. In February 1793 The Queen's Bays were increased by the addition of three troops.

Robert Long, who later gained fame as a Peninsular general and rose to the rank of lieutenant general, was commissioned into the KDG in 1791, and was the officer in charge of a recruiting party early in 1793; he received £5 for each recruit, out of which the recruit himself was at once given £2.18s, out a total bounty of £14.14s., but the poor recruit soon discovered that the balance was used to pay for his kit, and it seldom covered the cost. In the spring of 1793 the KDG assisted in a service at Norwich Cathedral as part of a national preparation for the war, but it was soon to prepare itself in a more practical way in Flanders.[3, 4, 7, 9, 10, 11]

Two squadrons of The Queen's Bays embarked for Flanders at Blackwall during May, and landed at Ostend. The Queen's Bays, together with the 3rd Dragoon Guards, marched to the vicinity of Tournai, where with some Austrian and Prussian cavalry and infantry, they formed part of a corps of observation during the siege of Valenciennes, and were camped between Lille and Tournai. Le Marchant of The Queen's Bays related in a letter home how some French prisoners had been brought into camp from an enemy outpost at Valenciennes. He asked two officers whether they had any profession before the Revolution, and received the answer that they were gentlemen. Their appearance led him to doubt this, and on questioning some of the other prisoners he discovered that one had been a barber and the other a private soldier in a line regiment. Le Marchant wrote, 'It is thus we find the Republicans prove their equality.'[3, 4, 5, 12]

Le Marchant's squadron was soon in action. It was detailed to accompany a Prussian and Austrian column, commanded by General Count Hohenzollern, in an attack on the French camp at Cassel, fifteen miles south of Dunkirk. The Queen's Bays squadron were the only British troops with the column, and Le Marchant on visiting his men the night before was astonished to find them all lying flat on their faces. When he asked the reason, he was told that they had all dressed their queues with the usual paste of flour and fat, and they wanted to avoid having to perform the chore again the next morning.

The attack on the French, carried out on 30 May, achieved surprise and the French infantry retreated, to be pursued by the cavalry. As the French rallied in a field of corn, the Austrian cavalry with The Queen's Bays on the right, charged and broke them. The French fled, leaving some sixty dead on the field. Writing home to his wife, Le Marchant commented,

I am just returned from a scene that, on cool reflection, makes my soul shrink within me; but it is one of the horrors of war. What gave me most pain was to see that the Austrians gave no quarter. Poor devils on their knees, merely begging for mercy, were cut down. My own people, thank God!, were as merciful as possible; and, I think, destroyed none in the pursuit, except such as would not give themselves up. Dive's party [his junior captain] had taken five men alive, but leaving them for an instant in pursuit of others, some Austrians came up and butchered them. My people behaved remarkably well in the face of the enemy, that is, for young troops.

Le Marchant received from Hohenzollern a warm letter of thanks and congratulation for the part played by The Queen's Bays.[13]

Three squadrons of The King's Dragoon Guards, commanded by Colonel Vyse, embarked at Blackwall on 1 July 1793 and after a swift passage landed at Ostend. They then marched to the relief of Nieuport, but on the approach of the British troops the French raised the siege and the regiment camped near the town. The diary of Lieutenant Colonel Russell, KDG, gives a graphic description of life in camp over a period when the cavalry did not have a great deal to do. During each month over the autumn and winter of 1793 and the spring of 1794, deaths from sickness among the men occurred with depressing regularity. On 18 August 'a detachment, consisting of officers, quartermasters, sejts, corpls, trumpeters and farriers with 120 privates, march'd to relieve the 37th Regt of foot in Ostend'; and the following day 'the remaining part of the Regt march'd into Ostend'. There are repeated references to the savage discipline exercised in the Army at that time: on 28 August, 'A court martial when Benjamin Major (servant to Captn Dawkins) was tried for breaking open a box and stealing to the amount of £28 sterling in cash, the property of Lieut. Balcomb, he was sentenced to 900 lashes – in the evening the above prisoner rec'd 300 lashes, vizt. 200 on his back and 100 on his backside.' On 20 September, 'a man tried by court martial for attempting to desert he was sentenced to 600 lashes but received only 350.' On the 14th, '9 deserters came in from the French army when after being examined by Col Vyse they were committed to the town gaol'. October is a tale of court martials for drunkenness, insulting sergeant majors and losing equipment, floggings, and men dying, mainly from Captain Serjeantson's troop. On 6 January 1794 'Troop marched to Ghistall', followed on the 7th by 'the Colonel's Troop marched from Ostend to Ghistall, Deynse and Pettingham', and on the 13th 'a general inspection of arms, accoutrements and horse appointments', while on the 14th '2 squadrons of cavalry, one of the Bays and one of the Blues, came through the town on their way to Courtray'. At Deynse, which was near Ghent, the KDG went into winter quarters and were

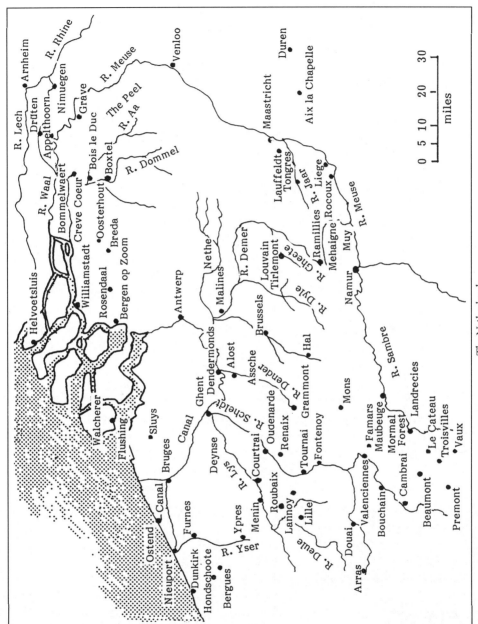

The Netherlands

further augmented by another twenty-six men per troop, as well as a second major and lieutenant colonel being added to the establishment.[3, 14]

The Queen's Bays were more actively engaged during the latter months of 1793. After the fall of Valenciennes, the regiment marched to take part in the operations covering the siege of Dunkirk. The Duke of York, commanding the army, divided his troops into two divisions, and The Queen's Bays were allotted to the division of the Hanoverian General Freytag, the only British among the Hanoverians being ten cavalry squadrons, including the two of The Queen's Bays. As the Allies laid siege to Dunkirk, Freytag established his headquarters at Hondschoote, but on 6 September the French, under General Houchard, advanced in five columns (two of them led by Vandamme and Jourdan, later to win fame under Napoleon) and, after some hard fighting, drove in the Hanoverians and wounded Freytag. General Walmoden replaced Freytag and concentrated his troops around Hondschoote, but on the 8th he was forced to retreat. The ground was so broken and marshy that The Queen's Bays, together with the 3rd Dragoon Guards, the Royal and Inniskilling Dragoons, fought dismounted as infantry. The Queen's Bays, however, lost only one trooper, shot through the body. Sadly, during the siege sickness and disease claimed many more than fell to the bullets of the French. On the evening of 8 September the Duke of York was forced to raise the siege of Dunkirk and the army retreated to Furnes. [4, 5, 14, 15]

Le Marchant described the situation:

> Our present position is divided from the enemy by the Yser, a river much smaller than we would wish, but wide enough to increase the difficulties of an attack. We are ranged along its banks, and the country is so well-wooded that our cavalry, in which the strength of our army lies, cannot at present be employed with advantage. On 6th September, they pushed hard to turn our right flank, where we were, and we were so pressed as to dismount to act as infantry. Our resistance checked their impetuousity, and towards four o'clock they began to retire. On the morning of the 8th, the action recommenced, and at two o'clock we were forced to give way, and retire. We learnt that almost all our baggage, ammunition etc had fallen into the enemy's hands.[16]

In October the Duke of York moved towards Tournai with 9,000 men, including The Queen's Bays amongst the cavalry, in order to strengthen the Allied forces watching Lille. On the 24th the Duke moved to drive in an advanced detachment which Vandamme had pushed out from Lille. As the French fell back, and were about four miles from Lille,

A [French] picquet of six officers and 150 men, which had been posted at the village of Saingain [Sainghin-en-Melantois], retreated across the plain towards Lezennes; they had nearly reached the last-mentioned village when a Squadron of the Second Dragoon Guards, led by Major Crauford, Aide-de-Camp to His Royal Highness, advancing with rapidity, gained their right flank, and charged them with so much vigour, that not a single man escaped; 104 prisoners were taken, and the rest killed upon the spot. The other Squadron of the Queen's Dragoon Guards, two Squadrons of the Royals, and a division of Austrian Light Dragoons came up in the pursuit.

It was the right squadron of The Queen's Bays that achieved this brilliant action, commanded by Captain James Hay, with fifty-six officers and men. Captain Hay had his horse shot under him but remounted a grey belonging to his farrier-major, which he rode for the rest of the action. The Queen's Bays lost three men killed, and four troop horses wounded. [4, 5]

As the weather worsened, further action was confined to isolated outpost skirmishes. The men of The Queen's Bays were quartered among the villages along the Belgian border, until in November they marched into their winter quarters at the cavalry barracks in Ghent.[4, 5, 12]

During February 1794 The King's Dragoon Guards at Deynse had a number of false alarms. On the 14th they 'rec'd orders to hold ourselves in readiness to march at a moments notice in consequence of which horses were immediately saddled — but we rec'd no further orders so that we unsaddled again in the afternoon — this was occasioned by the French making an attempt to cross the river near Menin.' On the 28th 'The Minute guns were fired at Menin (about 9 o'clock in the morning) in consequence of which the regt prepared to march, but was ordered to remain till further orders from His Ryl Hss. the Duke of York'. On the 6th 'a man rec'd 200 lashes for theft', and on the 10th another got '200 lashes for getting drunk when patrolling ... During the remainder of this month nothing material happened.' During February The Queen's Bays were moved from Ghent to the cavalry barracks at Tournai. They were held at this time in the highest regard, mainly because of the excellent training they had been given by Major Crauford. [12, 14]

On 3 April the KDG were reviewed by General Harcourt in watering order, and on the 8th the regiment marched with the 5th Dragoon Guards to Bellain near Valenciennes, joining the rest of the heavy cavalry on the heights at Valenciennes on the 14th. Two days later the whole of the army was inspected by the Emperor of Austria on the heights of Cateau. This was really a muster of all the available troops, and the twenty-eight squadrons of British cavalry were formed into four brigades. The King's Dragoon Guards were in Harcourt's brigade together with the 5th and 6th Dragoon Guards; The Queen's Bays were brigaded with the Greys and Inniskillings under General Laurie; Mansel's brigade consisted of the Blues, 3rd Dragoon Guards and the Royals; and Dundas had the 7th, 11th, 15th and 16th Light Dragoons. Russell relates that after the review 'we proceeded on our way for Catteau [sic], where we arrived about 11 o'clock at night, and halted on the heights above the town during the remainder of the night. The army began to move soon after daybreak and proceeded to the village of Fremont [sic] (then occupied by the French) from whence we drove them without any material loss on our side.'[3, 4, 7, 14, 15]

The advance after the review was made in two columns, The King's Dragoon Guards being in the column commanded by Lieutenant General Sir William Erskine, and the Queen's Bays in the one under the Duke of York. It was Erskine's force which attacked and captured Premont, inflicting some 2,000 casualties on the French and capturing twenty to thirty guns; The King's Dragoon Guards lost Captain Carleton, who was wounded by a cannon ball and died the next day, when he was acting as brigade major. The affair was marred because 'the Austrians after plundering the village, burnt it to the ground'. The King's Dragoon Guards 'after the engagement patroled with part of the 16th Light Dragoons towards Cambray the greater part of the night, but returned to Fremont about 3 o'clock in the morning'. The Duke of York's column attacked the village and entrenched post of Vaux, where The Queen's Bays with the Greys and Inniskillings charged the badly trained French Republican horse and helped to carry the Star redoubt. Major Hay had his horse shot under him, but this was their only casualty. The Duke of York was so infuriated by the conduct of some of the British that he had two looters hanged on the spot, without even the form of a drumhead court martial.[3, 4, 5, 7, 15]

Over the next few days the KDG and The Queen's Bays manoeuvred around Cateau, and on the 23rd 'marched to the grand camp at Cateau, leaving the 15th Light Dragoons in possession of the advanced post'. The regimental records of The King's Dragoon Guards state that the regiment 'on the 24th formed part of the force which made a successful attack on a body of French troops posted at Villers en Couche [sic], when the enemy were defeated and driven back into Cambray, with the loss of twelve hundred men and three pieces of cannon'. All other accounts of the brilliant action at Villers-en-Cauchies rightly give the glory for that occasion to the 15th Light Dragoons and the Austrian Leopold Hussars, and there is no other record of the KDG having been actively involved in that affair. Indeed the 15th King's Own Hussars are the only British regiment to carry Villers-en-Cauchies as a battle honour.[3, 14]

Early on the morning of 26 April General Chappuis led two strong French

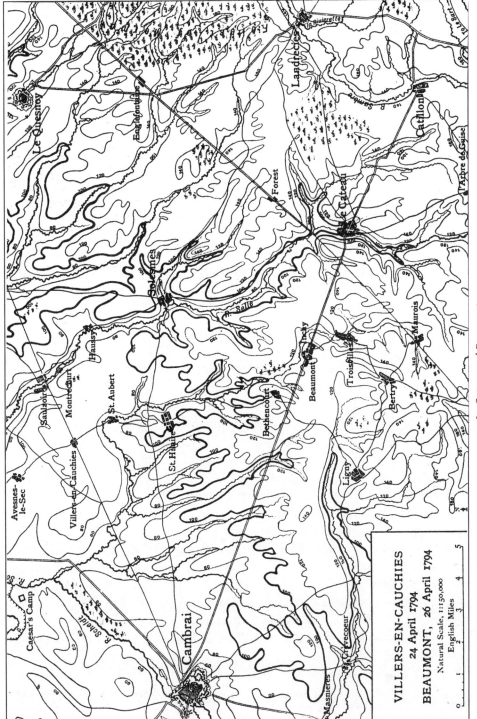

Le Cateau and Beaumont

columns, totalling some 30,000 men, from Cambrai to attack the Allied position at Cateau. Under cover of a dense mist the French managed to drive in the Allied outpost line, capturing the villages of Inchy and Beaumont. The French then began to form up for the main assault on the ground below the ridge on which the villages stand.[3, 7, 8, 14]

As the mist cleared the Duke of York brought up his artillery and made a great show of a feigned attack on the French front, sending a few light troops to keep their right occupied, as he had observed that their left was in the air. Colonel Russell, KDG, noted in his diary:

> Soon after daybreak the French army consisting of 28 thousand with 36 pieces of cannon made an attack upon our camp, when after driving in our advanced piquets, the following regts of cavalry were ordered to engage them, vizt 1st Dragoon Guards (6 troops), 5th Dragoon Guards (4 do), Oxford Blues (3 troops) and a regt of Austrian Zetchwitz Cavalry (cuirassiers) — our whole force did not exceed 2,500 and without cannon, we marched from the camp.[8, 14, 15]

The Duke of York brought the cavalry to his right and formed them up, out of sight of the French, in a fold in the ground between the villages of Inchy and Bethencourt. His first line consisted of the six squadrons of the Austrian Zetchwitz Cuirassiers under Colonel Prince Schwarzenberg, the second was Mansel's brigade of the Blues, 3rd Dragoon Guards and the Royals, and the third line was composed of The King's Dragoon Guards, the 5th Dragoon Guards and the 16th Light Dragoons; the whole force being commanded by General Otto.[3, 6, 15]

Otto advanced with caution, using every fold of the ground to conceal his movements, then came face to face with a body of French cavalry, General Chappuis among them. The French were immediately charged, overthrown and scattered; General Chappuis was taken prisoner. As the final ridge was cleared, the Allied force saw in front of them more than 20,000 French infantry drawn up in two lines with their guns all facing east, without any thought for an enemy approach from the north. Prince Schwarzenberg was an impetuous leader: the trumpets sounded and 'the British cavalry were not in the temper to conform too nicely to regulations; and with the British cheer, which had so disagreeably impressed the French at Dettingen, they swept down on the enemy's left flank — totally regardless of the furious fire of grape and musketry which was opened on them.'[3, 6, 7, 8, 14, 15]

General Mansel, who had been suffering from depression as a result of an imputation of cowardice for his conduct at Villers-en-Cauchies, suddenly shouted out as he led the British at a trot, 'I'll not get back from this alive!' Spurring his horse, he dashed into the French and was at once cut down. His son, who was acting as his aide-de-camp, went to his father's assistance and was taken prisoner. Colonel Vyse of The King's Dragoon Guards immediately took over command, and led the thundering squadrons straight into the French. Major Long and his leading squadron of KDG found themselves confronted by an extremely deep ditch, which unseated some troopers; while Long's charger, Conqueror, 'received a wound in his off foot near the hoof, by a musket shot which a scoundrel, whom I had pardoned the instant before on the condition of his laying down his arms, fired at me. My sword soon gave him the reward which such Criminal Ingratitude justly merited.'

Russell relates in his diary: '. . . after a short resistance they fled with the greatest precipitation (leaving behind them a great number of killed and wounded, among the former was a general, we like-wise took 1 general prisoner together with a number of men and 35 pieces of cannon with a great quantity of ammunition).'[3, 6, 7, 8, 14, 15, 17]

The King's Dragoon Guards lost seven men and twenty-nine horses killed, with Cornet Betson, twenty-three men and sixty-four horses missing; and three more men later died of their wounds. Cornet Caulfield was taken prisoner early in the engagement, but was retaken later in the day. The Duke of York issued a general order which said, 'The Austrian Regiment of Cuirassiers of "Zetchwitz", The Blues, 1st, 3rd, and 5th Dragoon Guards, The Royal Archduke Ferdinand's Hussars, and the 16th Light Dragoons, who attacked and defeated the principal column of the enemy on the right, have all acquired immortal honour to themselves.' The KDG received a reward of £500 by the Duke's order, on the basis of £20 for each cannon captured, £10 for each pair of colours and for each ammunition waggon, and £12 for each horse. Russell writes:

> The loss sustained by our Regt did not exceed ten men [among which] 3 were taken prisoner, our loss in horses was more considerable as we had 95 killed, wounded and missing in the course of three hours. The Oxford Blues and 3rd Dragoon Gds suffered considerably. In the evening, the different regts recd the thanks of his Royal Highness the Duke of York (Commander in Chief) in person for their gallant behaviour in the field – and likewise 4 pence per man to drink his Highness' health. A feu de joy [was] fired by the army (in honour of the victory gained over the French) beginning with 38 pieces of cannon and answered by all the regts in camp, the cavalry firing on horseback with pistols.[3, 6, 7, 14]

Papers found on General Chappuis gave the whole of the French plans. The Duke of York, taking advantage of this information, sent The Queen's Dragoon Guards [Bays], together with the Greys and Inniskillings, to operate towards St Armand.

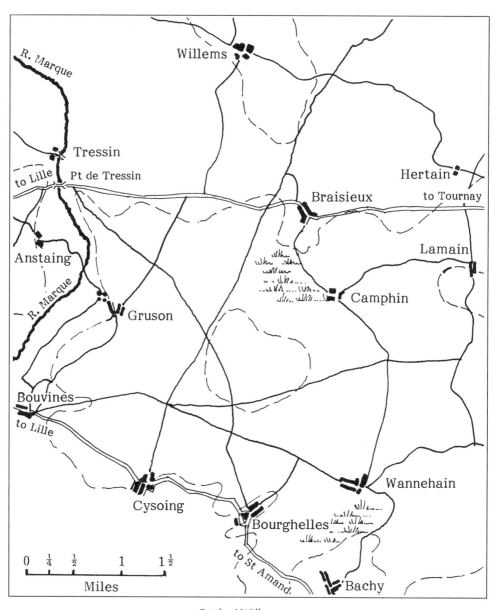

Battle of Willems

The Duke followed with the rest of the army towards Landrecies, which surrendered on 30 April. The Queen's Bays and The King's Dragoon Guards then
marched with the rest of the army towards Tournai to threaten the flank of the new
French armies which were now thrusting into Flanders. Tournai was reached on 3

May, and on the 10th the French, under Pichegru, attacked the Duke's entrenched position between Hertain and Lamain, advancing with two columns. One column was checked, the other captured the advanced Allied posts, but in so doing opened up a gap in their line. The Duke of York at once ordered sixteen squadrons, including The Queen's Bays, to advance across low ground to the south of Lamain and to turn the French right flank. Lieutenant General Harcourt led the cavalry over some difficult going, whereupon the French infantry formed squares and were able to beat off the first charges, which were delivered with insufficient speed. As the French retired towards the village of Willems, the cavalry followed and, reinforced by six more squadrons, charged and routed the accompanying French cavalry; but they were still unable to break into the infantry.

> At last, however, a little to the south of Willems, the battalion guns of the British infantry came up and opened fire, when the French, after receiving a few shots, began to waver. The squadrons again charged, and an officer of the Greys, galloping straight at the largest of the squares, knocked down three men as he rode into it, wheeled his horse round and overthrew six more, and thus made a gap for the entry of his men. The sight of one square broken and dispersed demoralised the remainder of the French. Two more squares were ridden down, and for the third time British sabres had free play among the French infantry. Over four hundred prisoners were taken, thirteen guns were captured, and it was reckoned that from one to two thousand men were cut down.

The Queen's Bays casualties were two men killed and two missing, with three horses killed, two wounded and two missing. The King's Dragoon Guards in Vyse's brigade were kept in reserve. Colonel Russell commented that 'During the whole engagement it rained very heavily'.[4, 5, 14, 15]

SOURCES

1 *Journal of the Society for Army Historical Research*, vol. 29, 1951, p. 85.
2 *Journal of the Society for Army Historical Research*, vol. 38, 1960, p. 141.
3 R. Cannon, *Historical Records of the British Army, 1st Dragoon Guards*, 1837.
4 R. Cannon, *Historical Records of the British Army, 2nd Dragoon Guards*, 1837.
5 F. Whyte and A. H. Atteridge, *The Queen's Bays*, vol. 1, 1930.
6 William Clowes, *King's Dragoon Guards, 1685-1920*, 1920.
7 Records of The King's Dragoon Guards, Regimental Museum, 1st The Queen's Dragoon Guards, Cardiff Castle.

8 Sir George Arthur, *Story of the Household Cavalry*, vol. 2, 1909.

9 E. Belfield, *The Queen's Dragoon Guards*, 1978.

10 The Marquess of Anglesey, *A History of the British Cavalry*, vol. 1, 1973.

11 G. Granville Baker, *Old Cavalry Stations*, 1934.

12 R. H. Toumine, *Scientific Soldier*, 1968.

13 *Journal of the Society for Army Historical Research*, vol. 30, 1952, p. 116.

14 Diary of Lieutenant Colonel James Russell, KDG, Flanders 1793-96, National Army Museum.

15 Sir John Fortescue, *History of the British Army*, vol. 4, 1910.

16 *Journal of the Society for Army Historical Research*, vol. 3, 1924, p. 117.

17 J. H. McGuffie, *Peninsular Cavalry General*.

16

Retreat to Bremen, England and Ireland
1795-1806

Early in May 1794 both The King's Dragoon Guards and The Queen's Bays marched with the army from the area around Le Cateau to Tournai. On 10 May the French 'began an attack on our advanced posts, which soon became general, with General Vyse's brigade kept in reserve. During the whole engagement it rained very heavily.' On 22 May, 'The French made a general attack on our camp, but were gallantly repulsed. The engagement lasted till 11 o'clock at night.' Again The King's Dragoon Guards and The Queen's Bays were kept in reserve.

> A column consisting of five or six thousand men made its appearance towards our left, on which account the brigade of Guards and British Heavy Cavalry remained ready for action on their camp ground all that day; but the French observing our advantageous situation, and dreading the thought of meeting the British Cavalry a second time on an open plain, thought proper not to make any approaches.[1,2]

From June onwards the Allies were in retreat. Belgium was abandoned to the French, and as the army fell back to Holland, discipline deteriorated. No remounts had arrived, the men's clothing and equipment were disintegrating, rations and forage failed, looting became commonplace. 'This day, 4th June, 4 prisoners recd the sentence of a general court martial for plundering and drawing their swords on Sir William Erskine. 3 recd each one thousand lashes, one five hundred.' More men had died from sickness and bad rations than from contact with the enemy, and both regiments were under strength. Gradually, however, things began to improve, and on 11 June 'The remount from England joined, consisting of 2 subalterns, 1 serjt, 40 men and 80 horses'.[1]

During the retreat The King's Dragoon Guards, in Major General Vyse's brigade together with the 8th and 14th Light Dragoons, marched to Malines on 21 May, where they encamped until June. They then moved on to Antwerp, staying there until August, when they marched to Breda, eventually moving to Bois-le-Duc (the 'boiled duck' of Marlborough's troops). The Queen's Bays were brigaded with the

6th Dragoon Guards, the Scots Greys and the Inniskilling Dragoons in the brigade of Major General Dundas, who had replaced Major General Laurie. Major Le Marchant of the Queen's Bays was offered an appointmenmt on the staff, but was unable to accept as the regiment was so short of officers.[6,7]

The Austrians were now retreating eastwards. The French having captured Antwerp, the Duke of York was forced to withdraw across the Dutch frontier. Then Valenciennes surrendered, and Le Marchant wrote to his wife, 'Perhaps it is fortunate that it happened so, otherwise we should have been attempting again to combat a Power we have unhappily found too much for us, and our allies, who would not act without subsidy, would have I fear not shown much exertion.' The Dutch, by now disheartened, looked upon occupation by the French as inevitable, and even welcome; so, as winter drew on, the British and Hanoverians, having parted with the Dutch at Breda, continued their retreat towards Germany. Le Marchant wrote on 9 September, 'God grant we may get well over the remainder of the campaign, and I trust the winter will effect a peace.'

During the retreat both The King's Dragoon Guards and The Queen's Bays were engaged on outpost duty, resulting in a number of skirmishes with the French. On 26 August Russell relates how the King's Dragoon Guards

> in the morning with a detachment from the Light Brigade consisting of about 100 men, marched from the camp in order to reconnoiter the enemy's advance post, when finding their force too powerful, we retreated, leaving the left Squadron and some of the Light Dragoons with some Hanoverian flying artillery. Soon after our leaving this detachment, they were attacked by the French, who were so much superior in number, that after a short resistance, they were obliged to retreat towards the . camp. In which we lost 1 serjt killed and 8 taken prisoner, (5 of which were retaken by the Hesse Darmstadt Light Dragoons). The enemy persued [sic] the piquet almost to the camp, while the remainder of the detachment sat on horseback, idle spectators, and could not go to their assistance. They were at last checked by a regiment of Hessian infantry; who were posted in a thick hedge, from whence, with two field pieces of cannon, they obliged the French to retreat in great disorder.[1]

The King's Dragoon Guards were again involved near the village of Ghilze on 15 September when they

> marched from the camp at 11 o'clock in the evening, and proceeded to the village of Boxtel, where we arrived in the morning, 15th, at about daybreak. We began an attack upon the advanced post of the enemy,

but were repulsed, they having taken up an advantageous position in the woods, likewise having a great superiority both in men and cannon. Which rendered it impossible with our small force to dislodge them, and not being able to bring them to a fair combat, which was much wished for by our men who were in great spirits, we were at length obliged to retreat, which we did in good order without any material loss – of our regiment we had 1 horse killed, and 1 man and horse wounded.[1, 3]

The retreat continued. The army crossed the Meuse at Grave, and The King's Dragoon Guards camped near Nijmegen before going into quarters on 19 October at Hensden. The cavalry went ahead of the main body of the army, and so escaped the worst of that terrible retreat. But as the regiments struggled over the ice and snow of north-east Holland in bitter weather, men and horses died at every bivouac, clothes were once more in rags, and rations so scarce that plundering became rampant and discipline collapsed.[5, 6, 7]

The Rhine was crossed on 13 November, and on the 21st the KDG 'were relieved by a detachment from the 3rd Brigade, consisting of the Bays'. The KDG then went into winter quarters at Eede and Benckam in Guelderland, but 'oweing to the badness of the weather we broke up camp, and went into the village, where we put our horses into the barns at the farmhouses, we likewise slept in the same places ourselves, not being permitted even to dress our victuals in the houses, so that we were obliged to make fires in the gardens and orchards'. On 14 January 1795 'kept our horses saddled the whole day, and turned out three different times in expectation of the French crossing the Rhine, as at this time the river was very hard frozen so that horses could pass over'. 'A party being ordered out to forage, at this time very scarce with us, they were attacked by a number of the farmers armed with muskets and pitchforks, who fired and wounded 2 of the party.' Eventually, on 18 May, The King's Dragoon Guards arrived in the vicinity of Bremen, but conditions were still appalling – as, in Russell's words, at 'Rhemels, by our men termed Miserable oweing to the scarcity of provisions, there being nothing to be had for money'. The Queen's Bays were billeted along the River Ems, from Rhein to Emden, with regimental headquarters at Osnabruck. On 28 February 1795 a Royal Warrant was issued to the Marquess Townshend, colonel of the Queen's Bays, in order to try to make good some of the shortages:

GEORGE R. These are to authorise you, by beat of drum or otherwise, to raise so many Men in any County or part of our Kingdom of Great Britain as are or shall be wanting to recruit and fill up the respective Troops of Our Regiment of Dragoon Guards under your Command to the numbers

allowed upon the Establishment. And for so doing this Our Order shall be
and continue in force for Twelve Months from the 25th day of March.

It was one thing to issue warrants from Whitehall, and quite another for trained
soldiers to arrive in the ranks.[1, 3, 8]

In March 1795 it was decided to bring back the troops to England, but it was 28
October before the six troops of The King's Dragoon Guards broke camp and
marched to Scharmbeck, where they remained until 5 November. They embarked
at Bremenlee on 7 and 8 November, arriving at North Shields on 16 December and
landing at South Shields on the 27th. The Queen's Bays embarked at Bremen and
landed at South Shields on 29 December 1795. There is no record of the casualties
suffered by The Queen's Bays, but The King's Dragoon Guards lost 59 dead from
sickness, 3 from their wounds, 8 killed in action, 4 prisoners of war, and 2 missing,
a total of 76. Of the horses, 247 died of disease, 30 were killed in action, 52 were
cast and 24 were missing or captured, a total of 353 horses lost.[3, 4, 5, 8, 9]

The Queen's Bays marched on landing to Ipswich, joining the depot there; later
they moved to Romford in Essex. The King's Dragoon Guards marched to Darlington, via Chester-le-Street and Durham, where 'we remained for 18 days to
refresh our horses, who were in a very low condition after being so long on the
water'. The march was resumed on 18 January 1796 and Romford was reached by
easy stages. There 'we joined the other part of the Regiment, after having, since
we landed, marched 251 miles from South Shields'. The 'other part of the Regiment' consisted of the six troops that had remained at home.[1, 3, 4, 5, 8]

With the return of the troops to England, a reduction of twenty-seven men per
troop was made. In December the KDG was further reduced by two complete
troops, but those officers and quartermasters were allowed to continue with the
regiment on full pay. When these reductions took place, two officers, a quartermaster, eight NCOs and fifty men from The King's Dragoon Guards volunteered
to serve in the 26th Light Dragoons. The officers were promoted on transfer, a
lieutenant to captain, and a cornet to lieutenant. In December three more officers,
Lieutenant Caulfield and Cornets Partridge and Townsend, volunteered for that
regiment. On joining, they all sailed for service in the West Indies.[3, 4]

Both The King's Dragoon Guards and The Queen's Bays marched to Wimbledon
Common to be reviewed by George III on 3 May 1796. The Queen's Bays returned
to Romford, The King's Dragoon Guards moved in June to a camp at Brighton,
and then in October to quarters in Salisbury. On 16 July 1796 Field Marshal Sir
George Howard died, and was succeeded in the colonelcy of The King's Dragoon
Guards by General Sir William Augustus Pitt, who came from the 10th Light
Dragoons.

During the summer of 1796 a number of changes took place in the appointments and establishment of the two regiments. A vetinary surgeon was added, and the position of regimental chaplain was abolished. Changes were made in the uniform, the length of the men's coats being considerably reduced, and the lace on their cocked hats discontinued. New hats were introduced, of black felt, turned up at front and back. Breeches were to be plush. Shoulder straps replaced epaulettes, and were yellow with red wings interlaced with brass plates. At this period the men's weapons were a musket and bayonet, two large horse pistols and a sword, with a cartouche pouch to carry thirty rounds of ammunition. In 1798 the musket, bayonet and two large pistols were replaced by a carbine and a single smaller pistol. In 1796 there were also changes in the saddlery of the heavy cavalry: housings and holster caps were dark blue with facings in the regimental colour, a rolled cloak was strapped behind the cantle, and a leather surcingle encircled the complete saddle and girth, which was the 1796 pattern with a crupper and breastplate. Also in 1796 new patterns of swords were introduced for both the heavy and light cavalry. The heavy cavalry sword had a broad straight blade, 35 inches long, ending in a hatchet point, the guard being a steel knucklebow. Brian Robson points out that it was 'useless either for slashing or thrusting. Indeed it must be almost the worst pattern ever issued.'[3, 4, 5, 9, 10, 11]

The ledgers of Cox and Co, Army Agents, provide some interesting information concerning the costs of clothing at that time. In 1797 The Queen's Bays paid £30. 15s for a standard, but two years later the price was £71. 16s. 6d. A trooper's hat cost 10s.2d, hat feathers for sergeants and trumpeters were 5s each, but those for the men only 1s each. The regiment was mounted on bay horses, with the trumpeters and farriers on greys. Recruit horses were purchased for £26. 5s each, but sergeants' horses cost the regiment £3. 3s more than any other regiment. The docking of horses' tails had been discontinued in 1764, but nag-tailed horses were still the civilian fashion and it was becoming increasingly difficult and expensive to buy long-tailed remounts. So on 10 August 1799 an Army Order directed that all regiments except the Life Guards and Royal Horse Guards were to be mounted on nag-tails. The King's Dragoon Guards, the Royal Dragoons and the 3rd Light Dragoons were the only regiments still allowed blacks, as they were becoming hard to purchase, but The Queen's Bays retained their bay horses; all other regiments were to have bays, browns or chestnuts.[3, 4, 5, 12, 13]

On 14 October 1796 The King's Dragoon Guards marched from Brighton to Guildford, and on the 26th to quarters at Salisbury. In June 1797 Robert Richards was appointed as the first veterinary surgeon to the regiment, and on 1 July the KDG went into camp at Weymouth until 28 September, when they marched to Croydon. The Queen's Bays left Romford and were quartered at Salisbury and Southampton during 1797. The following year they moved to Windsor and Croydon barracks, while The King's Dragoon Guards were nearby at Swinley. On 6 October 1798 The King's Dragoon Guards marched to Ipswich, and in August 1799 both regiments were augmented, the KDG by one troop, The Queen's Bays to an establishment of ten troops. The Queen's Bays were reviewed by the King at Wimbledon in the spring of 1799, later camping near Windsor, and then in October moving to Hertford, Ware, Hoddesdon and Hatfield. In August 1799 one sergeant and eight troopers of The King's Dragoon Guards transferred to the newly formed Wagon Corps, and on the 24th the regiment was ordered to have eight troops in readiness for foreign service. These eight troops marched to Canterbury on 12 September, but the order was cancelled and they countermarched back to Croydon.[3, 4, 5, 8, 9, 14]

The next few years saw a series of moves for both regiments:

	KDG	Queen's Bays
1800	Windsor, Croydon	Southampton
	Northampton, Leicester	Wellingborough, Peterborough.
	Exeter	
1801	Birmingham, Nottingham	Bristol, Bath
	Guildford	
	Northampton	
	Birmingham	
1802	Birmingham, Bristol	Scotland
		Piershill Barracks, Edinburgh
1803	Exeter	Ireland
		Dundalk, Navan, Lisburn
		Drogheda, Man of War
1804	Exeter	The Curragh of Kildare
	Arundel	Longford, Roscommon
	Norwich, Longford	Phillipstown, Tullamore
1805	Brighton	The Curragh of Kildare, Phillipstown
		Tullamore, Longford
		Dublin, to England
		Liverpool
1806	Brighton	Birmingham, Coventry
		Dorchester, Blandford
		Plymouth
		Salisbury
1807	Lewes	Chichester, Arundel

Over this period of garrison duties in England, Scotland and Ireland there were a number of changes to the establishments of both regiments. On 1 March 1800 an additional troop was added to the strength of The King's Dragoon Guards, and throughout June of that year ten troops of the regiment encamped in Windsor Forest, whilst the two remaining troops stayed at Croydon to look after the sick and the regimental baggage. On 25 March 1801 another ten men and ten horses were added to the strength of each troop of The King's Dragoon Guards, bringing up troop strengths to seventy-one men, with four sergeants, four corporals, one trumpeter, one quartermaster, and a captain in command with one lieutenant and one cornet. However, with the signing of the Peace of Amiens in 1802 the establishment of the regiment was reduced on 14 May 1802 to ten troops of fifty-six men per troop, plus four sergeants, four corporals and one trumpeter. A month later the full establishment was fixed at: 1 colonel; 1 lieutenant colonel; 1 major; 7 captains; 1 captain lieutenant; 9 lieutenants; 10 cornets; 1 paymaster; 1 adjutant; 1 surgeon; 1 assistant surgeon; 1 veterinary; 10 quarter-masters; 1 sergeant major; 1 paymaster sergeant; 1 armourer sergeant; 1 saddler sergeant; 30 sergeants; 30 corporals; 10 trumpeters; 570 privates. This strength was to include ten men per troop who were to be dismounted. A general order, dated from the Horse Guards 1 July 1802, had directed that an armourer sergeant was to be added to the strength of every regiment of cavalry and infantry. At the same time the establishment of The Queen's Bays was reduced from ten to eight troops.[3, 4, 5, 8, 9, 16]

The average height of troopers was to be five foot ten inches, and the horses were to be fifteen hands one inch and a half. The KDG officers wore scarlet, with blue facings to the cuffs and collar, and gold lace, but no facings 'on the ordinary uniform'. The troopers wore a red jacket, faced in blue with white lace, and the buttons marked K:D.G. The sergeants wore gold lace. The officers of The Queen's Bays at this period wore red, faced with black velvet and with silver lace, while the troopers wore a red jacket, with black collar and cuffs, royal lace, and the buttons of white metal marked Q.D.G. Their sergeants wore silver lace. An order from the Horse Guards dated 23 May 1803 discontinued the wearing of epaulettes and shoulder knots to distinguish the ranks of NCOs, and ordered the wearing of chevrons, which were

> to be formed of a double row of the Lace of the Regiment. The bars of the Chevrons are to be edged with a very narrow edging of Cloth of the Colour of the Facing of the Regiment. The number of bars as denoting the Rank of the Wearer are: Sergeant Majors 4 bars. Qr.Master Sergeants 4 bars. All other Sergeants 3 bars. Corporals 2 bars. They are to be affixed

on a piece of Cloth the Colour of the Coat, and worn on the Right Arm, at
an equal distance from the Elbow and the Shoulder.[3, 15, 17]

On 25 January 1803 Lieutenant Colonel Hawley of The King's Dragoon Guards
died and was succeeded by Lieutenant Colonel Elliott; this was a change that
was later to have a marked effect on the efficiency of the regiment. In July the
precarious peace with France had broken down, Napoleon was threatening invasion, and The King's Dragoon Guards were increased by twenty men per troop, of
which ten were to be mounted and ten dismounted. In September two additional
troops were ordered to be raised and added, and the two senior lieutenants and
cornets were charged with the task of recruiting the new men, with the reward of
promotion to captain and lieutenant if they were successful. Both The King's
Dragoon Guards and The Queen's Bays had three extra captains added to the
establishment, which released the field officers from the command of a troop,
though they retained the pay and allowances as if they were still in command of a
troop. [3, 4, 5, 8, 9]

Early in 1803 Robert Emmet attempted to seize Dublin Castle, and in the ensuing
scare The Queen's Bays were despatched to Ireland. In February 1802 Sir James
Erskine had been appointed as lieutenant colonel of The Queen's Bays; he was to
rise to the rank of lieutenant general in 1813, having commanded a brigade of
cavalry as a major general under Wellington in Portugal. On 3 December 1803 a
distinguished KDG was appointed to the lieutenant colonelcy of The Queen's
Bays. Lieutenant Colonel Robert Long had been commissioned into the KDG as a
cornet in 1791, and served with that regiment in the campaign in Flanders of 1793
to 1795. He was to attain the rank of major general, become adjutant general to
the ill-fated Walcheren expedition, and serve under Wellington in command of a
brigade at Vittoria. He was finally promoted to lieutenant general in 1821.[5, 8]

On 27 March 1804 two troops of The King's Dragoon Guards, under command
of Captain Craven, together with the 24th Foot [South Wales Borderers], were
reviewed near Norwich by Lieutenant General Sir James Craig. In December the
ten dismounted men per troop were ordered to be mounted. In April 1805 the
Duke of York approved 'the purchase of as many grey horses as are neccessary to
mount the Trumpeters [of the KDG] at present mounted on horses of a different
colour'. While stationed at Exeter The King's Dragoon Guards were commended
for the assistance they provided in training the local Yeomanry.[3, 20]

On 12 November 1804 the court martial of Lieutenant Colonel Elliott of The
King's Dragoon Guards opened near Brighton on three charges brought against
him by one of his own officers, Captain Sober. He was accused, first, of failing to
ensure that cast [unserviceable] horses were sold by public auction; secondly,

'carrying on an improper Traffic with the Cast Horses, highly disgraceful to the situation he holds in the Regiment, and thereby defrauding the Government'; and thirdly, 'for receiving Coals and Candles as a Barrack Allowance, whilst he lodged in the town and not in the Barracks, contrary to an express order on that head'. The paymaster, the adjutant, some officers, and rough riders gave evidence against their commanding officer. The main defence was one of malice by the witnesses, and of previous good character. Elliott was found guilty on all three charges, with some extenuating circumstances, and due note was taken of his previous service and character, but he was to be 'dismissed His Majesty's Service'. The sentence was confirmed by the King, and Lieutenant Colonel Henry Fane, from the 4th Dragoon Guards, was appointed to take over. This affair had gravely undermined the discipline of The King's Dragoon Guards, and there was discontent over the manner in which stoppages of pay were being made. Fane himself, between parliamentary duties and leave, spent only four months actually with the regiment. On 28 August 1805 Lieutenant Colonel Fuller from the 10th Light Dragoons arrived to be second lieutenant colonel. He was faced with a difficult position and acted firmly. For two years the KDG had had an absentee commanding officer or one who commanded no respect. Major Balcomb, the senior major, who had been serving continuously with the regiment, was made to take the blame for 'the relaxation of discipline'.[18, 19]

In August 1806 The Queen's Bays, having marched to Plymouth, embarked four troops on board HMS *Theseus*, *Malta* and *Captain* for service with a projected expedition to Spanish South America, commanded by Lieutenant General Whitlocke. When a French squadron was discovered nearby, the cavalry were immediately landed, the ships gave chase to the French and the project was abandoned. The Queen's Bays marched to Salisbury. In September 1807 Major General Charles Crauford was appointed colonel of The Queen's Bays in succession to the Marquess Townshend, who had died. The Duke of York reviewed twelve troops of The King's Dragoon Guards, together with the 3rd Dragoons, near Brighton early in June 1807, and noted the improvement in their discipline and movements.[3, 4, 5,]

SOURCES

1 'The Diary of Lieutenant Colonel James Russell, 1st King's Dragoon Guards', National Army Museum, Archive 8707-31.

2 *Brown's Journal.*

3 Records of The King's Dragoon Guards, Regimental Museum, 1st The Queen's Dragoon Guards, Cardiff Castle.

4 R. Cannon, *Historical Records of the British Army, 1st Dragoon Guards*, 1837.

5 R. Cannon, *Historical Records of the British Army, 2nd Dragoon Guards*, 1837.

6 R. H. Thoumine, *Scientific Soldier*, 1968.

7 Sir John Fortescue, *History of the British Army*, vol. 4, 1910.

8 F. Whyte and A. H. Atteridge, *The Queen's Bays 1685- 1929*, vol. 1, 1930.

9 W. Clowes, *The King's Dragoon Guards 1685-1920*, 1920.

10 M. Chappell, *British Cavalry Equipment 1800-1920*, 1983.

11 Brian Robson, 'The British Cavalry Trooper's Sword', *Journal of the Society for Army Historical Research*, vol. 46, 1968, p. 92.

12 *Journal of the Society for Army Historical Research*, vol. 17, 1938, p. 92.

13 J. M. Brereton, *History of the 4/7th Royal Dragoon Guards*, published by the regiment, 1982.

14 B. G. Baker, *Old Cavalry Stations*, 1934.

15 F. Grose and T. Egerton, *A History of the English Army*, vol. 2, 1801.

16 E. A. H. Webb, *A History of the Services of the 17th Regiment*, 1911.

17 *Journal of the Society for Army Historical Research*, vol. 16, 1937, p. 58.

18 W. Clowes, *The Court Martial of Lieutenant Colonel Elliott*, 1804.

19 W. O. 3/39 & W. O. 12/92.

20 *Journal of the Society for Army Historical Research*, vol. 30, 1952, p. 136.

17

Walcheren, England and Ireland, Flanders
1807-1815

During 1808 'the men's hair, which had been worn long, powdered and tied in a queue, was ordered to be cut short', and a hated practice ended for both The King's Dragoon Guards and The Queen's Bays, although the former did not comply until the following year. The King's Dragoon Guards, having marched from Arundel to York in June 1808, moved north in May 1809 to Berwick and then to Dunbar. On 31 July they were reviewed on Belhaven Sands by Lieutenant General Lord Cathcart, and in August marched to Piershill Barracks, Edinburgh. Before the regiment marched north, Lieutenant Colonel Fane, having been promoted to command a brigade, left the KDG and embarked at Cork for foreign service. Lieutenant Colonel William Fuller, who was to command the regiment at Waterloo, and who had been the second lieutenant colonel, took over as commanding officer.

The Queen's Bays left Arundel and Chichester in July 1808 and marched to Hastings, Piershill, Bletchington, Rye and Eastbourne, where they stayed until December, when they moved to Canterbury. [1, 2, 3, 4, 5]

Early in the spring of 1809 The Queen's Bays were warned to prepare for service overseas as part of an expedition to destroy the French-held port of Antwerp, which Napoleon claimed was 'a pistol pointed at the heart of England'. This town on the upper reaches of the Scheldt was the base for a gunboat flotilla and many privateers; its dockyard had been extended and was able to repair and build warships. The Queen's Bays provided six troops, which embarked at Ramsgate on 23 July. But the plans had escalated, and finally some 40,000 troops, escorted by a fleet of 37 ships of the line, 23 frigates and more than 100 sloops, sailed from the Downs in two divisions on 28 and 29 July, arriving at the mouth of the Scheldt on 1 August. The soldiers were under the command of the Earl of Chatham, the younger Pitt's elder brother, and the navy under Sir Richard Strachan.

Chatham decided that he could not risk sailing up the Scheldt to make a direct assault on Antwerp, and landed his army on the island of Walcheren in order to capture the fortress of Flushing, which guarded the approaches to the river. After being subjected to a heavy bombardment, Flushing surrendered on 16 August, but had held out long enough to give the French time to bring up strong reinforcements into Antwerp. By now, too, Walcheren fever was decimating the British regiments, helped by the appalling sanitation and bad water taken from the canals and shallow wells. The troops were dying by the hundred. There was not the best of communication between the naval and military commanders:

> The Earl of Chatham with sword drawn,
> Was waiting for Sir Richard Strachan;
> Sir Richard, eager to be at 'em,
> Was waiting for the Earl of Chatham.

But at least it was soon realised that there was little for cavalry to do among the polders and waterways of Southern Holland, and on the 4 September the six troops of The Queen's Bays landed back at Ramsgate, and joined the rest of the regiment at Canterbury on the same day.[2, 5]

In April 1810 the House of Commons had ordered one of its number, Sir Francis Burdett, Member for Westminster, to be sent to the Tower. There were widespread disturbances throughout the capital and in May The Queen's Bays were ordered to London in aid of the civil power, while the government enforced the authority of Parliament and had Burdett committed to the Tower under military escort. With calm restored, the regiment returned to Canterbury on 27 May and it was more than a year before they had reason to return to the metropolis – this time to be reviewed by the Prince Regent on Wimbledon Common on 10 June. There were some 20,000 troops on parade, including militia and volunteers; the other cavalry regiments included the Household Cavalry, the 3rd Dragoons, 10th Hussars, 12th Light Dragoons, 15th Hussars, and the 18th Hussars.[2, 5]

In 1809 Sir William Augustus Pitt died, and by a commission of 14 January 1810 Lord Heathfield was appointed colonel of The King's Dragoon Guards in his place. At this time the colonel of The King's Dragoon Guards carried out the court duty of being Gold Stick in Waiting to the Sovereign, since both colonels of the Life Guards were away on staff duties. The honour had hitherto been restricted to the colonels of the two regiments of Life Guards, but was later extended to the colonel of the Royal Horse Guards. On 19 May 1810 the regiment left Scotland, embarking at Port Patrick for Ireland, where it landed at Drogheda without incident. Regimental headquarters were at Lisburn until 28 June, when the regiment marched to Dundalk. It was reviewed by Major General Mitchell at Dundalk on 19 October 1810 and on 2 November 1811, before moving to Dublin on 9 December.

At the end of 1811 and during 1812 there were alterations to both the uniforms and horse appointments of The King's Dragoon Guards and The Queen's Bays.

Cocked hats were replaced with brass helmets with a black leather skull piece and a metal crest and plate bearing the name of the regiment with a reverse G.R. cypher. The helmet was surmounted by a long black horsehair crest, with a flowing tail, and in front a small tuft. This was the helmet that was worn by the KDG at Waterloo. A new jacket was issued, which did away with the bars of lace across the front, replacing them with broad lace on the collar and down the front, and around the skirt and cuffs. Buttons were dispensed with, the jacket being fastened by hooks and eyes. Parade dress consisted of white breeches and jackboots, but on active service cloth pantaloons and short boots were adopted (one pair of boots to be issued to each man every 3½ years). A narrow sword belt, fastened round the waist and without a bayonet frog, replaced the old broad belt, and short gloves replaced gauntlets. Officers were to have red cloaks and valises. All horse appointments were to be of brown leather, and the saddle was to have a divided pad. [1, 2, 3, 4, 5, 6]

In March 1812 there were changes to the establishment of The King's Dragoon Guards. The twelve troop quartermasters were replaced by a single regimental quartermaster and twelve troop sergeant majors, and at the same time a schoolmaster sergeant was added to the strength. The establishment of the KDG was now: 1 colonel; 1 lieutenant colonel; 2 majors; 12 captains; 12 lieutenants; 12 cornets; 1 paymaster; 1 adjutant; 1 quartermaster; 1 surgeon; 2 assistant surgeons; 1 veterinary; 1 regimental sergeant major; 12 troop sergeant majors; 1 pay sergeant; 1 saddler sergeant; 1 armourer sergeant; 1 schoolmaster sergeant; 48 sergeants; 12 trumpeters; 960 rank and file; 916 horses. TOTAL STRENGTH 47 officers, 1,037 rank and file. [1, 3, 4]

The King's Dragoon Guards were reviewed in Dublin on 12 June 1812 by Major General O'Loughlin, and they then marched on 7 September to Clonmel, where they were inspected by Major General Lee on 14 October. Whilst at Clonmel the regiment was pestered by thieves, and during April 1813 the post boy, carrying the Royal Mail from Clonmel to Limerick, was held up and robbed on two successive nights. As a result the postmaster applied for a mounted escort, and on the 27 April Privates Joseph Englefield and Abraham Cook of the regiment were ordered to escort the mails and 'succeeded in securing two men, part of a desperate and well-armed gang, who had long infested the neighbourhood, for which service the Dragoons received the thanks of the Magistrates and a reward of £10 each'. [1, 3, 4]

The Duke of York decided in 1813 to form four troops of military police, two of which were to serve in Britain and two with the Duke of Wellington in the Peninsula. All ranks were to be of exemplary character and would receive extra pay, a corporal getting 8d a day, a private 6d. On 28 May 1813 The King's Dragoon Guards provided one corporal, seventeen privates and eighteen horses for the new Cavalry Staff Corps, which embarked at Cork 'for a particular service in the Peninsula'. The two troops raised for service at home drew a further sixty-eight men from the KDG and seventy-six from The Queen's Bays. The other regiments providing men were the 7th Dragoon Guards, the 2nd Dragoons and the 7th Light Dragoons in England, and the 6th Dragoon Guards, 6th Dragoons and the 13th Light Dragoons in Ireland.[1, 3, 7, 8]

During 1812 and 1813 The Queen's Bays were actively employed in Yorkshire, Lancashire and Leicestershire in aid of the civil power. Throughout the industrial areas the Luddites had been destroying property and newly installed machinery in a vain attempt to halt progress and preserve jobs. It was a period of general unrest caused by the continual years of warfare and the distressed economic conditions. The troops were used to preserve property and to maintain peace. Throughout 1813, meanwhile, The King's Dragoon Guards remained at Clonmel, and on 4 October were again inspected by Major General Lee. In January Lord Heathfield had died and Lieutenant General Sir David Dundas was appointed colonel of The King's Dragoon Guards in his stead. Dundas had a long and varied career, being the author of an officially adopted book of drill, a strong advocate for the building of Martello towers, Quartermaster General from 1796 to 1803, and for two years Commander in Chief from 1809. In December 1813 The Queen's Bays were warned for service in the Peninsula with the Duke of Wellington, marching to transports moored at Deal and Ramsgate, but before they could embark the order was countermanded as a result of Wellington's success in driving the French back into France.[2, 3, 5, 10, 11]

The year 1814 saw the defeat of Napoleon and his exile to the island of Elba. The Bourbon family was restored to the throne of France and when Louis XVIII embarked for France from Dover on 20 April, he was escorted by two squadrons of The Queen's Bays. In June the regiment was stationed near London and took part in a grand review on 20 June in Hyde Park, when the troops were inspected by the Prince Regent, accompanied by the Tsar of Russia, the King of Prussia, and a retinue of foreign and British princes and nobles. A month later The Queen's Bays marched for Scotland. With the coming of peace the inevitable reductions in strength began. The Queen's Bays were reduced to eight troops, while the King's Dragoon Guards in Ireland were reduced to ten, with Captain Dawson's and Captain Bernard's troops being disbanded and the two junior lieutenants, Hawley and Brander, reduced to cornets, but allowed to remain in the regiment as lieutenants. On 30 August 2 corporals and 137 privates were discharged from the KDG, and on 28 September a further 12 sergeants, 3 corporals and 160 men were given their discharge, to be followed on 26 October by another sergeant. The total strength of the regiment was now 726 all ranks, with 565 troop horses. On 19

September The King's Dragoon Guards started their march to various Irish stations prior to embarkation for England. Headquarters went to Cork, two troops came from Fethard, one going to Banda, the other to Fermoy, another went from Carrick to Clogheen, and the last from Gort to Limerick. On 1 November the headquarters troop was inspected at Cork by Major General Forbes. In mid-November seven troops embarked at Cork, landing at Bristol on the 24th. On the next two days four troops marched for Coventry, two for Leicester, and one for Warwick. The remaining three troops of the regiment left Cork early in January 1815, landing at Bristol on the 12th and 13th, and marching, two troops for Northampton and one for Warwick. On 20 January 1815 the King's Dragoon Guards received five new standards and standard belts, the gift of the late Lord Heathfield, their deceased colonel, but the old standards had to be returned to Lord Heathfield's executors.[1, 2, 3, 4, 5]

Peace was short-lived. Napoleon escaped from Elba, landed in the south of France and began his triumphant return to Paris. The French troops deserted their monarch wholesale, returning to serve under the imperial eagles. Napoleon reached Paris by 20 March, Louis XVIII fled to Holland, and the nations of Europe again declared war on their old adversary. Great Britain, Russia, Austria and Prussia pledged to put 150,000 troops each into the field. Command of the British Army was given to the Duke of Wellington, but with the superb Peninsular Army now disbanded and many of its most experienced troops still engaged in America, it became a matter of putting together what troops were available.

The Queen's Bays immediately had their strength increased to ten troops, but remained in Scotland. The King's Dragoon Guards were increased to a strength of twelve troops, totalling 1,148 officers and men. In early April 27 officers and 505 men, with 537 horses, were ordered to the Low Countries under command of Lieutenant Colonel William Fuller, to join the British army assembling around Brussels. Eight troops marched and embarked at Tilbury, Gravesend and Purfleet on 16 April. After a night spent lying off Southend and another in Margate Roads, they arrived at Ostend on the 19th, but were not able to disembark until the 20th for lack of sufficient depth of water. The regiment marched by easy stages to Ghent, where it remained for a few days before proceeding to billets in the Dender valley. Colonel Fuller, who had been attending to affairs in London, joined the regiment on 27 April. Regimental headquarters was established at St Levens Asche with troops billeted around at Eygam, Liederkerke, Nyderhasselt, Aloste and Denderleur.[1, 2, 3, 4, 5, 12, 13]

The King's Dragoon Guards were brigaded in the Household Cavalry Brigade under the command of Major General Lord Edward Somerset. The brigade was made up of two squadrons each of the 1st and 2nd Life Guards and the Royal

Horse Guards, with the four squadrons of the KDG. Sir Henry Torrens, Military Secretary, in a letter to the Duke of Wellington of 21 April, wrote, 'Respecting the inefficiency in numbers of the Household Cavalry Brigade, four Squadrons of the 1st Dragoon Guards have been ordered to be attached to it.' So, of the 1,349 sabres of the brigade mustered at Waterloo, almost forty per cent of the total, 530 sabres, were KDG.[13]

Mid-April to mid-June was spent in the Dender valley between regimental duties and sightseeing. There were a succession of parades and watering orders, with field days on 20 May and on 4, 9 and 15 June. The first flogging since the regiment left Clonmel took place early in May, and two privates of Captain Sweny's troop were court martialled on 7 May. There were two more punishment parades on 12 and 14 June. Lieutenant Hibbert wrote:

> I got off an unpleasant duty, which is seeing four men flogged tomorrow for getting drunk on duty. One of them in a frolick loaded his carbine with three ball cartridges, one on the other, and very deliberately shot at the other three who were before in order, as he said, to let them hear the noise the balls made in passing. It was lucky that none of them was killed.[11]

Early in May a final inspection of arms, clothing and accoutrements took place. All deficiencies were made good, and old and excess kit was taken into store, packed, baled and sent back to Ostend, whilst the heavy baggage was stored at Ghent. 'The horses were living in clover, for their racks and mangers were full of it, and their stalls of clean straw up to their bellies, though this bounty was in some measure repaid by the manure, which was so valuable.' The unsatisfactory heavy cavalry swords were ordered to be ground 'to a point'. Private Charles Stanley wrote from 'Brusels Flemish Flanders' as 'Privert King's Dragoon Guards' to his cousin:

> Dear Couson,
>   I take this oppetunety of riting to you hoping this will find you all in good helth as it leaves me at present — I thank God for it — I have had a very ruf march since I saw you at Booton — We are onley 15 miles from Mr Boney Part Harmey wish we expect to have a rap at him exerry day — We have the most Cavilrey of the English that ever was none at one time and in gud condishon and gud sperrits — We have lost a few horses by hour marshing — I have the pleasure to say my horse is better every day whish I think im to be the best friend I have at presant — There is no doubt of us beting the confounded rascal — it ma cost me my life and a meaney

more that will onley be the forting of war — my life I set no store by at all
— this is the finiest countrey exer is so far before England — the peepel is so
sivel.

We have one good thing cheep — that is tabaco and everrything a
Corduley Tabaco is 4d per lb — Gin is 1s 8d per galland that is 2½ per
quart — and everrything in perposion — hour alounse per day is one pound
of beef and pound and half of bred — half a pint of gin — But the worst of
all we dont get it regeler and if we dont get it the day it is due we luse it —
wish is often the case I assure you — I hope you will never think of being a
soldier — I assure you it is a verry ruf consern — I have rote to my sister
Ann and I ham afraid she thinks the trubel to mush to answer.

I have not ad the pleasure of liing in a bed since in the cuntrey — thank
God the weather is fine wish is in hour faver — we get no pay at all onley
hour bed and mete and gin — we have had 10d per day soped from us wish
we shal reseive wen six months is expiered — I thank God I have a frend
with me.

I hope you will excuse my bad inditing and spelling.

Private Charles Stanley was killed in action at Waterloo. Pay for the men was
received on 5 June, with more coming on the 11th, which made up the men's pay
to 24 June.[13]

There were a number of reviews: the Earl of Uxbridge, commanding the British
cavalry, inspected the regiment on 6 May; on 24 May the Prince of Orange
inspected the two British brigades of heavy cavalry at Heldinghem, and following
this The King's Dragoon Guards were moved into billets around Ninove. On 29
May the Duke of Wellington and Marshal Blücher, accompanied by the Duc de
Berry and a whole train of lesser personages, inspected the whole of the British
cavalry near Grammont in the meadows by the River Dender, between the villages
of Jedeghem and Schendelbeke. The first line was composed of hussars, with a
battery of 9-pounders on each flank; in the second was the heavy cavalry, with a
battery of 24-pounder howitzers and 9-pounders in the centre, and another battery
of 9-pounders on each flank. The third line was composed of the light dragoons
with a 9-pounder battery on each flank.

The day was lovely and it was a splendid spectacle. The scattered lines of
Hussars in their fanciful, yet picturesque costumes; the more sober, but far
more imposing, line of heavy Dragoons, like a wall of red brick; and
again the serviceable and active appearance of the third line in their blue
uniforms, with broad lappels of white, buff, red, yellow, and orange — the
whole backed by the dark woods — formed indeed a fine picture.[1, 14]

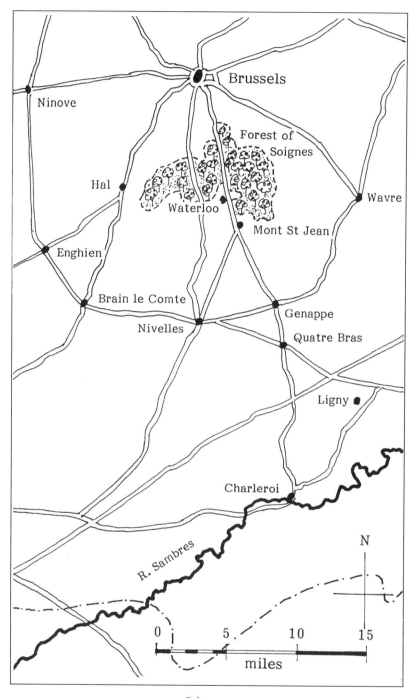

Belgium

Arriving on the ground covered with dust, the different corps had no sooner formed in their position, and dismounted, than off went belts, canteens and haversacks, and a general brushing and scrubbing commenced – for the Duke, making no allowance for dusty or muddy roads, expected to see all as clean as if just turned out: accordingly, we had not only brought brushes etc, but even straw to wisp over the horses. The whole line was in the midst of this business, many of the men even jackets off – when suddenly a forest of plumes and a galaxy of brilliant uniforms came galloping down the slope from Schendelbeke towards the temporary bridge. 'The Duke! the Duke! the Duke's coming!', rang along the lines, and for a moment caused considerable bustle among the people; but almost immediately this was discovered to be a mistake, and the brushing and cleaning recommenced with more devotion than ever; whilst the cavalcade, after slowly descending to the bridge and debouching on the meadows, started at full gallop toward the saluting point already marked out, the Duc de Berry, whom we now recognised, keeping several yards ahead, no doubt that he might clearly be seen. At this point he reined up and looked haughtily and impatiently around him; and as we were now pretty intimate with his manner it was easy to see that he was in a passion. The brushing, however, suffered no interruption, and no notice was taken of his presence. One of his suite was now called up and despatched to the front. The messenger no sooner returned than his Highness was off like a comet, his tail streaming after him all the way up the slope, unable to keep pace with him, for he rode like a madman, whilst a general titter pervaded our lines as the report flew from one to the other that Monnseer was off in a huff because we did not give him a general salute.[14, 15]

Wellington and Blücher arrived at one o'clock, and were greeted with a salute of nineteen guns. John Hibbert, KDG, wrote to his sister:

You may conceive what a sight it was when a line composed of sixteen regiments of cavalry, and all in the most beautiful condition imaginable. Lord Wellington did not dismiss us until six, so we had had quite enough of it by the time we got home, which was about nine. We were altogether about fifteen hours on horseback, and nothing to eat or drink; added to this it was the hottest day I ever felt and many men fainted in the ranks. The only accident we had, was in my Troop as we were on the march there. One of the privates being in a rage with his horse for not walking, hit him such a violent blow with his fist on the head that he broke his own arm. He deserved it most richly, and I told him by way of comfort that I was truly glad at what had happened.[11]

'We were comfortably situated in Flanders, in good quarters wanting for nothing, till the morning of June 16th, when at daylight we received a sudden order to march.' So wrote Troop Sergeant Major Page, KDG, in a letter home dated 3 July 1815. Naylor recorded: 'Before daybreak received an order to assemble at Ninove with all expedition.' The heavy cavalry under Lord Uxbridge were ordered to march on Enghien, a few miles to the west of Quatre Bras, but the orders did not reach the KDG until 3 a.m. on the 16th, and it took some time to concentrate the troops, scattered as they were in the various villages and hamlets. The regiment was paraded by 8 a.m. at Ninove, where it waited for an hour before moving off with the rest of Lord Edward Somerset's brigade.[12, 13, 14]

Progress was slow due to the many columns of cavalry converging, but Braine le Comte was reached by 4 p.m., where an hour's rest was granted for watering and food. The regiment pressed on towards Nivelles, and as they emerged from a forest, Naylor remembered, 'We heard a cannonade, and at times could distinctly see a smoke at some distance.' At once an order was given to advance at a brisk trot. So as to lighten their horses, the men untied the nets containing hay, and opened the mouths of the feed bags, 'which falling from them as they trotted on, the road was soon covered with hay and oats'. As the brigade entered Nivelles, the noise

> became more distinct, and its character no longer questionable – heavy firing of cannon and musketry, which could now be distinguished from each other plainly. We could also hear the musketry in volleys and independent firing. In the direction of the cannonade, volumes of grey smoke arose, leaving no doubt what was going on. The object of our march was now evident, and we commenced descending the long slope with an animation we had not felt before. It was now 7 p.m. and the brigade pressed on, 'expecting every moment to enter on the field of action'. Dusk began to close in as they trotted through the village of Hautain Le Val, and beyond it reached the edge of the battlefield after a march of nearly fifty miles.[12, 13, 14, 15]

SOURCES

1 R. Cannon, *Historical Records of the British Army, 1st Dragoon Guards*, 1837.
2 R. Cannon, *Historical Records of the British Army, 2nd Dragoon Guards*, 1837.

3 Records of The King's Dragoon Guards, Regimental Museum, 1st The Queen's Dragoon Guards, Cardiff Castle.

4 William Clowes, *The King's Dragoon Guards 1685-1820*, 1820.

5 F. Whyte and A. H. Atteridge, *The Queen's Bays 1685-1929*, vol. 1, 1930.

6 Captain R. G. Hollies Smith, 'A Short History of the Uniform of The King's Dragoon Guards', *The K. D. G.*, vol. 5, no. 5, 1958, p. 302.

7 Sir John Fortescue, *History of the British Army*, vol. 9, 1920.

8 *Journal of the Society for Army Historical Research*, vol. 47, 1969, p. 33.

9 *The Regimental Journal of 1st The Queen's Dragoon Guards*, vol. 1, no. 5, 1963, p. 339.

10 *Journal of the Society for Army Historical Research*, vol. 66, 1988, p. 67.

11 Lieutenant John Hibbert, KDG, MSS letters, Regimental Museum, 1st The Queen's Dragoon Guards, Cardiff Castle.

12 Captain James Naylor, KDG, MS diary, Regimental Museum, 1st The Queen's Dragoon Guards, Cardiff Castle.

13 Michael Mann, *And They Rode On*, 1984.

14 Cavalié Mercer, *Journal of the Waterloo Campaign*, 1927.

15 *The K. D. G.*, vol. 2, no. 5, 1936.

18

Quatre Bras, Retreat to Waterloo, Waterloo
1815

As The King's Dragoon Guards arrived at Quatre Bras, the tired horses stumbled from time to time over the corpses lying along the road and over the fields. It was 8 p.m. and the regiment formed a close column beside the road, before moving to bivouac in an open field of trodden down wheat, just behind the farmhouse of Quatre Bras. The horses were picketed and linked in column, saddled and bridled; it was a fine summer night and Sergeant Major Page noted, 'We marched this day 40 English miles and slept in the open corn fields, our horses being saddled ready to mount at a minute's notice, the French being in a wood close by us.'[1, 2]

As the troopers dismounted, they were sent to the farm well to draw water for the horses, but the crush was so great that the task took several hours. The horses were fed with what grain could be gathered from the field. The officers and men either stood by their horses or, lying down, wrapped themselves up in their long cavalry cloaks and tried to snatch some sleep. Page remembered that 'on the morning of the 17th at daybreak firing again commenced. So far it is what we call skirmishing.'

Soon after dawn the Duke of Wellington arrived, consulted Vivian, who commanded the hussar brigade providing the forward pickets, and sent out a patrol of the 10th Hussars to contact the Prussians. Wellington then dealt with some recently arrived despatches from England, lay down beside the crossroads with one of the despatch sheets across his face, and went to sleep. Waking up at 9 a.m., he remarked on the inactivity of the French, and then spoke to Von Massow, who had arrived with messages from General Gneisenau and Blücher. The Prussians had been badly mauled at the Battle of Ligny the day before, and in order to conform with their withdrawal, the Duke decided to fall back to the position he had earmarked at Waterloo. The infantry were ordered to march at once, while the cavalry with the horse artillery and some light troops were to cover their withdrawal. At the same time orders were despatched to Lord Hill, commanding the 2nd Corps, to march with the 2nd and 4th Divisions from Nivelles direct to Waterloo. By 11.30 a.m. the last of the infantry had disappeared from sight.

The King's Dragoon Guards had been standing to since daylight, when at 8 a.m. they were ordered to water their horses a little way to the rear. They then resumed their position. It was nearly two o'clock before a mass of the enemy were to be seen about two miles away. Lord Uxbridge pointed out to Wellington that with defiles in their rear and the infantry now too far off to be able to offer effective support, they were not in a good position. The Duke agreed and the order was given to retreat in three columns. The King's Dragoon Guards were with the centre column, and as they steadily retired down the Brussels road, they halted from time to time and formed up on either side of the road. The French ignored the two flank columns and concentrated their attention on the centre.[1]

Page recalled that as they moved off,

> There was one of the heaviest storms of rain ever known, accompanied by thunder and lightning. The fall of rain was so very heavy that in the fields, which were covered with corn, our horses sunk in every step up to near the hock. It is out of my power herein to express our situation – our boots were filled with water, and as our arms hung down by our sides the water ran off a stream at our finger ends.

Captain Naylor commented, 'We experienced the most severe fall of rain I ever beheld.'[1, 2, 3]

The two heavy brigades had moved off at the head of the centre column, and as they reached the narrow winding street through Genappe, there was not a soul to be seen; the windows were all shuttered, and the water cascaded from the roofs to rush in a torrent down the gutter in the middle of the road. The only other sound was the ring of the horses' shoes on the cobbles. Lord Uxbridge halted the two heavy brigades along the ridge which runs some 700 yards up the slope leading out of Genappe; the Household Brigade was on the left of the road facing the French advance, whilst the Union Brigade formed up on the right. Mercer, with his battery of horse artillery, described the scene:

> We suddenly came in sight of the main body of our cavalry drawn up across the chausee in two lines, and extending far away to the right and left of it. It would have been an imposing spectacle at any time, but just now appeared to me magnificent, and I hailed it with complacency, for here I thought our fox chase must end. Those superb Life Guards and Blues will soon teach our pursuers a little modesty! Such fellows! – surely nothing can withstand them.[1, 4]

Eighteen squadrons of French cavalry were now entering Genappe, and as their lancers came through the town, they halted for about fifteen minutes, facing the

British heavies. Those behind could not see that the front ranks had halted, and as they pressed on, the whole mass of French cavalry became jammed between the houses. Uxbridge, seeing their indecision, ordered the 7th Hussars to charge, which they did with great spirit, but with little effect on the dense masses. The French then advanced and drove back the 7th, and the contest relapsed into a seesaw. At this point the 1st Life Guards were ordered to charge. Kincaid of the 95th Rifles saw what happened:

> It did one's heart good to see how cordially the Life Guards went at their work; they had no idea of anything but straightforward fighting, and sent their opponents flying in all directions. The only thing they showed was in everyone who got a roll in the mud (and owing to the slipperiness of the ground there were many), going off to the rear, according to their Hyde Park custom, and being fit no longer to remain on parade![1, 5]

As the Life Guards charged, The King's Dragoon Guards formed up behind them as a second line; but their services were not needed. This left the KDG forming the rearguard as the British cavalry fell back onto the position at Mont St Jean, in front of Waterloo. The left half-squadron of The King's Dragoon Guards was in a sharp conflict with the French advance guard. Sergeant Major Page, in a letter of 12 January 1816, wrote how he 'took a horse on 17th June with a Frenchman's complete kit of arms, saddle bags etc on him. He had one of his ears nearly cut off. I gave him to a farrier to take care of for me while I was skirmishing with the French Dragoons, but he lost altogether.' After Genappe the French kept their distance, and although on two or three occasions they made as if to attack, they never pressed forward. So the retreat continued at a slow pace, the KDG retiring by alternate squadrons and with 'perfect regularity'.[1, 2]

As the column crossed the valley and climbed the slopes of Mont St Jean, Napoleon appeared on the ridge behind, mounted on his grey mare, Desirée. He at once supervised the placing of his guns as battery after battery, accompanying the French advance guard, arrived on the crest. All the time he cried out to the gunners, 'Fire! Fire! These are the English.' But the shot went over the heads of the cavalry as they climbed the long slope and wound their way over the ridge, dropping behind the skyline. At last the weary troopers dismounted and bivouacked near the farm of Mont St Jean. Uxbridge summed up the retreat, 'Thus ended the prettiest Field Day of Cavalry and Horse Artillery that I ever witnessed.'[1]

Trumpeter Samuel Wheeler of The King's Dragoon Guards put into verse his feelings of the past two days:

On the 16th June my boys that was the very Day
When we Received Orders for to March Away
To face the tyrants Army My Lads then we was bound
That on the plains of Waterloo Encamped was Around.

We espied our foes Next Morning As in a wood they lay
And like Britons we Advanced to Show them British Play
Our Grape Shot flew Among them to Put them to the Route
But still those Cowardly Raskels Refused to Come Out.

When Wellington Saw their Cowardlyness he Ordered a Retreat
Which Order Was Complied With His Design It was Compleat
We Retired through the Village of Geenap As you Soon Shall Hear
Followed by A Large Collum of the Enemies Lancers.

But In Our Retreat the Horrid thunder began to Roar
And Rain Like unto Rivuletts upon the Ground did Pour
But our brave English heroes Endured both heat and Cold
And Caused Our foe to Rue the Day the truth I soon unfold.

The first that Charged Was the Lifeguards the Enemy to Subdue
They Charged a Collum of the Lancers And Caused them to Rue
Till Half An Hours hard fighting those heroes Did Endure
And left 3 hundred Lancers A bleeding in their Gore.

When they Returned from their Work Our Regt. Was Called Out
For to face those french Dogs And Put them to the Route
But the Noble Earl of Uxbridge Some Danger Did Espie
Kings Dr Guards threes About he Loudly then Did Cry.

The Cunning french 3 field Pieces had Placed in the town
Thinking As we Advanced to Cut Our heroes Down
But Our brave Commander Soon Ordered us Away
And then brought up Some English Guns And on them began to Play.

Then We Again Retired and Inticed them on the Plain
And Gained a Good Position Wich we was Determined to Maintain
But the Night being fast Advancing we Could no Longer see
And Neither of the Armies Could Claim the Victory.[6]

The rain eased up during the early evening, but it returned with darkness and continued all night. Page wrote later, 'We remained in this situation the whole of the night halfway up to our knees in mud. Firing commenced the next morning, viz the 18th, at daybreak which made the third day. What seemed worst of all during these three days, we could draw no rations, consequently we were three days without anything to eat or drink.' The horses moved constantly to present their backs to the rain, and as the men moved around to attend to them, and to try to light fires, the whole area soon became a morass. The only rations were what individual troopers had managed to save, though some men dug up potatoes from a nearby field. Such fires as were lit were made with green wood, which gave off more smoke than heat, as the men huddled together for warmth. Some slept standing, others remounted and, wrapping themselves in their long cavalry cloaks, tried to sleep bending forward on their horses' necks.

The men were aroused before daybreak, and 'we began to get dry, and as the rain ceased we wrung out our clothes, put them on again, and very few of us have pulled them off since'. The men set to grooming their horses and cleaning their equipment, and by 6 a.m. the various regiments of the Household Brigade were assembled in brigade mass on their bivouac ground. They were soon moved forward to form a second line 200 yards behind the infantry, who were lying just below the ridge of Mont St Jean.[1, 2]

By 8 a.m. The King's Dragoon Guards, who mustered some 530 sabres, were formed up in the centre of the front rank of the Household Brigade, with two squadrons of the 1st Life Guards on their right and two of the 2nd Life Guards on their left. The two squadrons of the Blues were behind in reserve. The brigade was posted on the right of the Brussels-Charleroi road, with the Union Brigade formed up level with them on the left of the road. Lord Edward Somerset, commanding the brigade, sent one subaltern from each of the four regiments to ride forward to the crest of the ridge to observe and report to him on the French movements.[1]

Captain Naylor noted: 'At twelve a general cannonade commenced, by which we experienced some loss.' The men were ordered to dismount and lie on the ground beside their horses, so as to avoid the worst of the cannon fire. Sergeant Major Page commented,

> We lost many men and horses by the cannon of the enemy. While covering the infantry we were sometimes dismounted to rest the horses, and also when we were in low ground, so that the shot from the French might fly over our heads. Whilst in this situation, I stood leaning with my arm over my mare's neck when a large shot struck a horse by the side of mine,

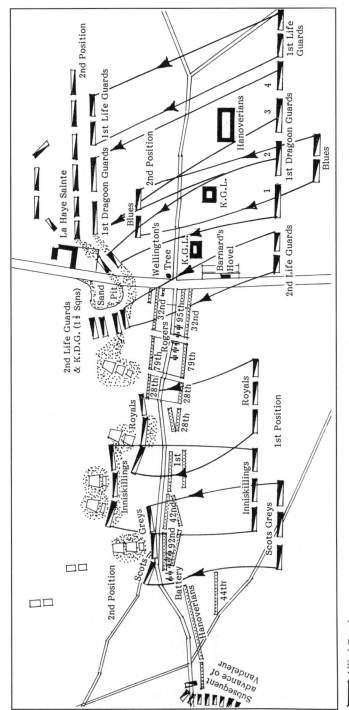

Battle of Waterloo: Daybreak until 3 p.m.

Allied Cavalry
French Cavalry
Allied Infantry in square
French Infantry en masse
Allied Infantry in line

2nd Position

2nd Position

1st Life Guards
1st Dragoon Guards
1st Life Guards

2nd Position

1st Dragoon Guards
Blues

Hanoverians

K.G.L.

Blues

La Haye Sainte

2nd Position

Wellington's Tree
K.G.L.
Barnard's Hovel

2nd Life Guards

Sand Pit

2nd Life Guards
& K.D.G. (1½ Sqns)

32nd
Rogers
95th
32nd

79th
79th

28th
28th

Royals
Inniskillings

1st

Scots Greys
42nd
92nd

Battery
Hanoverians
44th

Royals

Inniskillings

Scots Greys

1st Position

2nd Position

Subsequent advance of Vandeleur

killed him on the spot and knocked me and my mare nearly down, but it did us no injury.[1, 2, 3]

Napoleon had formed a simple plan of battle: he would open by attacking the right of the British position at the farm of Hougoumont, hoping to draw off Wellington's reserves to its support. The main assault, supported by a battery of eighty guns, would smash through the centre of the Allied position on either side of the Brussels road. The assault on Hougoumont was pressed by Jerome, Napoleon's brother, but the steadfast resistance of the Guards persistently foiled each attack, and more and more of the French reserves began to be drawn into what had only been intended as a spoiling attack.

At 1.30 p.m., in the centre, d'Erlon's corps of four divisions, comprising some 16,000 men, started to advance across the valley towards the left centre of the British position. The Emperor ordered Milhaud to support this attack with Dubois's brigade from the 13th Cavalry Division of the 4th Heavy Cavalry Corps. This comprised 1st and 4th Cuirassiers, who moved off at a trot, crossing the Brussels road. A hundred yards south of the farm of La Haye Sainte they formed into line, broke into a gallop, and caught a battalion of Hanoverians from Kielmansegge's brigade, who had been sent to reinforce La Haye Sainte, and wiped them out. Nine squadrons came past the farm on the right, and two on the left, rejoining as they passed by, to sweep on up the slope towards the very centre of the British line. 'The advance of the enemy's boasted invincible Cuirassiers was particularly imposing. Our first line was somewhat shaken, as this immense body of French Cuirassiers was advancing to force the centre of the position.'[1, 7]

On the right of the cuirassiers came d'Erlon's infantry. As they advanced, a Dutch-Belgian brigade, who had received a tremendous pounding from the French cannon, broke and fled before them, creating a gap in the Allied line which was quickly plugged by Kempt's 8th British Brigade. The whole weight of the French assault now fell on Picton's 3rd Division, and it seemed as though it was up to these 4,000 veteran British infantry to halt the 16,000 of d'Erlon's Corps, and the eleven squadrons of Dubois's cuirassiers, as they swarmed up and over the crest of the ridge.

The four subalterns of the Household Brigade, posted on the ridge, had reported back to Lord Edward Somerset on the French advance, and he had deployed his brigade into line. Uxbridge ordered Somerset to charge the cuirassiers, and then rode over to Ponsonby, in command of the Union Brigade, and ordered him to charge d'Erlon's infantry as soon as he saw Somerset move.[1]

Captain Naylor wrote, 'We deployed and, I think, about two o'clock a charge was made by the Heavy Brigade through a line of Cuirassiers and a reserve of Lancers.' Sergeant Major Page described the action:

> At this time the French seemed determined to get possession of a piece of ground where part of our line was drawn up: accordingly they brought up very heavy columns of infantry, and strong bodies of heavy cavalry, and our Brigade was ordered to form line immediately. Now comes the most bloody scene ever known – the French infantry and cavalry came boldly into the bottom of a very large field while we formed at the other end; they charged our infantry and as soon as they showed themselves to our front the word 'Charge' was given for our Brigade by Colonel Fuller.[1, 2, 3]

The brigade deployed outwards at 2.20 p.m., moving by threes to left and right, which took the left flank across the Brussels road. The KDG in the centre, and the 1st Life Guards on the right, wheeled left by threes and moved off at once. Owing to the urgency of the situation the 2nd Life Guards were still wheeling by threes to their right after the rest had started to advance.[1]

The King's Dragoon Guards and the 1st Life Guards descended into a sunken road athwart the ridge, crossed it, and scrambled up the bank opposite. On reaching the top they checked for a moment to steady the line, and charged. As they galloped forward, their right flank became advanced, so that the 1st and 4th Cuirassiers were struck obliquely.

> The Brigade and the Cuirassiers came to the shock like two walls, in the most perfect lines. I believe this line was maintained throughout. A short struggle enabled us to break through them, notwithstanding the great disadvantage arising from our swords, which were full six inches shorter than those of the Cuirassiers, besides it being the custom of our Service to carry the swords in a very bad position whilst charging, the French carrying theirs in a manner much less fatiguing, and also better for either attack or defence. Having once penetrated their line, we rode over everything opposed to us.[1, 8]

Uxbridge's charge, perfectly timed, hit the French at their moment of wavering, with their horses blown and winded by the long advance over slippery ground – much of which was uphill ploughland, into which the horses sank up to their knees. The opposing lines met with a crash, and the superior weight of the British heavies, both in men and horses, together with the advantage of the downhill slope, overthrew the French. Kincaid witnessed the scene:

> The next moment the Cuirassiers were charged over by our Household

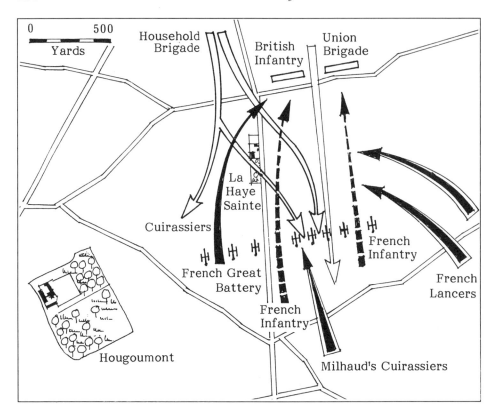

Battle of Waterloo: 3 p.m.

Brigade; and the infantry in our front giving way at the same time, under
our terrific shower of musketry, the flying Cuirassiers tumbled in among
the routed infantry, followed by the heavies, who were cutting away
in all directions. Hundreds of the infantry threw themselves down, and
pretended to be dead, while the cavalry galloped over them, and then got
up and ran away. I never saw such a scene in all my life.

Sergeant Major Cotton of the 7th Hussars also gave a graphic description: 'The
two cavalries dashed into each other: the shock was terrific; the swords clashing
upon the casques and cuirasses so that, as Lord E. Somerset humorously observed
to me – "You might have thought it was so many tinkers at work." '[1, 5, 9]

Individual KDG remembered their part. Corporal Stubbings was closely engaged with a cuirassier and narrowly escaped a downward cut from his sabre, which, however, took off the ear of his charger. Many years later Stubbings, then a pensioner, met his old troop horse, still serving with the regiment, when a detachment of the KDG passed through Market Worksop near Nottingham. 'The meeting between the two was most affecting, the detachment halted in front of the "Hare and Hounds", and villagers turned out to a man to see the "one-eared horse" on which John Stubbings rode at Waterloo.'[10]

Page remembered:

> We overturned everything, both infantry and cavalry that came our way, such cutting and hacking never was before seen. When the French lines broke and ran, our Regiment being too eager, followed the French cavalry while the cannon and musketry was sweeping our flank. Many fell and our ranks suffered severely — the Duke of Wellington, with tears, it is said, when he saw us so far advanced among the French, himself said he never saw such a charge, but he was afraid very few of us would return — his words were too true. However, of the 7,000 Frenchmen wearing armour, very few left the field. They were fine men but could not look us in the face, and dreadful was the havoc we made among them.[2]

The 2nd Life Guards on the left of the brigade were the last to form line, and when they came to cross the sunken road running along the ridge of Mont St Jean, they plunged down onto a mass of cuirassiers who had taken refuge there. They were followed by the left-hand squadron of The King's Dragoon Guards, and the whole surging mass of horsemen, cutting and thrusting as they went, poured across the main Brussels road opposite to the farm of La Haye Sainte. This confused welter of cavalry hit the remains of d'Erlon's infantry, which had just been broken by the charge of the Union Brigade on the left of the Household troops.[1]

The farm of La Haye Sainte had acted as a breakwater to the charge of the Household Brigade; most of the King's Dragoon Guards went with the 1st Life Guards, the Blues being in support, to the right of the farm, but at least a squadron and a half veered to the left with the 2nd Life Guards; there they joined the Union Brigade of the Royal Dragoons, Greys and Inniskillings, slaughtering d'Erlon's infantry as they went. 'Intoxicated with slaughter and inciting each other to kill, they pierced and cut down the miserable mass with glee.' Lieutenant Waymouth of the 2nd Life Guards 'more than once, during that advance, finding myself near Major Naylor of the King's Dragoon Guards, and to whom I spoke, leads me to suppose that some part of his Regiment may also have passed to the left' of La Haye Sainte. Naylor himself remembered that, at the advance, Colonel Fuller placed himself alongside and rode with him towards the left 'along with the current of our men, which was setting that way'.[1, 3, 8]

Lord Edward Somerset reported that 'the 2nd Life Guards, on the left of the Brigade, drove a portion of the Cuirassiers across the chaussée to the rear of La Haye Sainte, and down the slope. Here they were joined by the King's Dragoon Guards, who had crossed the road in front of the farm, and the two Regiments becoming mingled with Ponsonby's Cavalry, lost all regularity in the eagerness of the pursuit.' By now the men were out of hand, excited and paying no attention to shouted orders or trumpet calls as they galloped up the other side of the valley and through the Great Battery of the French guns. Several of the cannon were hurled down the valley and some troopers rode as far as the French artillery park, where the drivers, some mere boys, sat crying on their horses in the limbers.[1]

Cotton told how

> The enemy on the opposite heights were employed in destroying such of our cavalry as had ventured too far. In fact most of Ponsonby's Brigade, with a proportion of the Household Brigade, animated by their first success, pursued their advantage too far; they crossed the valley in disorder, and galloped up to the French position in twos and threes, brandishing their swords in defiance, riding along the ridge, sabring the gunners, and rendering about thirty guns useless; the bugles, or trumpets, sounding to rally, were unheeded.[9]

Napoleon now brought up Gobrecht's brigade of Jacquinot's division, consisting of the 3rd and 4th Lancers, who came in from the flank and chased any British cavalryman they could find, while from the centre of the French position the 5th and 10th Cuirassiers of Delort's brigade began to sweep the valley clear of the British stragglers.[1, 9]

John Hibbert records the scene: 'They were met by an immense body of Lancers who were sent for the purpose of attacking them. Our men were rendered desperate by their situation. They were resolved to get out of the scrape or die, rather than be taken prisoners, so they attacked them, and three troops cut their way through them; about a troop were killed or taken prisoner.' Naylor, who had charged down the hill and was still bearing to the left, found himself in a large field along with the French lancers. Turning away from them he managed to make his way back round the left of the Allied line. Captain Clark Kennedy of the Royals saw a small party of The King's Dragoon Guards making their way back. Mistaking them for some of his own men, for the uniform was very similar, he called out, 'Royals, form on me!' Back came the reply, 'We are King's Dragoon Guards – not Royals!' – and they rode on.[1, 3]

As the two and a half squadrons of The King's Dragoon Guards, with the 1st Life Guards and the Blues, re-formed on their original position, stragglers reappeared from the remains of the squadron and a half that had charged to the left of La Haye Sainte. Naylor remembered that 'Turner with about thirty men joined the Brigade'. More survivors trickled in as they made their way back, some in small parties, but mostly in ones and twos.[1, 3]

SOURCES

1 Michael Mann, *And They Rode On*, 1984.
2 *The K. D. G.*, vol. 2, no. 5, 1936.
3 Captain James Naylor, KDG, MS diary, Regimental Museum, 1st The Queen's Dragoon Guards, Cardiff Castle.
4 Cavalié Mercer, *Journal of the Waterloo Campaign*, 1927.
5 Captain Sir John Kincaid, *Adventures in the Rifle Brigade*, 1830.
6 'The Battle of Waterloo, by Trumpeter Samuel Wheeler, KDG', MS in possession of Lieutenant Colonel John Hibbert, Light Infantry.
7 R. Cannon, *Historical Record of the British Army, 1st Dragoon Guards*, 1837.
8 Major General S. T. Siborne, *Waterloo Letters*, 1891.
9 Sergeant Major Cotton, *A Voice from Waterloo*, 1849.
10 MS letters of John Stubbings, KDG, in possession of Ernest Shead.

19

Waterloo

1815

The losses to The King's Dragoon Guards from that first great charge had been heavy. The commanding officer, Lieutenant Colonel Fuller, was missing. John Hibbert commented, 'Poor Colonel Fuller and Major Graham are killed, I am afraid, without doubt, although they are returned missing.' There is confusion as to how Fuller met his death. Somerset thought 'Colonel Fuller must, on arriving at the farm of La Haye Sainte, have turned to his right, for I believe there is no doubt of his having been killed down the slope of our position, to the right of La Haye Sainte.' Another account tells of his dying at the south-west corner of the orchard of La Haye Sainte 'whilst gallantly leading the centre squadron'. A third stated that 'Colonel Fuller of 1st Dragoon Guards, as also Major Graham and Cornet the Hon H. B. Bernard and another officer, were taken prisoners; though their fate remains uncertain there can be no doubt that they were murdered by the enemy like many other brave unfortunate men.' Yet another account claimed 'Colonel Fuller was killed whilst pursuing the Cuirassiers: he boldly led his Regiment up the French height on the allied left of the Charleroi road.' A fifth version said, 'Many officers met their death in striving to rally their soldiers, and like the Colonels of the Scots Greys and the King's Dragoon Guards were slain within the French position.' John Hibbert of the KDG thought, 'In this affair poor Fuller lost his life, his horse was killed by a lance, and the last time he was seen he was unhurt but dismounted. Of course the Lancers overtook him and killed him, for our men were on the full retreat.'

Major Graham, the second in command, was dead, as was Major Bringhurst, who had served with the regiment since 1806. Captain Battersby and the adjutant, Lieutenant Tom Shelver, had been killed, and among the subalterns Lieutenant Brooke and Cornet Bernard, the eighteen-year-old son of Viscount Brandon. Hibbert wrote, 'You may conceive what a slaughter it was when we lost five entire troops out of eight.' The numbers lost in that one charge had reduced the effective strength of The King's Dragoon Guards to three troops.[1, 3]

The celebrated charge of the two heavy cavalry brigades has been much dramatised and criticised - the latter justly so, for there was a loss of discipline and cohesion with disastrous results, especially, within the KDG, among those who had veered to the left of La Haye Sainte. But the effects of the charge cannot be underestimated: Dubois's cuirassiers were annihilated, d'Erlon's whole corps was destroyed and — something that has often been overlooked — a major part of the Great Battery was put out of action for some critical hours. The losses suffered by both heavy brigades were crippling, but their part in the battle was by no means concluded. When Captain Turner, KDG, arrived back with his party, he took over command of the regiment, but a little later he was wounded in the arm by a cannon shot and Captain Naylor assumed command. There was at this point, about 3.30 p.m., a lull in the battle, and Wellington moved his infantry back off the crest and made them lie down. The cavalry was not so fortunate, for it had to continue in support of the infantry and remain mounted. There was, however, a good deal of movement to the rear, as the wounded and prisoners were escorted towards Brussels. The farriers of the KDG were used for this duty.[1, 3]

Marshal Ney was misled by Wellington's action in withdrawing the infantry, which, together with the movement of wounded to the rear, persuaded him that a massive cavalry charge could clear the crest and provide the *coup de grâce*. He accordingly gathered and personally led forty-three squadrons of magnificent cavalry against the centre of the British position.

> When the Cuirassiers had passed over the ridge, they were out of sight of the Lancers and Chasseurs, who immediately passed on to share in the contest. Our artillery received them in like manner; some of the artillery men rushing back to their guns, and after discharging them at the foe, taking shelter again within the squares, or under the guns. The firing produced a much greater effect upon such of the enemy's cavalry as were not protected by the cuirass or casque; consequently their ranks were much more disordered than were the Cuirassiers; still they pursued their onward course, passed the guns, raised a shout and swept round the squares. Some halted and fired their pistols at the officers in the squares; others would ride close up, and either cut at the bayonet or try to lance the outside files. No sooner had the broken squadrons passed the guns, than the gunners were again at their post, and the grape rattled upon the retiring hosts; but frequently, before a succeeding round could be discharged, the hostile cavalry were again upon them, and compelled them to seek shelter.[4]

As the French cavalry came surging forward, the remains of the Household Cavalry Brigade, drawn up in its original position, was given the order to charge,

Lord Uxbridge putting himself at their head. It was now 4.15 p.m. and they fell upon the advancing, and now disordered, cuirassiers, lancers and chasseurs and went for them hammer and tongs. The French horses were blown at the end of their charge, whereas those of the heavies had had time to rest. The French fought bravely. There was a spirited hand to hand cavalry contest, where both horsemanship and skill at arms decided the difference between death and survival. One slashing backhand stroke severed the head of a cuirassier, and both head and helmet were sent flying; the horse galloped on, obedient to the reins, with the headless rider sitting erect in his saddle until finally toppling to the ground. 'We galloped at the Cuirassiers and fairly rode them down; when they were unhorsed, we cracked them like lobsters in their shells.' Captain Wallace, KDG, remembered how most of his men slashed at all and sundry, including those of the enemy lying on the ground. Wallace himself, sparing a dismounted French trumpeter, recalled 'I did not slash at him but the trumpeter slashed at me!' But this second charge brought more casualties: Colonel Ferrier of the Life Guards and Major Packe of the Blues were killed, along with other officers and men.[1, 8]

As the French cavalry repeated their charges against the infantry squares, there developed a see-saw of counter charges by the ever-diminishing numbers of the Household Brigade. In all some eleven or twelve charges were made, with the men being kept well in hand, darting forward at the appropriate moment, then re-forming. But every time more men and horses were killed or wounded, and the brigade grew smaller and smaller. Some horses, having lost their riders, still formed alongside their comrades, and these riderless horses charged with their squadron. Even when the animals were wounded they kept in formation so long as they still had the strength to gallop.[1]

> At one time that memorable afternoon, the ridge and rear slope of our position were literally covered with every description of horsemen, lancers, cuirassiers, carabiniers, horse-grenadiers, light and heavy dragoons and hussars. The menacing approach of the French cavalry, who rode amongst and around our squares, was not quietly witnessed by our own horsemen: we made many spirited charges between and on every side of the allied squares. When at length the enemy's gallant but fruitless efforts became exhausted, our cavalry appeared and cleared the allied position.[9]

At 5.30 p.m. the French succeeded in establishing a battery of guns well forward of La Haye Sainte, which menaced the very centre of the Allied position. The King's Dragoon Guards and the Blues were moved up, supported by the Hanoverian Cumberland Hussars, but the latter began to give way. Uxbridge sent Captain Williams, his ADC, to bring them back, but the entire regiment fled the field and on reaching Brussels spread rumours of an Allied defeat. Lord Edward Somerset led the remaining King's Dragoon Guards and Blues against a new French attack, but their numbers were by now too few to do more than check the enemy's advance. Uxbridge then rode up to a Belgian brigade of heavy cavalry, but they refused to follow him. In the meantime the French attack had managed to capture the farm of La Haye Sainte.[1, 9]

The tattered remains of the Household Cavalry Brigade again re-formed in its original position, but not for long. Uxbridge moved it to the right of the Allied line, where a new threat was posed by a strong column of French infantry, supported by cavalry. The brigade formed and charged this fresh menace, but suffered badly from French musketry fire, even though the charge halted the advance and inflicted many casualties on the opposing infantry. Nevertheless the brigade was now too weak in numbers to be able to penetrate the column or to scatter it.[8]

The few men left had re-formed once more on their first position when Colonel Edward Lygon commanding the 2nd Life Guards had his horse wounded and left the field. The next senior officer was Captain James Naylor, KDG, who now took over command of the Household Brigade under Lord Edward Somerset. It was a sadly depleted command, consisting of a total of only a hundred men of all four regiments – 1st and 2nd Life Guards, Blues and King's Dragoon Guards.[3]

The other British heavy brigade, the Union, had suffered as badly and could only now muster the strength of a single squadron. It was brought across from its position on the left of the Brussels road to join the Household Brigade. So these two composite squadrons, all that remained in action of seven fine regiments, were extended into a single line of horsemen behind the equally battered remains of the infantry. A squadron of the 23rd Light Dragoons, which had become separated from the rest of its regiment, was attached as a welcome added strength to the now composite Household and Union Brigade.[8]

Uxbridge, seeing the weak state of these troops, advised Lord Edward Somerset to withdraw them. But a considerable space on the right of La Haye Sainte was now without any British infantry, and was covered by some Hanoverian infantry who were showing signs of great unsteadiness. These Hanoverians were supported by the Dutch-Belgian cavalry but, as Somerset remarked, there would be no holding the others if the Household Brigade moved off. So, as it was now 6.30 p.m., the composite brigade, in a single line to make as much show of force as possible, sat it out. They were exposed to a constant fire of both musketry and cannon, and more and more saddles were emptied. At 7 p.m. Captain James Naylor was wounded and forced to leave the field. He was taken to Brussels, where he was treated by William McAuley, the regimental surgeon of the KDG.[1, 3, 8]

By this time the Prussians were starting to arrive on the left flank of the Allied position. As they exerted more and more pressure, Wellington was able to move troops from his left to strengthen his terribly weakened centre. The two light cavalry brigades of Vivian and Vandeleur were brought across and posted behind some Dutch-Belgians and Hanoverians in the centre. As they crossed the front in columns of troops they passed the single line of surviving British heavies. Vivian shouted to Somerset, 'Lord Edward, where is your brigade?' 'Here,' replied Lord Edward. Vivian noted how

> Somerset pointed to the ground which was strewn with wounded, over whom it was hardly possible sometimes to avoid swerving. Wounded or mutilated horses wandered or turned in circles. The noise was deafening, and the air of ruin and desolation that prevailed wherever the eye could reach gave no inspiration of victory. Lord Edward Somerset with the wretched remains of the two heavy brigades, not 200 men and horses, retired through me, and I then remained for about half an hour exposed to the most dreadful fire of shot, shell and musketry that it is possible to imagine. No words can give any idea of it (how a man escaped is to me a miracle). It was this fire that had reduced the combined Household and Union Brigade to its sorry state.[1, 8]

The regimental sergeant major of The King's Dragoon Guards was Thomas Barlow, who rode with the regiment throughout the battle and was one of the fifteen KDG left in action at the end of the day, out of the 530 who had paraded that morning. In the first charge Barlow disabled a French cuirassier officer in single hand to hand combat. The officer was one of the finest swordsmen in the French army, and when he tendered his sword to Barlow 'in token of submission', Barlow presented it to Colonel Fuller, who complimented him on his bravery. As a result of this deed Barlow was one of the very few who was commissioned within a few days of the battle for a deed of valour.

The climax of the battle had now arrived. The attack by the Imperial Guard was defeated by Maitland's Brigade of Guards and by the quick thinking of Sir John Colborne, commanding the 52nd Foot, who wheeled his regiment onto the flank of the Guard, pouring in volley after volley until a combined bayonet charge routed these invincibles, and the cry was heard, 'La Garde recule!' The Prussian pressure on the French right was starting to turn their flank, and Wellington ordered the general advance. Uxbridge advised caution, suggesting that the troops should not go beyond the opposite heights, but Wellington's response was 'Oh, damn it! In for a penny in for a pound, is my maxim; and if the troops advance they shall go as far as they can.'[1]

The combined Household and Union Brigade, reduced to little more than a hundred men, were some 300 yards below the crest of the Mont St Jean ridge, but in spite of their small numbers they joined in the final pursuit. The French were by now a confused mass of cuirassiers, lancers, infantry and artillery, all trying to escape to the rear, whilst the fifteen KDG survivors of that terrible day rode among then cutting and slaughtering all in reach. Wellington and Uxbridge rode down the slope encouraging all in sight to press the enemy and not give them a chance to stand, when one of the last cannon shots of the day skimmed the neck of Copenhagen, Wellington's horse, and hit Uxbridge in the knee. 'By God,' Uxbridge cried, 'I've lost my leg!' 'By God,' exclaimed Wellington, 'so you have!' Wellington supported Uxbridge in the saddle until some soldiers and an ADC managed to carry him off the field.[1]

The Prussians now took up the pursuit, and the remnants of the combined heavy brigades halted and bivouacked for the night on the ridge by La Belle Alliance. They sat around warming themselves and trying to cook some dirty lumps of fat they had picked up, using the cuirasses of dead French cavalrymen in which to cook them. In memory of that evening on the field of Waterloo, the officers and sergeants of 1st The Queen's Dragoon Guards still dine together on Waterloo night in the sergeants' mess.[1]

John Hibbert wrote home, 'Altogether it was a most wonderful victory. No men but the English could have fought better than the French; they left two hundred and twenty pieces of cannon on the field.' Corporal Stubbings, now Sergeant Stubbings, KDG – promoted on the field – wrote to his father and mother:

> I take the first opportunity that lays in my Power of informing you that I ham in Good Health after the very sharp Ingagement Which tooke place on 18th June and Dear father it is a wonder that I escaped without receiving any injury for Whee was very much Exposed to Danger both on 17th and 18th June and a Most dreadfule Battle indeed and Ham sorry to say our Regiment Suffered severely in Killed and wounded. But thanks be to God that Spared My Life I came out of the field unhurt and I was in the hotest part of it and I Gained Great Praise for My Good and Coregous Beaver in the field and in Consequence of wich I Ham Maid Sergeant. It is dreadfule to relate the seens I saw on the 18th. The field for Miles around was covered with the wounded and Slain and in some places My Horse Could Not Pass Without Trampling on them and I am Sorry to inform you that Charles Stanley fell on that Ever Memorable Day the 18th June fiting Manfully in the Defence of his Country. French killed and wounded (and prisoner) on 18th June, 85,000 Men and the Duke of Wellington took from them 150 pieces of cannon. Martial Blucher tooke 60 Pieces

of Cannon and the french fled in all Directions those few that was left.[1, 10]

Sergeant Major Page, KDG, writing to his brother and sister, told them: 'Both the married men quartered with me at Romford were killed, one of them had three horses under him, and he lost his life on a fourth, which was a French horse.' Sir Hussey Vivian, commanding a light cavalry brigade, wrote: 'The Life Guards, Blues, with the 1st Dragoon Guards, gallantly met and repulsed the charges of the Cuirassiers in the very heat of the Action, and the losses of these Regiments afford evident proofs how severely they must have been engaged.'[1, 2, 8]

The cost to The King's Dragoon Guards had been frightful. The official casualty list published immediately after the battle gave three officers killed and four missing (all of whom proved to be dead), and another four officers wounded. The killed were Colonel Fuller, Majors Graham and Bringhurst, Captain Battersby, Lieutenant Brooke, Cornet the Hon. H. B. Bernard, and Adjutant Shelver. The wounded officers were Captains Turner, Naylor and Sweney, and Lieutenant Irvine. Forty rank and file were known to be killed, and 124 were missing (most of whom were later found to be dead), and more than 100 had been wounded. Total casualties were reported as 275 out of the 530 who had paraded that morning. Many men had had their horses killed under them, others had become dispersed in the general mêlée and had been unable to find the regiment on their return, others had attached themselves to other units. The final list of casualties, made several days later, showed that seven officers had been killed and four wounded. Two sergeant majors, eleven sergeants and 109 privates were dead; two sergeant majors, four sergeants, a trumpeter and 123 privates were wounded. Sadly, for any cavalryman, 269 horses had been killed. More than half the regiment which had been present had been lost, 129 all ranks killed, and 134 wounded.[1]

It is very difficult to estimate accurately the exact numbers that were present. Sir John Fortescue in his *History of the British Army* gives 29 KDG officers and 568 other ranks; Dwelley's Muster Roll gives 579 other ranks; Dalton's Waterloo Roll quotes 29 officers and a total of 608; Sir Morgan Crofton's account in the *Household Brigade Magazine* claims that there were 530 other ranks; George Jones's *Battle of Waterloo*, published in 1817, gives 529 all ranks, and Siborne gives 530. The Mint Medal Roll gives the total of Waterloo Medals issued to the regiment as 577, to which must be added a certain number of officers and men who were killed and whose relatives did not apply for a medal. If the rear party, the farriers and those left out of battle for one reason or another, is taken into account, the figure of about 530 all ranks on the field and in action seems the best that can be deduced from a conflicting mass of evidence.[1]

Trumpeter Samuel Wheeler, KDG, described the day in his own inimitable style:

> Then we Lay by for the Night With our horses by Our Sides
> To tell the Sufferings we Endured No Pen Can Never Describe
> For the Rain Descended In torrents the Lightening flashed so blue
> Still each Briton kept his Spirits up his foes to Subdue.
>
> Then Early the Next Morning Our Out Piquets did Espye
> Napoleon And his Army that was Advancing Nigh
> Our Army being Ready the Cannon began to Roar
> And Muskett Shot Into their Lines So Quickly we did Pour
>
> About ten O'Clock In the forenoon In A body Our Cavalry Lay
> Their Shots and Shells Into Our Collum So briskly they did play
> Two heavy Brigades then formed A line In Readiness to Advance
> To Charge the tyrants Curisears Wich was the Pride of France.
>
> Lord Somerset Commanded the Household Brigade
> And Led us On to victory through the hottest of Cannonade
> So well the Brigade then Played their Parts And Charged them so free
> Wich was the Cause my british boys of Gaining the Victory.
>
> O When we Came up With them the Slaughter It was Great
> Our Gallant troops So boldly had Soon Caused their lines to Break
> We charged them So boldly And Made them far to Run
> And Cut them down With Our broad swords Like Motes In the Sun.
>
> General Ponsonby's brigade Charged Next And I Am bound to Say
> They Done ther Duty Manfully upon that Glorious Day
> They Charged them So valiantly And Caused them to Rue
> That Ever they fought for boneypart on the Plains of Waterloo.
>
> But Now the Painful task Comes on I Am Sorry to Say
> We lost Many Noble officers Although we Gained the Day
> Beside Some thousands of Our Men Lay bleeding On the Plains
> And On the minds of Britons their Deaths Will Long Remain.
>
> But Now the battle Is over the Victory we have won
> Fill up A Bumper And Drink A health unto Duke Wellington
> Likewise the Earl of Uxbridge Lord Somerset Also
> That led us On Like heroes On the Plains of Waterloo.

On the field of battle lay 40,000 men and 10,000 horses. In among the dead lay
the wounded and the dying. For those who were still alive, it was not merely a

matter of enduring the long cold hours of the night: with darkness came the looters, who did not hesitate to murder in their search for booty. Anyone who attempted to resist was stabbed or shot. Some of the KDG wounded had found their way back to Brussels; Captain Naylor was at the Hotel Grand Mirror, and on the 19th he was put in command of all the wounded of The King's Dragoon Guards. Captain Turner was lodged at the Hotel de la Couronne d'Espagne. On 20 June Captain Sweney, KDG, and Lieutenant William Irvine, KDG, who had both been captured after the first charge, arrived in Brussels. Both had managed to escape when the French retreated, although Sweney had been wounded and now lodged with Turner. Naylor had recovered from his wound by 2 July and left Brussels to rejoin the regiment.

John Hibbert wrote home:

> This time last year in Ireland, before the reductions, we mustered in the field eleven hundred men – a thousand effective. Now the whole Regiment cannot muster three hundred including the depot, and about two hundred and fifty horses, and all our best and worthiest officers who were the most respected and liked in the Regiment, are all killed; in fact the Regiment will never be what it was. Colonel Fuller and the Adjutant, Captain Battersby, Majors Graham and Bringhurst and Lieutenant Brooke were all thrown into one grave and buried by the Thirty Second regiment. Young Bernard's body could not be found, but we know pretty well what became of him; he was taken prisoner after having been wounded, and not being able to keep up with the French, they killed him on the road. They served a great many English officers in the same way.

During the pursuit of the defeated French, further evidence of what had happened to the missing men came to light.

> What prisoners of our Regiment were taken, we certainly believe to have been murdered in cool blood by the French on their retreat. A sergeant major of ours was taken prisoner at Waterloo and the other day he was discovered lying dead with his head split open, and about seven men of our Regiment with him. He was the fattest man in the Regiment, and we suppose he must have knocked up, for they used them shamefully, driving them on with their bayonets and giving them nothing to eat or drink the whole day. Added to this they stripped them of everything they had except their overalls and shirts, and some had nothing but a blanket to cover them. No wonder our poor fellow knocked up with this usage; we

wonder how he got as far as he did, but he knew that there was but one alternative and that when he gave in, he was to die.[1]

SOURCES

1 Michael Mann, *And They Rode On*, 1984.

2 *The K. D. G.*, vol. 2, no. 5, 1936.

3 Captain James Naylor, KDG, MS diary, Regimental Museum, 1st The Queen's Dragoon Guards, Cardiff Castle.

4 Cavalié Mercer, *Journal of the Waterloo Campaign*, 1927.

5 Captain Sir John Kincaid, *Adventures in the Rifle Brigade*, 1830.

6 'The Battle of Waterloo, by Trumpeter Samuel Wheeler, KDG', MS in possession of Lieutenant Colonel John Hibbert, Light Infantry.

7 R. Cannon, *Historical Record of the British Army, 1st Dragoon Guards*, 1837.

8 Major General S. T. Siborne, *Waterloo Letters*, 1891.

9 Sergeant Major Cotton, *A Voice from Waterloo*, 1849.

10 MS letters of John Stubbings, KDG, in possession of Ernest Shead.

20

Paris, Occupation, England and Ireland
1815-1824

On the day after the Battle of Waterloo the Household Cavalry Brigade was moved to Nivelles, which gave the many stragglers, the unhorsed and the slightly wounded time to rejoin, so that very soon The King's Dragoon Guards mustered a hundred men. On 20 June the pursuit speeded up, and by 29 June John Hibbert was writing from within fifteen miles of Paris, 'We march from twenty to thirty miles every day, up always at three o'clock and sleep all night in the fields, for we lost all our tents and most part of our baggage. We shall enter Paris tomorrow.' On 9 July Sergeant Stubbings wrote home, 'Bonepart is completely defeted and Lewis 18th entered Paires yesterday and Whee expect all the fitting is over which I hope it is for it is Dredfule to relate the scens I saw on the 18th.' John Hibbert commented, 'Lord Edward [Somerset] ordered a sale today of the effects of the slain. I have purchased a good large cloak, erst the property of poor Colonel Fuller.'[1, 2]

The King's Dragoon Guards went into quarters at Nanterre. Hibbert wrote, 'We are comfortably situated within five miles of Paris and expect to remain here.' But the regiment was soon moved again to billets at Rouelle. On 25 October Hibbert was writing,

> We have the remnants of eight Troops – our wounded men have likewise joined us from Brussels, amounting to about sixty. We have about a hundred and eighty men including these, out of which there are not more than a hundred and twenty effective, that is with horses and clothing, but there are a hundred horses at the depot, which if we remain here will of course be sent out. We have not had any remounts yet, nor shall for some time, I mean recruits, for these are scarce at present, and the two Troops at the depot are but nominal ones, both together not exceeding twenty men.[1]

An order from the War Office, dated 29 July 1815, required 'That henceforth every N. C. O., Trumpeter and Private, who served in the Battle of Waterloo, or actions which immediately preceded it, shall be borne upon the Muster Rolls and pay lists of their respective regiments as "Waterloo men" and every Waterloo man to count two years service, in virtue of that victory, in reckoning his service for increase of pay, or for pension when discharged.'[3]

The Queen's Bays, stationed in Scotland, had been augmented to ten troops on the renewed outbreak of war, and were ordered after Waterloo to send six troops to France to reinforce the army, reduced by the terrible casualties suffered during the campaign.[5, 6]

The troops embarked at Ramsgate and Dover in three divisions on 27 June 1815 and landed at Ostend early in August. They marched from there to Paris, where they arrived on 26 August and were quartered alongside The King's Dragoon Guards.

On 2 September the 1st and 8th Brigades of Cavalry were reviewed by the Emperor of Russia, the Tsar Alexander; the 1st, or Household, Brigade was the same as at Waterloo, whilst the 8th was composed of The Queen's Bays, the 3rd Dragoon Guards and the 3rd Dragoons. These two brigades were again reviewed by the Tsar, the Emperor of Austria, and the King of Prussia, who were followed by a dazzling retinue of hangers on, and who duly complimented the cavalry on their appearance.[1, 3, 4, 5, 6]

During October and November 1815 the army was broken up, with many of the active regiments returning home or to other stations. Both The King's Dragoon Guards and The Queen's Bays were ordered to remain in France as part of the Army of Occupation. With the Life Guards and the Blues having returned to England, John Hibbert was writing to his sister on 13 December, 'We are now brigaded with the King's Own or Third Dragoons, and the Second Dragoon Guards, both regiments that have come from England within the last three months.' In 1816 The Queen's Bays were stationed at St Omer, and on 12 and 15 October 1816 they were reviewed by the Duke of Wellington, together with the 3rd Dragoons, the 7th Hussars, and the 11th, 12th and 18th Light Dragoons. On 22 October they took part in a 'sham fight' at St Germain, again in the presence of the Duke of Wellington.[1, 3, 4, 5, 6]

The King's Dragoon Guards spent the latter part of 1815 and the spring of 1816 in building up their strength. In July 1815 Lieutenant Colonel Acklom joined from England, and in August Lieutenants Bray, Hammersley, Brooke, Stephenson and Trevelliss arrived together with one sergeant, thirteen rank and file and twenty-seven horses, to be followed a few days later by Lieutenant Colonel Teesdale, Cornet Chalmers, Assistant Veterinary Surgeon Spencer and another thirteen privates. On 7 September 1815 Lieutenant Colonel Teesdale was appointed to command the regiment in succession to the late Lieutenant Colonel Fuller, killed at Waterloo. In December the losses in horses from Waterloo were made up by remount drafts from other regiments, thirty-seven horses being received from the 3rd Dragoon Guards, thirty-one from the 1st Royal Dragoons, thirty-four from the Scots Greys, and forty-seven from the 6th Inniskilling Dragoons, as the three latter regiments of the Union Brigade returned home. In January 1816 Lieutenant Polhill, Cornets Reed and Grierson, two sergeants and thirty-seven privates joined as further reinforcement from England.

John Hibbert wrote home: 'Nothing but bad news. Many dragoon regiments have received orders for marching homewards, and where do you think they are going to send our miserable remains? To a village called Catoff [Le Cateau] on the frontier between Flanders and France, where it is likely we may remain some years.' Sergeant Major Page was more explicit:

> It is finally settled for our Regiment to remain in France, the British Army (except that part which is now about returning to England) will remain on the frontiers of France three years, then, if nothing fresh happens, we shall return to England; our Regiment will be completed with horses from other Regiments which are returning to England; other things we shall have sent from England to us, such as arms, saddles etc etc to make us once more in fighting order. We have also many men and officers joined from England a few days ago, and about fifty men who are cured of wounds have joined us from Brussels.[1, 3, 4, 7]

Early in 1816 both The King's Dragoon Guards and The Queen's Bays were reduced in strength, the former from twelve to eleven troops, and the latter from ten to eight troops. Both regiments were now stationed in the Pas de Calais, and in March the KDG transferred thirty-seven men to the Mounted Staff Corps, while the regiment suffered the reduction of another troop. In April the KDG were warned for home service, and in preparation transferred ninety-five horses to the 3rd Dragoons, twenty-three to the 13th Light Dragoons and fifty-four horses to The Queen's Bays. The depot troop at Coventry received 118 horses from the 3rd Dragoon Guards in readiness for the regiment's return. On 3 April 1816 silver medals were received, 'to be worn suspended by a ribbon on the left breast, being one for each officer, N.C.O and private serving with the Regiment for their distinguished bravery at Waterloo in action with the enemy on the 18th June, 1815.' Prior to embarkation The King's Dragoon Guards were inspected by Lord Combermere, and the regiment was permitted to bear the word 'Waterloo' on its standards and accoutrements.[1, 3, 4]

The King's Dragoon Guards disembarked at Dover on 7 May 1816 and marched to the vicinity of Hounslow, while The Queen's Bays remained in France as part of the Army of Occupation. On 18 May the KDG were reviewed by the Duke of

York on Hounslow Heath, and two days later two squadrons marched to Colchester under command of Lieutenant Colonel Acklom as a result of some civil disturbances in the town. The remainder of the regiment joined the two troops at the depot now in Northampton. They had not been there long before they were sent to Lancashire and Yorkshire to deal with rioting weavers. Six troops went to Manchester, two to Sheffield, and one each to Bolton and Blackburn. Indeed the KDG were to be called upon so often to deal with civil disturbance following the long years of war that they earned the nickname of 'The Trades Union'. At the end of June one of the Manchester troops marched to Leeds, another was sent to Preston in August to deal with rioting, then in October another troop was sent to Wigan. In the same month the establishment of The King's Dragoon Guards was reduced to ten troops totalling 345 mounted and 290 dismounted men, and the blue-grey cloth pantaloons were replaced by kerseymere overalls. The King's Dragoon Guards were always dealt with, in the matter of reductions, on the same basis as the Household Cavalry, and differently from the rest of the cavalry of the line. John Hibbert commented, 'We are again changing our dress in every respect. This will all be very expensive and will come hard upon those who have nothing but their pay to subsist on.'[1, 3, 4, 8, 9]

During 1817 The Queen's Bays remained in the general area of Calais, and then moved to Cambrai where they were reviewed by the Duke of Wellington. They returned to Calais on 17 October and established headquarters at Guines. In 1818, on 29 June, the Duke of Kent reviewed the regiment on the sands near Calais, after which they marched to St Omer, and later to Cambrai. On 23 October they were again reviewed near Haspres, together with the rest of the cavalry of the Army of Occupation, by the Tsar of Russia and the King of Prussia. On 7 November The Queen's Bays embarked at Calais for England and landed at Dover the following day. Before they left France they received a brigade order from Major General Lord Edward Somerset in whose brigade they had served:

> Major General Lord Edward Somerset, in taking leave of the two Regiments, 2nd Dragoon Guards and 3rd Dragoons, which he has had the honour to command for nearly three years, begs to congratulate them on the approbation expressed at their conduct by the Lieutenant General commanding the cavalry, as well as by the distinguished officers who have lately seen them in the field. The Major General has great pleasure in assuring them of the sense he entertains of their good conduct. He requests the officers commanding the two Regiments and troop of Horse Artillery, and the officers in general, will receive his thanks for the zeal

they have manifested in the discharge of their duties, and he offers them his best wishes for their future happiness and prosperity.[5, 6]

On arrival in England The Queen's Bays marched to quarters at Egham and Staines, where they were reduced to an establishment of 439 men and 273 horses. In November 1818 the regiment marched to Bristol, and in December embarked for service in Ireland, where it was stationed in the south-east and west. A South-Eastern District order of 5 May 1819 stated:

> Major General Doyle is much pleased with the excellent appearance of the Queen's Bays made this day at their half-yearly inspection. The horses are not only of a superior description but the manner in which they have been trained, the condition they are in, the uniform seat of the soldier, and the riding in general, marked the attention which must have been paid by the Riding Master, Lieut Dyer, and every individual concerned. The manoevres in the field, the sword exercises, the marching and the several charges were performed with celerity and precision. The care of the sick, the state of the Hospital and of the Barracks, the good behaviour of the men since their arrival in the district, and the whole interior economy of the Regiment cannot be surpassed and reflect credit on Lieutenant Colonel Kearney, who appears to be ably supported by his command.[5, 6]

In 1820 The Queen's Bays marched to Dublin and on 23 September moved to the barracks at Newbridge; but before they left Dublin a garrison order put on record that

> Major-General Sir C. Grant cannot allow the 2nd Dragoon Guards to return to Newbridge Barracks without expressing to Colonel Kearney, to the officers, and to the men, his entire approbation of this excellent Corps. The appearance of the Regiment in the Field, the pointed attention of all classes to their respective duties, and the consequent precision of movement, have meritied and obtained his applause. The manner in which the Regiment is mounted reflects very great credit on Colonel Kearney, and strongly evinces the attention he has paid to this most essential part of the duty of a Commanding Officer.[5, 6]

In March 1817 The King's Dragoon Guards were called upon to deal with some disturbances in Manchester, when the unemployed of that city, known as 'the blanketeers' from the blankets they carried, tried to march on London to present a petition to the Prince Regent. The KDG, together with the Cheshire and Staffordshire Yeomanry, blocked their way and dispersed them. A trooper got a dent in his helmet from a brick and a marcher was mortally wounded by a sabre cut on the head. On 5 April the regiment received the thanks of the government 'for their ready and useful assistance during the disturbances'. In July the KDG moved two troops from Manchester and two from Sheffield to York, while one from Manchester and one from Wigan went to Sheffield. During the latter part of 1817 and the first half of 1818 various troops were subject to frequent moves:

August 1 troop from Blackburn to Leeds, then returned.

August 1 troop from Preston to York, then returned.

October 1 Troop from York to Beverley, then returned in January.

March 1 Troop from York to Tadcaster, then returned.

March 1 Troop from Huddersfield to Sheffield.

April 1 Troop from Wakefield to Sheffield, then returned.

May 1 Troop from York to Easingwold, and thence to Newcastle.

May 1 Troop from York to Carlisle.

In June 1818 The King's Dragoon Guards were ordered to Scotland, three troops marching from York and three from Sheffield, one each from Wakefield, Leeds, Carlisle and Newcastle. On arrival, headquarters and three troops were at Hamilton, two were at Ayr and two at Dumfries, with one each at Stirling, Kilmarnock and Glasgow, with a detachment at Dumbarton.[3, 4, 8, 9]

During 1817 a uniform system of riding was established for the cavalry, and there was a reduction of one lieutenant per troop. Another reduction took place during 1818 of 134 men, or 1 sergeant, 10 troopers, 2 boys and 7 horses per troop, and the five staff sergeants lost their horses. In May 1819 the five troops of the KDG in 'out-quarters' were ordered into Hamilton for inspection by Major General Hope; Captain Maxwell's troop from Linlithgow, Captain Sweney's and Captain Quicke's from Ayr, Captain Wallace's from Dumbarton and Captain Leatham's from Kilmarnock. After inspection, when they 'gained great credit in every respect', they returned to their out-stations preparatory to the whole regiment embarking at Port Patrick in June for Ireland and landing at Donaghadee. Only one horse was killed in transit. The KDG then came onto the Irish Establishment, and were quartered at Dundalk, Monaghan, Belterbet and Enniskillen, with detachments at Newtown, Magharafelt and Armagh. Shortly after their arrival in Ireland there was a change in the uniform, the short coats being replaced by long coats decorated with crossbar lace, and the material of the overalls being changed from blue to dark grey. At this period the dress sabretache became very elaborate, and the first Dress Regulations issued in 1822 described it as of a 'blue morocco pocket 12½ in. deep, 10½ in. wide at the bottom, 8½ in. at the top, face 15 in. deep, 13 in. wide at bottom, 9 in. at top covered with blue velvet and edged with 2¼ in. lace showing a light of blue velvet on the outer edges; a gold-embroidered GR

monogram surmounted by a crown, relieved in silver and encircled in oak
leaves; three rings at top for slings of belt.' To this should be added 'below
the GR monogram, a red velvet scroll with the battle honour "Waterloo"
in gold thread'. The helmet was described as 'Roman black glazed skull and
peak, encircled with richly gilt laurel leaves, rich gilt dead wrought scales and
lions' heads: bear skin top.' The great bearskin crest made it a most imposing
headpiece.[3, 4, 8, 10]

In February 1820 Sir David Dundas died and the new colonel was Sir Francis E.
Gwyn. In August The King's Dragoon Guards moved to the Connaught District,
with troops at Roscommon, Gort, Sligo, Loughrea, Portumna, Athlone, Dunmore
and Ballinrobe. Sir Francis Gwyn died in 1821 and General William Cartwright was
appointed colonel. In August 1821 the KDG were again reduced, by two troops,
giving a total establishment of 34 officers, 337 NCOs and troopers mounted, and
108 dismounted men. In May 1822 the regiment was relieved by the 12th Royal
Lancers and marched to Dublin, where it embarked in July, landing at Liverpool on
the 24th of that month to take up quarters at Manchester, Sheffield and Nottingham. Up to this date the standards of the cavalry had always been carried by
officers, but a general order of 30 November 1822 from the Horse Guards stated,
'His Majesty has been pleased to command that the Standards in the Cavalry shall
be carried in future by Troop Sergeant Majors.'[3, 4, 8, 12]

There are still in existence a number of regimental medals which were apparently
awarded over this period within regiments for bravery or for outstanding service. Five such medals 'For Military Merit' in The King's Dragoon Guards have
survived, the first awarded in 1816 to James Evans, the second dated 1817 to
W. Clarkson, and the third to Trumpeter William Hillier in 1818. Two earlier
examples, dated 1798 and 1799, were both awarded by Lieutenant Colonel W. M.
Hawley. The Queen's Bays awarded a medal 'For Military Merit' to W. Woodman
in 1815, and new examples for both regiments still come to light from time to
time.[13]

The Colonel of the Queen's Bays died in March 1821, and General Sir William
Loftus was appointed to the colonelcy. On 23 April 1821 the Lord Lieutenant of
Ireland, Earl Talbot, reviewed the Dublin garrison. The Queen's Bays were on
parade with the 3rd Light Dragoons and the 12th Royal Lancers, the Royal Artil-
lery, and the infantry regiments were the 23rd [Royal Welch Fusiliers], the 33rd
[Duke of Wellington's], the 43rd [Oxfordshire & Buckinghamshire Light Infantry],
and the 78th [Seaforth Highlanders]. The regiment was again highly complimented
for its turnout and performance. On 29 May The Queen's Bays embarked for
England, landing at Holyhead and Liverpool on 1, 2 and 3 June, and marching to
Sheffield, Burnley, Halifax and Huddersfield. In August the establishment was again

reduced to six troops, with a total strength of 335 NCOs and men and 253 horses. That month the regiment moved to Newcastle and Carlisle. In May 1822 there was a further move to York and Leeds.

During the next few years The Queen's Bays moved from station to station:

1823 Birmingham, Coventry and Nottingham

1824 Birmingham, Coventry and Nottingham

1825 Hounslow

1826 Manchester

1827 To Ireland

In 1823 The Queen's Bays had a troop under command of Major Charles Middleton stationed at Abergavenny. A dispute arose between Middleton and the local turnpike gatekeepers, who tried to levy charges against him, and against his batman, when they were out exercising their horses. The Mutiny Act contained a clause specifically exempting 'the horses of Officers and Soldiers on their march or on duty'. The turnpike men contended that when an officer was out riding without his troop, he was riding for pleasure and should therefore pay, and that the batman, not being an officer, should pay in any case. The local magistrates supported the turnpike keepers, and so Middleton appealed to the War Office for a ruling. There was then a lengthy correspondence both within the War Office and to Middleton, who claimed that it was his duty to make himself acquainted with the local roads, and that his horse had to be exercised, so this should count as duty. The War Office, after much deliberation and correspondence, decided that

> There cannot be a question on the point that Bat Men so passing for the purpose of exercising their horses are on duty, – in fact it must be assumed that Bat Men could not venture to ride their horses for their own pleasure, and the Act says nothing as to the men being in Regimentals when mounted. They would not dare to appear on their chargers out of Regimentals without the authority of their Commanding Officer and a Toll Keeper is not a fit person to decide whether soldiers are or are not on duty.

As to Middleton himself, the ruling was that

> His being in Regimentals is surely sufficient evidence that he is on duty and no other evidence should be required ... It would be extremely unpleasant on every occasion to be under the neccessity of either submitting to the toll or of summoning the Tollkeeper before a magistrate and thereof exposing the nature of his duty on which he had been employed, a disclosure that might perhaps not be politic nor proper at all times.

The exigencies of peacetime soldiering and petty bureaucracy do not change very much.[16]

While at Hounslow The Queen's Bays were reviewed on the Heath on 28 June, together with The King's Dragoon Guards, by the Duke of York, accompanied by the Dukes of Cambridge and Sussex. Both regiments were brigaded together with the Scots Greys in the Heavy Brigade; the other regiments were the 1st and 2nd Life Guards, the Royal Horse Guards, the 7th Hussars and the 12th Royal Lancers, together with a brigade of horse artillery. The Regimental Records of the Bays state that

> The condition of these Regiments, their martial appearance, the precision with which their several movements were executed, their uniform order and steadiness, combined with velocity in the several attacks, evinced a high state of discipline, and exhibited a splendid military spectacle which excited universal admiration and received the unqualified approbation of the Commander in Chief.[3, 4, 5, 6]

In Manchester in April 1826 the Luddites were destroying new machinery in the hope of saving their jobs. A meeting was held at St Peter's Fields (the scene of the Peterloo Massacre of 1819), when the speakers urged the crowd to destroy the power looms. The Cheshire Yeomanry were called out, but before they could be mustered there was serious rioting. Luckily The Queen's Bays and a company of the 60th Rifles were on hand and restored order by midnight. The next morning more trouble occurred. The Queen's Bays patrolled the streets throughout the day, dispersing the mob with the flats of their swords. The regiment was subsequently congratulated 'on the great forbearance shewn by them towards the populace in the performance of their duty'.[14]

In June 1823 The King's Dragoon Guards proceeded to Scotland and were stationed at Piershill Barracks, Edinburgh and at Perth, with detachments at Coupar Angus, Forfar, Queensferry and Haddington. Their stay in Scotland was short, and in 1824 they moved to Leeds, Newcastle and Carlisle. A painting in the Regimental Museum shows the regiment's baggage waggons outside St Nicholas's Church, Newcastle (now Newcastle Cathedral) in 1825. Four waggons are depicted piled high with baggage, and seated on top are numerous wives and children. Amongst items waiting to be loaded are two officer's trunks: the first bears the name of Lieutenant-Colonel G. Teesdale, the commanding officer, and the second that of Cornet Wilson, commissioned in 1823 and promoted lieutenant in June 1825.[15]

SOURCES

1 Michael Mann, *And They Rode On*, 1984.

2 MS letters of John Stubbings, KDG, in possession of Ernest Shead.

3 Records of The King's Dragoon Guards, Regimental Museum, 1st The Queen's Dragoon Guards, Cardiff Castle.

4 R. Cannon, *Historical Records of the British Army, 1st Dragoon Guards*, 1837.

5 R. Cannon, *Historical Records of the British Army, 2nd Dragoon Guards*, 1837.

6 F. Whyte and A. H. Atteridge, *The Queen's Bays, 1685-1929*, vol. 1, 1936.

7 *The K. D. G.*, vol. 2, no. 5, 1936.

8 *Digest of the Services of The King's Dragoon Guards*, Regimental Museum, 1st The Queen's Dragoon Guards, Cardiff Castle.

9 The Marquess of Anglesey, *A History of the British Cavalry*, vol. 1, 1973.

10 Captain R. G. Hollies Smith, 'Some Accoutrements of the King's Dragoon Guards', *Journal of the Society for Army Historical Research*, vol. 38, 1960, p. 47.

11 *Regimental Journal of 1st The Queen's Dragoon Guards*, vol. 3, no. 6, 1960, p. 47.

12 *Journal of the Society for Army Historical Research*, vol. 16, 1937, p. 139.

13 'Regimental Medals of the 1st & 2nd Dragoon Guards', *Regimental Journal of 1st The Queen's Dragoon Guards*, vol. 2, no. 5, 1971, p. 85.

14 *Journal of the Society for Army Historical Research*, vol. 19, 1940, p. 85.

15 *Regimental Journal of 1st The Queen's Dragoon Guards*, vol. 2, no. 6, 1971, p. 550.

16 Public Record Office WO43/217, folios 379-85.

21

Aid to the Civil Authorities, Canada
1824-1838

An inspection report of May 1824, when The King's Dragoon Guards were still in Edinburgh, noted that the regiment was still holding 419 bayonets. As a result, and after long deliberation in the War Office, it was decided in 1828 that heavy dragoons would no longer be issued with bayonets. With brisker step than the bureaucrats The King's Dragoon Guards moved down to London from the North in June 1825 in time for inspection by the Duke of York on Hounslow Heath, together with The Queen's Bays, and other regiments. 'Battle manoeuvres were executed on an imposing scale, and the precision of the troops received unqualified approval.' [3, 7, 8]

At the beginning of July the KDG marched from London to Canterbury, Shorncliffe and Deal, and in February 1826 two troops were sent to Norwich. In March 1826 the remaining six troops marched north again. Four troops and headquarters were at Leeds, and one troop each at Blackburn and Burnley. [3, 15]

Throughout 1826 The King's Dragoon Guards were used extensively to suppress rioting and preserve property from damage throughout Yorkshire and Lancashire. Immediately on arrival on 14 April they were called out by the magistrates to quell rioting in Lancashire; on the 25th Captain Skinner's and Captain Polhill's troops marched to Burnley and Blackburn on the requisition of the civil authorities; the Burnley troop had already been sent to deal with rioting at Clitheroe. On 26 April Captain Maxwell's troop was sent from Leeds to Otley and another troop to Addington to protect property; on 27 April all available troops left in Leeds marched to Manchester, where they were joined the next day by the two troops previously sent to Otley and Addington. On 1 May Captain Quicke's troop left Manchester for Wigan, and then moved on to Chorley on the 5th. On 2 May headquarters together with Captain Randall's and Captain Reed's troops marched from Manchester to Burnley, whilst Captain Polhill's troop left Burnley for Preston. Captain Maxwell's troop moved on 3 May from Manchester to Bury, and then proceeded to Bolton; Captain Skinner's troop went from Burnley to Blackburn. [1, 3]

During June 1826 The King's Dragoon Guards were used to keep the peace at elections at Kirkham and at Padiham, and to maintain calm at Huddersfield and Leeds. In July individual troops were used at Blackburn, Rochdale, Bradford, Huddersfield and Leeds. In October they were at Huddersfield and Bradford again, and at Sheffield. Then in February 1827, after a year of the most arduous duty in aid of the civil authorities, during which individual troops had been called upon to march as much as fifty to sixty miles in a day, The King's Dragoon Guards were again ordered north to Scotland.[1, 3]

While at Leeds Captain Frederick Polhill of the KDG performed a remarkable athletic feat.

> On November 9th, 1825, this officer, for a wager, rode 95 miles in four hours and seventeen minutes, and on the 17th April following, the same officer on the same ground walked fifty miles, drove fifty miles, and rode fifty miles, in the short space of nineteen hours and five minutes, and was afterwards drawn by the populace to the barracks in his carriage. The wager was for 100 sovereigns.[9]

The regiment spent only a year in Scotland, with troops stationed at Edinburgh, Glasgow and Perth, with detachments at Forfar, Coupar Angus, Witburn and Midcalder. In August 1827 a subaltern with twenty-one men was sent to Haddington in aid of the civil power, but the only other event of note was the death of General Cartwright and the appointment of Lieutenant General Sir Henry Fane as colonel of the regiment in his place.[1, 3]

In April 1827 The Queen's Bays were posted to Ireland, being initially stationed in Dublin and then moving to quarters around Limerick, Clonmel and Cahir. The regiment returned to England in April 1830 and was staioned at Manchester, moving to York and Leeds in 1831. In July 1831 General Loftus died and the new colonel was an old officer of the regiment, who was much loved and respected by all ranks, Lieutenant General Sir James Hay.[2, 4]

During April 1828 The King's Dragoon Guards moved from Scotland to York, Beverley, Carlisle and Newcastle; for the whole of 1828 and 1829 they were constantly sending individual troops and parties all over the North to help keep civil order. During April and May 1829 there were serious riots in Manchester, Macclesfield and Rochdale, and on 14 May 1829 Lieutenant Colonel Teesdale received the following letter from the Town Hall, Manchester:

> We cannot permit you to leave the town without tendering to you, the Officers, N.C.O's and Privates of the K.D.G., our grateful thanks for the very prompt and efficient services rendered to the inhabitants of

Manchester in the serious riots which occurred, but which were hap-
pily suppressed immediately preceding your departure, and we eagerly
embrace this opportunity of expressing the high sense we entertain of the
discretion and forbearance displayed by the Officers and men under your
command in the performance of this harassing and unpleasant duty.[3, 5]

It cannot have been easy at this period to exercise the command of a cavalry
regiment. Between 1828 and 1843 The King's Dragoon Guards were not once
together as a complete regiment; individual troops and detachments were sent off
here and there to cope with civil disturbance, to prevent smuggling and to round
up deserters. H. P. Parker, an artist of some renown, exhibited a painting in 1829 of
a KDG corporal and trooper securing a deserter, while a crowd of anxious relatives
look on. The painting is of interest in that it confirms regimental standing orders
for the cavalry, which say, 'On duties on foot and under arms Sergeants and
Sergeant Majors are dressed with sashes, the knots hanging on the right side,
pouch belt, carbine and Sword belt, with gloves hanging to the swivel. Corporals and Privates will parade with arms, flints and ammunition complete and in
trousers.'[6, 10, 11]

In May 1829 The King's Dragoon Guards embarked at Liverpool for service in
Ireland, and were stationed at Longford, Athlone and Gort. During 1830 and 1831
the regiment was constantly employed in trying to maintain the peace, and was
called upon to aid the civil authorities at Limerick, Ennis, Clonmel, Waterford,
Cashel and many other towns. The Regimental Record states, 'There was so much
excitement in Ireland, that the troops were constantly on the march, and more
harassing and painful services were even more frequent in 1831.' Between May
1829 and April 1832, when the regiment embarked at Dublin for Liverpool, no less
than sixty separate moves by elements of the regiment are recorded. While it was
in Ireland an increase of twenty-four horses was made to the establishment.[1, 3, 5]

The Queen's Bays moved to Scotland in April 1832, but returned to Nottingham
the following year. On 10 September 1932 the regiment was reviewed by the
Duke of Sussex at Bulwell Forest, and in 1834 the Bays moved to Ipswich, only to
return to Ireland in May 1835, where they remained until 1839 engaged in the
thankless task of maintaining the peace. A new colonel, Lieutenant General Sir
Thomas Gage Montresor, was appointed in February 1837 on the death of Sir
James Hay.[2, 4]

A picture in the Royal Collection by Dubois Drahonet, dated 1832, clearly
shows the uniform of Dragoon Guards at this time. A sergeant of The King's
Dragoon Guards has a black leather sabretache, and in the centre is a brass star
surmounted by a crown with the letters K.D.G. in the centre of the star. The helmet

was of black japanned tin with gilt oak leaves, the Royal Arms in front and gilt scales under the chin. The helmet had a high bearskin comb, which drew the comment from one officer, 'It was all very well to look at, but a beastly thing to wear, especially in a high wind.' The jackets were of scarlet cloth with facings of velvet in the regimental colour, and the cuffs had four loops with tassels arranged in pairs. The overalls were blue with a yellow stripe for both regiments. An inspection report of 1834 for the KDG states, 'The helmet worn by the band has a scarlet ornament above the crest, in lieu of bearskin.' Officers had to wear gauntlet gloves when in dress uniform, and white leather gloves in frock coat and undress. A general order of 2 August 1830 forbade Dragoon Guards to wear moustaches and ordered that the hair of officers and soldiers was 'to be cut close at the sides and the back of the head, instead of being worn in that bushy and unbecoming fashion adopted by some regiments.'[12]

On arrival at Liverpool in 1832, the troops of The King's Dragoon Guards were scattered, taking up quarters at Sheffield, Nottingham, Burnley, Derby, Leeds and Peterborough. In August and October 1832 two troops were moved to East Dereham in Norfolk; and men from one or both of these troops figure in a court martial held at the Cavalry Barracks at Norwich, 'into charges preferred against certain privates for disobeying the lawful commands of the Colonel of the 7th Hussars'. It was spring 1833 before the whole regiment marched south to be stationed at Canterbury, Brighton, Worthing, and in October it sent a reinforced troop to Chichester. As headquarters and four troops marched south they were halted en route at Chelsea, in order to keep the ground while the Household troops were reviewed by the Duc d'Orléans. In April 1833 the commanding officer, Lieutenant Colonel Teesdale, was made a Knight of the Royal Hanoverian Guelphic Order by William IV. In December there was a reduction of five men and three horses per troop, and in 1834 there was a change of helmet, which was now all of brass with a metal crest terminating in a lion's head, this being removable to enable a bearskin crest to be fitted for ceremonial occasions. Officers wore a gilt version when mounted, but at levees they wore a large cocked hat, and for undress a blue cloth forage cap with a band of gold lace, and a patent leather peak.[1, 3, 13].

In February 1834 William IV, with Queen Adelaide, reviewed the KDG in the Riding School of the Royal Pavilion at Brighton. Following this inspection the King ordered that the last remaining Waterloo horse still serving with the regiment, No 22 of 'H' Troop, and now twenty-five years old, was to be sent from Brighton Barracks and turned out to rest in one of the Royal Parks; the King presented the regiment with a very handsome cream-coloured horse from the Royal Stud in exchange. During 1834 'Experiments are now trying as to the practicability of arming both heavy and light dragoons with the lance, and lancers with the carbine, each will then have pistol, carbine, sword and lance.' In the event the lance did not come to the Dragoon Guards for another fifty years, and then only to the front rank. In April 1834 the KDG moved from Brighton to Dorchester, Exeter, Christchurch and Trowbridge, while still keeping a troop at Chichester.[1, 3, 14]

The regiment was involved again in 1835 in quelling civil disturbance. The first trouble came during contested elections to Parliament, when the KDG, having been moved to Birmingham, Coventry, Clifton and Abergavenny, had to send troops to South Staffordshire, West Bromwich, Wednesbury, Stourbridge, Wolverhampton, Great Barr, Dudley and Lichfield. The magistrates at Lichfield wrote thanking them 'for the promptitude with which they attended the requisition, to which the Magistrates attribute the prevention of serious injury to the property, if not the lives and limbs of His Majesty's subjects'. Captain Manning's troop at Wolverhampton on 27 May was called upon to disperse rioters; a local magistrate, the Revd Clare, read the Riot Act, and the KDG then 'executed this service with great firmness and forbearance, under every provocation and insult'. Manning's conduct came under investigation, and a letter to the Quarter Master General from the Wolverhampton magistrates read, 'I found it absolutely necessary to call in their aid, I think it proper to state that their conduct has been so truly excellent, that no language of mine can do them justice.' To reinforce this tribute the people of Wolverhampton presented Captain Manning with a silver snuff box, inscribed, 'Presented to Captain I. S. Manning, of the 1st, or King's Dragoon Guards, in testimony of their approbation and gratitude, by those inhabitants of Wolverhampton who witnessed the commendable forbearance and correct judgement which marked his military conduct in suppressing the riots in that town on the 27th May, 1835.' The second in command, Lieutenant Brander, was presented with a suitably inscribed cigar box.[1, 3, 5]

In the early months of 1836 The King's Dragoon Guards moved around the Midlands in aid of the civil power: Birmingham, Newcastle under Lyme, Wolverhampton, Kidderminster, Bridgnorth and Stafford afforded quarters for individual troops. Then, in May 1836, the whole regiment moved north to Manchester, Burnley and Derby. On the anniversary of the Battle of Waterloo a banquet was held at Apsley House in London, and on 18 June 1836 the gathering was commemorated by a painting of those present by William Salter. Among those portrayed is Lieutenant Colonel William Lionel Dawson Damer, captain and brevet major of the 1st Dragoon Guards. In May 1837 The King's Dragoon Guards marched to Liverpool and embarked for Ireland, being stationed at Dundalk, Athlone, Longford and Belfast. Throughout 1837 the regiment was employed in keeping the peace and quelling riots.[3, 15]

By 1838 the situation in Ireland had quietened down, but there was trouble in Canada, where open rebellion had broken out in 1837 in both Upper and Lower Canada. In England, too, the Chartist disturbances were growing more serious. During 1839 The Queen's Bays, together with the 1st Royal Dragoons and the 8th Hussars, were withdrawn from Ireland and posted to the troubled areas. The old custom of billeting troops in local inns was now resisted by commanding officers for fear of the men being suborned by the agitators, and local residents began to pay for the hire of large buildings to accommodate the troops.

During 1840, while The Queen's Bays were in Edinburgh, the regiment was returning from watering exercise on Portobello Sands when a led horse became restive, plunged suddenly, and knocked down and injured a bystander, Robert Carfrae. The troop sergeant major took the injured man to the barracks, where his wounds were dressed, and the orderly officer gave him 5s. This incident caused a considerable correspondence with the War Office, who had apparently never before received a claim for personal injuries involving a troop horse; and Carfrae, who was seventy years old, claimed he could never work again. Eventually, after much argument within the War Office and with the Treasury, an annual allowance of £20 a year was authorised to be paid to Carfrae.

In 1842 Major Le Poer Trench of The Queen's Bays was presented with a sword inscribed

> Presented to Major Le Poer Trench, of H. M. 2nd Dragoon Guards, as a small but sincere tribute expressive of the high esteem and respect which the inhabitants of the town of Burslem in Staffordshire, entertain for the gallant, firm, and temperate conduct, displayed by him, and the men under his command, on the eventful morning of the 16th August, 1842, when he successfully vindicated Social Order, Restored the supremacy of the Law, and effectually defeated the evil designs of a revolutionary mob.[4, 10, 17]

The King's Dragoon Guards had been put on notice for colonial service and along with the 7th Hussars were in readiness to embark. On 27 February 1838 Lord FitzRoy Somerset wrote from the Horse Guards: 'The period is now approaching when the troops under orders for embarkation will be put on board the ships appointed for their conveyance to British North America.' Trouble had been brewing in Canada for some years, and in the winter of 1837 it broke into armed rebellion. In Lower Canada the mainly French-speaking population felt that their culture was under threat, and that they were dominated by the English-speaking minority. In Upper Canada the agitation was for a more representative form of government, which drew a lot of support from the many Americans and Irish who had settled in the province; and because the frontier between Canada and the

United States was fluid, the claim arose that Upper Canada should become another American state.

Lord Hill wrote from London to Sir John Colborne, commanding in Canada, 'I have to add that the Commanding Officer of each of the regiments of cavalry has been ordered immediately to select a competent officer to proceed to New York for the purpose of purchasing horses for the regiments. They will be accompanied by a few rough riders of good character in plain clothes.' The King's Dragoon Guards embarked at Dublin on four transports, *Stentor, Maria, The Marquis of Huntley* and *Calcutta,* sailing for Canada on 3, 4 and 5 May. Some dismounted men were transferred to a transport, *The Prince Regent,* sailing from Cork on 12 May. Two troops of the K.D.G. together with one troop of the 7th Hussars remained behind, and proceeded to York to form a depot for both regiments.

The commanding officer of The King's Dragoon Guards, Lieutenant Colonel Sir George Teesdale, who had commanded the regiment since 1815, now decided to retire. The vacancy was offered to Lieutenant Colonel George Cathcart, who was already serving in Lower Canada as a staff officer on a particular duty. Cathcart wrote to his father on 24 May, 'I saw a rumour about my having the King's Dragoon Guards. I do not wish it, but if it should happen, I must make the best of it.' Lord Cathcart also received a letter from his son's wife, Georgiana, who was still in England.

> I received a letter from Lord FitzRoy Somerset requesting to see me at the Horse Guards. He told me that Lord Hill contemplated giving the command of the King's Dragoon Guards to George, if he is able to pay the difference between cavalry and infantry, £1,635. I answered that I was quite sure that George would like it very much. He then said the Regiment is now at Cork, embarking for Canada, that it would be as well it should be known who is to command it, therefore it should be settled immediately. If Sir George Teesdale is impatient for his money, Lord Somerset would get Cox to advance it.

Mrs Cathcart raised the money by means of gifts and loans, and on 24 June George Cathcart was writing in a different vein: 'As you may suppose, I am very much pleased at my appointment and very thankful to the kind friends who accomplished it.'[16]

The King's Dragoon Guards arrived in Quebec on 22 June 1838 and were immediately sent to Trois Rivières. On 27 June Lieutenant Colonel George Cathcart, who was on a particular duty at Ste Hyacinthe in Lower Canada, came over to look at his new command:

I came over here yesterday to see the horses they have bought, and have seen for us as nice a lot of 70 horses as I ever saw. They have been well and carefully selected, and many hundreds rejected. The state has been sent to me in due form from the K.D.G.s, by which I see we have 35 Officers' horses and 14 troop horses died on the voyage. We still want about 120 to complete. On 3 July the regiment was issued with swords of the new pattern and with new bearskins for their helmets. On the 7th Cathcart formally took over command.[16]

He reported developments in a letter home:

I found that from some alarm in the Upper Province, Sir John Colborne had decided to send a Squadron to Niagara. The first cavalry ever up there. I went down in a steamboat, and in two hours time I had my Squadron on board. By 9 a.m. the mounted men (for we only had 40 horses) were on their first march to Lachine, about 9 miles, where they embarked, and so on by the St Lawrence. The dismounted men and the heavy baggage are gone by the Rideau Canal. I have the Riding Master, Lieutenant Hammersley, and the Veterinary Surgeon already in Upper Canada to buy horses for the remaining 48 dismounted men. ['A' and 'E' troops made up this squadron, commanded by Captain Martin.] It will be something new on the landscape of Niagara to see a heavy dragoon in the foreground. The Niagara squadron 'obtained great credit immediately after its arrival from the exemplary conduct of some of the men in detecting and bringing to justice persons who attempted to tamper with them, and induce them to desert.[3, 16]

The day after taking up command, Cathcart wrote,

It is with equal pride and pleasure that I find myself most unexpectedly appointed to the command of this splendid regiment. We have 70 very good horses here, bought chiefly in the States, and many of them will soon be fit for duty as they have been ridden and are naturally of a very tractable sort. At present I have a good deal to do to put matters of interior economy in the proper equilibrium again, out of which they have been a little shook by the circumstances of their departure and frequent changes of command. But I have never had more willing or better disposed people to deal with in all ranks from the officer to the private. We are getting horses fast enough and very good and those brought out from England are splendid.

I have a remarkably nice corps of officers, among them Mr Turner, a very

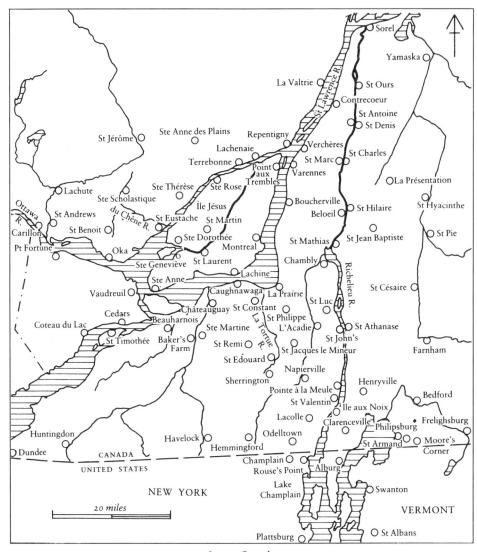

Lower Canada

gentlemanlike, as well as ornamental officer . . . I am getting on well with my new horses. I put them all into the ranks as soon as possible, taking care to put the best riders on the worst broke nags, and every morning from 5 a.m. to 7 a.m. they have been at Adjutant's drill. I am always there myself, and in three weeks of this steady work we have brought them to do everything essential for service, even to cantering in line. They now stand the sword exercises like old ones.[16]

Cathcart had placed orders for the troops' winter wear in Canada; they were to have

> a double-breasted pea jacket in regimental style, and a full length blue overcoat with regimental brass shoulder scales, held in by a white belt with percussion cap pouch, a white pouch belt over the shoulder, knee boots, fur gauntlets, a fur cap with a flap to cover the neck and ears, and fur leggings tied around the leg from below the knee so far as to cover the moccasins.[16]

Shortly after the KDG squadron had reached Niagara, it was felt that a show of strength was neccessary to discourage American freebooters from crossing the frontier. The occasion of a Governor's tour was used to hold an inspection.

> Old Niagara probably never did, and never will again, see such a gathering of cocked hats and radiant uniforms, when His Excellency was met by Sir John Colborne and Sir George Arthur with their respective staffs. An immense concourse, chiefly Americans, attended; the ground was kept by two companies of the 24th Regiment [South Wales Borderers] and a troop of Her Majesty's Niagara Lancers.

The 43rd Foot [Oxfordshire & Buckinghamshire Light Infantry] and the squadron of the KDG then held a field day with plentiful use of blank cartridge.[16]

On 16 October headquarters and the two squadrons of The King's Dragoon Guards at Trois Rivières moved to Chambly on the banks of the Richelieu river. They embarked on a steamer which had two vessels in tow and travelled down the St Lawrence as far as Sorel. They then marched, complete with forage nets, to St Ours, where they bivouacked for the night. On the 17th they reached St Charles, where they bivouacked for a second night. Early on the 18th the Richelieu was crossed in five ferry boats and Chambly was reached by lunchtime. The march was Cathcart's idea, 'as it will make a salutary sensation in the French districts which were the scene of last year's insurrection'. The heavy baggage was brought round by river, while the marching column had but one light cart for each of the four troops to carry the camp kettles, picket ropes and essential stores.[16]

SOURCES

1 R. Cannon, *Historical Records of the British Army, 1st Dragoon Guards*, 1837.

2 R. Cannon, *Historical Records of the British Army, 2nd Dragoon Guards*, 1837.

3 Records of The King's Dragoon Guards, Regimental Museum, 1st The Queen's Dragoon Guards, Cardiff Castle.

4 F. Whyte and A. H. Atteridge, *The Queen's Bays, 1685- 1929*, vol. 1, 1930.

5 William Clowes, *The King's Dragoon Guards, 1685-1920*, 1920.

6 *Regimental Journal, 1st The Queen's Dragoon Guards*, vol. 2, no. 6, 1972, p. 550.

7 War Office 3/257 of 1928.

8 B. G. Baker, *Old Cavalry Stations*, 1934.

9 *The K. D. G.*, vol. 5, no. 2, 1955, p. 108.

10 The Marquess of Anglesey, *A History of the British Cavalry*, vol. 1, 1973.

11 *Journal of the Society for Army Historical Research*, vol. 51, 1973, p. 3.

12 *Journal of the Society for Army Historical Research*, vol. 30, 1952, p. 123; vol. 28, 1950, p. 82; vol. 16, 1937, p. 187.

13 *The K. D. G.*, vol. 5, no. 5, 1958, p. 307.

14 *Journal of the Society for Army Historical Research*, vol. 28, 1950, p. 139.

15 *Journal of the Society for Army Historical Research*, vol. 49, 1971, p. 68.

16 Michael Mann, *A Particular Duty*, 1986.

17 Public Record Office WO 43/766, folios 50-65.

22

Canada
1838-1843

At 1 a.m. on the morning following the arrival of The King's Dragoon Guards at Chambly, a fire broke out in the officers' quarters of the 15th Foot [East Yorks]. The new barracks, built to accommodate the KDG, was adjacent, and all the buildings were of wood. Luckily the wind was in a direction which carried the flames away from the quarters and stables. The KDG turned out at once, saving several of the 15th from the blaze and managing to rescue some of their property. Two officers of the 15th died in the fire.

The area around Chambly was only quiet on the surface, the regimental diary commented, 'So secret were the plans of the rebels, so contradictory the information, and the unsettled state of things, no measures had been taken to prevent the projects of the disaffected.'

On 2 November Cathcart received information of a general uprising due to take place the following day. He considered the situation to be so serious that, handing over command to Lieutenant Colonel Charles Wellesley of the 15th Foot, he rode to Montreal to make a personal report, urging reinforcement of the area and, in particular, securing the bridge over the Richelieu at St John's. The 73rd [Black Watch] marched to the bridge at once, a move that was to prove decisive.

In the meantime the officer commanding at St John's decided to arrest the principal rebels, or Patriotes, as they were known. He rode to Chambly to secure assistance and Wellesley gave him twelve men of The King's Dragoon Guards under Captain G. D. Scott. They rode to an inn where the rebel leaders were meeting, but were made to wait for the arrival of some infantry. By the time the foot arrived, the Patriotes had fled. The KDG detachment then rode to Pointe-à-la-Meule to raid another rebel house, but without success. The regimental diary records, 'This was the first measure taken in the second insurrection.'[1]

The Patriotes had planned to set fire to St John's and to attack Chambly, but the arrival of the 73rd Foot, at the behest of Cathcart, had foiled their plans and thrown them into confusion. As soon as he returned, Cathcart took out a squadron of the KDG to round up the rebels gathering to attack Chambly, 'I trotted on, and by one o'clock had completely surrounded the premises where the meeting was held, with 80 loaded carbines, sprung and ready. The meeting had adjourned before we got there and we only got some arms and a few prisoners . . . The terror of our brass helmets was great, and the district never recovered from it.' Prompt and decisive action had paid dividends. Cathcart noted, 'Their plans seem already to have got into confusion, and I cannot find that they have any formidable leaders of note.'

The Patriotes gathered 3,000 men at Napierville, and on 9 November tried to reach the American border to secure arms and reinforcements, but were repulsed by a small force of Canadian Volunteers at Odelltown. Cathcart reported,

> On 8th November Sir John Colborne sent orders for me to bring all I could from Chambly to St John's. I raised a Corps of 60 Volunteers, left 100 of the 15th Foot and one weak Squadron of the K.D.G. I marched with the rest. By 5 p.m. the K.D.G. had arrived. The Grenadier Guards and the 71st Foot [Highland Light Infantry] and 7th Hussars was to attack from the direction of La Prairie. 4 Companies of the 15th Foot [East Yorks], and with the 24th Regiment [South Wales Borderers] and 2 guns was to attack from St John's through L'Acadie. The third column with the 73rd [Black Watch] and one company of the 15th, and supported by me with The King's Dragoon Guards, was to go up the River Richelieu – the infantry and guns by steamboat, and my Dragoons and gunhorses by land along the left bank. This move was to attack a strong post occupied by about 1,000 rebels, and then cut off all retreat to the south.
>
> On the 9th November, I moved down the Richelieu in time to cover the landing of the 73rd. The few rebels then disappeared and all was deserted. The moon rose at 4 a.m. and we marched, expecting to meet our enemy at dawn. We found him gone. We then marched to Napierville and all the three columns arrived at the same moment. The King's Dragoon Guards and the 7th Hussars met in the middle of the town. We took many prisoners, but all made themselves scarce as could.
>
> Sir John determined to go immediately to Odelltown to encourage the Volunteers and start them to cut off any runaways. He asked me to escort him with the K.D.G. I looked them over, and the hard frozen roads had broken and pulled off some shoes. I selected 50 good men and horses and two subalterns, and halted the remainder under Captain Thyssen. We set off at a good trot for the sun had softened the road, passed through Odelltown, where the battle had been the day before, and halted within a mile of the border. There were Volunteers there but no regulars, but for

my 50. My horses had been mounted from 4 a.m. to 5 p.m. with only one hour's halt. Our American horses, strange to say, beat our English ones at this work.[1]

On 12 November the KDG set off back to Chambly, arriving on the 14th. Cathcart commented,

We have not had a single sore back, and only a few sprains and broken feet owing to cast shoes. I have 144 horses of the two Squadrons fresh and fit for duty, and only 34 (mostly slight cases that I should leave in even for a field day) . . . On my return I immediately communicated with my friend the Curé, who told me that the rebels whom I went after on the 4th, had formed a camp in the Boucherville Mountain with three guns, and much ammunition and arms, and had been threatening Chambly. [They] had broken up and he had sent a young ecclesiastic to advise them to bring their warlike stores to him - to be voluntarily surrendered to me. As they did not arrive, I patrolled next morning with 20 Dragoons to the mountains about three miles off. We took possession of three guns, lots of artillery and other cartridges, twelve sacks of powder, and about forty muskets and several hundred pikes.[1]

On 21 November Cathcart went to Beloil, a centre of disaffection, with a troop of the KDG. He was to

show the people the inconvenience of free quarters and a state of war. I could not afford to relax my own discipline, and never let my Dragoons loose to help themselves, and never were men better conducted than mine are, and have been throughout. I began by arresting Dr Allard, a skulker. Having by means of some perambulations with my Dragoons and the free quartering made a great sensation, and caused sufficient alarm, I made known that all I then required was a voluntary surrender of all arms. If they were brought to the Curé to be surrendered by him to me as a peace offering, I would move. The priest started on an apostolic tour through his parish on his sleigh. By night the arms came in in sleigh loads, and we got about 30 or 40 stands.

Cathcart then went to Ste Hyacinthe. 'I found that all the principal people had been sworn to the rebellious oath.I required four officers and my three subalterns to enter seven dwellings simultaneously. This was done and the persons and papers secured. I marched the column of sleighs escorted by my Troop of the K. D. G. to St Charles and on into Montreal gaol.'

The regimental diary commented,

> Various similar services were performed by the King's Dragoon Guards which proves the utility of cavalry arising from the rapidity of their movements in the suppression of internal commotions, and although it became the duty of the Regiment to resort to active and vigorous measures, their good disciplne and exemplary conduct secured for them the respect and esteem even of those against whom they were called upon to act.[1]

Writing home on 2 January 1839 Cathcart summed up the situation in Lower Canada:

> I have had a very anxious and uncomfortable time, for when I went to Chambly we were led to expect the outbreak, and soon I was in a state of siege, and the secret oath was to exterminate all loyal subjects. My Regiment is in good order, very well conducted and really very happy. All is quiet here now, but the Canadians of Lower Canada know that they are in the natural course of things in progress of being supplanted by our increasing British population and the only way to keep them from openly resisting this now that they have once declared themselves, is by coercion and surveillance.[1]

American sympathisers, however, posed another threat. On 13 January,

> We had strong reason to expect an attempt to burn Phillipsburg, or Mississquoi Bay, by the sympathisers. To prevent it and to encourage the Volunteers, I was ordered to go and take command with one Troop of the K.D.G., the 66th Regiment [Berkshire] and two guns, and all the Volunteer cavalry and infantry on the border. Nothing occurred, but the reports of an attack in force gained ground. I patrolled and reconnoitred the border, and our visit had the good effect of encouraging the Volunteers, but we all marched back again without a chance of doing anything more.[1]

The KDG remained at Chambly, with one troop detached at St John's. There were still sporadic alarms and disturbances, and on 22 February there was an attempt 'to burn the barns occupied by our Squadron of the K. D. G. by night. Our sentries were on the alert and the picquet pursued them, but the rascals took to the ice, and two of our picquet who were on the point of cutting them off, got in and were in some jeopardy, on which the rascals cheered. This accident prevented their getting a shot at them.' But by the end of March the frontier had quietened down, and on 18 April Cathcart, who had been given command of a large area, 'resumed his regimental duties being at the same time still in command of the right bank of the St Lawrence'.[1]

The situation for the squadron at Niagara was much the same, with most of the trouble coming from incursions across the American border. Captain Martin, KDG, reported at the end of May:

> The burning of a barn and shed, and the plundering of the house of a Mr Taylor, near that of the late Captain Ussher who was murdered a few days previous. The barn and a shed of a Mr Miller were destroyed by incendaries who came across from the Yankee shore. The 43rd barracks, the Hotel at Drummondville and Sheriff Hammond's house were set on fire, and I am perfectly convinced that nothing but the utmost vigilance on the part of our sentries prevented it.[1]

Regimental duties were not confined to internal security, for a district general order stated: 'Captain Martin, Commanding the Squadron of King's Dragoon Guards at Fort George [Niagara], having handsomely proposed to give instruction in the Cavalry exercise and movements to the Militia Dragoons as may be ordered to attend him for that purpose, His Excellency the Lieutenant Governor directs that a Sergeant, Corporal and Private from each Troop of Cavalry and Volunteer Dragoons at Toronto, Hamilton and Niagara District with their horses be immediately sent to Fort George for a fortnight or even three weeks, should Captain Martin think it neccessary.' One such detachment came from Captain Denison's 1st West York Troop from Toronto, which was in later years to become the present Governor-General's Horse Guards, and the affiliated Canadian regiment to 1st The Queen's Dragoon Guards; this must be one of the earliest recorded instances of such a close link between affiliated regiments.[1]

With the situation becoming quieter, The King's Dragoon Guards settled down to a more routine existence. The accounts for 1 January to 31 March 1839 show that expenditure on the colonel's account amounted to £67. 4s. 1½d, which included the cost of running the regimental readquarters, the armourers', saddlers', and farriers' expenses. Cathcart noted, 'We have a very good brass band. Last autumn we completed our instruments by an importation of £250 worth, and as we have a very good master who can arrange for it, it makes very pretty music indoors or outdoors.'[1]

Cathcart was ever mindful of his men.

> I find in this sloppy muddy weather it is impossible to live without Jackboots.The boots made large enough to wear a long stocking over the sock, inside of the boot, and pulled over the knee. This was regimental for

the officers. The men had long stockings, but as they gave us 8/6d per man in consequence of the great wear and tear of last autumn, I have been able to give every man a pair of Jackboots to wear during the sloppy seasons of spring and autumn. They are all paid for, and the men very happy in them. Of course this is only for stable duties, watering order and drill order – or actually in the field. But if a General comes to see us, we must put on our Wellingtons, and make up our minds to wet feet. Our Jackboots are quite waterproof. They have the art of making them so here by putting the smooth side of cowhide leather outside, and then making a particular sort of composition of wax and boiled oil.[1]

On 24 May

We had a very grand review at Montreal. Sir John [Colborne] gave me the Cavalry Brigade with orders to play my own game and make movements and charges as opportunity offered. As soon as the 7th [Hussars] were through, we galloped, and on coming to some broken ground where the 7th charge had halted, we broke into files from the right of threes, and did the second division of the sword exercise at a gallop – then halted. These American horses have such good mouths that they never break away, and we can do these things at a gallop as closely and regularly as at a walk. We had not a single horse down that day, although the ground was in a very bad state, and many large ditches were in our way.[1]

In September 1839 Cathcart paid a visit to the Niagara squadron, and again in May 1840: 'I am just now relieving my Squadron at Niagara, one Troop at a time, with all their appointments, but leaving the horses at their respective stations. It is a long march, but the men are easily transported by steamer.' A month later Cathcart was writing,

I have completed the change, and the one comes down after having been out two years under the command of an excellent officer, Martin, and in so good a state did it come in, that all repairs of saddlery have been completed already, and with the help of some tailors, I have got it quite on a par with the rest. In the field too it has lost nothing but has only a few novelties and a little improvement in riding to learn to make it quite up to the mark.[1]

The state of the joint KDG and 7th Hussar depot at York was giving cause for worry:

I have reason to know that the system as regards the training of young officers at the Depot of the two Cavalry Regiments serving in Canada is as bad as is possible and that every description of boyish and unworthy irregularity is tolerated. The consequence is that young men are in danger of being irretrievably spoiled as officers and of contracting habits unbecoming gentlemen. I regret to say that both reports from home and specimens sent out confirm me in this opinion. I have therefore to submit my earnest wish that a proper Field Officer should be placed at the Depot.[1]

In June 1840 Cathcart was writing,

All is quiet and well enough here as far as can be expected. I have lost my much esteemed friend and Colonel [of the KDG], Sir Henry Fane, but have received a very civil letter from the new one [General Sir William Lumley]. I have got a new Riding Master, and a most zealous and efficient one he is, and so popular that NCOs and men come to ask to be put into a class to ride with him. I have now several excellent rides, who go through the whole business with lances, leaping bar etc, and with their stirrups of a right length, and hands in the right places.[1]

Cathcart had been experimenting with placing a blanket under the saddle, which caused wide interest. The board of cavalry officers at the Horse Guards asked for more information, and there were fears about additional expense, so the idea was not well received. Cathcart returned to the attack, pointing out that there would be no extra expense if the expensive shabraque were to be replaced by a good, but cheaper, blanket. He quoted from experience:

On 9th November 1838, one Squadron of the King's Dragoon Guards marched to St Valentine's in the evening, a distance of 10 miles and bad roads, at a trot so as to keep pace with a steamboat having troops on board, and arrived in time to cover their landing. The horses remained saddled that night. At 4 a.m. they marched, and one Troop, after the affair at Napierville, accompanied Sir John Colborne at a rapid pace to the frontier, as an escort, where they arrived at 6 p.m., and with the exception of perhaps one hour in that time (14 hours) never halted. Several long marches succeeded this day and the Squadron returned to Chambly without a single case of a sore back. This trial was in winter, partly in frost and partly in heavy rain. On 4th July a Squadron marched from Chambly at 4 a.m. and arrived after sunset at Phillipsburg, a distance of 41 miles, remaining saddled in bivouac, and marched a distance of 10 miles at daybreak. This march was under a powerful sun, and the Squadron returned without a single case of sore back.

At first the cavalry colonels remained unconvinced, but later Cathcart heard that there was no objection to the change proposed.

Cathcart drew up a manual, *Instructions for Cavalry in Canada*, which were a model of clarity, experience and commonsense. He also secured a change in the form of cavalry cloaks.

> The cloak with sleeves is preferable, the rain runs off freely and it is easily dried and rolled. It is a better thing for a man to sleep in on his guard bed or in bivouac, as it covers his legs and envelopes his whole body. It is a better covering for a man and horse on the march or as a mounted vidette. No active skirmishing can be performed with any cloaks on [without sleeves]. It admits of being rolled and easily carried.[1]

Cathcart also revised the standing orders of The King's Dragoon Guards, which were published on 1 September 1840 at Chambly.

> The Commanding Officer has thought fit to revise the Standing Orders of the King's Dragoon Guards, and embody the principles and regulations, to enable young Officers to learn their duty. The Commanding Officer is equally responsible for the maintenance of discipline and due subordination, whether on Parade, at the Mess, or in any other situation. No discussion on any point of duty is tolerated either at Mess, on Parade, or elsewhere.[1]

The majors and captains of the regiment were

> to attend particularly to the instruction and forming of young officers, not only as to their military duties, but as to their habits when off duty, and will not fail to call them, when neccessary, to that due sense of decorum worthy of officers of the King's Dragoon Guards, so essential to the maintenance of the reputation which the Regiment has so long upheld. [No officer] will ever be tolerated or countenanced, as hangers on, in the King's Dragoon Guards, with the ignoble motives of wearing the uniform, and enjoying the idle comforts and society of the Mess, without the pride and spirit to desire to distinguish themselves by the perfect knowledge and soldier-like performance of their duties. [Every officer] will frequently visit his men's rooms, and see everything is neat and clean, and kept so during the day. On the excellence of its Non-Commissioned Officers the state of a Regiment very much depends. Although familiarity, connivance or undue indulgence cannot fail to lead to contempt of authority, and a relaxed state of discipline, firmness, strict impartiality, friendly advice,

attention to the comfort of the men, and, above all, his own good example, cannot fail to ensure the Non-Commissioned Officer the respect and esteem of the Men of the Troop; and he will find his influence far greater, if thus supported, by fair means, than if asserted only by harsh language, uncalled-for severity, or a domineering manner.

The Corporal's position is a difficult one, but he will remember it is one of trial; and a man who shews himself capable of performing his duties properly as a Corporal, and preserves his character for sobriety, will be sure of promotion; and, as too many fail, those who are fit for it, and can take care of themselves, will not have long to wait for their advancement.

It cannot be too often repeated to the men that they are on no account to marry without leave, and their marrying at all must be discouraged as much as possible. It is impossible to point out in too strong terms, the inconveniences that arise, and the evils which follow a Regiment encumbered with women.[1]

Attitudes towards marriage may have changed, but the general good common-sense of Cathcart's standing orders pass the test of time.

Cathcart had some trouble with his second in command.

The Regiment under my command came out on this service without a Major. It was not until June 1839 that Major Slade arrived, and I was sorry to find that having been sixteen years on half-pay and labouring under bodily imfirmities, had not only forgotten all the details of duty, but could neither walk nor ride in a manner neccessary for the efficient performance of his duty. After he had been scarcely nine months doing duty and often on the sick list, he applied for leave of absence. I submit that a Military Medical Board be ordered to examine and report whether Major Slade is fit for the service.[1]

Another officer gave trouble. Captain Scott

left the barracks at half past five a.m., having been playing at whist and drinking more or less all night, having fallen into the hands of a most culpable medical friend, who administered 'the rather too large dose' [of laudanum] under the effects of inebriety himself. Captain Scott is now under medical treatment by the Surgeon of his own Regiment. He is an Officer of very irregular habits in respect of drinking instead of supporting me in the management of young Officers, and is to them a constant bad example and companion in irregularity.[1]

Early in 1843 The King's Dragoon Guards received orders to return to Britain. Before they left Chambly Cathcart arranged a special parade.

> On the parade ground I had spread a tablecloth on which were placed two kettledrums with their banners. The kettledrums were filled with excellent punch. The three Squadron standards with their bearers and escort of troop sergeant majors, with swords drawn, took post behind, and the kettledrummer with two silver soup ladles, instead of drumsticks, stood ready to bale out the precious contents. When each man had received his share in a goblet, the band played 'God Save The Queen', and then I called upon the Regiment to drink Her Majesty's health. I then sounded the dismiss, and as the men doubled to their quarters, they raised a cheer which could be heard a mile off.[1]

The people of Chambly, nearly all French-speaking, gave the regiment an address:

> 5 years nearly have elapsed since your arrival in the Province, and though the circumstances of your arrival afford no pleasure in the retrospect, and the interval which ensued has been often clouded, yet we contemplate with unalloyed pleasure the whole period of your presence here. In the beginning it afforded protection and allayed differences, and throughout its whole length it has been unmarked by violence, licentiousness, or disorder of any kind, and latterly it has exercised a benign influence, holding up an example of order, social enjoyment, and social improvement, which has been, and will hereafter be, beneficial to us all.

On 9 July 1843 the KDG began its march to Quebec for embarkation, arriving at Ramsgate at the end of September, and marching to Canterbury, having lost only nine horses on the voyage. Here it was joined by the depot squadron, received ninety-nine troop horses from the 7th Dragoon Guards, and by 20 October were pronounced fit for service with a strength of 445 NCOs and men and 374 troop horses.[1]

SOURCES

1 Michael Mann, *A Particular Duty*, 1986.

23

Crimea, Indian Mutiny
1843-1857

The Queen's Bays spent much of 1842 keeping the peace against the Chartists in Cheshire and Staffordshire. In the Potteries the rioting was particularly severe, and The Queen's Bays were constantly called upon, together with the Staffordshire and Cheshire Yeomanry, to charge with drawn swords. The disturbances continued from July until the middle of September and entailed much patrolling, especially at night, and the provision of mounted escorts for prisoners during the day.[1]

In June 1843, for the fourth time since the end of the Napoleonic Wars, The Queen's Bays were sent to Ireland, where they remained until May 1848. During that time they were stationed at Ballincollig, Longford Barracks in Dublin, and Newbridge. During 1843 there was a change in the helmet, when the unwieldy bearskin crest was replaced by a brass helmet surmounted by a long black horsehair tail. The cost of a commission in the regiment at this time was: lieutenant colonel, £6,175; major, £4,475; captain, £3,225; lieutenant, £1,190; cornet, £840.

There was yet another change in the pattern of the helmet in 1847, when the 'Albert' helmet was introduced. The *United Service Gazette* on 27 March congratulated the heavy dragoons on the adoption of the Prussian helmet as 'it is light, fits well to the head, produces an evenness of pressure, and undeniably offers the best kind of protection against a bullet or sword cut'. The Albert helmet had a black horsehair plume for all dragoon guard regiments. The shape has basically remained the same for full dress ever since. On 13 May 1847 an order came from the War Office instructing the officers of all regiments of dragoon guards to discontinue the use of a shabraque as a part of the horse furniture, and later that year another instruction ordered an alteration in the skirts of the coatees of both men and officers, which were to be shortened and squared and 'entirely divested of padding and stuffing'. In 1850 the number of chevroms to be worn by NCOs was laid down for heavy dragoons: the regimental sergeant major was to wear four chevrons with a crown above; troop sergeant majors three chevrons with a crown above; sergeants three chevrons; and corporals two only.[2, 3, 4, 5, 6, 7, 8]

While The King's Dragoon Guards were serving in Canada in 1840, Prince

Albert of Saxe-Coburg arrived in Britain to become the Consort to Queen Victoria. At this time there was a vacancy in the colonelcy of the KDG, following the death of Sir Henry Fane. Lord Hill, the commander in chief, had suggested to the Queen that Prince Albert might become colonel of The King's Dragoon Guards. Albert had, however, become attracted to the 11th Hussars, who had provided his escort on arrival. The Queen wrote to Lord Melbourne, 'Albert, I know, does not wish to have the Dragoons anyhow.' To satisfy Albert's wish a complicated arrangement was made: Sir William Lumley gave up the colonelcy of the 6th Dragoons and became colonel of the KDG; General Philpot relinquished the colonelcy of the 11th Hussars and got the 6th Dragoons, leaving Albert free to be appointed colonel of the 11th Hussars, which the Queen promptly ordered should be designated 'Prince Albert's Own'.[9]

The three squadrons of The King's Dragoon Guards, having returned from Canada in 1843, were stationed at Canterbury, where they were joined by the Depot Squadron from York, this being the first time the whole regiment had been together since 1828, a period of fifteen years. Later that year Lieutenant Colonel Cathcart retired on half pay and was succeeded by Lieutenant Colonel Hankey. Inspections during 1844 and 1845 found the regiment fit for service, and during February and March 1845 the KDG were called upon to police the election of a new Member of Parliament for Canterbury. In April the regiment left Canterbury for the West Country and was stationed at Dorchester, Trowbridge and Exeter.[10, 11]

The new pattern helmets were issued to the KDG during 1846, and each troop was augmented by four men. During April 1846 the regiment was once again scattered around the country, with troops stationed as far afield as Carmarthen, Brecon, Birmingham, Dudley, Trowbridge and Newport. From time to time individual troops were called out throughout the area to assist the civil authorities in maintaining order. During May 1848 The Queen's Bays returned from Ireland and occupied Piershill Barracks at Edinburgh, where they stayed until April 1850 before moving to York and Manchester. In April 1848 the old long-tailed dress coats of the dragoon guards were changed for a shorter coat of a similar length all round.[2, 10]

During May 1848 The King's Dragoon Guards marched to Liverpool from their various scattered stations and embarked for more service in Ireland. Landing at Dublin, they were stationed at Cahir, Clonmel and Limerick, and during 1848 and 1849 were employed in patrolling, escorting officials, and keeping the peace. In April 1850 eight troops marched to Portobello Barracks in Dublin.

The following year Cornet Marter joined the KDG.

A terrible ordeal awaited him. The Regiment was one of the finest in Her

Majesty's service, but among the Officers the so-called 'good old days' of hard drinking were in full swing, immorality was bare faced, practical joking of a serious nature was common, and the bullying of anyone who did not swim with the tide was carried on unchecked by senior Officers. Moreover most of the Officers were rich — the next poorest to him had £300 a year besides his pay, while Marter had but £100, and had undertaken not to get into debt. For a year or more his life was almost intolerable. Then a new Commanding Officer joined (Spottiswood from the 9th Lancers), who quickly brought about a much better state of things.[12]

In December 1850 General Sir William Lumley, colonel of The King's Dragoon Guards, died; he had also been the very popular commanding officer of The Queen's Bays, and his death was much regretted. Lieutenant General the Earl Cathcart was appointed to the KDG in his stead. During March 1851 all dragoon guard officers were ordered to wear a yellow stripe on their undress trousers. The Queen's Bays moved back to Ireland during 1852 and joined with the KDG in trying to keep the peace.

The troops were for six weeks constantly on duty in Belfast and the neighbourhood during the disturbances between Protestants and Roman Catholics. The call for them to turn out usually came about dinner time, and it was not uncommon for them to have to forego that meal, and patrol the streets till midnight or later, much firing taking place between the rioters.[2, 8, 10, 12]

The Duke of Wellington died in September 1852, and in November both The King's Dragoon Guards and The Queen's Bays sent a detachment to London, of three officers and eight NCOs and men, to march in the funeral procession. The KDG officers were Major Spottiswood, Captain Sayer and Lieutenant and Adjutant Bradbury. During August 1853 The King's Dragoon Guards provided a mounted escort under Captain Sayer for the visit to Ireland of Queen Victoria: 'The weather was persistently wet — the streets and roads deep in slush, which was whirled off the wheels of the Queen's carriage upon the "review order" uniforms of the officers riding by it. The difficulty of turning out day after day smartly under these circumstances can be imagined.'[10, 12]

From the raising of both The King's Dragoon Guards and The Queen's Bays the colonel of each regiment had been responsible for providing the clothing and equipment, and had been given a fixed sum, known as 'off-reckonings' for this purpose, for every man on the establishment. A board of general officers, appointed by the King, laid down the patterns, which were then 'sealed', and each colonel had to ensure that the clothing and equipment conformed to these 'sealed patterns'. As the payments were calculated according to the numbers on the establishment, rather than those who were actually serving, it was obvious that public funds were going into the colonels' pockets for men who did not exist. With the outbreak of the Crimean War, and the ensuing heavy casualties, the ancient system became a matter of public scandal. Consequently the Secretary of State, Sidney Herbert, abolished 'off-reckonings' and paid both colonels a fixed sum for clothing; this was set at £800 a year for The King's Dragoon Guards and £450 a year for The Queen's Bays.[13]

In July 1854 The King's Dragoon Guards left Ireland and were stationed in Scotland at Piershill Barracks in Edinburgh, where they provided an escort for Queen Victoria from Holyrood to St Margaret's Station. While in Edinburgh the regiment was attacked by a particularly virulent form of cow fever, which caused a number of deaths. Cornet Marter, with the assistance of one sergeant, was detailed to take a draft of forty-one recruits and eighty horses of the Scots Greys, which had been attached to The King's Dragoon Guards, for embarkation for the Crimea. The recruits were raw Glasgow boys, only partly trained and with little idea of discipline, but with a ready aptitude for drinking whisky. Each recruit had to lead a horse as well as ride, and having travelled by rail as far as Preston, they had two days' march to Liverpool. Marter and the sergeant had to drive the mob, hunt them out of inns, force them to look after their two horses, pull them out of bed in the morning, see them saddle up, and get them moving. With much relief the two KDGs got the party on board HMS *Assistance* at Liverpool and returned to Edinburgh as fast as they could.[12]

In May 1855 the KDG was warned for service in the Crimea, and in July embarked at Liverpool on the transports *Arabia*, *Himalaya* and *Resolute*, arriving at Balaclava in August with a strength of 2 field officers, 4 captains, 8 subalterns, 6 staff officers, 35 NCOs, and 318 rank and file, with 268 troop horses. A depot was left at Edinburgh, which moved to Exeter in August. The *Himalaya* was one of the first screw-propelled steamships, and when launched was the largest in the world. Built for the P & O Line, she was bought by the Admiralty as a troopship and was eventually sunk by bombing during the Second World War when acting as a hulk in Portland Harbour.[10, 11, 13]

The King's Dragoon Guards were encamped on arrival in the Crimea at Kadikoi, but within ten days they were moved further inland owing to an outbreak of cholera. The KDG, together with the Carabiniers, reinforced the original regiments of the Heavy Brigade (the 4th and 5th Dragoon Guards, the Royals, Greys and Inniskillings). The Cavalry Division, which included the Light Brigade (4th and 13th Light Dragoons, 8th and 11th Hussars, 17th Lancers, now reinforced by the

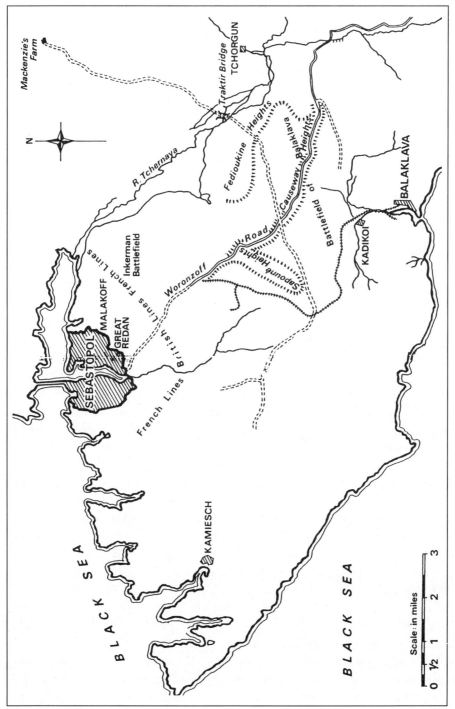

The Crimea

10th Hussars and 12th Lancers), was now commanded by General Scarlett. A KDG officer, Captain H. D. Slade had acted as ADC to General Scarlett until he was invalided on medical grounds in September 1854.[10, 11, 14]

One wing of the regiment under Major Briggs was present at the Battle of the Tchernaya. The Russians managed to surprise the Sardinians under cover of a thick mist, and then attacked the Fedioukine Heights. The French supported the Sardinians and the Russians were driven back. The wing of the KDG was brought up in support, but although present was not actively engaged. Lord George Paget described the day: 'This may be called the poetry of battle to us — to ride out nearly three miles from one's house in the morning; see a general action; return home to a comfortable wash and breakfast, and then spend the afternoon on the battlefield.'[10, 13, 15]

While Sebastopol was under siege and being bombarded, the cavalry had little to do. The summer of 1855 passed pleasantly for them: there were the routine patrols, but apart from that the officers organised race meetings, to which they invited their French allies, although the French officers declined to ride. Major Forrest of the 4th Dragoon Guards noted, 'There is the most remarkable contrast between the get-up of the French and English Officers, the former very smart and with their waists drawn in, trousers well strapped down etc., whilst the generality of our Officers throw off the soldier as much as possible and attempt the sporting get-up.'[14]

In September 1855 the troops received the issue of the Crimean War Medal, described by Colonel Hodge of the 4th Dragoon Guards as 'a vulgar looking thing, with clasps like gin labels', which were quickly dubbed 'Port', 'Sherry' and 'Claret', 'How odd it is that we cannot do things like people of taste.' The men of the KDG earned but one of the clasps for 'Sebastopol'.[13, 14]

During the first week in November The King's Dragoon Guards, along with the rest of the cavalry, embarked on troopships to go into winter quarters at Scutari, where they remained until June and July 1856 before returning to Britain. Although peace was not officially sealed until the signing of the Treaty of Paris in March 1856, hostilities had terminated with the winter. The KDG had lost twenty-five men killed or died of disease during the campaign, with a number sent to recuperate in the better climate of the Bosphorus, among those being Paymaster W. Smith, Adjutant D. Wall, Quartermaster J. Bradbury and Assistant Surgeon J. Jephson. It is interesting to note that ill-health seemed to affect the headquarters officers more than those with the service troops.[10, 11, 13]

The King's Dragoon Guards arrived home with a strength of 2 field officers, 4 captains, 8 subalterns, 6 staff, 23 sergeants, 12 corporals, 5 trumpeters, 4 farriers and 236 troopers, and they brought back 203 horses. On landing in mid-August they went into billets at Farnham, but on 1 September the regiment marched into camp at Aldershot. Three weeks later they moved, partly back to Farnham, but with troops at Yorktown, Blackwater and Bagshot. They were not there for long, for at the end of October 1856 they moved to Exeter, where they heard that the regiment was to bear on its standard the battle honour of 'Sevastopol'. At Exeter the Crimean contingent rejoined the depot of six troops, and the strength of the regiment was established at 8 troops, with 34 officers, 57 NCOs, 8 trumpeters, 8 farriers and 461 privates, with 334 troop horses. As the regiment was fifty men over strength, these were allowed to remain as supernumeraries until they could be absorbed by natural wastage.[10]

The Queen's Bays had been in Ireland since 1852. In 1856 they were stationed in Dublin, when there was an alteration in the regiment's uniform, with the facings being changed from black to buff. On 11 July 1857 the news reached London of the mutiny of the Indian sepoys at Meerut on Sunday, 10 May, and their capture of Delhi the following day. A telegram was at once despatched to The Queen's Bays, informing them that they were to proceed to Liverpool immediately en route for Canterbury, before embarking for India as part of the reinforcements. One troop was to remain at Canterbury as the depot, while the other nine troops embarked on 25 July in the transports *Blenheim* and *Monarch*, under command of Lieutenant Colonel Hylton Brisco and with a strength of 28 officers, 47 sergeants and 635 other ranks.[2]

The King's Dragoon Guards left Exeter for Aldershot in June 1857 and on 4 August they were also ordered to prepare for Indian service, with an augmented strength of 43 officers, 58 sergeants and 679 other ranks, and 703 horses. The regiment marched for Canterbury on 7 August, having been inspected by the Duke of Cambridge two days earlier. A depot troop was formed of 3 officers, 9 sergeants and 74 other ranks. The regiment was carried by rail to Gravesend and embarked on 24 August in the transport *City of Manchester* with 31 officers, 49 sergeants and 462 other ranks. The regiment was issued, on embarkation, with the new breech-loading Sharpe's carbine. The *City of Manchester*, a long and narrow four-masted vessel built for trade in temporate climes, was completely unsuited to the tropics. The mouth of the Hooghly leading to Calcutta was reached on 7 November, after two stops for coaling at the Cape Verde Islands and Cape Town.[10, 11, 12]

In the meantime The Queen's Bays had had a longer voyage, taking four months, and did not reach Calcutta until 25 and 27 November. It was a difficult journey; one of The Queen's Bays described the conditions:

> She [the troopship *Hanover*] was a two-decker vessel of 1300 tons burthen.
> The number on board, exclusive of crew, totalled 595 men and 2 females.

On quitting England we found our daily rations were 1 lb of biscuit, 12 ozs of meat – always salt meat – never once during the voyage seeing a bit of bread or a bit of fresh meat. We had tea and sugar, of each a small quantity. Water we found was a very valuable asset. The cook claimed three pints per man, this reducing the quantity available for other purposes to two pints per man per diem. The rust from the iron tanks caused the water to become yellow, and it eventually got so bad we had to strain it through a towel before drinking it. The biscuit was like biting a deal board, it was hard I can assure you. The ship was rationed for four months' voyage, and should the voyage last longer, half-rations was the order of the day. Each man was served out with a three pound stick of tobacco to last the voyage. Smoking was allowed one hour a day on the forecastle, the bugler on duty sounding 'Light Up' and 'Out Pipes'. A breach of this order was severely dealt with for fear of fire. Drill of any kind was impossible on account of the crowded state of the vessel. There was no canteen or other place where neccessaries could be obtained. When twenty eight days out from Gravesend, we sighted Madeira, the only land sighted between England and India. In sight of that island we lay becalmed for twenty-eight days. There the monotony of a troopship was manifest. To air the messing deck, which served as sleeping room, all hands were turned up on the upper deck. Then one-half had to sit down to allow the other half to walk about. After three months at sea the Captain announced that we were just over half of our journey. Consequently half-rations became the order of the day – three pints of water, half-pound of biscuit (and it was hard more ways than one), and half-quantity of everything else. Every man did his own washing with sea-water, for which purpose salt-water soap was supplied, but if you rubbed for a week, no sign of lather appeared. The drying-ground was over the side of the ship, and on many occasions some villain cut the ropes, allowing the clothes to drop in the sea. Mangling was left to your own consideration. Most of us folded and sat upon it. I may state that from first to last our food was not of the most appetising kind. On Sundays, boulle soup was given for a change, but it wasn't generally approved of, as one mess found a dead mouse in theirs, and another mess found a man's finger with a rag wrapped round it. However there's an end to all things, and after a voyage of 141 days we arrived at Calcutta.[2]

Both The King's Dragoon Guards and The Queen's Bays found, on arrival in Calcutta, that there were only enough horses to mount one heavy cavalry

Southern India

regiment. Although the KDG had arrived first, Colonel Campbell of The Queen's Bays had travelled overland, in advance of his regiment, and as he was a cousin of the Governor-General, he had managed 'to obtain the promise of the horses' for the Bays. In addition the men of The King's Dragoon Guards were considered to be 'the heaviest of the heavies' at that time, so there was said to be difficulty in obtaining suitable mounts for them. The KDG were not pleased, and one officer wrote, 'The Bays did not arrive for a fortnight, and then in a sickly state, and so, when cavalry was urgently needed up the country, the KDG were ordered down to Madras.' They were then posted on to Bangalore and Arcot. Lieutenant Colonel Charles Foster was the commanding officer, having just joined the KDG as the regiment embarked, and he had a reputation for being cold and austere.

On the outbreak of the Mutiny there had been four British cavalry regiments in India, of which only two were in the Bengal Presidency. With the arrival of the reinforcements there were by March 1958 a total of eleven regiments, the KDG, The Queen's Bays, the 3rd and 7th Dragoon Guards, the Carabiniers, the 7th and 8th Hussars, the 9th, 12th and 17th Lancers, and the 14th Light Dragoons.[2, 10, 11, 12, 13]

The KDG who were at Arcot were stationed in a local barracks in order to watch the Madras Light Cavalry, who were suspected of an inclination to mutiny. 'This, though called a fair specimen of an Indian quarter, is as wretched a place as I could have imagined – good barracks are a few ruinous looking bungalows.' In November 1858 a squadron, under command of Captain Slade, marched to Bellasy with the object of capturing some rebel chiefs in the Darwar and Belgum jungles. The countryside was extremely disturbed: camp followers were constantly attacked and beaten, tent ropes were cut, and kit was stolen so cleverly during the night that the sleeping soldiers were not awakened. One day a small party that had been left to clear the camping ground was attacked by a mob and had to draw their swords and use them. The squadron was out on this duty until February 1859, when it was ordered back to Bangalore. But this was typical of the frustrating internal security work throughout the Madras Presidency to which the KDG were condemned, while their sister regiments were dealing with the mutineers. In April 1859 Lieutenant Colonel Pattle arrived in Bangalore to take over command from Lieutenant Colonel Foster.[10, 11, 12]

Captain W. H. Seymour of The Queen's Bays wrote to his mother:

> Until now, from the time that we disembarked, we have been leading such a worrying, disagreeable, busy life this is the first time I have been able to sit down. We have all been more or less indisposed since coming off board ship, and this I believe invariably the case after a long voyage, and I, along with the rest, was unwell for four or five days, but by steady living and a little physicking from the Doctor, I was soon mended up, and now never felt better in my life.
>
> Our troubles now commenced, for we received orders on our arrival, to the effect, that we were to be mounted immediately, on horses that had been procured for us, and sent up the country to Sir Colin Campbell with the utmost speed. On the 28th, received orders to proceed on the following day with my Squadron and 160 horses to Raneegunge. Paraded on the morning of the 29th before daylight, took over our horses, crossed the Hooghly by ferry, and reached this by railroad, 130 miles from Calcutta, the following morning, marching into camp and picketing our nags. Two other Squadrons came up the two following days, and we have now here

of the regiment some 500 men and 400 horses. The remainder of the regiment we expect every day, but with how many more horses we shall march off, I can hardly say, they being exceedingly hard to get, notwithstanding the government are paying 1000 and 1500 rupees for our troupers. We are so far capitally mounted, and are very anxious to continue our march up the country, but we shall not get away from this for another week, notwithstanding we are using almost superhuman efforts to get our saddlery, etc, fitted and ready. Among our horses we have 120 from the Governor General's Bodyguard, and I believe we are to get the whole of the remainder.

The route of our march has arrived. By it, it seems that we are to march the usual distance a day, as far as Benares, and then proceed by forced marches to Allahabad, and so on. We are living here in a camp in tents, and though this is the cold season, and old Indians say the weather is delightful, the thermometer in my tent is over 80 Fah.

SOURCES

1 *Journal of the Society for Army Historical Research*, vol. 19, 1940, p. 140.
2 F. Whyte and A. H. Atteridge, *The Queen's Bays, 1685- 1929*, vol. 1, 1930.
3 *The K. D. G.*, vol. 5, no. 5, 1950, p. 301.
4 *Regimental Journal of 1st The Queen's Dragoon Guards*, vol. 4, no. 4, 1983, p. 1102.
5 *Journal of the Society for Army Historical Research*, vol. 52, 1974, p. 229.
6 *Journal of the Society for Army Historical Research*, vol. 53, 1975, pp. 1, 2; vol. 55, 1977, p. 110.
7 War Office 3/138 of 13 May 1847.
8 *Journal of the Society for Army Historical Research*, vol. 16, 1937, pp. 200, 201.
9 Royal Library, Windsor Castle.
10 Records of The King's Dragoon Guards, Regimental Museum, 1st The Queen's Dragoon Guards, Cardiff Castle.
11 William Clowes, *King's Dragoon Guards, 1685-1920*, 1920.
12 Marter Family Diaries, Regimental Museum, 1st The Queen's Dragoon Guards, Cardiff Castle.
13 The Marquess of Anglesey, *A History of the British Cavalry*, vol. 2, 1975.
14 J. M. Brereton, *History of the 4/7th Royal Dragoon Guards*, published by the regiment 1982.
15 S. G. Calthorpe, *Cadogan's Crime*, 1979.

24
Indian Mutiny, Lucknow
1858

The Queen's Bays were, by the end of December 1857, nearing the troubled areas. On 24 December Captain Seymour wrote:

> Here we are, en route for Allahabad, where we are to be formed into a brigade, along with the 7th Hussars and two troops of Horse Artillery, under our Colonel, Brigadier Campbell. We get to Allahabad about the 15th of next month, and very hard marching we shall have had, as the last ten days are forced marches. This, on a new regiment, is trying. We are marching without our heavy baggage, yet we have no less than 2000 paid camp followers! We march at 4 a.m., always in the dark for two hours, and this adds much to the ordinary confusion of a march. The worst part of my tale, however, is to come – cholera, a disease from which we hitherto have been free, has made its appearance, and we lost our first man, since leaving England, yesterday, and have a sergeant now dying. I believe every new regiment has had it but ourselves, so we have been hitherto fortunate. The poor fellow's case yesterday was the most rapid case of cholera I have as yet seen: he was taken with it on the march in the morning, and died at sunset, a fine, strong, hale old dragoon. My establishment is complete, and consists of a bearer, a kitmutgar, a doby, a clashy, 4 syces, 4 grass cutters. My campaigning stud is also complete – two chargers and two baggage ponies. My tent is a government one, and in it I 'hang out' with one of my subalterns. Regimental mess we have none, but we mess by troupes [sic], i.e. captains with their subalterns, and very comfortably too – though off our rations and without tables and chairs! The more we get up country, the colder it becomes, and although the mid-day is very hot, yet marching before daylight is very cold work, shiveringly so![1]

On 11 January 1858, on nearing Allahabad, Seymour wrote from twenty miles beyond Benares:

We have been continuing our march daily, unparalleled I believe with a young cavalry regiment in India: no halt having been named in our route, and we are finishing up with forced marches to Allahabad. I, thank God! have kept my health and spirits perfectly, but we have had much sickness – cholera, fever and dysentery among our men, and owing to the unsafeness and usual inconvenience of this country, have lost more men than we otherwise should have done, for we are obliged to carry the sick daily with us – poor fellows! Out of 278 dhooley bearers that accompanied us from Raneegunge, not one is now with us, nor have we had one for days, all having bolted! Our native servants are behaving better than we had reason to expect, though we are quite at their mercy, and are obliged to put up with their roguery, idleness and drunkenness. Our commissariat is wretched, and there is no help for it. Our carriage is wretched in the extreme, breaking down daily; and owing to our baggage, in consequence, not being up, one has to shake down in one's cloak on mother earth. The elephants with our tents and our camp followers are foot-sore with the forced marches, and the latter generally come just before we march, if at all! Notwithstanding all this, I am as bonny as possible – really happy – and I am certain from this, and my former campaign, that I am quite cut out for a rough life.

We reach Allahabad on Thursday next, and, I hear, shall barely have breathing time there before we shall flesh our swords. Our men will go at the scoundrels with a will, I know, for they are as keen as they can be – quite exasperated at the tales we hear. All the police stations, and most of the European bungalows that we have passed for some days, are in ruins, and frightful must have been the havoc. Scarcely fifty perfect yards of the electric telegraph, but it has all been mended now.

Our horses are turning out very well, in fact, we are capitally mounted, but it is a nuisance not knowing whether they will stand fire or not – the 'sine qua non' of a trooper – as they never carried a dragoon before. I hear the Carabinier troopers have got the regiment into several scrapes on this account, having turned tail with their riders. I trust this may not happen to the Bays, who have not been on active service for forty years. I can tell you nothing as to the state of affairs in this country, as we seldom, if ever, see an Indian paper, and one can believe nothing that one hears, but everyone agrees out here that we have some roguish work before us, and that India will never again be what it was, so entirely is confidence shaken.[1]

The Queen's Bays were first in action as a regiment on 23 January 1858, but one

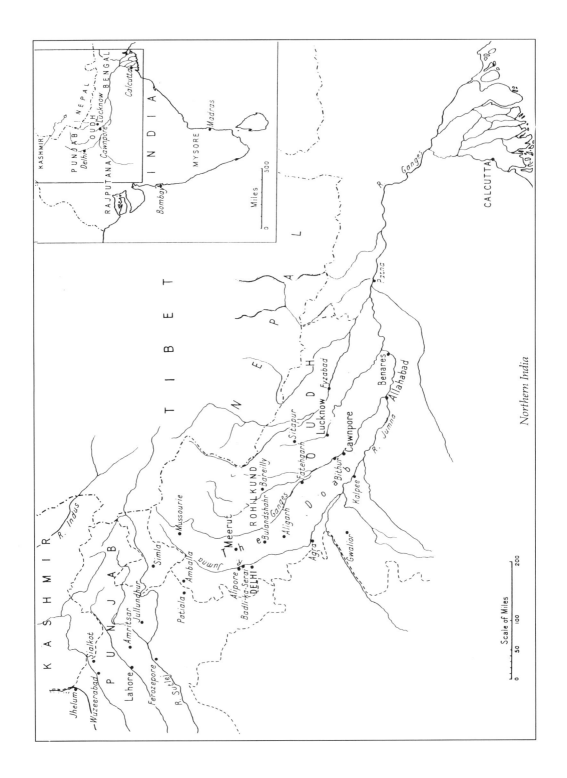

Northern India

of their officers had already won the coveted Victoria Cross, while serving with the 9th Lancers before Delhi. Lieutenant Robert Blair had been commissioned as a cornet into the 9th Lancers on 16 December 1853; he then purchased a lieutenantcy on 2 November 1855 and exchanged into The Queen's Bays on 20 December 1856, but was still serving with the 9th Lancers in India on 28 September 1857, waiting for his regiment to arrive. On that day a brigade of mutineers was encountered entrenched in the village of Bulandshahr. The horse artillery opened fire, but the infantry could not be persuaded to move. Captain Anson of the 9th Lancers wrote, 'They could not be got to look round a corner or to advance in any way.' Lieutenant Colonel Ouvry of the 9th Lancers decided to rush the position: 'I ordered them to charge through the main street. I went through with them myself. We passed through a shower of musketry from both sides of the houses. We met with no loss till we got to the other side of the city. There the enemy made a stand for the moment, but the head squadron charging, the rebels took to flight.'[2] Blair's citation reads:

> A most gallant feat was here performed by Lieutenant Blair, who was ordered to take a party of one serjeant [sic] and twelve men and bring in a deserted ammunition waggon. As his party approached, a body of fifty or sixty of the enemy's horse came down upon him, from a village, where they had remained unobserved; without a moment's hesitation he formed up his men, and, regardless of the odds, gallantly led them on, dashing through the rebels. He made good his retreat without losing a man, leaving nine of them dead on the field. Of these he killed four himself; but, to my regret, after having run a native officer through the body with his sword, he was severely wounded, the joint of his shoulder being nearly severed.[3]

In this action the 9th Lancers won four other VCs, and Lieutenant Blair became the only officer of The Queen's Bays to be awarded the Delhi Clasp to the Indian Mutiny Medal in addition to that of Lucknow.[3]

On 23 January 1858 two squadrons of The Queen's Bays, commanded by Major J. Percy Smith, who had been attached to the regiment from the 10th Hussars on embarkation, were part of a force under the command of Brigadier-General F. H. Franks. They encountered a body of mutineers at Nusrutpore and at once attacked them, when one of the Bays squadrons, led by Captain F. G. Powell, made a brilliant charge. The enemy lost 1,800 men and some guns were captured. The British had only eleven casualties, of which five were wounded men of The Queen's Bays, who also lost seven horses killed or wounded.

Seymour wrote home on 30 January describing the action:

I have however been in action. On the night of the 22nd, after having been in bed and asleep for about an hour, I was awoke by the adjutant saying that I was to march immediately with my own and another Squadron of the regiment, under our junior major, on a secret expedition of a few days' duration. We got our saddles on our horses, and were very soon ready to start, but after awaiting further orders for more than two hours at our horses' heads, were turned in again. However we were not doomed to be disappointed, for we (3rd and 4th Squadrons) left Allahabad the next day at 11 a.m. accompanied by a troop of Royal Horse Artillery, the whole under command of Lieutenant Colonel D'Aguilar, of the Artillery, and after marching by a circuitous route of upwards of 30 miles, joined a division of infantry and foot infantry, under Brigadier-General Franks, at a place called Secundra. We soon learned for what purpose we had been sent there, viz., to attack the next and the following day two strong positions of mutineers in a jungle within a few miles of the place, but with uncertain and conflicting intelligence as to the precise whereabouts. Paraded the following morning at daylight, having roughed it without tents or roof the previous night. Our force mustered about 3000 of all ranks, viz., Bays 198, Horse and Foot Artillery 400, the remainder consisting of Infantry (5 companies H. M. 97th Regiment [Royal West Kents] and Goorkhas). Although we paraded very early next day (24th) we did not leave our camping ground until about 9 a.m. and did not come upon the enemy until the afternoon. We found them entrenched in a strong position, in thick jungle, and their fire opened on us somewhat suddenly, our artillery returning their fire pretty fiercely, theirs making no impression on us, only one (artilleryman) being hurt. After this had been going on for some time, our junior major, with our 3rd Squadron, thought he could take some of their guns – being on the extreme right – and very pluckily charged, with however no effect, being unable to get over deep nullahs which had not been apparent even from a short distance off; however, we got off cheap, losing but three horses killed and five men slightly wounded, our men managing to take some 25 of the enemy. It had the effect of making the enemy budge, for they soon after retired from their position under a heavy, though distant, fire from our artillery and infantry, leaving two of their guns in our possession, better certainly than none, though I think we ought to have managed to get some more, for we hear they took away nine with them. It was impossible to pursue them in their flight, for the jungle was so dense that no cavalry could act in it. I nearly got into a scrape, for being towards the end of the action on the left flank with my Squadron, under the orders of the officer commanding the artillery, and supporting the guns, and being far in advance with my trumpeter, I discovered a large body of the enemy retiring across our left front, so I rode back to my chief as hard as I could to report this, but could not get him to move up quickly, but at last when he did, I obtained permission from him to move on further with a subaltern and about thirty men, which we did as hard as we could go, for about a mile. We caught up the enemy, but could not follow them, they having literally buried themselves in the jungle, where we could not follow them, and whilst trying our best to get on, under a heavy fire of musketry, we were ordered back. Luckily and marvellous to relate, we sustained no loss, as their balls came pouring on us pretty thick.[1]

In February 1858 The Queen's Bays were ordered to march north from Allahabad to form part of a force under the commander in chief, Sir Colin Campbell, which was to move on Lucknow. Since the recapture of Delhi, Lucknow, the ancient capital of Oudh, had become a centre of resistance for the mutineers. At the start of the Mutiny, the rebels had seized the city and besieged the British, who were cooped up in the Residency. Havelock with a small relief force fought his way through to the beleaguered garrison, reaching them just in time, but both the besieged and the relievers then had to endure a second siege. In the autumn of 1857 Sir Colin Campbell, with a second relief column, managed to raise the siege, but had not sufficient strength to capture the town, and so retired with what remained of the garrison and its imprisoned Europeans. Now, in February 1858 he was assembling a force of 20,000 men to march on Lucknow, which was held by 130,000 mutineers, many of whom were regular sepoys, whilst others were followers of their hereditary chieftains.

The country was so unsettled that on their march to join the Lucknow force, part of The Queen's Bays were diverted to deal with another threat. Seymour wrote on 6 February from 'Maharajpore, one day's march from Cawnpore':

The left wing (mine again) were turned out at a moment's notice, and by another telegraphic message, were off in an hour's time, on a three days' trip. Our force consisting of the left wing, 'Bays' (mine and another Squadron), and two guns of R. H. Artillery, the whole under command of our junior major, Percy Smith. Our orders, to go in search of a Madras force of one infantry regiment and four guns, sent from Futtehpore two days previously with Brigadier Carthew, and to co-operate with them against a force of the mutineers — strength and whereabouts as usual uncertain. Off we got at 3 p.m. and marching until 9 p.m., bivouacked in the open (as before on our last expedition) at Binkee. We could hear no tidings of our friends, who had however been here the night before, so the next morning at four o'clock off we went again, and marching until midday, had to give up the chase. Failing to find the Madrassees, who however we heard had driven the rebels across the Jumna, and had returned to Binkee, we rested our horses and selves for a couple of hours, and retraced our steps to Binkee, which place we reached at 10 p.m. and where we found our friends, who had not even had a chance of firing a shot, the Sepoys, etc., having cut and run at their approach, leaving their camp for loot and crossing the Jumna by boats, where, for want of means, it was impossible to follow them. This account consoled us a little, as we should have been preciously disgusted if the Madrassees had 'cut us out'. The following day we all returned to Futtehpore, much fatigued, having roughed it for two nights, besides being in the sun and our saddles all the daytime. On the 12th we continued our route to Cawnpore, where we hope to make no stop, continuing our march to Lucknow. We are all well, and having left our sick behind at Allahabad (where we have a depot for sick, recruits and young horses) have hardly an invalid with us. Our men have certainly become wonderfully acclimatised and are in great spirits at the idea of taking Lucknow, etc, etc. Six weeks from hence will have decided all, and will, I trust, see us en route for cantonment at Meerut, Umballah, or some good quarter for the summer months. I suspect, however, that we shall be wanted next cold season, as the country will be for a long time in an unsettled state, even after the fall of Lucknow.[1]

Seymour found the moves distressing:

I never was a grumbler, but really this life is miserable; everyone is disgusted. India must have been a luxurious, pleasant place, but I am sure it is that no longer. You cannot imagine the wreck of the European bungalows throughout this part of the land. Everything that seemed to appertain to the English or Europe has been rased to the ground. Dozens of bungalows are in ruins, in fact, none remain, and yet what strikes the newcomer as most wonderful is, that with the voice of their country upraised against us throughout the length and breadth of the land, we should have such a hold on them as we again have. We arrive at Cawnpore, I believe, in eight days' time, and Cawnpore is 45 miles from Lucknow.[1]

From Cawnpore Seymour wrote:

We marched in here yesterday morning. All is desolation around, not an

European bungalow or edifice standing, churches and all destroyed. One has, of course, visited the 'slaughter house', 'well', 'compound defended by General Wheeler', etc, etc. – all is desolation. One scarcely knows what regiments are here, but I have discovered the 23rd Fusiliers [Royal Welch Fusiliers], and the 3rd battalion Rifle Brigade. We move across the Ganges en route for Lucknow in a day or two – perhaps tomorrow, our brigade consisting of Bays, 7th Hussars and two regiments of irregular Cavalry, under Brigadier Campbell. Another cavalry brigade consists of 9th Lancers etc, etc.

The cavalry division was commanded by Hope Grant, a 9th Lancer, and consisted of two cavalry brigades; the 1st under Brigadier Little had the 9th Lancers, 2nd and 5th Punjab Cavalry, Wale's Sikh Horse and the 2nd battalion Military Train; the 2nd Brigade, under Campbell of The Queen's Bays, had in addition to the Bays and 7th Hussars, the 1st Punjab Cavalry and Hodson's Horse with Barrow's Volunteer Cavalry.[1, 4]

The force moved on 1 March. The Queen's Bays, joining the army on the 3rd, were attached to Sir James Outram's column, composed of all arms, and given the task of forcing the crossings of the River Goomtee. On 5 March Seymour wrote to his mother, from the Dilkoosha (Deer Park) Camp, before Lucknow:

As I foretold in my letter to you, posted the day before yesterday, we marched that night at 10 o'clock. A long march so far as time went, and the most exciting one I have had, 15 miles, but 12 hours in accomplishing it. The Dilkoosha had been taken the previous day, and for this place we departed. We arrived at Alumbagh during the night, our force consisting of 'Bays', Artillery, 2 battalions Rifle Brigade, and 23rd Fusiliers and a long train of carts of 'Material'. After passing through, we had to feel our way through Jellalabad (a mud fort), taken by us some hours before, I believe, and daylight showed us our friends the Pandies, around the doomed city of Lucknow. The gentlemen, however, took great pains to keep at a respectful distance, in fact, would not let us get near them; however, taking advantage of our passing through jungle, they managed to set fire to two of the carts, which fortunately instead of containing ammunition, merely contained chopped sugar-cane for the cattle. Reached our camping ground, as above, at 10 a.m. having made a circuitous march round Lucknow. Dead-beat, we were glad to get our horses picketed, and I slept like a top from breakfast until dinner time – not so fortunate however were all of us, for nearly half the regiment went off on picquet duty. The duties of the cavalry are very hard, our horses being always saddled up, and we have all kinds of outpost duty to perform. This morning I have again been under fire. Rising with the sun, some of us went off to have a look at Lucknow. We got two capital views of the city, the one from just in front of our camp, where are some of our batteries, though at the risk of several bobs of head. The place seems as strong as John Sepoy can make it – very strong indeed, and unless they 'bolt' we shall have a severe tussle for it. Of large extent, full of large strong buildings. After taking a good synopsis of the town from a palace in our possession, we cantered off to see our friends of Nusseratpoor who have just arrived at Sultanpore, and whilst inspecting a pontoon bridge we are making across the Goomtee, the Pandies came out of the city in great force to drive us off, but a shot or two from us sent their Sowars galloping off. These fellows are the greatest cowards you can conceive; the 9th Lancers say that ever since the Mutiny first broke out they have not killed 100 of them, though there are 25,000 of them. The weather becomes sensibly hotter every day. Whilst I am writing the order has come for us to be off this evening along with the 9th Lancers, on a two days' dowr (excursion), 'cooked provisions, tents left standing, ' so away we are![1]

The next day, 6 March, two squadrons of The Queen's Bays made a spirited charge under command of Major Percy Smith, suffering nine casualties. Major Percy Smith, Corporals Hunt and Nicholls were killed and six men wounded. Like the KDG at Waterloo the enthusiastic but inexperienced Bays got out of control in broken ground over which they should never have been led. A squadron of the 2nd Punjab Cavalry, led by Captain D. M. Probyn, charged on their right and had only three men wounded. Seymour described the action:

About 10 a.m. we came on bodies of cavalry and infantry of the enemy. 'Bays' were ordered to the front to 'charge and pursue'! Away we went as hard as possible, Major Percy Smith and I leading. We did not stop for three miles, cutting down, pursuing and cutting up the Pandies right up to Lucknow, and across the river. We are told the most gallant, smartest, though somewhat rash thing that has been done before Lucknow. Mr Russell (Times Correspondent) perhaps will tell you how many we cut down, though he has not been with us! Alas! however, we lost our best officer, shot dead alongside and within five yards of myself, the nearest of all, with some fifteen men, to Lucknow. Poor Percy Smith, our Junior-Major! The 'recall' had just been sounding all over the place for us, and we had just been polishing off some 50 infantry that we had got in a body. He fell without a groan, and I and four of my Troop tried to bring his body off, but their cavalry bore down on us in such numbers that it was impossible. We, however, Heaven knows how, got his helmet, sword, pistol, watch and medals, which I cantered off with, the last but one to leave the spot; the remaining one being a corporal of my Troop, whose horse would not let him get up again – this poor fellow was cut into ribbons! I returned without a scratch, though my charger got a nasty sabre wound on his off foreleg, which will prevent my riding him for some time. Our vetinerary surgeon says that another eigth of an inch would have done for him. I cut three fellows down, besides hitting one or two others. All this seems perhaps like egotism, but I cannot help telling it to you. We roughed it in the open that night, and had nothing to eat all that day or night – no baggage being up.

The charge captured an elephant, and killed between sixty and eighty mutineers. [1, 4, 5, 6]

On 8 March Seymour wrote:

Yesterday I was out all day in command of the largest party I have yet commanded, protecting a working party clearing a road from the Dilkoosha here, and we got up some heavy guns last night. My force consisting of one Squadron 'Bays', one Troop 9th Lancers, Watson's Horse, two Horse Artillery guns, and 100 Sikh infantry. Some of my Troop horses are frightfully wounded. I never saw such cuts. My Troop alone, we fancy, cut up over 70 Sepoys, and I had only 45 men mounted in the morning.[1]

On 14 March Seymour was in camp.

North side of Lucknow. On the 9th, it happened that I was cavalry captain in camp, and having wandered to the front, got up just in time to become a spectator of the attack on the N.E. side of the city on that day, when the 'Yellow House' and several mosques, and a considerable portion of this part of the town was taken. It was the prettiest attack I have ever seen made, and the Pandies were sent flying in all directions, and it ended in our coming into more than we wanted that day. Our camp, consequent on the operations in front, being left with but few troops, was in some danger, for large bodies of the enemy's cavalry kept threatening us; we could however only get near them for a moment during the day. These Pandies hovering about caused us to strike our tents, etc., so we had to rough it in the open that night. The next day we made an extensive reconnaissance, and cut up a lot of fellows in various places, though I am sorry to say, with the loss of one of our best officers, Sandford of the Sikh Irregulars, who was shot through the head approaching the armed village. The following day we had again a hard day's work, working close round the city, cutting up as we went. The extent of Lucknow would surprise you – it seems endless.

On 16 March the mutineers were finally cleared out of Lucknow, but some 20,000 sepoys managed to escape, and lived to fight another day. On 15 March Sir Colin Campbell had despatched the two cavalry brigades to chase the rebels fleeing from the city, but the cavalry scattered all over the countryside and had little success; but, more importantly, they were not available on the 16th to block the retreat of the bodies of mutineers, and did not return to camp until the following day.[1, 4]

SOURCES

1 F. Whyte and A. H. Atteridge, *The Queen's Bays, 1688-1929*, vol. 1, 1930.

2 E. W. Sheppard, *The 9th Queen's Royal Lancers, 1715-1936*, 1939.

3 W. A. Williams, *The V. C.s of Wales and the Welsh Regiments*, 1984.

4 The Marquess of Anglesey, *A History of the British Cavalry*, vol. 2, 1975.

5 *History of the 2nd Punjab Cavalry, 1849-1886*, 1888.

6 G. W. Forrest, *Selections from the Letters, Despatches and Other State Papers Preserved in the Military Department of the Government of India*, 1893.

25

India
1858-1869

With the capture of Lucknow and the dispersal of the rebel forces, operations were concerned with the pacification of the countryside and the elimination of the various separate bodies of mutineers. On 13 April 1858 Seymour wrote from The Queen's Bays' camp outside Lucknow:

The Headquarters Staff march tomorrow. I fear we shall not have a quiet hot season of it, a Squadron of ours having just gone off with the expedition to Fyzabad. They will, I am afraid, continually work the unfortunate 8,000 left to take care of Lucknow.

April 14th. An extra six months' batta has been granted us for the taking of Lucknow – 1095 rupees a Captain's share. Prize money we are to get – perhaps years hence. They dig up and find rupees, jewels, valuables, etc, every day, in and around Lucknow; five per cent being given to natives who give information. There is, and has been for some time, and will be all the hot season, a daily sale of prize things. Some of our people have spent small fortunes at it. It is the lounge of the morning, and a dangerous one too, for one's pocket. Hitherto I have expended but thirty rupees at it. One of my purchases being a 'tulwar' or native sabre, with a gilt handle and velvet case, for which gaudy toy I gave eleven rupees; it turns out that the huge handle is silver; and it, alone, is valued at fifty rupees. This morning I gave ten rupees for twenty yards of fine Scotch cambric (brand new) one and a quarter yards in breadth. Is this dear? – and what would it cost in the 'ould countrie'?

Although we have been through the roughest campaign I believe ever known in India, and we are still roughing it, and are likely to rough it until this time twelve months, still I confess I do not dislike the country. I am perfectly happy here, and enjoy better health than I did in Ireland, barring the fever that I have just had. Everyone tells us that we have not seen India – things are so topsy-turvy; and this, I believe, is the case. I am at

present living on half my Captain's pay, or at any rate a very little more
than half. Weather, very hot! – old Indians say so, and sleep out of doors. I
do not find it so very, very warm – newcomers do not, their first hot
season, I believe. We are left with 8,000 troops, if so many.[1]

On 15 April The Queen's Bays were again on the move.

We (two Squadrons) left our camp at 5 a.m. yesterday, and form part of a
Moveable Column under the command of Brigadier Eveleigh – comprising
some 1,000 Cavalry, Artillery (heavy and light), and 2,400 Infantry from
20th [Lancashire Fusiliers], 23rd [Royal Welch Fusiliers], 53rd [King's Own
Shropshire Light Infantry], and 3rd Battalion Rifle Brigade and Sappers.
News has arrived that General Hope Grant has been surprised; and we
may be off to his relief, but I do not think it likely. We are encamped not
far from the Dilkoosha. The thermometer in my tent yesterday was 112
Fahrenheit, and a punkah-wallah at work!

[23 April.] One of our Officers has gone home sick, and two are going
off to the Hills on medical certificate for six months. One of them is Price,
our Junior Major. He and the youngster who accompanies him to the Hills
both caught my fever; and my servant Smith has also had it, as also many
of our men. It leaves one very debilitated; and I thought I should have
been sent away somewhere; but, all of a sudden, one morning I felt a
change come over me, became stronger and stronger, and now never felt
better, stronger or healthier; in fact, it has done me good. We are shock-
ingly exposed where we are now, for the sake of regularity, our camp
being in Brigade order; whereas, where the rest of the Regiment is, at the
Moosabagh, we were somewhat sheltered under a tope of mango trees.
We all therefore wish this Moveable Column at an end, although they will
find some other amusement of a like nature for us soon after we go back to
the Moosabagh, I suppose, as they say the Cavalry are to be out all the
hot season. It will be a long time before 'India is quiet again'. I look
forward to being on active service for the next twelve months; I mean
until the commencement of the next hot season. I have been living on the
contents of saddle-bags since leaving Calcutta; and shall not see my heavy
baggage, I suppose, for a year or a year and a half. I have, however, nearly
everything I want; in fact, in writing to Calcutta the other day, I could
only think of tea and a sponge. The townspeople are coming back to
Lucknow fast, having all fled; the consequence is, that we shall have capital
bazaars. If they had remained, or merely kept away for even some hours,
they would not have been nearly ruined, leaving all their dwellings to be

plundered by our hordes of camp-followers, who are pretty handy at loot.[1]

On 26 April Seymour was writing from Bunnee Bridge between Lucknow and Cawnpore:

What between the Sepoys and our Generals, it is being 'taken out of us' uncommonly. The Moveable Column at Dilkoosha dispersed to their quarters in and around Lucknow the day before yesterday, returning with many sick. We brought in out of our Wing three sick Officers – two with fever, caused by the exposed position we were in. Well! we fully expected to have a little rest under our mango tope at the Moosabagh – a rest, however, of but twenty four hours for me, for at 1 p.m. yesterday down came an order at a gallop for the other Wing to jump into their saddles immediately and be off – that a large force of Sepoys had attacked Bunnee Bridge, where we have a small Madras force, burnt villages, and cut the telegraph. This latter is true. The Colonel, Brisco, and both Majors being on the sick list, I had to go off in command of the Wing. Immediately after we had left, and when at the place of meeting the remainder of the Moveable Column under Brigadier Eveleigh five miles off, we heard that the 'immediate' order had been countermanded, arriving at the Moosabagh too late for us. Such was our smartness in getting off! Consequently we had been without a morsel to eat from breakfast yesterday until now – midday today. Marched all last night, and reached this place at sunrise – found a Sepoy force of 3,000 had only shown themselves yesterday, and are now eighteen miles off. Our tents, etc, etc, have come up, and we are now making up for lost time. We are off, I suppose, tonight after these Sepoys; but whether we shall catch them or not remains to be proved. As we are all as keen as mustard, if we do happen to come across them I fancy they will fare badly. My men are behaving capitally, and I have no trouble whatever with them. I hear we are to be out and about for about a fortnight. Brigadier Horsford's Column joins our camp here in a day or two. This savours of work![1]

On 8 May, back at the Moosabagh Camp at Lucknow, Seymour wrote his next letter.

My last letter was posted at Bunnee Bridge, the day prior to an expected 'affair' which, however, never came off. On the 27th April the Cavalry and Artillery went out in search of the enemy; but, reconnoitring over some extent of country, no enemy could be found, so we returned to our camp.

At Bunnee we remained for a week, when we were relieved by another Column under General Grant, and came back here in two marches. My party was particularly fortunate in bringing back a few sick, some of our united force bringing in a great many, and losing men of sunstroke – now, I regret to say, by no means an uncommon termination to the British soldier's existence here, owing to the immense exposure he is obliged to undergo.

We are the healthiest body of troops in or near Lucknow, and yet we have or had over 60 in hospital, out of our strength of 450. Nine of our Officers are struck off duty, sick; in fact, I am one of the few who are in a perfect state of salubrity, for all our Cornets are at the depot, or on their way out there. Thank God, I am holding out famously, and really never felt better than at the present moment; and I am certain that with a moderately sound constitution, and leading a regular life, not exposing oneself unneccessarily to the sun, it is an easy matter to get on in this climate; but it is not a climate to trifle with oneself in, as one immediately feels it to one's cost. Within the past week we lost several men of sunstroke – a few hours illness. Our Regimental Sergeant Major, a fine young Dragoon, and a man who would have held H. M.'s commission before long – died of it three days since. I have not a single Sergeant left with my Troop, all being in hospital with fever. This Troop Sergeant Major of mine has never been near a hospital during his service of fifteen years, until now. I am grieved to say that we fear our Junior Major – Price-is dying. He had a relapse of fever, and has been delirious for three days. He is in great danger, and I fear he will not get over it – a right dear fellow, for whom I have the greatest regard. Our Senior Major is still ill. I am Second in Command of the Regiment, owing to the two Regimental Majors being sick, and Colonel Campbell, a Brigadier, at Cawnpore. You never saw a set of fellows so disgusted as ours are with this country, with but one or two exceptions. Some swear they would sooner break stones in England than remain here. We are certainly far worse off than any troops here, being the only ones in tents! – thermometer upwards of 113 Fahr. in them.

[On 9 May] We had a dust storm last night about bed-time, and having had a rehearsal of it at Bunnee, I at once turned into bed. It was as bad a dust storm as our 'savans' had ever seen; and on awaking this morning, I found myself and my worldly goods a quarter of an inch deep in sand. This morning the thermometer has fallen to 84 Fahrenheit, but it will be as hot again tomorrow. Poor Price is a little better this morning – not much.

The doctors were up with him all night, as he was in convulsions. This temporary change has rallied all our sick today in a wonderful manner. We are in an uncertain state of mind; fearing we may be sent on another dowr; but I do not think it likely, as there is every chance of our being paid a visit here by some of the thousands of the enemy hovering about Oude. Unless a vast body more of troops are sent out here, we shall never decimate these fellows, and decimate we must, before we can have rest. I fear in England a strong feeling is rising in favour of these villains. Ah! that invariably mistaken muffish 'British Public', always led astray.[1]

A letter written on 15 May relates:

The weather here is terribly hot, and we are still in tents. A Wing of the Regiment returned from a some-days dowr today; they have, of course, brought in several sick. The Headquarters Wing (remaining here, and with which I have been) buried four men the day before yesterday; and we have daily cases of sunstroke – yesterday no less than seven. Among our losses are poor Price and our Riding Master: the former of fever and the latter of sunstroke. Price's death gives me my majority without purchase. I am the third Major who has held this majority within three months. Poor Price died three days since – unconscious for several days previously. Kirk (the Riding Master) died at Cawnpore (where we have a depot of un-trained horses, and recruits and young officers) after a few hours' illness – promoted from the ranks seven months since. The Sepoys are all around us here, and as they know our garrison is weak and sickly, intend to take a mean advantage of our plight, and attack us; and as this is the period of a great Mohammedan Feast, and from spies, we expect an attack almost hourly ... [22 May] I have out of my Troop of 48 men, 17 sick. This is pretty dreadful.[1]

The Queen's Bays were attached during March to a column commanded by General Grant, and made up of 3,000 British and Sikh troops, the cavalry being the 7th Hussars, five squadrons of irregular horse and one squadron of The Queen's Bays, with the task of dispersing and destroying a well-equipped force of 20,000 mutineers gathered in the area between the Rivers Goomtee and Ganges. Grant moved out on the road to Sitapur and engaged in a number of actions across Oudh, but by the middle of May the heat was such that even hardened troops were dying in scores, and Grant returned to Lucknow. During these operations the Bays squadron, commanded by Captain Hutchinson, was in action at Koorsee and was involved in an affair at Barree.[1, 2]

On 13 June 1858 Grant again left Lucknow with some 3,500 men, the cavalry element consisting of two squadrons of The Queen's Bays, under Major Seymour, together with the 7th Hussars, Hodson's Horse and the 1st Sikh Horse. The column advanced to attack a large body of 15,000 mutineers strongly entrenched at a river crossing, about eighteen miles east of Lucknow at Nawabganj on the road to Fyzabad. After a twelve-mile night march the enemy were surprised, but fought back bravely. The action lasted for three hours, and 'at one time our small force was completely surrounded and fought in every direction'. Towards the end of the action,

> A very large body of Ghazis and two guns advanced against Hodson's Horse. The onslaught of the fanatics, led by standard bearers, was not to be checked by the small body of troops before them, but Sir Hope Grant, seeing the danger in time, brought up the rest of the battery and the 7th Hussars, and at length the enemy retreated, having lost nine guns and 600 men killed.

In this mêlée the 7th Hussars and Hodson's Horse were supported in their charge by the two squadrons of The Queen's Bays. The action was over by 8 a.m. but 33 men of the British force had died of sunstroke, while by the end of the day 250 were in hospital, whereas Grant's battle casualties were but 67. At this period the Bays had only two of their own captains fit for duty, H. M. Stapylton and O. P. C. Bridgman.[1, 2, 3, 4]

On 6 July Colonel Campbell died, following his promotion to the rank of brigadier general to be in command of the Cawnpore Division, a post which he was never able to assume. During September Colonel Hylton Brisco retired, and Lieutenant Colonel W. H. Seymour took over command of the regiment. During July and August Grant tried to rest his troops as much as possible, and although two sorties were made to disperse mutineers and prevent them from gathering, the force remained at Lucknow until the hot weather and the rains abated.[1, 2, 3]

Operations started again in September, and on the 21st a squadron of The Queen's Bays, led by Captain F. Graham Powell met the enemy at Dawah. The squadron charged and dispersed the mutineers, who put up a strong fight in hand to hand combat; in the charge and pursuit Corporal Walls and Troopers Hart and McGuinness were killed, and several men were wounded. A few days later the same squadron, as part of Brigadier General Purnell's column, drove out a force of sepoys from Selimpur.[1]

In October 1858 The Queen's Bays were transferred to Brigadier Sir George Barker's column, and on the 8th of that month were engaged in a smart action at Jamo. Lieutenant Colonel Seymour led The Queen's Bays in person against a party of the 42nd Bengal Native Infantry, who were concealed in a dense jungle of sugar cane.

> The mutineers were between 30 and 40 in number. They suddenly opened fire on Lieutenant Colonel Seymour and his party at a few yards' distance, and immediately rushed in upon them with drawn swords. Pistolling a man, cutting at him and emptying with deadly effect at arm's length every barrel of his revolver, Lieutenant Colonel Seymour was cut down by two sword cuts, when Trumpeter Thomas Monaghan and Private Charles Anderson rushed to his rescue, and the Trumpeter shooting a man with his pistol in the act of cutting at him and both Trumpeter and Dragoon driving at the enemy with their swords, enabled him to rise and assist in defending himself again, when the whole of the enemy were dispatched. The occurrence took place soon after the action near Sundeela Oudh.

So read the citation for the two Victoria Crosses which were awarded to Trumpeter Thomas Monaghan and Private Charles Anderson. In the same action a boy trumpeter of the Bays, John Smith, taking on an enemy sepoy in single combat, cut down and killed him.[1, 5]

The main body of the regiment continued under command of Brigadier Barker in the pacification of Oudh. They took part in the storming of the fort at Birwah on 21 October, and one squadron, commanded by Lieutenant Francis Lavington Payne, was detached to General Grant's column and was engaged in the passage of the River Gogra near Fyzabad on 24 October. The 2nd Gurkhas, then known as the Sirmoor Battalion, were operating in the area.

> A cloud of dust was observed in the distance and a halt was called, as it might have proved to be a body of rebels, and the actual whereabouts of any of our columns was unknown to the Commandant. The Goorkhas formed for attack and loaded, when with glasses they made out a European Officer, who turned out to be a Major Carnegie with a party of Irregular Horse moving to take up ground for the camp of General Barker's column following in the rear. This column shortly afterwards passed the Goorkhas, and at Raniganj Colonel Reid received orders to return and join General Barker's force, which consisted of a Troop, Royal Horse Artillery, a Heavy Battery, the 2nd Dragoon Guards [Bays], 8th Irregular Cavalry, 3rd Battalion Rifle Brigade, and Boileau's Police battalion. With this force the Sirmoor Battalion did duty during the cold weather, aiding in quelling disturbances, dismantling forts, etc, and were

present at the capture of Kyrabad and Biswah, which, however, did not entail much fighting.

In the spring of 1859 Colonel Seymour and Colonel Walker participated in two actions near Bungdon in Oudh between 22 and 27 April.[1,6]

As many mutineers had sought refuge in the hills and mountains of Nepal, Jung Bahadur, King of Nepal, begged the British that they might be hunted down. During the cold weather of 1858-59 two squadrons of The Queen's Bays, under Major Hutchinson, chased the remnants of these mutineers, and at the Jowah Pass they expelled a force of sepoys with a charge. The only casualty, slightly wounded, was Cornet Torrens, acting as ADC to General Grant. The last flames of the Mutiny were now being extinguished, and on 5 June 1859 The Queen's Bays entered cantonments at Lucknow, having been in the field for twenty consecutive months.[1,2]

There were awards for the services of The Queen's Bays during the Mutiny: Colonel Seymour was made a Companion of the Order of the Bath, Major Hutchinson was given a brevet lieutenant colonelcy, and Captain Bridgman would have received a brevet majority had he not died before it could be awarded. 464 Bays who served with the regiment during the Mutiny received the Indian Mutiny Medal with the Lucknow clasp, and 196 received it without the clasp. As previously mentioned, Lieutenant Blair, VC, was the only member of the regiment to receive the two clasps of Delhi and Lucknow. Colonel Walker left the regiment to proceed to China as assistant quartermaster general under Grant, where he served in close contact with The King's Dragoon Guards before retiring after the successful conclusion of that campaign.[1,7]

When Lord Canning, the Governor General of India, retired, he was escorted on his farewell tour of the North West Provinces by a squadron of The Queen's Bays. In 1860 the regiment was inspected by Field Marshal Sir Hugh Rose, the commander in chief, who was extremely complimentary, and during 1861 The Queen's Bays provided an escort of a squadron to the new Governor General, Lord Elgin, at the inaugural installation of the newly created Star of India.[1]

Many members of The Queen's Bays anticipated that the regiment would soon be sent home, but they were to remain in India until the winter of 1869. In 1861 a false alarm raised hopes of impending embarkation, but regiments that had served in the Crimea were given precedence, and The Queen's Bays were moved to Benares early in 1862. During 1862 the overall stripes and cap bands of The Queen's Bays were changed from yellow to buff, in order to match the regiment's facings, and velvet replaced cloth for the facings. There was a further change in 1864 when 'the present wide loops of yellow lace on the front of the collar entirely hiding the facing now that the collars are cut so much lower, Her Majesty is pleased to order its discontinuance and the substitution of an edging of white braid all round the collar'. At the same time the loop and button on the cuff were replaced by an Austrian knot of white braid, and the loops of yellow lace on the tunic skirt were done away with. In 1865 the Albert helmet was modified as the result of Crimean and Mutiny experience, and strengthened into the pattern which is very similar to the one worn in full dress today. During 1863 the establishment of the Regiment was reduced to: 1 lieutenant colonel; 2 majors; 7 captains; 7 lieutenants; 7 cornets; 1 paymaster; 1 adjutant; 1 musketry instructor; 1 riding master; 1 surgeon; 2 assistant surgeons; 1 vetinerary surgeon; 37 sergeants; 28 corporals; 7 trumpeters; 7 farriers; 434 rank and file; 490 troop horses. In 1863 the regiment was allowed to carry the battle honour 'Lucknow' on its standard.[1]

On 6 January 1865 The Queen's Bays moved to Muttra to relieve the 7th Dragoon Guards, and were inspected by the new Commander in Chief India, General Sir H. R. Mansfield, who commented in an order: 'His Excellency desires to express his thanks for a very excellent field day. In the whole course of his service he has never seen anything better. The Regiment moved with great steadiness, rapidity and precision, the horses well broken and managed, and strict attention must evidently be shown to every detail.' In 1865 The Queen's Bays provided the escort to the new Viceroy, the title that had now superseded that of Governor General. He was the famous Mutiny hero, Sir John (now Lord) Lawrence and the regiment attended him during the great durbar which he held at Agra. On 1 April 1867 the establishment of the regiment was again reduced to 21 sergeants and 378 rank and file, although a paymaster sergeant, armourer sergeant, bandmaster sergeant, saddler sergeant, farrier major, hospital sergeant, sergeant-instructor of fencing, sergeant cook, trumpet major and orderly room clerk were added.[1] During their time at Muttra, The Queen's Bays were inspected by Major General Troup on 25 March 1868 and by Brigadier General Stannus on 24 October. The regiment was ordered to move to the Bombay Presidency on 15 November 1868, making an exchange en route of camp equipment with the 11th Hussars at the River Parbutti, who then relieved them at Muttra. The Bays arrived at Mhow on 12 December; there were three further inspections during 1869, and also orders to return home. Nearly a hundred NCOs and men elected to stay in India, transferring twelve to the 5th Lancers, nine to the 16th Lancers, eight to the 3rd Hussars, seventeen to the 4th Hussars, two to the 11th Hussars, twenty-two to the 18th Hussars and twenty-one to the 21st Hussars. The Queen's Bays embarked in the troopship *Malabar* on 31 December 1869.[1] The divisional commander wrote:

As the 2nd Regiment of Dragoon Guards will have left the Mhow

Division before the Major General purposes to return to Mhow, the Major General is particularly anxious to assure Colonel Seymour, CB, the officers, non-commissioned officers, and men of The Queen's Bays of his admiration of the appearance, high tone, and general good conduct evinced by that Regiment on all occasions since they have served under his command.[1]

The journey back to England was swifter than the long journey out to India round the Cape, for the regiment was landed at Suez, travelled by railway from Cairo to Alexandria, and arrived at Portsmouth on 31 January 1870, having taken one month instead of four.[1]

SOURCES

1 F. Whyte and A. H. Atteridge, *The Queen's Bays, 1685- 1929*, vol. 1, 1930.

2 Sir John Fortescue, *History of the British Army*, vol. 13, 1930.

3 The Marquess of Anglesey, *A History of the British Cavalry*, vol. 2, 1975.

4 F. G. Cardew, *Hodson's Horse, 1857-1922*, 1928.

5 W. A. Williams, *The V. C.s of Wales and Welsh Regiments*, 1984.

6 L. W. Shakespear, *History of the 2nd King Edward's Own Goorkha Rifles*, 1950.

7 Indian Mutiny Medal Roll, 2nd Dragoon Guards [Queen's Bays].

26

China

1860

In 1860 the British and French Governments went to war with China for the third time to secure satisfactory trading facilities. The trade consisted in great part of an opium traffic, which undermined the health of the Chinese people and which the Chinese authorities were anxious to stop. In 1858 a British naval squadron with Royal Marines attempted to capture the Taku forts at the entrance to the Pei-ho river, and were ignominiously defeated. Such a blow to European prestige could not be allowed, and the British and French Governments assembled a combined force of 10,000 British and 7,000 French troops to be despatched to China.[1]

The King's Dragoon Guards were stationed at Bangalore in the Madras Presidency. On 1 March 1860 the strength of the regiment was increased by fifty-eight men from the 12th Lancers, who were returning to Britain, and on that day orders arrived for a wing of the regiment to proceed on service to China. On 2 March 14 officers and 312 men, together with 313 troop horses, marched to Madras for embarkation. The colonel, Lieutenant Colonel Pattle, was appointed to command the cavalry brigade being prepared for China, and so the four KDG troops were commanded by Lieutenant Colonel Sayer. The other regiments forming the brigade were Probyn's and Fane's Horse. On 19 April the KDG embarked on sail transports, 'A' Troop on board the *Frank Flint*, 'G' Troop and regimental headquarters on the *Sirius*. The headquarters officers were Lieutenant Colonel Sayer, Captain Wingfield, Lieutenant Greaves and Veterinary Surgeon Thacker. 'B' Troop embarked in the *Trimountain*, sailing on 6 May, and 'F' Troop followed in the *Eastern Empire* on 15 May. On 12 February 1860 the regimental families of the service troops of the King's Dragoon Guards had arrived in Madras from Britain, just when the men selected for service in China were preparing to embark. So after one long separation, the families were condemned to another. These women and children stayed in Bangalore making their lives on their own, having had a brief welcome from their menfolk.[1]

The first two troops of the KDG arrived in Hong Kong on 15 May, but the last two did not reach there until 19 and 29 June, and the following KDG details until the middle of July. Luckily the three cavalry regiments had all brought their horses with them from India, all in excellent condition, for it proved extremely difficult to secure adequate horseflesh in China. General Sir Hope Grant, commanding the British troops, wrote, 'The two irregular cavalry regiments were really magnificent. The King's Dragoon Guards was also one of the finest regiments in the service, and altogether I had reason to be proud of my little cavalry force. It was commanded by Brigadier Pattle of the King's Dragoon Guards.'[1]

The troops re-embarked at Hong Kong, sailing for North China, where the British force was assembling at Talienwan Bay on the Manchurian peninsula, which afforded an excellent anchorage for the fleet. The cavalry made their camp at Odin Bay a few miles to the south, where there was a supply of fresh water. Robert Swinhoe of the Consular Service travelled from Hong Kong with some of the KDG:

> The *Sirius* had on board Brigadier Pattle and fiftyt-five troopers of the King's Dragoon Guards, with their horses. My fellow passengers were mostly officers on the General's staff. And a curious group we were: there was just that amount of disagreeableness that usually occurs among Englishmen who are strangers to one another, and yet are fully aware of the appointment and position that each holds; in a word, there was no conviviality. At sundown on the 21st June we arrived at Talienwan.

Three more KDG troops arrived at Odin Bay on 4 and 5 July, with the fourth troop disembarking on the 14th. Grant reported, 'On my arrival at Talienwan, I found the whole of the British troops anchored in the harbour, with the exception of about 120 men of the 1st Dragoon Guards, all in good health and ready for active operations.'[1]

There was a delay while the French, who were ill prepared, tried to assemble an adequate supply of horses. Thomas Bowlby, the *Times* correspondent, 'sailed to the Cavalry Camp, and saw all the camp, consisting of all the Artillery, the KDG, Probyn's Horse and Fane's Horse. The horses are in excellent health; there are but 40 sick out of 1775.' On 13 July there was an inspection for the benefit of the newly arrived Lord Elgin of the Foreign Office, and the French General de Montauban. Bowlby described the scene:

> We found the troops drawn up on the beach, a battery of Armstrong guns on the left, the K.D.G., and Probyn's and Fane's Horse. It was a lovely sight, with the vast varieties of hue and colour – the dark blue of the Artillery contrasted well with the light blue tunics and red turbans of Fane's Horse; whilst Probyn's blue black tunics and turbans were relieved by the scarlet uniforms of the King's Dragoon Guards.

Grant himself wrote,

> I was glad to show off the other day to General Montauban our Artillery and Cavalry force, and I am sure you would have been pleased with their appearance. There were about 1,000 men on parade; handsome, fine look-ing fellows, well dressed and turned out, and their horses in beautiful condition. General Montauban stated that it was a sight to see in Paris or in London, but one he did not expect to see so far from home.[1]

The King's Dragoon Guards embarked on 23 July, leaving at Odin Bay some details under command of Major Slade. The fleet gathered in the Gulf of Pechilli and approached the Chinese coast. The landing was to be made by the 1st Divi-sion, followed by the cavalry brigade, opposite the town of Pehtang; the KDG in the *Sirius, Frank Flint, Trimountain, Eastern Empire* and *Harry Moore*, each carrying between 60 and 80 horses and men, landed 13 officers and 313 NCOs and men, with 339 horses, on 5 and 6 August; an officer of the Royal Scots was struck by 'the workmanlike way in which the blue- jackets landed the horses of the cavalry brought in by the gunboats, with whips and slings on their little foreyards. A horse fully accoutred was hoisted up, swung over the jetty, and dropped ashore on its legs before it knew what was being done to it.' Once ashore the cavalry had to make the best of the appalling conditions around Pehtang, made worse by persist-ent rain. Bowlby commented, 'The cavalry horses are standing nearly knee deep in the rain, which has completely flooded the fort.'

On 9 August Grant sent out a reconnoitring party of fifty men of the KDG, 'B' and 'F' Troops, and two squadrons of Probyn's Horse, under command of Major Probyn. They were to report on the state of the country to the north of a causeway which carried the road from Pehtang to the Taku forts, since the country on either side of the causeway was proving to be very marshy. Probyn had strict orders not to be drawn into action, and although several bodies of Tartar horse were seen hovering in the distance, they were left alone. After a long detour the cavalry got to within a mile of the Chinese positions and found that the going was practicable for all arms, and that water was plentiful.[1] Wolseley, the assistant quartermaster general to the force, noted:

> The 10th was rainy, and on the 11th we had some slight showers, but when day broke [on the 12th], although the weather was looking threaten-ing, yet the rain did not come down. So the exodus from Pehtang, with all its detestable odours, began. It was arranged that the 2nd Division should move out along the track reconnoitred by the Cavalry on the 9th, and turn the left of the enemy's position, whilst the 1st Division

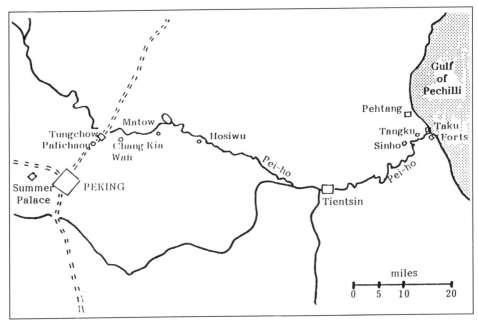

China, 1860

and the French advancing along the causeway towards the enemy's front
should take their works. All the Cavalry were to accompany the 2nd
Division . . . Immediately the 67th [Hampshire] has cleared the ground, the
cavalry will follow in the following order; K. D. G., Fane's Horse, Probyn's
Horse, and 3 guns of Stirling's Battery.

The waggons stuck in the mud and progress was slow, particularly by the heavy
cavalry. The advance was delayed. Wolseley commented, 'It was painful to see the
cavalry horses struggling on knee deep with their heavily accoutred burdens.'
Grant reported,

> This force at last got over. It was attacked by a large body of Tartar
> cavalry, some 3,000, who in the most daring way came up to the guns.
> Our cavalry were let loose upon the Tartars, and they had a hand to hand
> fight, cutting down about 70 or 80. The enemy behaved very gallantly
> and under better rulers would make excellent troops.[1]

As soon as the infantry deployed, a mass of Tartar cavalry trotted forward in
loose order. Sweeping right round the cavalry brigade, they began to threaten the
rear. With a loud yell the KDG, Probyn's and Fane's charged and scattered the
enemy, cutting down numbers of them. As the 1st Division advanced 'a large force

of Tartar cavalry, certainly some 2,000 or 3,000, rode with wild cheers' straight at them. The KDG, Probyn's, and Fane's again charged, but the Tartars were no longer prepared to stand and beat a hurried retreat.

> The guns opened fire upon the enemy's entrenchments. The enemy stood well for a few minutes, and then, after a few rounds more, bolting as fast as their little active ponies could carry them, so that by the time our infantry had reached the place, the only occupants were dead and dying horses and men. The whole army now advanced and passed through the village of Sinho.

The King's Dragoon Guards had one trooper slightly wounded.[1]

The next Chinese position was at Tangku, on the Pei-ho river, which was carried by the infantry. On 15 August some commissariat officers went out, escorted by a few KDG and Indian sowars, to gather supplies from the countryside. As they neared a village, they were fired on by a Tartar cavalry picket. The KDG and Indians at once gave chase, killing three Tartars and wounding three others. All six were brought in with their horses and equipment.[1]

Grant now closed up on the Taku forts, which were assaulted with great bravery on 21 August by the 44th (Essex) and 67th (Hampshire) Regiments, when six Victoria Crosses were won by the two regiments. The casualties had been heavy, 17 men killed, and 19 officers and 153 men wounded. The Chinese retreated up the Pei-ho river towards Tientsin and Peking. The King's Dragoon Guards set out from their camp at Sinho for Tientsin on 25 August, marching along the north bank of the Pei-ho through open flat country. Reaching Tientsin on the 27th, they encamped on open ground around the city.[1]

It seemed to Grant that with the capture of the Taku forts, the opening up of the Pei-Ho river and the signing of the terms of capitulation, the campaign must be at an end. But it soon became clear that the peace negotiations were merely a Chinese ruse to gain time, and to entice Lord Elgin, the Foreign Office plenipotentiary, to Peking. Bowlby wrote: 'Received an invitation from Lord Elgin to go to Peking. We only take up to 1,000 men, of whom it is arranged that the K.D.G.s shall form 300, Probyn gives 100, Fane 100, and fifty picked men from each regiment, with a Battery.' With the collapse of the negotiations, it was decided to advance on Tungchow.

On 8 September The King's Dragoon Guards and Fane's Horse moved out in advance of the infantry. Bowlby noted, 'The King's Dragoon Guards with their white helmets and scarlet coats were followed by Fane's Horse in their blue tunics and red turbans, the pennons on their lances streaming gaily in the wind.' The march was not easy for the cavalry because the roads had not been maintained.

'The country was excessively difficult for troops to advance along, and very dangerous for horses, as the fields were full of millet, which when cut, formed very sharp stalks in the ground, and injured them severely.' Grant commented, 'I cannot say how thankful I am that the Cavalry was sent from India with this force. I scarcely know how we should have got on without it, from the clouds of horsemen which encircled our small forces. And right gallantly the Tartar horse came at us and charged up nearly to our guns.' The cavalry became victims of their own success, however, as more and more demands were made on them for escorts and detachments, dissipating their strength. The KDG provided escorts for the various diplomatic missions, and orderly dragoons for the staff.[1]

By 13 September twenty of the KDG and some fifty sowars had reached Hosiwu, about halfway to Tungchow. On the 17th Mr Parkes of the Foreign Service, Mr Loch, Private Secretary to Lord Elgin, and Colonel Walker of The Queen's Bays, and quartermaster general of the cavalry brigade, set out for Tungchow to prepare the way for Lord Elgin. They were escorted by Lieutenant Anderson of Fane's Horse with twenty-five of his sowars, together with five KDG troopers. The same day the Allied force marched for Matow, and on the 18th left there for Chang-Kia-Wan. Advancing three miles, the troops encountered a picket of Chinese cavalry, who immediately fled, but a few miles further on the force came up against a large Chinese army, stretching across a front of some five miles and barring the way forward. Grant at once halted the troops, ordered the baggage to concentrate in the last village, and sent a KDG officer to close up the rearguard for its protection.[1]

Loch suddenly galloped back into the lines with the news that Parkes's party with the five KDGs and four of Fane's sowars had left Tungchow at 5 a.m. that morning to mark out the area of the Allied camp in front of Chang-Kia-Wan, and that they had passed large bodies of Chinese troops. Parkes, arriving at Chang-Kia-Wan, had left Colonel Walker to watch these movements and returned to Tungchow, taking with him only Private Phipps, of The King's Dragoon Guards, to demand an explanation from the Chinese. [1]

Grant was in a difficult position, with his envoys and their escort behind the Chinese lines. He sent out his cavalry to both flanks, with strict orders not to become involved. He took up a position on a mound, some 400 yards from the Chinese lines, where he could clearly see the red coats of Colonel Walker's KDG escort moving about in the midst of the greyclad Chinese. After two or three hours of waiting, there was a stir in the Chinese lines and Colonel Walker and his escort were seen to be galloping through the surrounding Chinese and heading for the Allied lines. Immediately the Chinese opened fire with their artillery along the length of their position. General de Montauban, whose French troops were on the right of the line, charged with the 2nd Chasseurs de Vincennes, while the Spahis of his escort supplemented by a squadron of Fane's Horse swept round the position, charging three times and doing great execution.[1]

Colonel Walker arrived with a commissary and only two of the KDG troopers. He related how having been shown an unsatisfactory camp site without water in the midst of the Chinese troops, he was left to wait. The attitude of the Chinese around the party started to change. A group crowded around Walker and one man suddenly tilted Walker's scabbard so that the sword came out, which was immediately seized. A Chinese officer had the sword returned. Then Walker observed a French officer, with a deep sabre cut on his head, crying out for help. He held out his hand to lead him away and the surrounding Chinese rushed at them, dragged the Frenchman down, pulled Walker's sword out of its scabbard and tried to pull him off his horse. Walker, trying to rescue his sword, cut his hand badly, and shouted to his companions to ride for their lives. As the party spurred their horses, the Chinese opened fire from every side, wounding two of the men and killing one horse.[1]

The Allied artillery, returning the Chinese fire, soon began to silence the Chinese batteries and scatter their infantry. The Tartar horse, meanwhile, had gathered in great numbers on the Allied left flank. Probyn's Horse, numbering only 106 sabres, charged them without hesitation, soon backed by The King's Dragoon Guards, who chased the fleeing Chinese for several miles. [1]

On 20 September the cavalry probed forward and found that the Chinese had taken up a strong position in front of the Yang-Liang canal, which connected Peking with the Pei-Ho river, and which could only be crossed by two bridges, one wooden, the other — at Palichao — marble. While the infantry attacked frontally, the cavalry were to make a wide sweep to the left to drive the Chinese right flank onto its centre, forcing them to retreat over the two bridges. The King's Dragoon Guards set out from Chang-Kia-Wan at daybreak on 21 September, but after two miles they had to halt to give the French time to come up. Grant, surveying the Chinese position, was joined by Lord Elgin with his escort of KDG. As the troops arrived and took up position, in the words of Grant's report,

> The King's Dragoon Guards and Fane's Horse, with Probyn's regiment in support, now advanced to the charge; the first-named taking a bank and ditch on their way, and attacking the Tartars with the utmost vigour, instantly made them give way. Fane's men followed in pursuit, and on reaching the margin of a road jumped into it over an interposing high bank and ditch. The front rank cleared it well; but the men in rear, unable to see before them owing to the excessive dust, almost all rolled into the ditch.[1]

Wolseley also witnessed the charge.

Our cavalry which had been slowly moving forwards, went straight at them, Fane's Horse and the King's Dragoon Guards in the first line, Probyn's regiment in support behind. The Tartar cavalry had halted behind a deep wide ditch, upon seeing our troops advancing towards them, from which position they delivered a volley as our cavalry reached it. The horses of the irregulars are always ridden in short standing martingales, which effectively prevent their jumping well; so, when our line reached the ditch, but very few of the irregulars got over it at first, many of their horses, unable to pull up, tumbling in, one over the other. The King's Dragoon Guards, however, got well in among the Tartars, riding over ponies and men, and knocking both down together like so many ninepins. The irregulars were soon after them, and in the short pursuit which then ensued, the wild Pathans of Fane's Horse showed well fighting side by side with the powerful British Dragoons. The result was most satisfactory. Riderless Tartar horses were to be seen galloping about in all directions, and the ground passed over in the charge was well strewn with the enemy. At no time subsequently during the day would they allow our cavalry sufficiently near for a second charge.

Later Wolseley remembered that the Chinese 'were mounted on small ponies, our men on great troop horses. The men of the King's Dragoon Guards were then about the biggest of our cavalry of the Line, and as they went thundering forward with loud shouts, their opponents may well have thought that their last hour had come.'[1] Sidney Herbert, in reporting this charge to the Queen, wrote:

The charge of the King's Dragoon Guards was an act of horsemanship most remarkable. The Tartars were posted on an elevated mound with a deep ditch in front, and the Horse had not only to clear the ditch, but to leap up the height at the same time. Only one man was unhorsed. The Sikh cavalry tried to do it, but upwards of thirty saddles were immediately empty. On looking at this and another obstacle with a deep drop which the Dragoon Guards passed, he [Grant] says it is impossible to conceive how cavalry could do it.[1]

Trumpeter John Goldsworthy, KDG, was orderly trumpeter to the general commanding the 1st Division.

On the morning of the 21st September, 1860, I was ordered to rejoin my Regiment which was about to charge the enemy. This we carried out in excellent style, notwithstanding the difficult nature of the ground, and I had disposed of some six or seven of the enemy, when I noticed that

Lieutenant W. S. McLeod, of the Madras Cavalry, who was attached to my Regiment for duty, was surrounded by seven or eight Tartars, one of whom was preparing to give him a final stroke, when I pierced him through the neck, killing him instantly. I then turned my attention to the others, and succeeded in killing, or mortally wounding, them all, thus saving the Officer's life. I followed in the pursuit and managed to slay three Mandarins, and, catching the eye of Lieutenant Marsland, remarked to him: 'That's the way to polish them off, Sir!' He replied: 'Well done, Trumpeter, go on and polish some more off.' I did so, but on coming to a halt, as soon as the smoke had cleared away, found that I was simply in the midst of the Chinese Army. I resolved to clear myself, so, putting spurs to my horse, I made a dash and rode straight through them about a mile and a half in the direction of my Regiment, which I found formed up for the roll-call. I had to cut my way right and left to get back through the Tartar army as I did, and my trumpet was shot off my back, but I calculated that forty six fell to my sword that day. I recovered my trumpet the next day, it having been found by one of Probyn's Horse, and on unrolling my cloak from my saddle four bullets fell out.

The reason for this was probably because the Chinese soldiers, having difficulty in ramming the bullets down the barrels of their matchlocks, filed down the slugs to make them fit more easily, with the result that, when fired, the bullet had no force behind it.[1]

An eyewitness of the charge related how the cavalry

were withdrawn from view by the cloud of dust that enveloped them, and nought could be seen of the encounter save an occasional gleam of an uplifted sword, or puffs of grey smoke from the discharged carbine or pistol. In a minute, as it were, the cloud of dust was swept away, and the gallant Dragoons appeared drawn up in line, as if nothing had happened. We soon moved over the ground. One Private lay dead of a matchlock ball wound through the heart, and a Captain of Dragoons dropped to the rear with a bad cut on his arm: but the ground was strewn with the dead and dying Tartars. I stopped with the Commander in Chief's Doctor to look at the dead Private. As his face was turned upwards, one Dragoon that stood near remarked to another, 'Why, Bill, if it ain't old Charley! Poor fellow! He has gone to his long home!'[1]

The wounded KDG captain was Captain Bradbury, and the dead trooper was Private Webster. Privates Napier and Davis were severely wounded, and Privates

Lawrence, Hughes, Ductat, Mason and Pollett slightly wounded, although Pollett later died of his wounds. Eight horses were missing. Grant, in reporting to the Duke of Cambridge wrote, 'I am happy to say in the last fight the 1st Dragoon Guards did very well and charged into a troop of Tartars riding over a ditch and a bank on the way and killing a great many.' Trumpeter Goldsworthy was recommended for the Victoria Cross, but the recommendation was turned down. Lieutenant Marsland, who became a general, tried for many years to gain recognition for Goldsworthy, and thirty-six years later, as a saddler sergeant with the 3rd Hussars, Goldsworthy was at last awarded an additional pension of 6d a day. In 1862, when the regiment was back in India, Lieutenant McCleod presented Goldsworthy with a gold watch and chain in gratitude for saving his life.[1]

Grant paused at Palichao in order to concentrate his force for an advance on Peking. Each day the KDG were sent out to reconnoitre the Chinese positions, and Wolseley commented:

> Our cavalry was, indeed, of the utmost use to us throughout the whole campaign. Our two regiments and a half of cavalry there rendered most valuable service. With even that small force we were enabled to scour the country all round our camps to a great distance. Our cavalry inflicting almost all the loss which the enemy sustained.[1]

SOURCES

1 Michael Mann, *China, 1860*, 1989.

27

China, India, England and Ireland
1860-1877

On 6 October 1860 the cavalry resumed the advance on Peking. General Grant reported:

> We halted the night about three miles east of the north east angle of Peking, and the following morning resumed our march in a north westerly direction, so as to pass along the northernmost face of the city, out of gunshot of the walls, and attack Sankolinsin's army, which was supposed to be encamped directly in our front. The country about here is not good for cavalry, from the great number of trees, villages and hollow roads: I accordingly despatched the Cavalry Brigade, with two six pounders with mounted detachments, with orders to advance on the road leading to the Emperor's Palace at Yeun Ming Yeun, with a view to cutting off the enemy's retreat in that direction.[1]

Swinhoe of the Foreign Service was attached to the cavalry brigade.

> I found the Brigade halted and awaiting the signal for a general march. I reported myself to Brigadier Pattle, commanding, and rode by his side as the troopers advanced through the pretty wood-abounding country of this neighbourhood. The Cavalry were advancing to the northwest, when the vedettes reported large bodies of Tartars moving north. The Brigade was halted, and a Squadron was sent, but the Tartars sighted its approach, and made off.

Grant in his report said,

> The French, anxious to join us in our advance, struck off to their right, finished on the [Summer] Palace without meeting any opposition, and occupied it about nightfall. The Cavalry Brigade had reached the Palace about two hours before this, and were there waiting for us to join them; on their way they saw a body of the enemy's cavalry, but were unable to come up with them.[1]

The French General de Montauban offered to show Pattle and his officers over the Summer Palace, and they were astonished to see how thoroughly the French were looting the place, although de Montauban kept on insisting that he had forbidden it. As the cavalry brigade spent another night in the vicinity of the Summer Palace, they were able to turn their position to considerable personal advantage.[1]

The hostages whom the Chinese had seized around Chang-Kia-Wan had been abominably treated. Now that the Allies were before Peking, the Chinese began to release some of the survivors, who reported their treatment. Private Phipps of The King's Dragoon Guards, one of the escort, had been seized, and on arrival in Peking he was paraded through the streets and then taken to the Summer Palace. He was thrown onto his face, and his hands and feet were tied together behind him; the ropes were pulled as tight as possible, and then soaked in water to make them shrink further. He and the others were placed in a kneeling position and kicked over onto their backs. If they tried to move they were beaten and forced back so that all their weight rested on their bound hands, which, with the circulation cut off, soon became swollen and black. They were then carried into a court-yard, with a Chinese soldier allotted to each prisoner, where they were left exposed to the sun and rain for three days and nights without any food or water. If they moved they were kicked and beaten. When they pleaded for food, dirt was forced into their mouths and they were kicked about the head. At the end of the third day, a little food was handed out, and irons were placed on their hands and feet.

On the afternoon of the fourth day, the prisoners were split up into four parties, Private Phipps being taken in a cart with Bowlby, the *Times* correspondent, a French officer and four Sikhs. Phipps spoke a little Hindustani, but the Sikhs spoke no English; however, two of the Sikh sowars survived and testified:

> Another cart with us containing Daffadar Mahomed Bux, a French Officer, very tall and stout, with a brown beard, and a Dragoon named Pisa [Phipps]. We were taken into the fort, and for three days were out in the open and in the cold. They then pulled us into an old kitchen and kept us there for three or four days. Mr Bowlby died the second day after we arrived. The next day the Frenchman died. Two days after Jawalla Singh died. Four days afterwards Phipps, King's Dragoon Guards, died; for ten days he encouraged us in every way he could.

On 14 October a cart brought the coffins of the dead prisoners. 'They were found to be in such a fearful state of decomposition that not a feature was recognisable, and it was only by the tattered garments that the doctors made them out to be the remains' of eleven dead prisoners, including Private John Phipps. Wolseley commented:

> Up to the day of his death, he [Phipps] never lost heart, and always endeavoured to cheer up those about him when any complained or bemoaned their cruel fate. Even to his last moment of consciousness he tried to encourage them with words of hope and comfort. All honour to his memory: he was brave when hundreds of brave men would have lost heart. Nothing except the very highest order of courage, both mental and bodily, will sustain a man through the miseries of such a barbarous imprisonment and cruel torture as that which Private Phipps underwent patiently, his resolute spirit living within him up to the very last moments of his existence.

Private Phipps's China Medal is now displayed in the Regimental Museum of 1st The Queen's Dragoon Guards.[1]

Grant and Lord Elgin determined to bury the prisoners with full military honours and to burn the Summer Palace as a punishment for the prisoners' treatment. On 18 October The King's Dragoon Guards marched to the Summer Palace, where separate buildings were allotted to parties for destruction. Sidney Herbert informed Queen Victoria: 'The [Peace] Treaty would never have been obtained had the Summer Palace not been burned, but its destruction greatly alarmed them, and gave additional force to the threat to destroy the Palaces at Peking.'[1]

Grant decided to evacuate Peking and get his troops back to Hong Kong before the Chinese winter set in. The King's Dragoon Guards set out for Tientsin on 7 November, where they embarked their dismounted men in the steamer *Atlanta* on the 21st, losing Private Offendale, who fell overboard and drowned. On the 22nd the mounted men marched thirty-five miles in eight hours in atrocious conditions, arriving at Taku that evening. The Regiment embarked on 23 November; 'G' Troop reached Madras on 14th January, 'B' Troop on the 16th, 'A' Troop on the 18th, and 'F' Troop on the 24th. All four troops marched to Bangalore, arriving between 24 January and 3 February 1861. The China Medal was issued to all who had taken part, with two bars inscribed 'Taku Forts' and 'Pekin', and the regiment was allowed to bear both engagements as battle honours on its standard. Brigadier Pattle and Lieutenant Colonel Sayer were awarded the CB for their services.[1]

In April 1862 a committee set up by the War Office decided to divide the cavalry into heavy, medium and light regiments, a move described by the *Army and Navy Gazette* as 'the disorganisation of the Cavalry Service'. The King's Dragoon Guards and The Queen's Bays were designated as medium cavalry, although in practice it is difficult to see what difference that made. It was at this time too that the Australian Waler became the standard remount for cavalry regiments serving in India. Standing between 15 and 16 hands, the Waler had endurance and stamina, apart from being able to carry the weight of a laden cavalryman.

In 1864 all regiments of dragoon guards were issued with a new pattern sword, whose steel hand guard was pierced with a Maltese cross, and which had a curved cutting blade of some 35 inches. Also in 1864 the regiments of dragoon guards on home service were given a new carbine, the Westley Richards, weighing only 6½ lb, but this weapon was not issued either to the KDG or Bays as both regiments were serving at the time in India. [2]

On 22 April 1863 The King's Dragoon Guards was reduced to seven troops, resulting in the break up of 'D' Squadron. On 17 November 1864 the KDG left Bangalore, marching under command of Captain Edelmann for Secunderabad, which they reached on 24 December. On 7 March 1865 the regiment was inspected by Brigadier General Grant, and on 15 August it was told to hold itself in readiness to return to Britain. At once the process of men volunteering to transfer to other regiments remaining in India started, with 150 NCOs and men going to The Queen's Bays, 3rd Dragoon Guards, 7th Hussars, 16th Lancers and the 18th, 19th, 20th and 21st Hussars. On 4 January 1856 the KDG marched from Secunderabad for Bangalore under command of Lieutenant Colonel Sayer, arriving on 12 February and encamping at the Arab Lines, where 454 horses were handed over to the 16th Lancers. On 5 March regimental headquarters with 'B', 'E', 'F' and 'H' Troops went by rail from Bangalore to Madras, where they embarked on the *Lady Melville*, sailing on the 8th with a strength of 1 field officer, 1 captain, 5 subalterns, 2 staff officers, 1 schoolmaster, 21 sergeants, 11 corporals, 6 trumpeters, 5 farriers and 202 troopers; in addition, there were 24 women and 40 children. The left wing followed on 19 March, with 1 field officer, 2 captains, 3 subalterns, 1 staff officer, 8 sergeants, 9 corporals, 2 trumpeters, 1 farrier, and 58 troopers, embarking on the *Ivanhoe* with 15 women and 29 children. The *Lady Melville* arrived at Gravesend on 24 March 1866, four men having been lost during the voyage. The main party were joined at Gravesend by the left wing, who had arrived at Portsmouth, and the united regiment proceeded by rail to Colchester.[3]

On arrival at Colchester The King's Dragoon Guards received 355 horses from the 11th Hussars, and were joined by the depot troop, bringing the regiment's numbers to 1 field officer, 4 captains, 9 subalterns, 2 staff officers, 8 sergeants, 8 corporals, 2 trumpeters and 190 troopers; a total strength of 208, with 15 women and 28 children. On 25 October 1866 a squadron, consisting of 'B' and 'G' Troops under command of Major Alexander, marched to Norwich to provide the mounted escort for the Prince of Wales who was attending a musical festival. The city of

Norwich passed a resolution, 'The Mayor, Aldermen and Citizens resolved unani-
mously that the thanks of this Council be presented to Major Alexander, K. D. G.,
for the kindness and assistance in carrying out the arrangements during the visit of
H. R. H. the Prince and Princess of Wales to the City.' In November 'D' Troop was
called out to Chelmsford to aid the civil power.[3]

On 27 March 1867 the KDG moved from Colchester to Aldershot, and in June
the establishment was augmented by an additional major. In July 1867 there was
another move to a camp on Hounslow Heath, from whence the regiment was to
march to Hyde Park for a review by Queen Victoria. This was cancelled, however,
and the KDG returned to Aldershot. In September the regiment was issued with
the new breechloading Snider carbine, and during January 1868 at his special
request a bearer party of one sergeant and twelve men carried the coffin of General
Sir Thomas Brotherton, colonel of the regiment, to the grave. Brotherton was
succeeded as colonel by General Sir James Jackson. The strength of the Regiment
was now established at: 1 lieutenant colonel; 2 majors; 8 captains; 10 lieutenants; 7
cornets; 1 paymaster; 1 adjutant; 1 riding master; 1 quartermaster; 1 surgeon; 1
assistant surgeon; 1 veterinary; 556 NCOs and men; 368 troop horses. On 31 May
1868 the KDG received a new standard; the old one was presented by the officers
to Colonel Pattle on his retirement.[3, 5]

In August 1868 The King's Dragoon Guards was stationed at Sheffield, with
squadrons at Coventry, Birmingham and Preston. On 1 April 1869 regimental
headquarters moved to Manchester, the mounted men marching, and the dismounted party remaining a day longer to hand over the barracks, then travelling
to Manchester by rail. Captain Marter, who had been away from the regiment on
staff duties, rejoined and took command of 'A' Squadron at Birmingham. Two
months later the new commanding officer, Colonel Slade, who had taken over from
Colonel Pattle, inspected the squadron, and 'was much pleased with the turn out
and working of the men, condition of the horses, and the state of the barracks and
quarters', even though the drill field was four miles from the barracks. On 28
August 'A' Squadron at Birmingham marched via Lichfield, Uttoxeter, Leek and
Macclesfield to rejoin the main body at Manchester. Later that year, in October,
Lord George Paget, Inspector of Cavalry, inspected the regiment and commented
that the officers' ride was the best that he had ever seen.[4, 5]

During November 1868 a troop commanded by Captain Hatfield went from
Birmingham to Brierly Hill, while 'C' Troop under Captain Benthall marched to
Blackburn, both to the aid of the civil power. A later resolution from the Blackburn
Council read:

That the thanks of the Mayor and Corporation be conveyed to Captain

Benthall, the Officer Commanding and the men under his command, comprising the detachment of Cavalry which visited Blackburn at the Municipal Election, for their impartial conduct during their visit, under such painful and at all times difficult circumstances, as it is at all times painful for soldiers to be called out in their own country. This Corporation also couples with that Vote of Thanks a prayer that the Military Authorities will look as leniently as possible upon any indiscretion the men may have been guilty of through the too lavish hospitality of our fellow townsmen.[4]

In May 1869 the old system whereby a number of troops made up each cavalry regiment was changed to an establishment of four sabre squadrons. The amount of weight that a cavalry horse should carry was a constant source of argument, between the optimum needed for efficient action and what the rider actually required; with the latter nearly always taking precedence. The Director of Veterinary Services considered that fifteen stone should be the maximum weight, but most troop horses were expected to carry eighteen and even twenty stone. Queen's Regulations stated that 'the horses of Regiments of Cavalry are not to be allotted to Troops according to colour', but The Queen's Bays always ignored this regulation, and on returning from India in 1870 were 'according to traditional practice, remounted on bay horses'. For this to be made possible forty-four troop horses were acquired from each of the following regiments: The King's Dragoon Guards, the 4th, 5th, 6th and 7th Dragoon Guards, and the 1st Royal Dragoons.[3, 6]

On 6 December 1869 The King's Dragoon Guards received a telegram ordering the regiment to be ready to embark at Liverpool for service in Ireland, and to move in two days' time, the normal length of an English posting being cut short by two years. The regiment marched on 9 December, arrived at Liverpool the next day, embarked in two ships and reached Dublin on the 11th, after 'a fearful night – a gale came on – everyone was ill – horses struggling, seas drenching all'.[3, 4]

In the meantime The Queen's Bays, home from India, were stationed on arrival for a few weeks at Chichester, but soon moved to their new depot at Colchester, where they received a new establishment: 1 colonel; 1 lieutenant colonel; 1 major; 7 captains; 7 lieutenants; 3 cornets; 1 paymaster; 1 adjutant; 1 riding master; 1 quartermaster; 1 veterinary; 1 RSM; 1 bandmaster; 1 RQMS; 7 troop sergeant majors; 1 pay sergeant; 1 armourer sergeant; 1 saddler sergeant; 1 farrier major; 4 farriers; 1 hospital sergeant; 1 cook sergeant; 1 fencing instructor; 1 orderly room clerk; 1 trumpet major; 7 trumpeters; 21 sergeants; 21 corporals; 9 shoeing-smiths; 2 saddlers; 1 saddle-tree maker; 374 troopers. TOTAL 483 all ranks. 300 troop horses.

Her Majesty Queen Elizabeth, The Queen Mother, Colonel in Chief 1st The Queen's Dragoon Guards. (By gracious permission of Her Majesty Queen Elizabeth, The Queen Mother.)

ABOVE: Inspection of The Queen's Horse (KDG) and Peterborough's Horse (Bays) by King James II on Hounslow Heath, 1686. Painting by Joan Wanklyn, property of the Officers' Mess, 1st The Queen's Dragoon Guards. BELOW: The Queen's Horse (KDG) capturing the French drums at Ramillies, 1706. Painting by R. Hillingford, property of the Officers' Mess, 1st The Queen's Dragoon Guards.

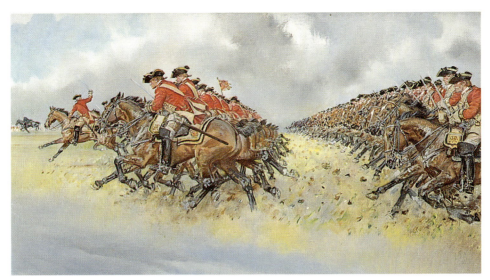

ABOVE: Following the Marquess of Granby at Warburg, 1760: The King's Dragoon Guards and The Queen's Bays. BELOW: The Duke of Wellington with The King's Dragoon Guards at Waterloo, 1815. Paintings by Joan Wanklyn, property of the Officers' Mess, 1st The Queen's Dragoon Guards.

Officer of The Queen's Bays, 1825. Lithograph by L. Mansion and L. Eschauzier.

Sergeant Fogarty, King's Dragoon Guards, on Hercules, escorting French-Canadian prisoners, Canada, 1838. Painting by M. A. Hayes, property of Captain A. C. McCallum, QDG.

ABOVE: Charge of The Queen's Bays at Lucknow, 1858. Painting by H. Payne, property of the Officers' Mess, 1st The Queen's Dragoon Guards. BELOW: Private J. Doogan, King's Dragoon Guards, winning the Victoria Cross by rescuing Major Brownlow, KDG, at Laing's Nek, 1881. Lithograph in possession of the author.

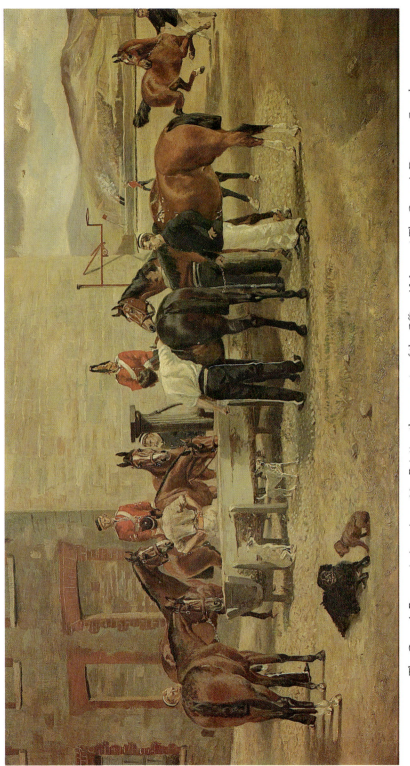

The Queen's Bays watering, circa 1880. Painting the property of the Officers' Mess, 1st The Queen's Dragoon Guards.

ABOVE: The Emperor Franz Josef, Colonel in Chief of the 1st King's Dragoon Guards, inspecting the regiment. Painting the property of the Officers' Mess, 1st The Queen's Dragoon Guards. BELOW: The leading scouts of The Queen's Bays, circa 1895. Painting by H. Payne, property of Major General Sir Desmond Rice, QDG.

Lieutenant Quested's patrol of The Queen's Bays at Fampoux, the Battle of Arras, April, 1917. Painting by Lionel Edwards, the property of the Officers' Mess, 1st The Queen's Dragoon Guards.

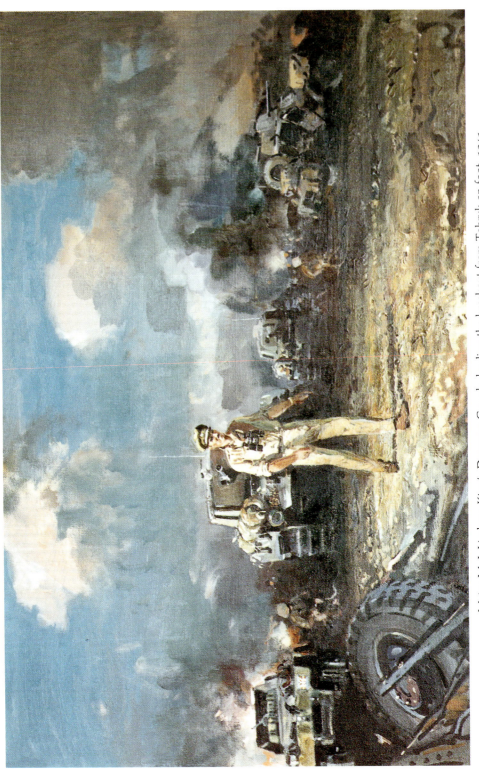

Major M. J. Lindsay, King's Dragoon Guards, leading the breakout from Tobruk on foot, 1941.
Painting the property of the Sergeants' Mess, 1st The Queen's Dragoon Guards.

Lieutenant Colonel T. Draffen commanding The Queen's Bays at Knightsbridge, Battle of Gazala, 1942. Painting by Joan Wanklyn, property of the Officers' Mess, 1st The Queen's Dragoon Guards.

1st The Queen's Dragoon Guards in Borneo, 1965: Saracen and Ferret armoured cars. Painting the property of the Officers' Mess, 1st The Queen's Dragoon Guards.

On arrival at Colchester the regiment was allowed to change its title to that by which it was commonly known, and received the offical designation of 2nd Dragoon Guards (Queen's Bays) in place of the old 2nd (Queen's) Dragoon Guards. During August and September 1870 The Queen's Bays were inspected by the Inspector General of Cavalry, and by the commander in chief, the Duke of Cambridge, who said that the regiment was in better order and condition than any other cavalry regiment he had seen since the regiment's return from India. The following year Major General Freeman Murray, commanding the Eastern District, was equally complimentary. In August 1871 the regiment moved from Colchester to Aldershot, occupying the East Cavalry Barracks and taking part in the autumn manoeuvres.[7]

Colonel Seymour, who had commanded the regiment with such distinction, was promoted to general in June 1872. Before Seymour gave up command to his successor, Lieutenant Colonel H. H. Steward, he made himself even more popular with all ranks by donating a sum of two hundred guineas for the purchase of warm clothing for the wives and families of the men of The Queen's Bays, to commemorate his fourteen years of command. Colonel Steward only remained in command until 1874, when he was replaced by Major Thomas William Snede.[7]

During 1875 The Queen's Bays moved to Ireland, where they were to spend five years. Whilst in Ireland, they were inspected by their old commanding officer, now Major General Seymour, who received a very warm welcome from his old regiment and was able to congratulate them for having had fewer courts-martial for two successive years than any other mounted corps stationed on Irish soil. This drew from the Duke of Cambridge yet another congratulation on the regiment's excellent continuing record. During 1879 and 1880 there were widespread civil disturbances throughout Ireland over the issue of land, exacerbated by a general failure of the crops in Western Ireland. During 1880 The Queen's Bays were used in aid of the police and civil magistrates throughout the west of Ireland, and at one time there were 7,000 cavalry, infantry and police under arms in County Mayo alone. At the end of 1880 the regiment returned to England and was stationed at Manchester and Liverpool, until in 1884 The Queen's Bays moved to Aldershot.[7]

The King's Dragoon Guards arrived in Ireland on 18 December 1869 and was stationed at Caher, Waterford, Clonmel, Tipperary and Carrick. The march from Dublin and Holyhead to the stations was carried out in appalling weather, with rain, hail and gales. The first public duties performed by the regiment were to help with the elections at Waterford and Tipperary in February 1870, when 'every available horse was mounted'. In July 1870 the regiment was concentrated at Newbridge, this being the first time it had been together as a whole since August 1868, and the following month its strength was augmented by an additional 83 men and 50 troop horses. In October, however, detachments were again sent out to Carlow and the Curragh. During 1870 the KDG was reorganised back into the troop system, and in early 1871 an additional troop was authorised, bringing the strength up to eight troops, each troop having 4 officers and 63 men with 40 troop horses.[3, 4]

Over the period there had been a number of changes in the uniform and equipment. In 1855 the shabraque had been abolished, but in 1874 it was reintroduced and was worn until 1897. In 1857 the KDG had changed the black horsehair plume on the Albert helmet for the red one, which has been worn ever since. Paintings by Orlando Norie dated during the 1870s show officers and men wearing the leathered overalls, which were discontinued by the regulations of 1874, although they may have continued to be worn for some time afterwards. In 1873 a new pattern helmet was introduced, very similar to the one worn in full dress up to the present time.[8, 9]

During 1871 The King's Dragoon Guards moved to the Curragh, and the following year to Dublin. On 12 January 1872 General Hankey was appointed as colonel in succession to Sir John Jackson, who had died. Various troops were despatched to Galway, Banbridge, and Belfast during 1872 in aid of the civil power. During August 1873 the regiment moved to Ballincolig, just to the south of Dublin, until in January 1875 the KDG marched to Dublin, embarked at the North Wall and sailed for Glasgow, where on disembarking they marched to their new station at Edinburgh.[3]

Trooper R. Smith describes the journey:

> We embarked on a troopship at Dublin for Glasgow en route for Edinburgh. After packing our saddles and arms belts in our corn sacks, which we always carry in marching order, our horses went on board, and then made a start for Glasgow. We had a very good crossing. The next morning we had to unpack our saddles, and saddle up again for another march towards Edinburgh, which took us two days and one half day. I was very glad we found Edinburgh a very nice station. The Scotch people were so very civil, a great contrast between Irish and Scotch.[10]

Trooper Smith's diary gives a good account of the training of a KDG recruit in the 1870s:

> Reveille sounded at 5 a.m. Stables from 5.30 until 6.45. Breakfast at 7 a.m. Riding drill at 7.45 a.m. until about 10. Come in and change into stable dress and if you could steal a few minutes clean a few things for dismounted drill as well. But there would soon be a Sergeant or Corporal

come yelling out 'Forage', meaning to draw hay, corn and straw. And it would be very nearly 11 o'clock then. At 11.15 stables until 1 p.m. then dinner. After dinner swords, guns, boots and clothing wants cleaning for drill again at 2.15 p.m. Come off drill at 3.30, almost as soon as you get off drill it sounds school, and then it is schooldays again. No matter what scholar we have to go until an examination comes off. We leave the school at about 5 minutes to 5, and then tea; at 20 past 5 it sounds stables again for an hour, and perhaps you may be put on night guard at 7 at night until 5 next morning, and start the same again until we come off guard. That life lasted me about 8 months before I was dismissed my drills. It used to be after dinner then old soldiers to bed and recruits to drill. After going through that course of drill, I had another to do and that was musketry, and as we had no butts at Ballincolig we were sent to a place called Youghal by the sea. We were there about a fortnight. Then returned to our station again.[10]

On 23 February 1874 the Patent Office issued Patent No 685 to Major Walter Clopton Wingfield of The King's Dragoon Guards in respect of a game invented by Major Wingfield, which he called 'Sphairistike', but which is now universally known as lawn tennis. Wingfield, who lived in Pimlico, claimed twenty-five years later that

> No living soul has ever discovered what Sphairistike means! The object and intention of my invention consists in constructing a court by means of which the ancient game of tennis, as played by kings for crowns and kingdoms, is much simplified. It can be played in the open air, on a lawn, on ice, or any suitable level area, and dispenses with building special houses for the purpose costing thousands of pounds.

Wingfield also invented a 'Cyckhana', where the contestants, mounted on bicycles and armed with sabres, rode round a track slashing off as many 'Turks' heads' as they were able within a given time.[11]

While the KDG was in Edinburgh it furnished escorts during 1876 for the various ceremonies associated with the General Assembly of the Church of Scotland, but in June the regiment moved to Manchester, Liverpool and Bury. At this period a patrol jacket replaced the old stable jacket. In June 1877 the establishment was reduced to 601 NCOs and men, and 379 horses.

SOURCES

1 Michael Mann, *China, 1860*, 1989.

2 J. M. Brereton, *History of the 4/7th Royal Dragoon Guards*, published by the regiment, 1982.

3 Records of The King's Dragoon Guards, Regimental Museum, 1st The Queen's Dragoon Guards, Cardiff Castle.

4 The Marter Diaries, Regimental Museum, 1st The Queen's Dragoon Guards, Cardiff Castle.

5 W. Clowes, *King's Dragoon Guards, 1685-1920*, 1920.

6 The Marquess of Anglesey, *A History of the British Cavalry*, vol. 2, 1975.

7 F. Whyte and A. H. Atteridge, *The Queen's Bays*, vol. 1, 1930.

8 *Journal of the Society for Army Historical Research*, vol. 50, p. 197.

9 *Journal of the Society for Army Historical Research*, vol. 40, p. 55.

10 R. Smith, KDG, 'Sketches of Home and Abroad', Regimental Museum, 1st The Queen's Dragoon Guards, Cardiff Castle.

11 *The K. D. G.*, vol. 1, p. 55.

28

Zulu War
1878-1879

In May 1878 The King's Dragoon Guards moved from the North to Aldershot. Smith, now a lance corporal, relates:

> I was told off to march a man for turning out late. Not a very pleasant billet, but I had to do it without cribbing; it was a long march 22 miles I believe. I had my horse, but his was taken with the Troop, but we got on capital, as I kept giving him a lift, and me walking for a few miles. We were treated well on the road, having a few shillings given us, as the people thought that this man must have done murder or something bad, but only for not being in time on parade, it is not much trouble for an officer to say let him walk.[1, 2]

Captain Marter, promoted to the majority, described 'My first Field Day in command of the King's Dragoon Guards. There was a scrimmage between parts of the opposing forces, and some men were injured. I carried off a Colour-Sergeant's rifle, which he had deliberately thrown away, and brought it home.' In August the regiment moved from North Camp to the East Cavalry Barracks, where Marter and his family took over the commanding officer's quarter, as Colonel Alexander was a bachelor. Smith, now promoted to full corporal, thought, 'And a vey lively place Aldershot is for a soldier, plenty of work and plenty of pleasure and life.' In July 1878 the experimental patrol jacket was discontinued, and the men reverted to wearing stable jackets.[1, 2, 3]

On 11 February 1879 an unexpected order was received for the regiment to go to South Africa on active service against the Zulus. Smith, now a sergeant, described the resulting feverish activity:

> One morning, as we were going out on reconnaissance duty, we had got about a mile away from our camp, [when] we received the command to halt, and found out we were for the Zulu campaign, as soon as we could be got in order. So we had to return to camp and commence work, for we had

only a week to do it in, and a very busy week it was. When a Regiment is ordered out, there are so many different things wanted, such as volunteers and horses. We had men and horses from several regiments, some from the Scotch Greys, 3rd Dragoon Guards, 2nd Dragoon Guards, 6th and 7th Dragoon Guards, to make up our strength, saddlery and accoutrements wanted besides cork helmets for brass. Every man had to be inspected by the doctor to see if he was fit for service, and too many inspections to mention.

The left wing, under command of Major Marter, left Aldershot on 27 February, followed the next day by the right wing under Lieutenant Colonel H. Alexander, the commanding officer. The KDG embarked at Southampton in two steamships, *Spain* and *Egypt*, hired by the Government from the National Line, with 634 all ranks and 545 horses.[1, 2, 3, 4, 5]

Smith gives a graphic account of the embarkation and journey:

There were a great many friends to see us go away from Aldershot, particularly the female class. Married women and suchlike pulling all sorts of faces, but they had to be left, and away we went. We did not sail on the day we embarked, but the next morning, on account of a very dense fog. We get plum pudding and salt junk (or pork) and pea soup; other days bully beef, or tinned meat, bread three times a week and crackers [biscuits]. People without a good set of teeth gets on rather badly, we get a kind of porter every day that made it better.

The King's Dragoon Guards arrived at Durban and disembarked on 9 April 1879. 'When all were landed, the Cavalry were despatched to the camp prepared for them at Cato's Manor, and on Saturday, the 12th were inspected by Lord Chelmsford, the Commander in Chief, attended by Major General Clifford and Colonel Crealock.' Smith recalled, 'We stayed in camp for two or three days to get our horses used to solid footing, for they were like drunken men on landing, having stood up for six weeks.' The horses looked like 'tucked up whippets', and having always fed on cut fodder, they had never supported themselves by grazing, and so refused, at first, the rank grass of Natal, which, around Durban, was in any case worked out. Marter noted that the 'men and horses were tormented by ticks'. On 14 April General Marshall, commanding the cavalry brigade, made up of the KDG and the 17th Lancers, ordered the two regiments to parade through Durban. The troopers rode in light marching order, 'and the scene as they passed through the crowded streets will never be forgotten. The Dragoons all carried the Martini-Henri carbine, in addition to their swords.' General Marshall issued an order,

Col Alexander, Officers, NCO's and men of the K.D.G. I have the greatest pleasure in telling you that the appearance of the Regiment as it marched was most satisfactory. I feel confident that the Regiment is in a most efficient and satisfactory state. I shall make it my special duty to report to the highest authorities in this country and back at home the high state in which I find it to be, and I feel that I cannot make use of terms too strong in so doing. You will publish what I have now said in Orders, with a view to its being entered in the Regimental Records.[1, 2, 3, 4, 5, 6, 7, 8]

The march inland started on 17 April, but the horses were still so weak and overloaded that they only managed ten miles a day, and even then required to rest every third day. Pietermaritzburg was reached on 23 April,

A very pretty little town [noted Smith]; after getting all our camp equipment we started upcountry towards Dundee, where we again halted for a day or two. From Dundee our Regt. the 17th Lancers and two Batteries of Artillery were ordered to go to Rorke's Drift and Isandula [sic]. We did not take any tents with us, and had to have the sky for a roof above us, and very cold it was then with only an oil cloth, one blanket and our cloak. This is what we call bivouacking.

Marter records that 'General Marshall decided upon making a reconnaissance over the Isandhlwana battlefield where the 24th Foot had been massacred four months earlier and breaking the spell that seemed to hang over it.' The column arrived at Rorke's Drift on the evening of 20 May and bivouacked near Fort Melville, which had been built about a mile from the mission station.[1, 2, 3, 8, 9, 10]

Miles Gissop, a 17th Lancer, noted on 21 May, 'Revellie at 2.30 a.m. General Parade at 3 a.m. Marched from camp with K.D.Guards.' The King's Dragoon Guards, with a wing of the 17th Lancers, set out at 4 a.m., passing Sihayo's burnt kraal and working through the hills by the valley of the Bashee to the Nqutu Plateau, descending to the battlefield of Isandhlwana, arriving there at about 8.30 a.m. Vedettes were at once posted. Marter wrote:

Before daylight we forded the Buffalo River, and made our way along a track between hills covered with scrub jungle, in which it was very difficult to keep a lookout. As daylight broke, the waggons of the ill-fated force could be clearly seen in the distance against the sky. On arrival there was the camp, the oxen inspanned in the waggons, the horses at their picket posts, the Officers' Mess and their baggage, the Quartermaster's Store and supplies, and officers and men lying about in their uniforms — dead — but singularly lifelike, as from the state of the climate the bodies had only

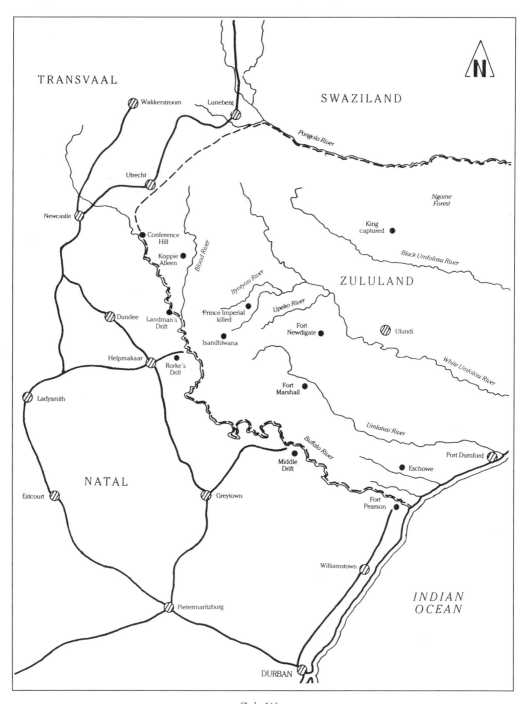

Zulu War

dried. Many were recognisable. They had not been mutilated. Birds and beasts did not seem to have molested them, and the Zulus had removed nothing but arms and ammunition, and part of the canvas of the tents.

Marter made a sketch map of the scene, which is displayed in the Regimental Museum.[1]

Sergeant Smith had a different view:

> Such a sight as I saw going over that hill, I shall never forget, some of them laying on the ground as if looking up at us, others in other ways, but the way they had mutilated their bodies were painful to see. I believe there were about six hundred lay dead on the field, tents and all torn up, and everything left, only fighting gear which they took away. The General in command did not keep us long there, for the sight was dreadful to see, and I cannot tell all that they had done to our comrades bodies.

Archibald Forbes of the *Daily News* wrote:

> At the top of the ascent beyond the Bashee, which the Dragoon Guards crowned in dashing style, we saw the steep, isolated, and almost inaccessible crag of Isandhlwana. Dead men lay thick, some were almost wholly dismembered, I forbear to describe their faces. Every man had been disembowelled. Some were scalped, and others subjected to yet ghastlier mutilations. The clothes had lasted better than the poor bodies they covered, and helped to keep the skeletons together.

The difference between these accounts may lie in the fact that it was the Zulu custom to slash open the bellies of their dead opponents in order to release their spirits. Marter goes on, 'With such light tools as we had, we buried some of the bodies, Colonel Durnford among them, and, having brought every spare horse and all tackle procurable, dragged about 40 waggons back to Rorke's Drift.'[1, 2, 3, 4, 6, 8, 10, 11, 12]

On 22 May The King's Dragoon Guards left Rorke's Drift and advanced to Landman's Drift and Koppie Allein on the Blood River. 'We marched at 8 a.m. for Landman's Drift after 24 miles we arrived in camp at 4 p.m. The nights and mornings were dreadfully cold. So much we could scarcely saddle our horses our hands being so numb. During the day it was very hot indeed.' Marter was sent by Lord Chelmsford to make the first reconnaissance into Zululand, in order to determine the line of advance of 2nd Division, with two squadrons – a ride of about fifty miles. On the way back, while Marter was occupied with the rearguard, Captain Benthall, KDG, took it upon himself to trot off with the main body back

to camp. This left three officers and nineteen men on their own in enemy territory with broken country and fading light. Marter with another officer and six
men managed to reach camp some two hours after dark, but the remainder were
stranded without food or cloaks for the night. The remaining two officers and all
the men, bar three, came in at midday the following day. Two of the missing men,
whose horses had died, were brought in later that day, whilst the final man was not
found until the third day.[1, 9, 12]

It had been the intention for The King's Dragoon Guards to move to Standerton,
but in view of the difficulty in procuring forage for the cavalry brigade, it was
decided that the KDG should remain on the frontier. On 26 May the right wing of
headquarters and four troops, under Colonel Alexander, left Landman's Drift for
Conference Hill, and the left wing under Major Marter was to return to Rorke's
Drift. Marter commented,

> It was now decided that the K. D. G. should remain on the frontier, whilst
> the 17th Lancers accompanied the column advancing into Zululand. In
> vain did General Marshall remonstrate, but he was not on good terms
> with Lord Chelmsford, and his arguments were disregarded. In vain did
> Marter endeavour to persuade his Colonel to make a stand against the
> heart of the Regiment being thus broken. At length, almost beside himself,
> Marter went to Lord Chelmsford's tent, and notwithstanding repeated
> rebuffs, did not leave until he had brought him reluctantly to agree that
> Marter, with one Squadron, should accompany the column.

The remaining two troops of the left wing, under Captain Douglas Willan,
returned to hold the post at Rorke's Drift. Sergeant Smith reported: 'On account of
us losing our convoy of provisions, we were in the cart, and had to stay on the
borders, and do convoy duty.'[1, 2, 10]

Marter continues:

> With the Colonel's permission Marter picked officers, men and horses,
> about 186 all ranks, and on the 1st June in high spirits, [they] crossed the
> Blood River into Zululand with the Column. That day the Prince Imperial
> was killed, and early next morning, Marter and his Squadron, with some of
> the 17th Lancers, was sent out to find his body. This was done and the
> body brought into camp.

It was on 2 June that 'D' and 'H' Troops discovered the remains of the Prince
Imperial at the Ityotosi River, and one troop escorted the body, covered with a
blanket and on a bier made up of lances, to the camp on Itelezi hill.[1, 3, 4, 10]

On 4 June intelligence came in that a body of Zulus were in the area of the

Ityotosi river. General Marshall with the 17th Lancers and Marter's squadron of
the KDG reconnoitred the track forward as far as the Upoko river, where they
joined forces with Buller's flying column. Buller had discovered some 300 Zulus
near some kraals on the far side of the Upoko river, and had driven them into some
thorn bush on the lower slopes of the Ezunganyan hill. He had then burnt the kraal.
He was about to withdraw, with the loss of two men wounded, when Marshall and
the cavalry arrived. Three troops of the 17th Lancers advanced and, coming under
fire, dismounted and engaged the enemy without much effect, as the Zulus were
well concealed in the long grass. Marshall moved the KDG forward in support and
ordered the 17th Lancers to withdraw. As they fell back their adjutant, Captain
Frith, was killed by a Zulu bullet. Marter described the action:

> The enemy were strongly posted in a wood intersected with dongas
> behind four kraals. Buller's men managed to set fire to the kraals, but
> having had several horses shot, and men wounded, found it neccessary to
> retire. Colonel Lowe [17th Lancers] then, against General Marshall's or-
> ders, advanced with the 17th to within 150 yards of the wood, and dis-
> mounted some men. I supported him, placing a Squadron in echelon on
> either flank, and we were potted at for about twenty minutes. Frith, the
> Adjutant of the 17th, was shot dead and Martini-Henry bullets flew high,
> and others were more dangerous.

As the KDG retired across the Upoko river, they were closely followed up by the
Zulus, who opened fire, but without effect. The cavalry re-formed column, returning to a new camping ground on the banks of the Nondweni river.[1, 3, 4, 6, 7, 8, 9, 10, 13]

On 6 June Marter again led out his two troops to reconnoitre the area, when
large numbers of Zulus were seen and some kraals were shelled and burnt. On 8
June the KDG squadron at Rorke's Drift under Captain Willan marched to the
Upoko camp and returned the following day, keeping open the lines of communication. On 16 June, in Marter's account,

> I was ordered back with my Squadron to Fort Newdigate, a fort which had
> been formed a few miles back. I still longed for the front, and begged to
> go on. The 17th was a most wearisome day, and I tried to the last to
> get off going back to Fort Newdigate, trudging backwards and forwards
> from one Staff Officer to another. We marched at about 3.30 and took
> up my new command. The garrison was two Companies of the 21st
> Fusiliers [Royal Scots Fusiliers], my Squadron, two Gatling guns, a Com-
> pany of Bengough's Contingent [Natal Native Contingent], and about
> four mounted Kaffirs and Basutos. The fact is that Lord Chelmsford and

General Marshall did not agree. The former therefore decided to break up the Cavalry Brigade, and General Marshall was relegated to the lines of communication.

Lord Chelmsford had decided to command the main body himself, and he placed the security of the lines of communication under Major General Clifford, giving him three infantry battalions, The King's Dragoon Guards and some of the Natal Native Contingent. The squadron of the KDG at Fort Newdigate was placed in charge of a convoy which was pushed through to Koppie Allein, in support of General Wood's flying column. Another was attached to Baker Russell's column; while the two remaining squadrons were stationed at Conference Hill on the lines of communication.[1, 4, 9, 10, 13]

The squadron at Fort Newdigate had to meet all convoys and pass them from the frontier on to the next fort, and on the intervening days they were charged with raiding to the left rear of Lord Chelmsford's column, burning kraals and devastating the countryside so as to clear out any Zulus. As the Zulus had burnt most of the grass, there was little grazing for the horses, and since Lord Chelmsford had ordered the supply of corn to be restricted, many horses died from pressure of hard work and lack of provender. As the infantry in the fort were forbidden to go out on escort duties, the handful of dragoons were hard worked. After the war was over the Zulus were asked why they had not attacked the slenderly guarded convoys, and they said that the scouting was so good that they could never get near enough to see what troops were with each convoy. Sergeant Smith remembered, 'During our stay in Rorke's Drift we had some very hard days, for our duty there was to keep up communication with troops marching to Ulundi, and had some days to march 15 to 20 miles out until we could see them.' General Marshall said that he had never seen such scouting, and made a special report to that effect to the commander in chief.[1, 2, 4, 7, 8, 9, 10, 13]

While at Fort Newdigate Marter took out two men to the spot where the Prince Imperial had been killed. They carted white stones seven miles on an ox cart and built a cairn. Queen Victoria later sent out a cross to be placed at the head, but specially requested that the work of her soldiers should not be disturbed.[1].

On 28 June, Sergeant Smith recalled, 'Our orders was to go to Isandula [sic] to commence to bury our comrades of the 24th [South Wales Borderers], Royal Artillery and volunteers that was killed there.' Colonel W. Black was in command and took 'thirty King's Dragoon Guards mounted, fifty on foot, 140 2-24th Regiment, 360 natives, and fifty native horse'. They 'were working on the 20th, 23rd and 26th, and have completed the burial of the remains of those who fell at the battle of Isandhlwana'. Smith continues;

We marched at 3 o'clock. Half of us had to walk on account of having more men for working. It was not a very comfortable duty. I for myself did not like it, as we had to dig holes by the side of each, and the ground rocky to make it worse for us, but it must be done, so we had to settle down to it the best we could. On account of us having to dig so close to the bodies was because when they were moved, not only the stench but they fell to pieces, and would have been worse, but their clothes held them together a great deal.

Colonel Black ended his report: 'I cannot close without calling attention to the hard work so cheerily undergone by the fifty dismounted King's Dragoon Guards, who each day trudged twenty two miles to and fro, besides their labours on the widespread field.'[2, 8]

By the beginning of July the main column approached the royal kraal of Cetewayo, the Zulu king, at Ulundi. On 4 July the troops, which included the 17th Lancers with a troop of twenty-four men of The King's Dragoon Guards under Lieutenant Brewster, were roused before dawn and advanced in an open square, the flanks and rear being covered by the 17th Lancers, the troop of the KDG and the Mounted Irregulars. Colonel Drury Lowe, commanding the 17th Lancers, had Major Brownlow, KDG, as his orderly officer. At about 8.45 a.m. the Mounted Irregulars became engaged with 'all that remained of the young manhood of the [Zulu] nation', and as the Zulus advanced, the Mounted Irregulars slowly retired to the cover of the square, 'firing their carbines as close to the advancing warriors as they dared', followed by the whole Zulu army. The Zulus advanced on the square with the greatest bravery. 'It was now about 9 a.m. and the guns were already hard at work. Before many minutes, the infantry volleys began tearing into the attacking black masses.' As soon as Lord Chelmsford saw signs of the enemy wavering, the 17th Lancers and the KDG troop, who had been standing to their horses' heads, were ordered out in pursuit of the Zulus. As the troopers prepared to mount, the Zulu reserve suddenly charged and the troopers dismounted again until this final charge was broken by the relentless fire from the square. The cavalry then emerged through the rear face, the 21st [Royal Scots Fusiliers] and the 94th [Connaught Rangers] opening their ranks to allow the horses to trot out, form line and, advancing from the trot to the gallop, charge, sweeping around upon the Zulus and putting them to flight. The chase was pressed for some three miles before the troopers were held up by dongas and, being rallied, were recalled. The whole action lasted no more than three quarters of an hour, and the military power of the Zulu nation had been broken. The KDG suffered one horse killed and one wounded.[3, 4, 5, 6, 7, 8, 9, 10, 13, 14, 15]

Lord Chelmsford had been replaced as commander in chief by Sir Garnet Wolseley, who, on arrival, claimed Captain Douglas Willan with 'A' Troop KDG as his personal escort, retaining them for the whole of his time in Zululand, and subsequently as far as Pretoria. Sergeant Smith related,

> On Sir Garnet coming up country, we were taken from Rorkes Drift by him as escort to Ulundi, which was about six or seven days' march, and not short marches either. The day before we got to Ulundi, which was called Fort Victoria, we had a terrible storm at night, having blown nearly all the tents down, officers as well as ours. Horses running loose, donkeys and bullocks, men shouting, such a row was never heard. Thunder and lightning besides pouring with rain, for myself, I crept under a waggon out of it until morning, drenched to the skin, everything we had was wet, having been exposed after the tents went down. We had a dram of rum given us next morning, and it was a God send I can tell you, for it put a bit of life into us.

Marter was amazed by the ferocity of the storm: '236 of the transport bullocks lay dead in the camp, and many others that had broken away, were found dead in the bush afterwards. The force could not advance for lack of transport animals, and the troops were formed into burying parties.'[1, 2, 3, 9]

Leaving most of the troops to guard the stranded store waggons, Wolseley pushed on with 'A' Troop, and on 13 August reached the headquarters camp at Ulundi. 'Our next place', wrote Smith, 'was Ulundi, which was a ruin, and Sir Garnet commenced to collect in the Zulu's arms. One day we found an artillery gun in a donga, which we brought into camp, belonging to the artillery at Isandula [sic].' 'A' Troop found both of the guns which the Zulus had captured from 'N' Battery of 5th Brigade at Isandhlwana, and brought them into Ulundi. Marter writes: 'The main Zulu army had been beaten by the forces under Lord Chelmsford, but had not been followed up, the British troops retiring immediately after the battle, as if they had met with a reverse. It was not therefore known whether the Zulus had re- united to any extent, and at all events peace could not be made, or any arrangement for settling the country effected, until the King either surrendered or was captured.[1, 2, 3]

On 4 August 'C' and 'F' Troops KDG, with Captains Brownlow and Watson, joined Colonel Baker Russell's flying column and remained with it until it was broken up at the end of the month. On 9 August Captain Gibbing's troop made a reconnaissance of the Ulundi plain with orders (1) to trace the reputed waggon trail from Ulundi to Conference Hill, and to report on its condition, (2) to make contact with Baker Russell's column, thought to be fifty miles away near the Intabinkulu

Mountains, (3) to return by an alternative route, and report on its possible use for the army marching out of Zululand. Gibbing's troop fulfilled all its tasks and the route reconnoitred was later used. On 23 August Colonel Alexander with headquarters marched from Conference Hill out of Zululand to Utrecht in the Transvaal.[3, 9, 10, 14]

SOURCES

1 The Marter Diaries, Regimental Museum, 1st The Queen's Dragoon Guards, Cardiff Castle.

2 R. Smith, KDG, 'Sketches of Home and Abroad', Regimental Museum, 1st The Queen's Dragoon Guards, Cardiff Castle.

3 Records of The King's Dragoon Guards, Regimental Museum, 1st The Queen's Dragoon Guards, Cardiff Castle.

4 W. Clowes, *King's Dragoon Guards, 1685-1920*, 1920.

5 The Marquess of Anglesey, *A History of the British Cavalry*, vol. 3, 1980.

6 D. R. Morris, *The Washing of the Spears*, 1966.

7 D. Clanmer, *The Zulu War*, 1975.

8 C. L. Norris Newman, *In Zululand*, 1988.

9 *Regimental Journal of 1st The Queen's Dragoon Guards*, vol. 3, no. 6, 1978, p. 546.

10 *Narrative of the Field Operations Connected with the Zulu War of 1879*, 1989.

11 F. Emery, *The Red Soldier*, 1977.

12 'The Recollections of Miles Gissop', *Journal of the Society for Army Historical Research*, vol. 58, 1980, p. 85.

13 M. Barthorp, *The Zulu War*, 1980.

14 Sir Henry Everett, *The Somerset Light Infantry, 1685-1914*, 1934.

15 J. Laband, *The Battle of Ulundi*, 1988.

29

Zulu War, First Boer War, Laing's Nek
1880

The power of the Zulu nation had been destroyed at Ulundi, but the Zulu king, Cetewayo, was still at large. Until he either surrendered or was captured, it was felt that there could be no peace. A number of abortive searches were made by various parties, but intelligence was hard to come by as the Zulus tried to protect their king by giving misleading information. In particular a party led by Lord Gifford showed great perseverance, following a number of false leads.[1, 2, 3, 4, 5]

On the morning of 27 August 1879 Major Marter rode out with his squadron and some men of the Natal Native Contingent and Lonsdale's Horse, to search for Cetewayo on information that the king was hiding in the vicinity of the 'Ngombe forest. The squadron marched twenty-four miles and camped, having had a very rough ride, with the men's tunics torn to shreds by thorns and overhanging branches. When they moved off the next morning, a Zulu wandered by and made an apparently chance remark, which Marter determined to follow up. With two local guides they reached the summit of a mountain range, when

> the guides made a sign to the party to halt, where tall forest trees on the left hid the men and horses from being seen from below; and when they had dismounted, they called Major Marter to the edge of the precipice, where the trees opened out a little, crawling along on hands and knees. Stopping there themselves, they told the Major to go on to a bush a little further round the edge, and to look down. He saw a kraal in an open space 2,000 feet or more below, at the bottom of a basin, three sides of which were precipitous and clothed with dense forest.[1, 11]

Sergeant Smith remembered how Marter saw

> the kraal down at the bottom of an hill, about a stone's throw off to look at, but it took about three hours to get to it, having to lead our horses all the way through a forest. When we got through our Major says to us, 'When I say "Gallop", I want you to gallop, ' which we did up one hill,

down another, and then on a level with stones as big as wheelbarrows to get over, through, or anyway we liked. Hear one man and then another calling out 'Stop that horse, ' as the horse had fallen down with them, but it was everyone for himself, until we surrounded this kraal. The inmates were quite surprised, as none of them knew where we had sprung from. [The king] was a long time before he would surrender to us, but he was told that if he did not come out, we should burn him out, so he quietly came out, and looked as stately as a General coming to review a few thousand men on parade.[6]

The return to Ulundi was not uneventful, as Cetewayo and his attendant wives and followers were uncooperative. During the course of one night lions killed three of the horses; on the second day's journey from the kraal,

there was only a single track, and us leading our horses, some in front and some behind. All at once there was a cry of escape, and no mistake about it, after counting up we had eleven left. And the Major says, 'Is the king gone?' so I told him not. He tells me to bind his hands, which I did by a piece of raw bullock's hide, that we tied our horses up with, and led him, with the Major behind him with loaded revolver. As we advanced at the head of our column, there was another of our prisoners escaped among the long grass, and a few of our mounted men tried to cut him down with their swords, but the grass so strong, and not able to see him, for he was like a snake dodging about, but one man drew his carbine and shot him at arm's length. Again on the track for Ulundi, which we reached on Sunday morning, as the troops were dismissed from Church Parade, and of course they all gave us a good greeting, as all now was over.[6, 11]

There was, however, more trouble.

On arrival with the king at Ulundi, it was found that Lord Gifford had generally stated in camp that Marter had obtained the information, which enabled him to take the king, from the note [sent by Gifford to Marter]. Marter, taking one of his officers with him, found Lord Gifford, and told him plainly that he had no right to make the statements he had made, as the note had given no clue whatsoever to the king's whereabouts. Gifford said 'I thought I had.' Marter replied, 'If you only thought you had, you should not have made positive statements calculated to damage another officer.'

The *Army and Navy Gazette* of 6 September 1879 read: '1st Dragoon Guards. The most satisfactory accounts reach us from Natal as to the general behaviour of the Regiment and its efficiency. Our correspondent says, 'Notwithstanding the serious disappointment suffered by the Officers and men at finding themselves deprived of the honour of marching on Ulundi, all ranks have worked in a spirit which does them no little credit. Though some have never tired of maligning this regiment, it has since its arrival in Natal shown that it is not behind-hand when work is to be done.'[1, 10]

'C' and 'F' Troops under Captain Brownlow were still with Baker Russell's flying column, who moved against a Bapedi chief, Sekukuni, known as 'Cetewayo's Dog'. They were engaged in the attack on Manganobi's stronghold and Sekukuni's town, both of which were completely successful. Two KDG officers, Lieutenant C. J. Dewar and Cornet E. L. Wright, were attached to the 94th (Connaught Rangers) when Dewar was severely wounded by a bullet in his thigh. Some natives of the Natal Native Contingent started to carry him down, but then abandoned him when about forty Bapedi appeared. Two privates of the 94th (Connaught Rangers), Privates Flawn and Fitzpatrick, came to his rescue, one carrying him and the other holding off the Bapedi with rifle fire. Both of these men were awarded the Victoria Cross. During October Marter heard that he had been promoted to brevet lieutenant colonel in recognition of his services in capturing Cetewayo. During the same month the 17th Lancers sailed for India, leaving their horses for The King's Dragoon Guards; both regiments had been issued with a new pattern saddle for the campaign, known as the 'Universal Pattern, Angle Iron Arch', but both regiments reported unfavourably upon it.[1, 4, 7, 12, 13]

The King's Dragoon Guards returned to the Utrecht District of Natal. Lieutenant Colonel H. Alexander was put in command of the district when Sir Garnet Wolseley moved his headquarters to Pretoria, taking with him Captain Willan's 'A' Troop. On 9 September 1879 'E' and 'H' Troops rejoined headquarters at Utrecht, having marched from Ulundi, and on 12 September 'C' and 'F' Troops came back from Baker Russell's flying column; 'D' Troop commanded by Captain Thompson was at Middelburg. Sergeant Smith did not think much of Utrecht: 'We went into a store, a place where almost anything you want you can get for money. Four of us had a bottle of beer each, which cost a pound, 5/- a bottle and we thought this a licker, I can tell you, and very soon moved off to camp again!' Smith was with 'A' Troop. 'We had now about nineteen days march to do before we got to Pretoria, the Capital of the Transvaal, and myself not in the best of health, as I had dysentery, but I went with my Troop, as I did not like Headquarters.' While still at Utrecht, 'We were changing horses, as ours were about done up, and these at Headquarters had only just arrived from England with a draft of men, and were in very good condition.' On the march to Pretoria, 'We halted at Standerton for one day, and we let our horses loose to graze by the side of our camp, and they had a general stampede; about thirty of them galloped away from camp, but our old ones that we had not changed at Utrecht, stayed in camp as they were used to it. We got all back again, but three, which we never got again as the Boers had collared them.'[1, 6, 7, 13]

The KDG expected to follow the 17th Lancers to either England or India, but in the autumn of 1879 the Boers of the Transvaal started to hold mass meetings demanding a return of their independence, and Sir Garnet Wolseley telegraphed the War Office in October that he would have to retain the regiment. The Transvaal had been annexed in 1877 by a handful of police, and the British administration considered that the Boers were too disorganised and unsoldierly to pose any real threat. The King's Dragoon Guards were moved up into the Transvaal with headquarters at Wakkerstroom, and troops at Pretoria and Heidelburg.

> On 21st September Marter was sent across the hills to Wesselstroom [sic] to see if there was camping ground for the Regiment, and grazing for the horses in that District. When about halfway Marter and his two orderlies came to a Boer farm, and asked for a drink of water. It was refused. On 6th October [1879] the Headquarters of the Regiment arrived [at Wakkerstroom] and camped on the ground chosen. Colonel Alexander being appointed to command the District, Marter took command of the Regiment.

'A' and 'H' Troops in Pretoria were called out on a number of occasions. Smith recalled that 'The Troop came in for Sir Garnet had to go down to Pretoria to read a proclamation to the people, as we expected some disturbance from the Boers.' On two occasions the KDG were called out following attempts by the Boers to seize arms from the town stores. The Pretoria squadron, camped on the Potchefstroom road outside Pretoria, was ordered to draw swords, should the Boers approach them, and, if neccessary, charge them, and then return to Pretoria. Wolseley then formed a movable column to deal with any trouble, consisting of the KDG, 120 mounted infantry, 4 guns of N/5 Battery Royal Artillery, and 13 infantry companies.[1, 6, 7, 8, 12, 13]

The King's Dragoon Guards received orders in September 1880 to move to India. Headquarters and the various troops assembled at Newcastle, where their horses were sold to the Boers, and they marched from there at the end of September to Camperdown, and thence by train to Durban, the men embarking on HM Troopship *Orontes* for Bombay. A detachment of two troops was left at Pietermaritzburg, under the command of Major Brownlow, with the intention of proceeding to England to form a depot. Rider Haggard later described this move as 'a piece of economy that was one of the immediate causes of the [Boer] revolt.' During the regiment's time in the Transvaal two men died of sickness at Heidelburg and nine at Wakkerstroom, but there had been a much more serious loss through desertion. Living conditions were primitive, the cost of any amenities was extremely high, and morale was low. A sergeant of the KDG deserted with £200 of troop and canteen money; a KDG officer, on his way to the races at Harrismith in the Orange Free State, stopped at a forge to have his horse shod, only to find that the blacksmith was a deserter from the regiment. The Boers encouraged anyone with a rifle or a horse to desert, and a Boer woman used her brothel to lend deserters civilian clothes and money, until she claimed that if only she had started earlier 'there would not have been a private left in the King's Dragoon Guards'. Of the 260 men who deserted from the British troops in the Transvaal, 67 were KDG. General Colley, who took over command of the troops in the Transvaal, wrote: 'The desertions among the King's Dragoon Guards have been exceptionally heavy, especially since the regiment was put under orders for India.'[1, 7, 8, 12, 13, 14, 15]

In December 1880 the Boers in the Transvaal proclaimed an independent republic and seized Heidelburg; the small British garrisons soon found themselves beseiged and isolated. Major General Colley, who had succeeded Wolseley, decided not to wait for reinforcements, but to march to the relief of the stranded garrisons. Major Brownlow with the two KDG troops was ordered to Newcastle, where a composite squadron of 120 men was formed from convalescents, drivers from the Army Service Corps and volunteers from two infantry battalions, who had little or no knowledge of horsemanship. Against the Boers who were natural horsemen, Colley decided that his meagre cavalry would have to be carefully nursed; half the squadron was to act as advance guard, and half as rearguard. Private Venables, KDG, wrote to his brother:

> So we are just on the border now and advance into Boer country tomorrow and first go to Wakkerstroom to relieve two companies of the 58th Regiment [Northamptonshire], our little force consists of twenty five of our depot, all old soldiers, me about the youngest and 100 mounted infantry. The Boers are all good shots; they won't be like the Zulus; you will see we shall lose a good few before it is settled. ·

Colley set out from Newcastle on the morning of the 27 January 1881. Two days' slow progress brought his small force, numbering 1,200 men, in sight of the majestic Drakensberg Range, guarded by the square-topped hill of Majuba from which, across their road, ran a low ridge called Laing's Nek. The Boers, numbering probably about 1,000, were under Joubert, and had entrenched themselves along

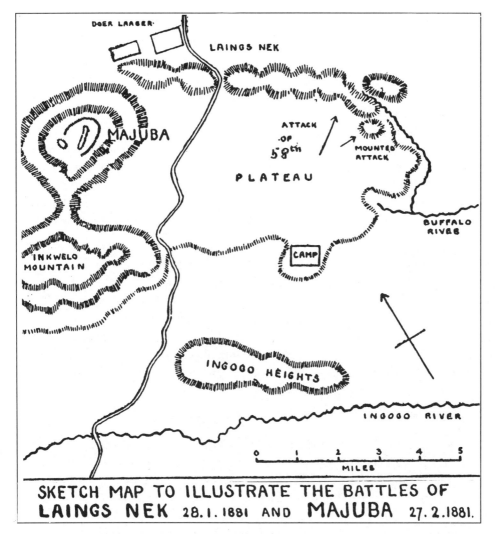

SKETCH MAP TO ILLUSTRATE THE BATTLES OF
LAINGS NEK 28.1.1881 AND MAJUBA 27.2.1881.

Laing's Nek

the ridge barring the way. The road over the Nek lay in the centre of a rough crescent of stony hills some six miles in length, and was only open to a frontal approach.[7, 8, 12, 13, 15, 16, 17]

On 28 January 1881 Colley moved out to attack the Boer position. The Naval Brigade with one company of the 60th went to the left to shell the right of the Boer entrenchments with rockets, the 58th (Northamptonshire Regiment) was to assault the centre, while Major Brownlow with his squadron was to seize an undefended spur which ran out from the Nek on the left of the Boer position. As the shelling started the Boer horses could be seen bolting in every direction, giving the impression that they were evacuating the Nek. In fact the burghers were shaken, although most of the shells and rockets had gone over the position, but they merely gritted their teeth and dug in. The 58th advanced in column of companies, with their colours carried unfurled in the centre (the last occasion when this was to happen).[7, 8, 12, 13, 15, 16, 17]

Major Brownlow led his squadron against the spur on the right of the Boer position. The spur was about 500 feet high and the approach to it was across three quarters of a mile of open ground. In the meantime the Boers had occupied and entrenched the spur, and as the squadron advanced at the trot, they came under heavy and accurate fire. This caused Brownlow to mistake his direction and, as the squadron broke into a gallop, to veer to the right. Colley reported afterwards:

> The charge was splendidly led by Major Brownlow, who with Sergeant Major Lunny, K. D. G., was first on the ridge. Major Brownlow's horse was shot under him and Sergeant Major Lunny was instantly killed; but Major Brownlow shot the Boer leader with his revolver and continued to lead his men who now crowned the ridge. Could he have been supported the hill was won, for the Boers had begun to retreat; but the fire was heavy while many of the horses were quite untrained to stand fire. The support troop was checked, the leading troop, fatigued with all their leaders down, could not push on and the whole gave way down the hill.[7, 8, 12, 13, 15, 16, 17]

In a private letter to Wolseley Colley was less guarded.

> Brownlow bore more to the right than I had intended and came under fire and, drawing up his men facing the steep part of the hill, charged right up to it before the infantry had begun their ascent. Of course in action the man on the spot must often decide the ground and the moment of the charge, and Brownlow's was gallantly made. Brownlow and a part of his leading troop, consisting principally of K.D.G.s, actually crested the ridge, his sergeant-major and corporal being shot dead. Brownlow, who was on foot, got off by a miracle; the whole lot went headlong down the hill and although the losses were not heavy (four killed and thirteen wounded, almost all out of the leading troop) the mounted men were practically out of action for the rest of the day.[13]

Major Brownlow was indeed fortunate, for his horse had been shot dead under him and he found himself dismounted among the Boers. Private Doogan, KDG, Brownlow's servant, was charging with the troop when he saw the Major dismounted among the Boers. At once he rode up and, although himself severely wounded, dismounted. He urged Brownlow to mount his horse, and while he was so doing he was himself wounded a second time. Both Brownlow and Doogan managed to retreat down the hill after their men. For this gallant action Doogan was awarded the Victoria Cross. [18]

The KDG losses had been heavy in proportion to the numbers engaged. Sergeant Major Lunny, who killed one Boer with his revolver and wounded another before he fell with six bullets in his body, and Corporal Stevens were killed among the Boers. Four KDG were wounded – Major Brownlow, Private John Doogan twice, Privates William Brown and George Coles. Another KDG, Private Venables, was captured by the Boers. The three privates were all dangerously or severely wounded, Brownlow only slightly. Thirty-two horses were hit. [12, 13, 18]

The leading troop, composed mainly of the KDG, had won the hill, and the Boers were actually running for their horses, but the second follow-up troop, made up of mounted infantrymen, thought that all their leaders had been killed and, without a man being hit, they turned tail and galloped back down the hill. One of the retreating Boers saw what was happening and shouted to his fellows, who promptly reoccupied their positions. Colley observed that if the mounted squadron had been regular cavalrymen on trained horses 'the position would have been ours, without much loss'. 'Poor Brownlow, who behaved splendidly, is quite broken hearted, and when he came down the hill refused to speak to his men or go near them. I have been telling him we must make allowances for untrained men with untrained horses.' [7, 8, 12, 15, 17]

In the meantime the 58th (Northamptons) were advancing on the main ridge. They came in sight of the Boers just as the cavalry charge failed, and the Boers on the spur were able to add their flanking fire to that of their comrades on the main position. As the 58th scaled Laing's Nek and arrived at the crest of the ridge, breathless and disordered by their climb, they were met by a fusillade of accurate fire from the Boers lying in cover. The colonel ordered the 58th to charge and was immediately shot from his horse. Other officers encouraged their men forward, only to be killed one after another. The 58th at once responded, but the fire from the Boers was too deadly for the charge to be effective, and ammunition was starting to run low. The only course left was for the regiment to withdraw. This was carried out with great discipline and bravery, Lieutenant Hill winning the Victoria Cross for his gallantry. The 58th re-formed at the foot of the hill, in spite of having had three officers and seventy-five men killed and two officers and ninety-one men wounded, but Colley had had enough and the action was called off. [15, 16, 17, 19]

The Boers followed up their victory by threatening Colley's lines of communication, intercepting a mail detachment as it was crossing the Ingogo river. A thousand Boers were reported to be mustering on the Schuinshoogte, the heights which dominated the Ingogo crossing. Colley decided to make a demonstration in force to clear the road, and a column set out, consisting of thirty-eight mounted men, of whom five sergeants and ten privates were KDG, under Major Brownlow, together with five companies of the 60th Rifles and four guns. In fact the Boers only numbered about 250 men, but they disputed the crossing of the Ingogo and pinned down the British force in an area where there was very little cover of any sort. The horses were easy targets and by 6 p.m. only eight artillery and four staff horses were still alive. Colley ordered Brownlow to clear a party of particularly troublesome burghers, but the men were caught in a murderous fire as soon as they broke cover going down to the river. Nearly all the horses were hit as well as many of the men; the survivors took shelter behind the dead horses and gun limbers. Farrier Sergeant Davis, KDG, received a gunshot wound in the head. Afterwards Colley wrote: 'With a moderate force at my disposal, it would not have been difficult to have rolled up the Boer right, which was dangerously extended and exposed, but the small mounted detachment under Major Brownlow, K.D.G., was too weak for such an attempt.'

The KDG suffered one further casualty during the First Boer War – Private Alfred Jessop, KDG, who had been released from prison on the outbreak of hostilities for return to active service. He was wounded in an attack on the Boers at Elandsfontein Ridge on 16 January 1881, when he was attached to the 21st Foot (Royal Scots Fusiliers). He was then again confined to the cells from 26 to 31 March! The remains of the mounted squadron stayed at Mount Prospect during the Battle of Majuba and the following peace negotiations, and by the end of March 1881 the KDG troop was on its way back to England to form the belated depot.[7, 13]

The main body of The King's Dragoon Guards had embarked at Durban for India. Sergeant Smith, who was now again Private Smith, having been reduced for being rude to an officer of another regiment, recalled,

> On sailing out of harbour the band of H. M. Ship Bordeica [sic] played us out with Old Lang Syne as well as our own band. On this voyage we had a capital time, singing every night and other amusements during the day, if not on duty. We arrived in Bombay on the 15th October, and proceeded at once to Deolali Depot. We stayed a few days to get our Indian kit. Then we started for Meerut, which was our station. When we arrived in Meerut the band of the 15th Hussars played us into camp, which is a very pretty barracks. By this time we began to understand about their coin, as a rupee was as two shillings, four annas as sixpence. We had plenty of coolies

about us now, dhobies which we paid twelve annas a month for washing as many clothes as we had to do, and there were plenty for we had six suits of white clothes each, and wanted a clean one every day in summer, and if on duty very often two. We had cooks as we paid the same amount to. Barber four annas a month for a shave every morning. Water carriers, sweepers, besides any amount of others about us.[6]

The *London Gazette* of 18 April 1882 carried the appointment of Major Marter to be the lieutenant colonel commanding the regiment. He

quickly discovered that the difficulties with which he would have to deal were far more numerous and important than he expected. He took command with full intention of taking time in reforming abuses, avoiding friction, and endeavouring gradually to lead into better ways, but he found such a state of affairs, more particularly among the Officers, that, from his point of view, delay could not be tolerated. There was no thoroughness in duty anywhere, and a system of bribery had crept into Non Commissioned Officers and others for carrying out ordinary regimental matters. He worked as much as possible through his Second in Command, Major Thompson, a man possessing good temper and tact. Some two or three Officers supported him, but the rest were sadly troublesome, and continued so throughout the period of his command. The Non Commissioned Officers and men came round to the new order of things, and gave no trouble whatsoever.

There was little recreation of a wholesome kind beyond the limits of their lines, while the Canteen had the sole attraction of drink, Marter formed a Regimental Institute, containing Refreshment Tavern, Recreation Room, Library and Writing Room. The Regiment congregated there in the evening, hardly any men leaving the lines. Crime almost ceased. After much trouble and delay, Marter acquired a good piece of ground, well turfed, and close to the lines, as a cricket ground, and enclosed it with posts and chains.[1]

SOURCES

1 The Marter Diaries, Regimental Museum, 1st The Queen's Dragoon Guards, Cardiff Castle.
2 D. R. Morris, *The Washing of the Spears*, 1966.
3 M. Barthorp, *The Zulu War*, 1980.

4 C. L. Norris Newman, *In Zululand*, 1988.

5 *Narrative of the Field Operations Connected with the Zulu War of 1879*, 1989.

6 R. Smith, KDG, 'Sketches of Home and Abroad', Regimental Museum, 1st The Queen's Dragoon Guards, Cardiff Castle.

7 Records of the King's Dragoon Guards, Regimental Museum, 1st The Queen's Dragoon Guards, Cardiff Castle.

8 W. Clowes, *King's Dragoon Guards, 1685-1920*, 1920

9 *Sheffield Daily Telegraph*, 7 October 1879.

10 *Army and Navy Gazette*, 6 September 1879.

11 *Regimental Journal of 1st The Queen's Dragoon Guards*, vol. 3, no. 6, p. 549.

12 The Marquess of Anglesey, *A History of the British Cavalry*, vol. 3, 1982.

13 Elizabeth Cox, 'The First King's Dragoon Guards in South Africa, 1879-1881', *British Military Review*, vol. 6, no. 3, June 1985.

14 H. Rider Haggard, *The Last Boer War*, 1899.

15 L. Maxwell, *The Ashanti Ring*, 1985.

16 *Journal of the Society for Army Historical Research*, vol. 30, 1952, p. 167.

17 J. Lehmann, *The First Boer War*, 1972.

18 *The K. D. G.*, vol. 5, no. 2, 1955, p. 107.

19 R. Gurney, *History of the Northamptonshire Regiment, 1712- 1934*, 1935.

30

Sudan, Abu Klea, India
1880-1893

During the early part of the 1880s trouble had been brewing in Egypt and the Sudan, and in 1882 Wolseley carried out a brief but brilliant campaign, culminating in the Battle of Tel-el-Kebir and the occupation of Cairo. The Army of Occupation in Egypt required reinforcing, and The Queen's Bays were ordered to send drafts to a number of cavalry regiments. Volunteers were called for, with twenty-eight men going to the 4th Dragoon Guards and fifty-seven to the 7th Dragoon Guards. In 1884 Gordon was besieged in Khartoum, and public opinion forced a reluctant Government to mobilise a relief expedition. The advance up the Nile from Egypt was to consist of two columns, the River Column comprising the infantry, and a Desert Column made up of a hastily raised and trained Camel Corps, which Wolseley in Egypt had insisted upon being formed. Contingents were drawn from all branches of the cavalry, and the Bays provided a party of two officers, Captain Gould and Lieutenant Hibbert, with forty-three non-commissioned officers and men.[1, 3, 5, 6, 7, 8, 9, 10]

The Queen's Bays, being dragoon guards, went to the Heavy Camel Corps, and were joined with the Royal Horse Guards contingent in No 2 Company; No 1 Company was made up of the 1st and 2nd Life Guards, No 3 the 4th and 5th Dragoon Guards, No 4 the Royal Dragoons and Scots Greys, No 5 the 5th and 16th Lancers. The men were dressed in grey tunics, cord pantaloons, blue puttees and brown boots. They carried the 1882 type bandolier carrying fifty rounds, a brown leather waistbelt with a twenty-round pouch, and a haversack. Individual regiments were distinguished by insignia sewn onto the sleeve. The camel was laden with a rifle bucket, another 100 rounds in saddlebags with the man's kit, a three-quart leather waterbottle, a six-gallon water-skin, a blanket, shelter tent, greatcoat, waterproof sheet and cornbag. The men carried the Martini-Henry rifle with a sword bayonet. Officers were similarly dressed, but generally wore boots and Sam Browne belt for a sword and revolver.[1, 2, 3, 8]

The men of the Camel Corps sailed from Portsmouth on 26 September 1884 in the P & O *Deccan*. The Duke of Cambridge wrote,

They are a very fine body of men, and will no doubt form a valuable body
of soldiers from the special object had in view, but they are the cream of
every corps to which they happen to belong, and I must confess I regret
the necessity which has arisen in sending them so constituted.The principle is unsound. It is most distasteful to Regiments and officers and men;
especially the Commanding Officers are disgusted at seeing their best men
taken from them, and themselves being disbarred from sharing in the
honours and glories of the service upon which they may happen to be
engaged.[1, 6]

On arrival in Egypt Wolseley moved the men of the Heavy Camel Corps by
boat up the Nile to Korti, where he had formed his 'fighting base'. It was at Korti
that the men were given their new steeds. 'Generally speaking, the camel gets up
just as the man gets his foot in the stirrup, and the results are curious. Woe betide if
you try to throw your right leg over before the beast is up; you will infallibly come
a hideous cropper. The only thing is to stand in the stirrup, let him rise with you,
and then get into your seat.' Another variation was

Mounting a frisky camel is exciting work for the beginner, and nearly
always ends in a cropper. The mode of procedure should be thus: having
made your camel to kneel by clearing your throat loudly at him and
tugging at his rope, shorten your rein till you bring his head round to your
shoulder, put your foot in the stirrup, and throw your leg over. With his
head jammed like that he cannot rise, and must wait till you give him his
head. Unless you do as directed, he will get up before your leg is over, and
you will infallibly meet with a hideous catastrophe. . .

A camel's hind legs will reach anywhere – over his head, round his chest,
and onto his hump; even when lying down an evil-disposed animal will
shoot out his legs. His neck is of the same pliancy. He also bellows and
roars at you whatever you are doing, saddling him, feeding him, mounting
him, unsaddling him. To the uninitiated a camel with his mouth open and
gurgling horribly is a terrifying spectacle; but do not mind him, it is only
his way. He hardly ever bites, but when he does you feel it for some time![1,
4, 6, 7]

Gordon had been besieged in Khartoum for ten months. On 31 December 1884 a
message was received at Korti written on a tiny piece of paper, 'Khartum all right.
14.12.84. C. G. Gordon.' The messenger, however, sounded a more urgent note:
'Fighting goes on day and night. The food we have is little. We want you to come
quickly.' Wolseley at once decided that the Camel Corps should strike across the

Bayuda Desert, taking the shortest route to Khartoum, 150 miles to the south. The main force would continue by the much longer route along the course of the Nile.[6]

The Desert Column set out on 14 January 1885, with 3,000 camels – the Heavy Camel Regiment, of which the Bays were part, the Guards Camel Regiment, and the Mounted Infantry, rode in a solid phalanx of forty camels abreast, stretching over half a mile in length. General Sir Herbert Stewart, in command of the column, intended to reach the wells at Abu Klea in two days' march, but on the afternoon of the 16th the advance guard of the 19th Hussars reported that the Dervishes were in force around Abu Klea. General Stewart decided to camp for the night within a hastily constructed *zariba* of thorn bushes, rocks, ammunition boxes and camel saddles, and to continue the advance in daylight. [3, 6, 7]

On the morning of the 17th the baggage, most of the camels and the sick and wounded were left in the *zariba* with a guard, and Stewart formed a square for his advance to the wells. The Bays were stationed with the Naval Brigade and the Gardner gun in the centre of the rear face of the square; a line of skirmishers was pushed out on the flanks. This unwieldy formation made slow progress down a valley leading to the wells, the centre of the valley being filled with clumps of long grass. The square soon came under fire from a line of Dervishes whose banners could be seen running across the valley to the left. In order to avoid the scrub grass, to attend to the wounded and to get rid of sick and wounded camels, the square had to make frequent halts, which tended to throw the formation into confusion, especially at the rear. When the square was some 500 yards from the line of Dervish flags, a halt was made to allow the rear face to close up, as it had been delayed by the baggage and ammunition camels lagging behind. Before this could happen, 5,000 Arabs leapt from the cover of the scrub and charged the left front of the square in two columns.

The flanking skirmishers raced back to the square, thus masking its fire until the Dervishes were barely 200 yards away. Colonel Burnaby of the Blues ordered the dragoons to wheel outwards, presumably to give covering fire for the skirmishers, whereupon the Bays and Blues started to edge to the right in support of their comrades, weakening the rear face further. In the meantime the fire from the 7-pounder guns and the mounted infantry posted on the left front and side of the square had been so effective that the great mass of Dervishes veered to their right into more broken ground, joining up with three more columns of Arabs, to charge at a breakneck speed against the left rear of the square. Burnaby saw the danger and ordered the dragoons to re-form, but he was too late. A gap of sixty to eighty yards had opened between the Lancers and the Greys and Royals.[3, 6, 7, 8]

Lord Charles Beresford of the Naval Brigade had the five-barrelled Gardner gun manhandled twenty yards outside the left rear corner of the square, and opened fire

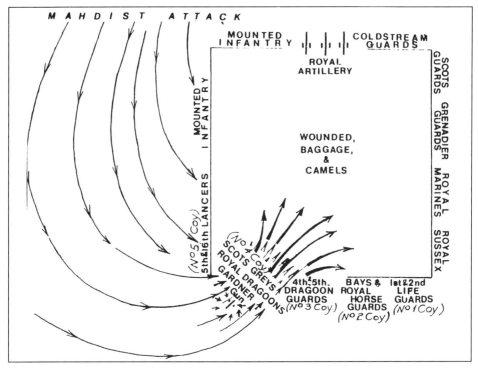

The Attack on the Square at Abu Klea

on the oncoming Arabs. As the fanatical horde neared, the Gardner jammed. The bluejackets serving it were speared, Beresford only escaping by throwing himself under the gun. Burnaby rode out to assist the last of the skirmishers, and while defending himself had a heavy spearpoint thrust into his throat, cutting his jugular vein. The mass of Arabs poured into the gap behind the Gardner, and the rear of the square became a confused mêlée of savage hand-to-hand fighting, the Bays and their fellow cavalrymen pitting their unfamiliar rifles and bayonets against the spears and swords of the Arabs. The Dervish charge was only stopped by the inert mass of tethered camels in the centre of the square. Gleichen of the Guards recalled: 'I found myself lifted off my legs amongst a surging mass of Heavies who had been carried back against the camels by the impetuous rush of the enemy.' The rear ranks of the mounted infantry and the Guards, on the front and sides of the square, turned about, pouring volleys into the packed Dervishes. As the following Arabs saw their fellows falling in great numbers under this fire, they began to lose heart, wavering and falling back. A body of Dervish horse made a final charge, but were driven back by the steady fire of the Bays and Household Cavalry. The whole action had lasted only ten minutes.[3, 4, 6, 8]

Burnaby lay dying thirty yards outside the square. 'A young private of the Bays, a mere lad, was already beside him, endeavouring to support his head on his knee; the lad's genuine grief, with tears running down his cheeks, was touching, as were his simple words: "Oh! sir; here is the bravest man in England, dying and no one to help him."' The Bays' contingent had themselves lost four men killed in the scrimmage. The total loss was 9 officers and 66 men killed, and 9 officers and 72 men wounded, but the Dervish casualties were enormous, 1,100 bodies being counted around the square. Corporal Middleton wrote home, 'After the scrimmage was over we had plenty to do. Frightful sight after smoke cleared away. Wounded men, dead men, dead camels, all of a heap. After clearing all wounded marched into wells, stayed all night.'[1, 6, 7]

It was late the following day before the march was resumed, and it was 19 January when the column came in sight of Metemmeh and the Nile. Again the Dervishes lined a ridge by the village of Abu Kru. Lieutenant Hibbert of The Queen's Bays described the action:

> Our Squadron Leader was wounded at the commencement and my Captain was in the laager disabled by gout, so I was the senior officer in the squadron. At last the enemy charged the left and front faces of the square, but not one of them got within 150 yards. We then marched down to the river and drank it nearly dry out of our helmets. Under arms all night in the bitter cold and nothing to eat. Next morning half the force went to bring up the camels, and the rest of us fortified the village. Next day we went out to reconnoitre Metemmeh. It is a big sand town a mile long and all loopholed. As we advanced they fired on us with three guns, but did very little damage. Suddenly we saw the Khedival flags of some steamers coming down the river. They came right abreast of us and landed, and it turned out to be Gordon's four steamers which had been on the look-out for us ever since September. We shelled the town a bit and then retired, and established ourselves strongly on the river. We have indeed had a very rough week of it. In the heavy division we lost 25 per cent killed and wounded. The bravery of the enemy is simply wonderful, and the worst of it is they are such good shots, people say as good as the Boers.

On 1 February the news came that Khartoum had fallen, and Wolseley recalled both the River and the Desert Columns.[3, 4, 7]

In the autumn of 1885 The Queen's Bays were again detailed for service in India. The regiment embarked in the troopship *Serapis* on 21 November, under the command of Lieutenant Colonel Hemming Lee, and finally reached the new station at Ambala on 3 January 1886. The Bays suffered a number of changes in command at this time, Colonel Lee giving up in favour of Lieutenant Colonel Leslie French, who himself only commanded for two years before making way in 1887 for Lieutenant Colonel J. J. S. Ashburner. Ashburner retired on 15 August 1888, to be succeeded by Lieutenant Colonel F. C. L. Kay.

On arrival at Ambala The Queen's Bays took over the horses and camp followers of the 9th Lancers, and among the latter was the Jemadar syce of 'B' Troop, Karim Baksh, who had served with the regiment at Lucknow during the Mutiny, and had helped to bring in the body of Major Percy Smith.

During 1886 and 1887 the Bays held the record of having less crime than any other regiment, not only in India, but in the whole of the British Army. They also had a distinguished sporting record. During their ten years in India the Bays nine times carried off the annual Lloyd Lindsay Mounted Competition, which was open to all British mounted regiments. Regimental Sergeant Major Greenwood and Sergeant Kent won the first and second place respectively in the Inspector General of Gymnasia's competition for the best 'man at arms' in India. With the exception of 1889, The Queen's Bays entered the inter-regimental polo tournament from 1886 to 1894, and for the three successive years of 1892, 1893, and 1894 won the tournament, defeating in those finals the 7th Hussars, the 16th Lancers and the 5th Lancers. The teams were: 1892 Captain V. G. Whitla, W. H. Persse, Captain Kirk, C. K. Bushe; 1893 and 1894 W. H. Persse, H. W. Wilberforce, Captain Kirk, C. K. Bushe.1894. The tournament had only started in 1877, and polo ponies were strictly limited to a height of 13.3 hands. In 1894 the Bays fielded twenty-four ponies, who were allowed to be played in the tournament, and of these eighteen were Arabs, four were countrybreds and two were Australian Walers. Every pony had to be under five years old and each had to be measured the day before the opening of the tournament and registered each year in the Polo Calendar. Whilst the Bays were at Sialkot and Rawalpindi they competed in the Punjab Polo Tournament at Lahore, and each year, bar one, were defeated in the final by the Maharaja of Patiala's team, which in those days included the famous polo player Colonel Hira Singh. While at Sialkot Captain Clark kept a small pack of hounds at Phooklian on the banks of the River Chenab, and hunted jackal with them. Captain Kirk owned a very good mare, Engadine, and an Australian pony, Teddy, which he raced in partnership with Captain Sherston of the Rifle Brigade, winning a number of races, including the Kurnal Stakes for horses bred by General Parrott at the Kurnal Stud.[1, 11]

Lieutenant W. H. Wilberforce of The Queen's Bays saw service on the staff of Major General W. K. Elles during the Hazara operations of 1891 on the North West Frontier; also in 1891 Lieutenant N. M. Smyth served on the survey staff of the Zhob Valley Expedition, gaining promotion to the rank of captain. In 1895 the

Bays heard that their old commanding officer, Major General Seymour, had been promoted to lieutenant general, and had been nominated as the regiment's colonel in chief in succession to General Walker, who had retired. In the autumn of 1895 the Queen's Bays embarked in the *Malabar* for Egypt, arriving on 28 October.[1]

The King's Dragoon Guards at Meerut were joined in 1882 by a nineteen-year-old, who was destined to become a famous explorer and mystic. Francis Younghusband was welcomed as a young officer, partly because Sir Dighton Probyn, VC, had recommended him to Colonel Marter, and also as his arrival allowed a colleague to go on leave before the hot season ended. Younghusband found that the regimental day started at 5 a.m. and ended at 10.30 a.m. when all ranks retired to their quarters until sundown. Adjutant's Parade lasted from 5.30 a.m. until 7 a.m., to be followed by Stables until 8 a.m. and then Orderly Room, after which a sharp ride to the swimming baths before returning to the mess for a breakfast at 10.15 a.m. of porridge, quails or curry. At 6 p.m. the officers emerged to play polo or tennis, to ride or drive, until formal mess at 8.15 p.m., followed by billiards and cards.[14]

Younghusband shared a bungalow with the adjutant, Captain Hennah, whom he admired almost as much as he did Colonel Marter, but in April 1883 Hennah died and Marter was promoted. The keen young intellectual, out of his depth in a cavalry regiment, discovered that

> The art of warfare was the last topic of conversation that was likely to rise. I found that these same men, as soon as there was any active service in sight, would move heaven and earth to get there. Keenness for sport did not mean indifference to service in the field. All it meant was disinclination to the monotony of preparation. On the other hand I was surprised at the friendship I experienced. To my surprise I found my brother officers excellent fellows; and in my heart of hearts I envied them their good nature. They never went to church except when paraded for service. Their talk was of little else than ponies or dogs. Their language was coarse. And yet they were a cheery lot, always ready to do each other, and even me, a good turn, and secretly possessing an ideal of their own to which I would have been thankful to attain: it was simply to be a 'good fellow', and a good fellow in their eyes was above a good Christian or even a good soldier.[14]

In 1882 Mr Lowry, KDG, won the Kadir Cup for pig-sticking on Moorcock. The Kadir had been started in 1867, and Lord Charles Beresford commented, 'Pig-sticking is a rough rude sport, but a mighty enjoyable one.' The hog hunter, it was said, had to have the seat and judgement of a fox hunter, the eye of the falconer, the arm of the lancer and, above all, a horse fleet, active, bold and well in hand. This was the only time that the cup was won by either a KDG or a Bay, although in 1914 Alexander, KDG, on Iced Water came near to winning.[15] The King's Dragoon Guards left Meerut on 13 June 1883 and marched to Rawalpindi, not arriving there until 19 December. Marter wrote, 'The Regiment marched not a man showing a sign of irregularity.' Younghusband remembered, 'The Regiment marched to Rawal Pindi all the way; in their scarlet; carbines and swords; jackboots and white helmets.' Private Ashton died at Ambala whilst en route. The strength of the regiment was 13 officers and 325 other ranks, but it was accompanied by 1,304 camp followers, with 43 bullock carts, 227 camels, 184 ponies and 157 bullocks, 1 mule, 3 ambulances and 8 dhoolies.

> The Regiment marched as in an enemy's country, with Advanced and Rear Guards extending, when the country permitted, and with flankers out. An Officer was detailed daily to furnish a sketch, and to report on the road, and on the country on either side to a distance of about three quarters of a mile.[12, 13, 14, 18]

In April 1884 Lieutenant Colonel W. H. Thompson took over command. Younghusband, who was now adjutant, did not find life easy.

> Punctilious to a degree, both in personal equipment and behaviour, he expected a clockwork precision – that and nothing more. He looked for a perfectly timed mechanism, in which the eye of the most inquisitive inspecting General would be unable to find a flaw ... [During an exercise] the Regiment was moving over rough ground, and it was certainly not moving like a machine, but with considerable speed and like a live being for all that. There was much apparent but no real disorder, and the Colonel was furious. 'Mr Younghusband, you have been working the Regiment as if we were on active service. We are nothing of the kind. We have the General coming to inspect us, and we must have it orderly. Go to the rear and another Captain will take your place.'

From then on Younghusband was not the first adjutant to resort to guile, 'Most tactfully, therefore, after [each] parade I would inform the Colonel of any irregularities, and ask him if he had observed any others, so that I could correct them at Adjutant's parade. Gradually I weaned him from his dreadful habit! And a ceremonial parade became a true work of art.'[12, 14]

The Colonel later told Younghusband, 'Get going with games, sports, entertainments, theatricals, sing songs and all the rest of it.' The help of a young subaltern was enlisted, who became so enthusiastic as to give up shooting and polo, and the result was the setting up of an athletic club, dramatic club and boxing club, together with football and cricket matches, sports, a gymkhana, and a regimental magazine. Before Marter's time of command and the ensuing period in India, the men had been left very much to their own devices.[14]

The KDG at Rawalpindi were formed into six troops: 'A' Troop – Major Willan, TSM Foulkes, roans and greys; 'B' Troop – Major Veitch, TSM Moran, dark chestnuts; 'C' Troop – Captain Willett, TSM Barry, greys; 'D' Troop – Major Forbes, TSM Alderson, bays; 'E' Troop – Captain Hardy, TSM McGill, blacks; 'F' Troop – Captain Le Veta, TSM Nix, light chestnuts. During 1886 the regiment was augmented to a strength of 29 officers and 602 other ranks, with 526 horses. In January 1888 Lieutenant Colonel Thompson retired on half-pay, and was succeeded by Lieutenant Colonel G. V. C. Napier. During 1888 three officers and nine other ranks took part in the operations against the Black Mountain tribes, and were awarded the India General Service Medal with the Hazara Clasp.[12, 16]

In December 1889 The King's Dragoon Guards left Rawalpindi, and marched to Muttra to relieve the 3rd Dragoon Guards. In June the following year Colonel Napier, a most capable and popular commanding officer, was killed when he fell from his horse out pig-sticking; Lieutenant Colonel Douglas Willan assumed command in his place. During 1890 the Tsarevich of Russia was touring India and paid a visit to the regiment. He was so well entertained, and formed such a favourable impression of the men and horses, that he presented the officers' mess with a valuable set of gold cups and a jug, the whole inlaid with beautifully carved enamel (now to be seen on display in the Regimental Museum). The King's Dragoon Guards embarked at Bombay in the troopship *Euphrates* on 17 October 1891. On reaching Suez one squadron disembarked to form a part of the Army of Occupation in Egypt, relieving a squadron of the 17th Lancers. The rest of the regiment, on reaching England, marched to Shorncliffe, arriving on 2 November 1891.[12, 13, 17]

One of Marter's sons, William Marter, had been commissioned into the Royal Fusiliers in 1887, but exchanged into The King's Dragoon Guards in April 1882 and joined the regiment at Shorncliffe. During October 1892 the KDG were moved to Windsor, where they relieved the 1st Life Guards. In April 1893 Major Spencer returned to the regiment with 'C' Squadron, comprising 5 officers and 165 other ranks who, having completed their tour in Egypt, had been relieved by the 7th Dragoon Guards. In May a squadron of 100 men marched from Windsor to London for the opening of the Imperial Institute by the Queen, which drew a message from Her Majesty that she had 'been pleased to express her gratification at the appearance of the troops, and at the excellent manner in which they performed their duties.' In July 1893 The King's Dragoon Guards left Windsor for Norwich, 'A' and 'C' Squadrons marching via Richmond, Romford, Chelmsford,

Colchester, Ipswich, Eye and Scole; while 'B' and 'D' Squadrons took the route, Watford, Hertford, Bishop's Stortford, Haverhill, Bury St Edmunds and Diss. William Marter was sent ahead by rail to take over the barracks from the 8th Hussars.[12, 13, 18]

SOURCES

1 F. Whyte and A. H. Atteridge, *The Queen's Bays, 1685- 1929,* 1930.

2 M. Barthorp, *The British Army on Campaign, 1806-1902,* no. 4, 1988.

3 M. Barthorp, *War on the Nile,* 1984.

4 P. Warner, *Dervish,* 1973.

5 L. James, *The Savage Wars,* 1985.

6 M. Alexander, *The True Blue,* 1957.

7 F. Emery, *Marching over Africa,* 1986.

8 H. Keown-Boyd, *A Good Dusting,* 1986.

9 Sir Evelyn Wood, *British Battles on Land and Sea,* 1915.

10 Winston Churchill, *The River War,* 1899.

11 J. R. C. Gannon, 'Polo', *R. A. C. Journal.*

12 Records of The King's Dragoon Guards, Regimental Museum, 1st The Queen's Dragoon Guards, Cardiff Castle.

13 W. Clowes, *King's Dragoon Guards, 1685-1920,* 1920.

14 G. Seaver, *Francis Younghusband,* 1952.

15 J. R. C. Gannon, 'History of the Kadir Cup', *R. A. C. Journal.*

16 *The K. D. G.,* vol. 5, no. 1, 1954, p. 64.

17 *King's Dragoon Guards,* 250th Anniversary, 1935.

18 The Marter Diaries, Regimental Museum, 1st The Queen's Dragoon Guards, Cardiff Castle.

31

England, Ireland, Second Boer War
1893-1901

The pattern of life for The King's Dragoon Guards in Norwich was typical of a cavalry regiment serving at home during the late Victorian period. The *Eastern Daily Press* reported on 19 July 1893 that 'the first Squadron of the 8th Hussars marched from the Cavalry Barrcks, Norwich, and the remaining Squadrons left on the 20th. The 1st King's Dragoon Guards, commanded by Lt. Col. H. P. Douglas Willan, marched in on the same dates.' On 7 September the paper carried the news that 'in consequence of a telegram received from the War Office, the K. D. G. left Norwich for the scene of the colliery riots in the Midland counties. The Regiment entrained at Trowse, the horses being conveyed in bullock trucks. 'A' Squadron proceeded to Mansfield, 'B' to Rotherham, 'C' to Wakefield and 'D' to Dewsbury.' Marter's son, William Maurice, now a captain in the regiment, reported:

> The Regiment received unexpected orders to send about 200 men to Yorkshire and Nottinghamshire in aid of the Civil Power – the prolonged coal strike having resulted in riots. We marched out of barracks at 8.45 p.m. and got off by train at midnight, reaching Wakefield at 6 a.m. in command of 50 men. During that day we rode round a number of collieries dispersing crowds. I and my men were hooted at, and a little coal was thrown. We were billeted for two nights in a farmhouse. Being then relieved in command of the party, I proceeded to Pontefract, and did duty with a detachment there for ten days. On 21st September I rejoined my own Squadron at Mansfield, marching the 46 miles in the day on my second charger 'Jenny'. On 4th October the Squadron returned to Norwich.

By 22 October all the squadrons had returned by rail to Norwich.[1, 2, 3]

There was at that time at Norwich a resident pack of staghounds, which each cavalry regiment hunted and then handed on to its successor. Marter enjoyed excellent sport with this pack, his mare Jenny carrying him particularly well over the big Norfolk ditches. In July 1894 a military tournament was held in the Agricultural Hall, in aid of the clothing fund for the local cadet corps, in which the

KDG took part, together with the Depot of the Norfolk Regiment and the Norwich Artillery Volunteers. In October the city was visited by the Duke and Duchess of York who opened the Castle Museum and Fine Art Gallery. 'The streets on the route from the station to the Castle were lined by troops, consisting of the 1st K.D.G.' At Thorpe Station

> a procession was formed, the escort being furnished by the 1st King's Dragoon Guards under Captain Birkbeck and Lieutenant Langton. At the Castle public excitement increased when distant cheering announced the progress of the procession, and when at length the glittering helmets and scarlet uniforms of the Dragoons forming the advance guard were sighted, followed by the leading sections of the escort, the cheers of the crowd became more and more hearty.

Later, at the Cathedral, 'during the time occupied by the assembling of the company, the band of the 1st King's Dragoon Guards played selections of music. As the head of the procession advanced along the aisle, the band of the Dragoons played the National Anthem.' In May 1895 the KDG also provided the escort for the Prince of Wales when he visited Norwich Cathedral.[1, 2, 3, 4]

Whilst at Norwich the officers indulged in a series of amateur dramatics, giving three performances of *The Area Belle* and Burnand's burlesque, *Black-Eyed Susan*, at the Norwich Theatre in aid of the Soldiers' Widows Fund; and five performances were given of *Easy Shaving* together with a burlesque of *Faust* in aid of the Norfolk and Norwich Hospital.

In December 1894 Captain Bogle Smith married a local girl at Woolverstone Park, Ipswich, and Major Forbes, the second in command, attended a promotion examination at Ipswich. During 1894 the colonel of The King's Dragoon Guards, Lieutenant General James Sayer, who had commanded the regiment in China in 1860, visited Norwich, when a group photograph was taken of the officers in a variety of dress. The group included two future colonels and no less than seven commanding officers. On 14 June 1894 Lieutenant Colonel Douglas Willan resigned after four years in command, and was succeeded by Lieutenant Colonel R. C. B. Lawrence. During August 1895

> The 1st King's Dragoon Guards, who have been quarterd in Norwich for about 2 years, have received orders to march to Colchester today. In consequence of this, Lieutenant Colonel Lawrence and Officers of the regiment regret that their regimental sports cannot take place. It has already been announced that this popular regiment will be succeeded by the 7th Dragoon Guards.[1, 2, 3, 4, 5]

In February 1896 lances were issued to the front rank of dragoon guard regiments, and the magazine Lee Metford carbine came into general cavalry use. Marter described a typical exercise:

On the 12th May, 1896 General Luck [then Inspector of Cavalry] saw us in the field, and a rare doing he gave us. On the 13th we marched out of Colchester at about 9 a.m., reconnoitred to Ipswich, being opposed by a skeleton force found by the Gunners, and a distance of 26 miles. On the 14th we reconnoitred north on a front of about 7 miles and came into contact with the 7th Dragoon Guards from Norwich, and threw out outposts along the river. We marched 37 miles. On the 15th there was a general concentration of the outpost line at 3 a.m., but I was left watching my bridge till 7 a.m., when the war was over, and we marched 30 miles back to Ipswich. On the 16th we marched back to Colchester, and the next day I went round the sorry horses, and there were 30 casualties.[1, 3]

Early in 1896 Queen Victoria was staying at Nice, where she met the Emperor of Austria, Franz Josef, and offered him the appointment of colonel-in-chief of the 1st King's Dragoon Guards. The Emperor accepted with pleasure, writing three years later to the Prince of Wales, 'I am delighted with this beautiful present, which will always remind me of its kind donor, but also of the gracious disposition of The Queen, your noble Mother, who did me a singular honour when she appointed me Colonel-in-Chief of the King's Dragoon Guards.' The appointment was announced in the *London Gazette* of 24 March 1896. By the Emperor's command the KDG adopted as their regimental badge the Habsburg coat of arms, the Austrian double-headed eagle, this being sanctioned for use as cap badge, collar badge and sergeants' arm badge by the Army Clothing Department in 1898. In April 1896 Colonel Lawrence with Captains Briggs and Bell-Smyth were received by His Imperial Majesty Franz Josef in Vienna, where they presented him with the old regimental standard. The Emperor, in turn, conferred the 2nd Class of the Order of the Iron Crown upon the colonel, whilst the two captains received the 3rd Class of the Order. Franz Josef had a set of band parts sent to the regiment; this turned out to be the *Radetzky March* composed by Johann Strauss in 1848 in honour of Field Marshal Count Joseph Wenzel Radetzky, and it has been the Regimental Quick March ever since. In addition, on guest nights in the KDG officers' mess after dinner the National Anthem was followed by the band playing the tune 'Austria' in honour of its colonel-in-chief.[1, 6, 7, 8, 9]

The Queen's Bays reached Egypt in the troopship *Malabar* on 28 October 1895 and were posted to Cairo. There they received their magazine Lee Metford carbines, and the front rank was equipped with lances. In November 1895 Lieutenant

Colonel Lambert retired and was replaced by Lieutenant Colonel J. E. Dewar. While the regiment was serving in Egypt a number of officers saw active service. Captain W. H. Persse, in command of a squadron of Egyptian cavalry with the rank of *bimbashi* (major), took part in all the campaigns from the advance to Dongola in 1896 to the capture of Khartoum in 1898. During the reconnaissance of the Dervish *zareba* at the Atbara on 29 March 1898, Persse showed exceptional leadership when the Egyptian cavalry met a counter-attack by a large force of the Dervish cavalry, made up of the best fighting men of the Baggara tribe, who were the finest of the Sudanese horse. For his conduct Persse received a brevet-majority.[10]

Captain N. M. Smyth of The Queen's Bays was attached to the intelligence department of Kitchener's staff during the advance on Khartoum. On the day of the Battle of Omdurman, Smyth was awarded the Victoria Cross. The citation reads, 'Captain Neville Maskelyne Smyth, 2nd Dragoon Guards – At the Battle of Khartoum, on the 2nd September 1898, Captain Smyth galloped forward and attacked an Arab who had run amok among some camp-followers. Captain Smyth received the Arab's charge and killed him, being wounded with a spear in the arm while so doing. He thus saved the life of one at least of the camp-followers.' Farrier-Sergeant Escritt, Queen's Bays, also attached to the Army Veterinary Department during Kitchener's advance on Khartoum, was awarded the Distinguished Conduct Medal for his gallantry.[10]

In 1896 The Queen's Bays were ordered to return to England, arriving home in 1897 to be stationed at Aldershot. At this time the drum horse was a white or grey, the kettledrummer a corporal, who wore a white helmet plume, as did all members of the band, in contrast to the full dress black plume of the rest of the regiment. The bandsmen wore a white aiguillette from the left shoulder. The drumhorse had a shabraque, surmounted by a black lambskin with white edging. The drum banners bore the regimental crest on a buff background edged with gold lace.   At the outbreak of the South African War in 1899 the regiment was required to furnish drafts for other cavalry regiments from time to time. The regimental strength was increased in February 1900 to 32 officers, 72 sergeants, 10 trumpeters, 782 other ranks and 609 horses. On 2 February 1902 The Queen's Bays provided 100 men under the command of Lieutenant Colonel Dewar to march in front of the gun-carriage carrying the coffin at the funeral of Queen Victoria.[10, 11]

During August 1901

> Sudden orders were received to furnish a strong mounted detachment to proceed direct by rail to Bangor, North Wales, to assist the local authorities in connection with the strike of quarrymen at Bethesda. Captain R. D. Herron was in command. 'A' Squadron received orders at

4.30 p.m., whilst the Squadron was at Evening Stables, and within ninety minutes they had paraded in full marching order − sword, lance and carbine − marched to Aldershot Town Station and entrained. The Regiment had only returned two days previously from the Cavalry Manoeuvres at Salisbury Plain after experiencing seven days' continuous rain. In fact, red jackets, blue patrols, jack-boots, blankets, etc, were all out being dried. Saddles had to be packed, kits to be collected from the drying-line, and men to be relieved off guards. After an uneventful journey, the Regiment arrived at Bangor early one morning, being met at the Station by the Civil and Police officials. It was soon discovered that its presence was not popular. The men of the detachment had great difficulty in making themselves understood, as the inhabitants only spoke the Welsh tongue. Strangely, the inhabitants did not have a liking for the Military; however, the good example and behaviour of the men was beyond question, and the people of Bangor simply loved the Regiment, and our three months' stay was made a pleasure. The whole three months the detachment carried out strong patrol work, and on two occasions turned out to assist the Police. The men displayed great coolness, and no unpleasant incident occurred.[10]

In 1898 The King's Dragoon Guards moved to Ireland, and were stationed in Dublin. At this period the drum horse was either a piebald or a white, the kettledrummer a trooper. He, like the rest of the band, wore a white plume to his helmet, in contrast to the regimental colour of red, and an aiguillette from his left shoulder. The drum banners were blue with the Royal Arms in the centre, surrounded by scrolled laurel leaves carrying the regiment's battle honours. The drum horse had a scarlet throat plume and a black lambskin edged in red over the saddle, but no shabraque was worn until 1904, when a scarlet tuft plume was also added to the brow band.[11]

An issue of Regulations in 1898 stated that 'Standards and Guidons of cavalry will be carried by Squadron Sergeant-Majors'. Those regulations still apply to the present time. The standards of both The King's Dragoon Guards and The Queen's Bays were of crimson silk damask, edged by a gold fringe, with cords and tassels of gold and crimson mixed. Each standard carried the battle honours of each regiment, the regimental title and seniority in abbreviated form, and the White Horse of Hanover; The Queen's Bays also carried the cypher of Queen Caroline of Anspach, Consort of George II. Dress Regulations of 1900 laid down that officers' cloaks were to be of blue cloth, reaching, when dismounted, to the ankle. The KDG lining was scarlet and that of The Queen's Bays white; collars were blue, but in the case of the KDG lined in red. Officers' mess staff at this time numbered sixteen, including two coachmen.[11, 12, 13, 14, 15, 16]

In April 1900 Lieutenant Colonel H. Mostyn Owen, commanding the KDG, wrote from Marlborough Barracks, Dublin to a young officer about to join the regiment:

> I have received your letter about a horse, at £10 a year – there will be no difficulty about that, but I shall not try to find a horse till I know whether you are a big man or a small man – nor shall I take any steps to mount you on this valuable animal till I have dismissed you from your Riding School. As regards your kit on joining, I would like you to go to the Regtl. tailors if you have not given your orders already. I will pass no article, whoever it is made by, if it is not absolutely correct and of the best quality and workmanship.
>
> S & L know this – one man I will not allow to make anything is ----, they turned out a disgraceful kit for a young officer recently. I hope you will join as soon as possible – we are very short, owing to so many officers being in S. Africa. Anything you want to know write to me, or if you can spare a day, come over here and see me. Punchestown is next week and we can, I am sure, find you a room somewhere.[17]

One of the officers who had gone to South Africa was Marter, who had been appointed as brigade major to the 14th Brigade in 7th Division. He took with him his favourite mare, Jenny, and his KDG soldier servant, Burt. Soon after arrival his brigade was in action against a strong position held by the Boers and he was wounded in the stomach – a wound, which in spite of initial hopes, proved to be mortal. The news came to the regiment in Dublin on 5 April, while the officers were at dinner. 'The band stopped, and a Funeral March was played.' Regimental orders stated, 'The band will not play in public or at Mess for one month, and Officers will wear mourning for a period of one month.' It is of interest that the Aldershot Military Cemetery for the year 1900 contains the graves of four KDG troopers, and between 1901 and 1903 there are ten graves of men of The Queen's Bays.[3, 18]

Another officer of the KDG, Colonel R. C. B. Lawrence, who had recently given up command of the regiment, had become chief assistant inspector in the Remount Department at the War Office. This was situated in a fourth-floor flat in Victoria Street which was so small that it was

> a constant source of trouble, because we have crowds of visitors; one officer has to see a visitor, and naturally the interview disturbs the work of

the man sitting next to him, and the visitors overflow into the passages; there is no getting rid of them; every sort and kind; stockbrokers who want to get a contract for Russian horses, Germans, Frenchmen, and any number of horse dealers, ladies who want to know when their sons are going to arrive at the Cape.[21]

On 3 April 1900 Queen Victoria paid a visit to Dublin, and she herself described the day in her journal:

The procession, consisting of four carriages, then started, mine coming last. Arthur [Duke of Connaught] rode near my carriage all the way, and I had a travelling escort of The King's Dragoon Guards. The whole route from Kingstown to Dublin was much crowded, all the people cheering loudly, and the decorations were beautiful. The drive lasted two and a half hours.[13]

In January 1901 The King's Dragoon Guards returned to Aldershot prior to embarking on the 8th of that month in the *Maplemore* for South Africa, under command of Lieutenant Colonel Mostyn Owen. But a detachment of the KDG commanded by Lieutenant Wayte Wood had gone to Australia to take part in the Centennial celebrations, as part of the Imperial Representative Corps. On 1 January 1901 they 'processed through the streets of Sydney — received great reception — unparalleled enthusiasm of people — crowds'; on the 17th they were in Brisbane, where they received a telegram from the War Office, 'Detachment K.D.G.s proceed Cape direct as soon as possible.' However, from Brisbane they went to Hobart and then Melbourne, where they boarded the *Orient* on 15 February, arriving at Cape Town on 8 March.

It was July before this detachment caught up with the main body of the regiment up country. Ten months later, in November 1901, The Queen's Bays sailed in the troopships *Orotava* and *Fortunatus* from Southampton for South Africa. They landed at Cape Town on 6 December with a strength of 24 officers and 513 men, 43 chargers and 445 troop horses. Lieutenant Colonel Dewar was in command, with Lieutenant Colonel Fanshawe, Major J. A. Walker, and Captains T. Ward and R. D. Herron as the senior officers.[10, 12, 19]

The King's Dragoon Guards on arrival, like all other reinforcements, had to move north by rail to join the army. The first sign that they were nearing the field of operations was their coming to the railhead of De Aar, described by one officer:

At its best De Aar is a miserable place. Not made — only thrown at the hillside, and allowed by negligence and indifference to slip into the nearest hollow. Too far from the truncated kopjes to reap any benefit from them.

Close enough to feel the radiation of a sledge-hammer sun from their bevelled summits – close enough in summer to be the channel of every scorching blast; in winter every icy draught. Pestilential place, goal of whirlwinds and dust-devils, ankle-deep in desert drift, as comfortless by night as day.

Here the KDG were formed into a cavalry brigade with the Prince of Wales's Light Horse, and 'G' Battery of the Royal Horse Artillery, who had come down from Pretoria; the whole under command of Colonel E. C. Bethune. The brigade, later joined by the 3rd Dragoon Guards, was to be employed in chasing the elusive Boer Commando commanded by De Wet.[12, 19, 22]

The KDG concentrated at Naauport in the Orange Free State, with two squadrons detached to reinforce the column commanded by General Plumer. De Wet had invaded Cape Province with from 2,500 to 3,000 men plus some artillery, and Plumer had located him at Philipstown, scaring him into a retreat towards De Aar and the Orange River Railway. Kitchener thought that he had managed to surround the area in which De Wet was operating, but the Boer leader, doubling back on his tracks, was too wily to be caught. Late in the afternoon of 13 February Plumer found De Wet laagered at De Put. The weather had broken and both sides halted knee deep in water. The rain continued to pour down all night. The following morning Plumer attacked the Boers, holding them to their position, with The King's Dragoon Guards and the Imperial Light Horse containing them in front while he turned the Boer line from the west. The Boers then sent off their waggons, following them round to Bas-Berg; but the pursuit was bogged down by yet another storm and progress on the veldt hampered by mud two feet deep. On 16 and 17 February the chase was continued, but De Wet had doubled back north towards the Orange Free State, where he was briefly contacted by Plumer as he rested his tired commando at Gras-Pan. The contact was made by a small party of the KDG, who attacked a Boer rearguard position unsupported and were all captured. The pressure on De Wet was such, however, that once he had disarmed the prisoners and taken their horses, he let them go.

On the 21st intelligence was received that De Wet's guns were just ahead of the column at Slyp-Steen, and that their animals were exhausted. The troop horses of the KDG were also near foundering, but the advance was pushed on, with the KDG in the lead. After a chase of three hours Colonel Mostyn Owen sighted two guns ahead on the road to Disselfontein, surrounded by a body of Boers. In spite of the fact that the pace of the pursuit had foundered many of the regiment's horses, at the sight of the column bearing down on them as they prepared to laager, the Boers panicked and fled. By nightfall, when Plumer's column had joined

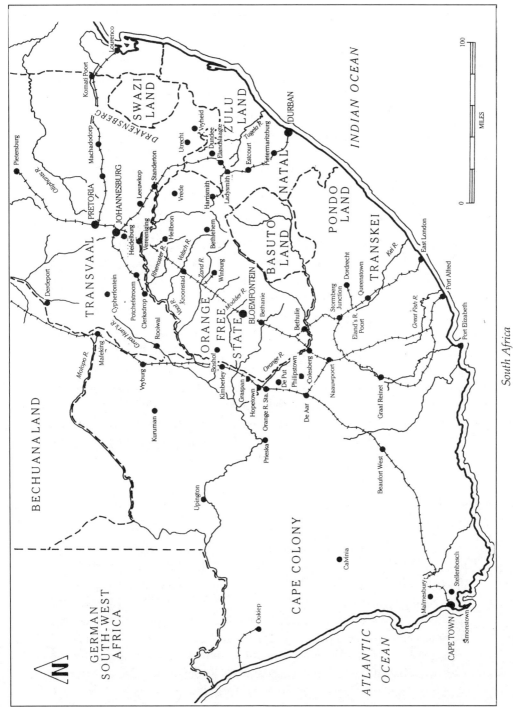

Bethune's at Disselfontein, two guns and their ammunition carts and 102 burghers had been captured.

Kitchener in his despatch wrote:

> The close pursuit of the various columns had the effect of driving De Wet north to the Orange River, west of Hopetown, where, being hotly pressed by General Plumer, his 15-pounder and a pom-pom were captured by our mounted troops under Lt-Col Owen, 1st Dragoon Guards. De Wet eventually crossed the river, but over 200 prisoners, all his guns, ammunition, and waggons fell into our hands. He undoubtedly quitted Cape Colony with great loss of prestige.[12, 19, 22]

An officer of the KDG was not quite so sanguine:

> Almost before the last of the horses had been detrained, the whole nature of, and necessity for, the movement had changed. In short, everything had turned out as the Brigadier had anticipated. Plumer, with the tenacity for which he is famous, had clung to the rearguard of De Wet's column, snatching a waggon here and a tumbril there, until he himself could move no further.
>
> De Wet had outlasted him, and had, moreover, seen that it would be useless to carry out his original programme. So he doubled and doubled again, with the result that the cleverly devised scheme of relays of driving columns was out of joint, and a dozen units were uselessly spread out over the veldt a hundred miles from the place in which the invader was catching his breath, within jeering distance of the column which had run itself stone-cold in his pursuit. So within forty eight hours of the start the whole plan had to be reconstructed.[22]

The tedium and exhaustion of the endless treks across the barren veldt were graphically described.

> The column swung out into the great dry Karoo prairie. It was a comfortless trek. Earth and sky seemed to have forgotten the rain of the preceding days; or it may have been that the storms which had distressed us had been purely local, for we have struck a great waterless plain which showed not the slightest sign of moisture. The shuffling mules and lumbering waggons churned up a pungent dust; a great spiral pillar of brown cloud mushroomed out above the column; no breath of air gave relief from the vertical rigour of the sun; the great snake-like column sweated and panted across the open, reporting its presence to every keen-sighted Dutchman

within a radius of fifteen miles. At midday the Brigade came to a halt at some mud holes, which furnished sufficient clayey water to allow the sobbing gun-teams and animals to moisten their mouths. Neither was there shade from the merciless sun. Men crawled under waggons and watercarts if they were fortunate enough to find themselves near them, or, unrolling their blankets, extended them as an awning, and burrowed underneath. The oppression of that still heat![22]

SOURCES

1 Records of The King's Dragoon Guards, Regimental Museum, 1st The Queen's Dragoon Guards, Cardiff Castle.
2 *Eastern Daily Press, Norwich Gazette,* 1893.
3 The Marter Diaries, Regimental Museum, 1st The Queen's Dragoon Guards, Cardiff Castle.
4 *Norwich Chronicle,* 1894.
5 *Regimental Journal, 1st The Queen's Dragoon Guards,* vol. 3, no. 5, 1977, p. 462.
6 *Journal of the Society for Army Historical Research,* vol. 46, 1968, p. 43.
7 *Regimental Journal, 1st The Queen's Dragoon Guards,* vol. 1, no. 8, 1966, p. 667.
8 By gracious permission of Her Majesty The Queen, The Royal Library, Windsor Castle.
9 E. Belfield, *The Queen's Dragoon Guards,* 1978.
10 F. Whyte and A. H. Atteridge, *The Queen's Bays, 1685- 1929,* 1930.
11 *Journal of the Military Historical Society,* vol. 13, 1962, p. 54.
12 W. Clowes, *The King's Dragoon Guards, 1685-1920,* 1920.
13 Michaela Reid, *Ask Sir James,* 1987.
14 *Regimental Journal, 1st The Queen's Dragoon Guards,* vol. 3, no. 6, 1978, p. 520.
15 *Journal of the Society for Army Historical Research,* vol. 42, 1964, p. 110.
16 *The K.D.G.,* vol. 4, no. 5, 1949, p. 116.
17 *Regimental Journal, 1st The Queen's Dragoon Guards,* vol. 4, no. 4, 1983, p. 1104.
18 Aldershot Military Cemetery Registers, Aldershot Military Museum Trust.
19 J. Stirling, *Our Regiments in South Africa,* 1903.
20 Diary of Lieutenant Vere Wayte Wood, KDG, Regimental Museum, 1st The Queen's Dragoon Guards, Cardiff Castle.
21 The Marquess of Anglesey, *A History of The British Cavalry,* vol. 4, 1986.
22 'The Intelligence Officer', *On the Heels of De Wet,* 1902.

32

Second Boer War
1901-1902

The King's Dragoon Guards remained with Bethune's column during the months of March, April and May 1901. The force was strengthened by the addition of six squadrons of the Imperial Yeomanry, when they moved to the north east area of the Orange River Colony to take part in a series of drives and mopping up operations under General Elliott. In June the KDG were formed into a separate column, together with the 4th Imperial Yeomanry, under command of Colonel Lowe of the 7th Dragoon Guards; another column, consisting of the 7th Dragoon Guards and the 12th Imperial Yeomanry, was commanded by Brigadier General Broadwood, who also had overall command of both columns. Lieutenant Wayte Wood rejoined the regiment from Cape Town on 16 July, and his diary gives a graphic account of each day's work, with both its boredom and occasional excitement.

> July 18th. Commenced trekking from Vredeford Road, marched about 12 miles.
> Greyhounds killed several hares en route. July 19th. Marched through very pretty defile. Destroyed farms en route. Struck the Vaal River, camped in open ground with small copje [*sic*] overlooking us. One man 7th D.G. taken prisoner, and one man 'B' Squadron. As we entered the defile saw a grey horse of the 7th D.G. which had been shot in the shoulder. Guns fired off 4 or 5 shots up the valley when we halted at midday. July 21st. 'A' and 'B' [Squadrons] with convoy. 'C' Squadron, Quicke, Grey and Langton went out at midnight, captured 5 waggons and 4 Cape carts. July 22nd. Right Flank Guard, started 6 a.m. through beastly drift. July 23rd. Arrived Clerksdorp in time to water and feed before dark. Entered Transvaal from Wolman's Drift.[3, 4, 7, 8]

At the end of July General Elliot formed The King's Dragoon Guards into a separate column commanded by Colonel Mostyn Owen, who was given two guns in support, in order to carry out a sweep with other columns west of the

Kroonstad Railway. 'On 2nd August near Gras Pan Captain Quicke, King's Dragoon Guards, of Colonel Owen's column, with two squadrons of his regiment, effected the capture of a laager of 65 waggons and 4,000 cattle.' Gras Pan, close to Wonderfontein, was to be Broadwood's bivouac for the night of 2 August and Wayte Wood's diary records the action:

> Big day, 19 hours in the saddle, no food, no drink. Captured large number waggons and few Boers. My horse fell twice heavily through holes in the ground – under fire for a short time – got waggons to camp at midnight. 'A' and 'B' Squadrons turned out to assist us, but 7th D.G. arrived first. Took us for Boers! 55 waggons and 12 Cape carts captured.

Next day 'A' and 'B' Squadrons brought in a further seven waggons and some more Cape carts. On 4 August, 'Trekked to Besters Rust Camp doing rear guard. 4 more waggons and one or two Cape carts captured by 'A' and 'B' Squadrons – 'A' Squadron took 21 waggons and 17 Cape carts.'[1, 3, 4, 9]

So the tiring and mainly tedious work of drives and trekking continued. On 19 August Wayte Wood reported,

> To Rietfontein, Sgt Baker and my Troop right flank. Crossed bad drift – ground very difficult – high stoney copjes [sic]. 'A' Squadron advance flushed Commando of 50 Boers. We got into camp at 2 p.m., having started at 5 a.m. August 20th. To Worago with main body. Boers fired on flank guard – about a dozen of them came galloping in. Pom Pom and one squadron Yeomanry went out. August 22nd. Arrived Brandwater Basin after 5 1/2 hours' march with General Broadwood. Posted 24 men on two Neks and copjes – with Pom Pom and a few 15 pounders. Charlton shot in arm, one man wounded severely – died. Halted at Nek 5.30.p.m. Looking, thought stones on hill looked like Col Beard's detachment of the Black Watch Mounted Infantry. Marched to their relief. August 23rd. 2 Officers and 65 men of the Black Watch [arrived] at evening. We arrived at scene of Black Watch position 7 a.m. Beard's detachment surrendered yesterday about 9 p.m. They had one man killed, 2 or 3 wounded. August 25th. Chased Boers, Pom Pom cleared them. Yeomanry went out in afternoon, lost two horses. Boers buried man found at farm. August 26th. Mealie Mill blown up with gun cotton – destroyed contents three farmhouses, burnt some of the outhouses – started out 10 a.m., returned 2.30 p.m.[4]

The drives and cordons went on through September, with the Boers for the most part eluding the columns, who methodically laid waste the countryside in order to destroy the commandos' means of support. Wayte Wood records:

1st September. Reveille 6 a.m., started 8 a.m. Held Nek to Moolman's Hoek – watched battle in valley – guns fired three or four times. Boers fired 75 to 100 rounds with no effect. September 3rd. Reveille 5.30 a.m., started 6.46 a.m. as Advance Guard. Climbed copje, and led horses down bad bridle path. 'A' Squadron exchanged shots with Boers on left flank, guns fired 5 or 6 rounds.

Early in September the KDG were transferred to De Lisle's column.

12th September. De Lisle's column 5.30 a.m. went over old grounds, crossed drift, and two farmhouses examined. Captured 50 waggons and 20 Cape carts. Interviewed 12 prisoners. Quicke received Brevet Major, Champagne but damned bad dinner. September 22nd. Reveille 5.30 a.m., marched 8 a.m., crossed drift into a hollow, and we were no sooner over than the Boers opened fire. Quicke and self occupied ridge on left rear. Boers made good shooting at us, bullets coming uncomfortably close! I fired about 35 rounds, guns pooped off something like 40 rounds. At this spot we have lost one man killed, one wounded and six taken prisoner. September 24th. Boers sniped until driven off by guns. We ran into Colonel Brigg's [KDG] column and heard his Pom Pom working hard. September 25th. Met another convoy on its way to Bethlehem. Guns pooped off a few rounds. 'A' Squadron captured one prisoner – soft-nosed bullets on him. September 26th. Met man who told us 17th Lancers had been mauled, 3 Officers killed.[4, 7, 8]

October 8th. Marched to meet convoy from Harrismith. Quicke, Gray and Longworth with 100 men marched at 10.30 p.m. to surprise some Boers in farmhouses, and almost succeeded, brought back with them eight beautiful horses from one farm, and 70 to 80 good trek oxen. October 14th. Yeomanry advance, Boers fired volleys at them – 60 Boers. 15 pdr and Pom Pom came into action and pooped off a few rounds at the retreating Boers, who scattered in all directions. They were followed up by the Yeomanry, but they did not wait. October 27th. At 1 a.m. two Troops under Quicke and Gray surrounded a farm, and were fired at from the farm. The inmates had had warning through the dogs barking. Quicke, whilst rounding the corner of a kraal, was shot dead through his left hand, both lungs and heart. All deeply regret his untimely end. Renton attacked an outpost at 11.30, killed two Boers, and wounded some five or six at least.

October 28th. Reveille 3 a.m., marched 4 a.m. to same place as yesterday, where we again met with some resistance. Guns fired 40 to 50 rounds. Bullets coming unpleasantly close. My Troop (13 men) fired 450 to 500 rounds. I fired a good many too. October 29th. Started out at 4 a.m. Long copje shelled for some time, Boers in strong position on a hill like the side of a house. A few men from 'A' Squadron got up, but had to retire post haste! We were with General Broadwood and the guns.[4]

Wayte Wood wrote home:

We are kept on the move the whole time, no rest at all, but it does us no harm. About every third night we are on some night enterprise rounding up farms or getting into position to try and catch the slim 'Bohee' — it is a most difficult task for many reasons. In the first place he has naturally got the best horse, since he can rest him as soon as a column goes by, and wait for the next, whereas our poor horses never have rest, and many have to be shot from sheer exhaustion. The wear and tear now is immense. From the papers you will have seen that the Regiment has been most unfortunate losing two good Officers and two wounded. We have been fighting every day. Poor Major Quicke was a very sound fellow. He was my Squadron Leader and got his brevet for capturing the 58 waggons and 12 Cape carts on 2nd Aug. last, a part in which I am pleased to say my Troop took, and was first in at the finish. It was exciting and my horse fell twice through putting his foot in a hole galloping through the long grass. I rode him to a standstill and the poor thing had to be shot a day or two later. We were 19 hours in the saddle without food or water.[10]

In a letter home in January 1902, Wayte Wood described the November activities of the KDG:

We were in the Squadron, Captain Quicke, Captain Williams, Gore, Langton, Harris and myself. In the three weeks poor Quicke and Williams were shot dead, and Harris wounded in the foot — it is all so sad — the latter it is feared injured for life. On the 28th November we came in touch with De Wet about 4 o'c in the afternoon, having galloped 5 or 6 miles — we had been trekking all the morning. The Boers waited for us to come into range, and then let us have it! They were in a very strong position with a deep wide donga between us and them. De Lisle counted the shells that they fired at us — 40 x 15 pdrs. They also fired between 300 and 400 rounds of pom-pom, and made excellent shooting, the shells coming right in amongst us. One man had the butt of his rifle blown away, another shell dropped right under the horses — one heard in the distance a faint pom-pom and wondered where the shells would drop. Eight rounds in line ripped the ground up behind me and the next lot I heard whistle over my head – the shots drop a little way in front of me. This happened whilst we were retiring for cover behind a ridge . . .

One day we saw quite 1500 Boers. We disturbed them in holding a conference, in which it was said at the time that they voted 5 to 2 to continue the war for two years, and if they met with no success in that time, they would surrender. Unfortunately they have met with some success. The capture of Col ---'s [sic] column has given them new life, but we hope it won't last long. As I write, a report has come that 900 Boers are just outside Kroonstadt. The 4.7 gun has fired half a dozen rounds at them. So far none of the Boers have been seen from the blockhouses, and I suppose they want to break through the line.[10]

Wayte Wood's diary of 9 November described Captain Williams's death.

'C' Squadron advance, my Troop right flank, Gray left, Longworth and Harris with Hulton centre. Day ended badly. Boers in strong position, and guns nowhere. Williams shot through stomach, also Hughes of my Troop; former succumbed after about an hour. Harris shot through the ankle. My Troop had a narrow squeak of being badly mauled, had we not taken cover. Got right onto top of copje, seven horses shot. Dill narrowly escaped, his horse shot through the saddle amd back, but still alive. Boers stripped Williams of almost everything. November 10th. Buried poor Williams. Hughes died. More scrapping today, guns used. 'C' Squadron down to 50 men. November 11th. Advanced on nasty looking ridge, killed a Boer and wounded another mortally. Guns killed two or three in rear. November 13th. 'B' Squadron sent on, lost 4 or 5 horses shot. Sgt Allen shot in foot. 'B' Squadron in nasty corner. November 27th. Fighting on left flank, Eastwood wounded in shoulder and arm, and three others in 'B' Squadron.[4]

A court of inquiry described how Hughes, KDG, met his death. Corporal Burbidge, KDG, stated:

Mr Wood sent me out in charge of a patrol of three men in advance of the Troop. When I was approaching the kopje, I saw one or two Boers on it, and opened fire on them. Seeing them retire, I advanced onto the kopje, sending Private Gunthorpe to the right, myself and Private Brittain working up the centre, whilst the late Private Hughes went to the left. We were all well extended. As we ascended Hughes disappeared round the left corner, and reached the summit first. He came on some kraals, which were occupied by the Boers, who opened fire on him, and on me as soon as I showed myself. There were also some Boers to my right, high up, whom we could not outflank. The fire being severe and having no cover, we retired thinking that Hughes would do the same. Mr Wood covered my retirement, and when I was sent forward again, I found Hughes, lying wounded by a kraal. He told me that 7 or 8 Boers sprang up, and said 'Hands Up', an order which he could not help complying with. After they had taken his arms and ammunition, and rifled his pockets, a young Boer, about fifteen years of age, shot him through the stomach.[11]

The Queen's Bays arrived in South Africa in December 1901. Shortly afterwards Lieutenant Colonel Dewar was invalided home, to be succeeded by Lieutenant Colonel Fanshawe. Based in Colesberg, the men were armed with rifles and bayonets, with 150 rounds of ammunition each. The regiment moved in January 1902 to Winburg in the Orange River Colony, where Captain Smyth, VC, rejoined from Egypt. The Queen's Bays were at once engaged in chasing De Wet as a part of a column, which included the 7th Hussars, and which was commanded by Colonel Lawley. After several skirmishes, and having suffered a number of casualties, the column captured several hundred prisoners, some thousands of head of cattle and large quantities of stores.[2]

In an incident on 28 February 1902 Lieutenant Ing was wounded and thrown from his horse. Trooper Roberts, Bays, seeing his plight, galloped up, dismounted and placed the wounded officer on his horse, remounted and brought Ing back safely in spite of being pursued by a party of Boers.[2]

In March the regiment was moved up to Springs in the Transvaal, where Lawley's column was to operate against Piet Viljoen's Commando, which had possibly been joined by the Heidelburg Commando under Alberts. On 31 March the column was at Boschmanskop, when information was received that two small groups of about 200 Boers each were encamped near Enkeldebosch and Steenkoolspruit farms, some twelve miles away. At 1 a.m. on 1 April The Queen's Bays were to make a night march to surprise the Boers at Enkeldebosch, while next morning the 7th Hussars were to attack Steenkoolspruit, where the Bays would join them. As there was the possibility of a third Boer laager further on, the Scouts would reconnoitre, and, if neccessary, the column would deal with that later. The Bays marched at 1.30 a.m. with a strength of 284 NCOs and men. At 3.15 a.m. a farm was reached, which was surrounded and searched, but found to be empty; the Scouts in advance then crossed a stream and found some horses hobbled near a laager. On receiving this information, Colonel Fanshawe decided to attack. 'B' Squadron was sent to surround the laager, while the other two squadrons would go straight for it. As the Bays attacked the Boers were completely surprised: many of them fled and a lot of their horses were stampeded, though some Boers fought for a short while, firing from their waggon line. One of the prisoners said that there was another laager higher up the spruit, and Colonel Fanshawe, thinking this to be the second laager, decided to attack it. Major Vaughan, 7th Hussars, the column intelligence officer, had been with the flanking squadron and had captured Commandant Pretorius as he was trying to escape in a Cape cart; but later that night Pretorius managed to slip away.[2]

As The Queen's Bays moved to attack what they thought to be the second small laager, they came under fire from the front and flank. The flashes from the Boers' Mausers soon showed that they were faced by a very much larger force than their intelligence had indicated, and the order was given to retire. The squadrons went files about and galloped back some 300 to 400 yards up a slope, where they dismounted and opened fire on the Boers, who were advancing from their laager. The position was unsatisfactory, as it was overlooked and had an open flank, so Colonel Fanshawe determined to make a stand around a rocky kopje about a mile away, where the regiment could hold out until reinforcements could reach them from Boschmanskop. 'C' Squadron was ordered to seize the kopje and duly occupied the south and north-east crests, and then the other two squadrons retired and took up positions prolonging the line to the north west, with the horses being left to the west. It soon became apparent that the new position was not as good as had been hoped, for the line was too extended and the ground allowed the enemy to come too close. Added to this, in the darkness and the rain it was not easy to organise the defence.[2]

The first attack came in against 'C' Squadron on the right, who were already under heavy fire from the direction of the Boer laager to the north east. A mounted Boer Commando then charged from the south east, firing from the saddle and shouting as they came on. The Bays at the point where the charge was made were very thin on the ground, but they kept up a steady fire and managed to beat off more than one attack. However, the Boers renewed their pressure, and as more and more men were hit, the enemy managed to gain a foothold at the south-east corner of the kopje. By this time the adjutant, Captain Mullens and Major Vaughan of the 7th Hussars had all been wounded, but they continued in action.

As dawn broke it became clear that the Boers were in great force, and that they were working around both flanks of the Bays' position. Colonel Fanshawe decided to retire to a dominant crag some three miles back, called the Leeuwkop, or Lion's Head, which was close to the Pretoria-Standerton road. 'B' Squadron fell back about a mile to a ridge from where it could cover the withdrawal, followed by 'A' Squadron, leaving 'C' Squadron as rearguard. 'C' Squadron's retreat was conducted by Lieutenant Allfrey, as the squadron leader, Captain Ward, and Lieutenant Hill had both been wounded, and twenty-three killed and wounded Bays were left on the ground.

A small party of eight NCOs and men under Captain Smyth, VC, had been isolated on the kopje, their horses having been shot. The Boers repeatedly called on them to surrender, but they refused and held out for another twenty minutes, until all but Smyth were shot. He then managed to crawl away through the long grass, catch a riderless horse, and gallop back to the regiment.

By the time that the three squadrons had reached the intermediate ridge, the Scouts brought the news that the Boers had already occupied Leeuwkop. Fanshawe then ordered the regiment to retire to some small kopjes near Boshof's Farm, and on to some rising ground to the west and to the north west of Leeuwkop. As they retired, the Bays drove off some Boers on their left flank, reached the rising ground north of the Pretoria road, and there made a stand. A further withdrawal was then made to a height to the west of their position and some six and a half miles from Boschmanskop, where a long front was adopted to prevent the Boers from out-flanking the regiment, who by this time were outnumbered by four to one. 'B' Squadron, under Major Walker and Captain Mullens, was in a strong position on the right, near the Leeuwkop-Boschmanskop road. 'C' Squadron was about a mile to their left, commanded by Lieutenant Allfrey, and reinforced by Major Vaughan and some of the Scouts; while Colonel Fanshawe, with 'A' Squadron, was still further to the left. The Boers soon outflanked 'A' Squadron, forcing a further retirement, when Major Walker was killed.

As the squadrons retired, halting in turn to check the pursuit, they came at last in sight of Boschmanskop, but at this point Captain Herron, commanding 'A' Squadron, was killed. When they were two miles from Boschmanskop, they saw the 7th Hussars riding towards them. The Bays formed up once more and stopped a final charge of the enemy, and then two British guns opened up from Boschmanskop. A 7 a.m. the Boers started to retire towards Leeuwkop. A squadron of the 7th Hussars pursued the retreating Boers, and a party of The Queen's Bays followed to assist the wounded.

During the retreat some thirty Bays had been taken prisoner when their horses were shot under them, and the Boers returned these men the following day, having stripped them of their arms and equipment and some of their clothing. It then transpired that the Boer force was a concentration of ten Commandos under General Albrechts, numbering between 1,000 and 1,200 Boers. The Boer losses were variously estimated at between 30 and 75 killed and 40 wounded; among the killed were Commandant Prinsloo, Barend Prinsloo, Field Cornet Niekirk and two other field cornets; and Commandants Scheepers, Farel and Viljoen, with Hans

Botha's son, wounded. The Queen's Bays lost 2 officers and 13 men killed, and 3 officers and 59 men wounded (8 of whom subsequently died); 120 of the Bays horses were killed or badly wounded.[2]

December 1901 saw The King's Dragoon Guards in De Lisle's column together with the 3rd and 7th Dragoon Guards and some mounted infantry. They were given the task of patrolling the blockhouse line between Kroonstad and Harrismith, a distance of 130 miles. The objective was still to catch De Wet, but he proved too wily and elusive. On 8 December Wayte Wood wrote:

> Saw more Boers than I ever have, 600 to 700 travelling to our left, and 300 to our front towards the hills, besides numerous small parties – about 1500 altogether. Colonel Bing, Broadwood and De Lisle all had a smack at them. December 26th. Sniped at, a few shots as soon as we appeared on the skyline. Marched all night, arrived at farm and captured 7 Boers, two more caught on flank by 7th D.G. December 28th. De Wet and his crowd located – came under shellfire and pom-pom from him, strange sensation![2, 7]

On 16 February 1902 a party of 500 KDG, 7th D.G. and 6th Mounted Infantry were sent out on a forced march to capture De Wet, who was supposed to be at Elandskop. The column marched forty-two miles in less than six hours, to capture six elderly Boers and one helio, and to find that De Wet had been nowhere near. Peace negotiations were now in process, but the KDG were employed in a number of drives right up to the end of hostilities.[1, 3, 7, 8]

Two officers of The King's Dragoon Guards achieved distinction during the war, serving away from the regiment and in very different fields. Lieutenant Colonel W. H. Birkbeck became the Assistant Inspector of Remounts in South Africa. The going had been so hard, and the wear and tear on horseflesh so tremendous, that the supply of remounts became crucial to the success of the campaign. Haig commented, 'No one could have done this remount work as well as Birkbeck has.' Major C. J. Briggs commanded the 1st Regiment of the Imperial Light Horse from early in 1901 until the end of the war, gaining great distinction. On 22 March 1901, at Geduld Farm, near Hartebeestfontein, Briggs with 175 men was suddenly charged by 400 of De La Rey's Commando. The ILH were dismounted, but quickly remounted and then slowly, squadron by squadron, retired over a distance of four miles, holding off the enemy and fighting their way back to their mountain camp. Briggs, whose horse had been shot under him, conducted the whole operation 'armed with a cigarette and a knobkerrie'. For the loss of two officers and five men killed, and some sixteen wounded, Briggs inflicted about two dozen casualties on the Boers and earned from General Jan Smuts, who was with the Boers, the verdict:

'The rearguard action fought by the I. L. H. was the most brilliant one I have seen fought by either side during the whole campaign.' Briggs's leadership at Cyferfontein in January 1901 and at Rooiwal on 11 April 1902 increased his reputation.[5]

Between 18 and 20 April The Queen's Bays took part in their last operation of the war, a drive across the veldt, which proved to be abortive except for some Boers captured at Palmiefontein on 6 May. By now peace negotiations were in hand. Between 8 April and 10 May the regiment had marched 900 miles, arriving on 20 May at Heidelburg. Peace was signed on 31 May 1902.[2]

SOURCES

1 W. Clowes, *The King's Dragoon Guards, 1685-1920*, 1920.

2 F. Whyte and A. H. Atteridge, *The Queen's Bays, 1685- 1929*, 1930.

3 John Stirling, *Our Regiments in South Africa*, 1903.

4 Diary of Lieutenant Vere Wayte Wood, KDG, Regimental Museum, 1st The Queen's Dragoon Guards, Cardiff Castle.

5 The Marquess of Anglesey, *A History of the British Cavalry*, vol. 4, 1986.

6 'The Intelligence Officer', *On the Heels of De Wet*, 1902.

7 J. M. Brereton, *A History of the 4/7th Royal Dragoon Guards, 1685-1980*, published by the regiment, 1982.

8 E. Belfield, *The Queen's Dragoon Guards*, 1978.

9 Lord Kitchener's Despatches, 8 August 1901.

10 *The Regimental Journal of 1st The Queen's Dragoon Guards*, vol. 4, no. 4, 1983, p. 1104.

11 'Evidence before a Court of Enquiry,' South Africa, Regimental Museum, 1st The Queen's Dragoon Guards, Cardiff Castle.

33

South Africa, England, India, France
1902-1914

With the peace signed, The King's Dragoon Guards formed a part of the Army of Occupation and was stationed at Potchefstroom in the Transvaal. The Queen's Bays were ordered to Middleburg in June 1902, being present at the surrender of Louis Botha's Commando at Kraal Station on 5 June, and were then posted to Pretoria in August. The King's Dragoon Guards had eight other ranks killed, whilst twenty-seven other ranks died of disease; twenty-nine other ranks had been wounded; two officers were captured and five other ranks were posted as missing. The officers killed were Major F. C. Quicke, and Captains W. M. Marter, P. R. Denny and E. A. Williams, and those wounded were Colonels A. H. M. Edwards and J. C. Briggs, Major H. de C. Eastwood, who was wounded twice, Captain Rasbotham and Lieutenants L. S. Denny, Harris and F. H. Charlton. The Queen's Bays lost seventy-eight other ranks killed, fourteen other ranks died of disease; four officers and fifty-one other ranks were wounded. Two officers were killed, Major Walker and Captain Herron. The wear on the horses was terrible: out of 775 with which the Bays landed, 748 had been lost by the end of the campaign.[1, 2]

Five officers of the KDG were awarded the CB, Colonels A. H. M. Edwards, and S. Bogle Smith, Lieutenant Colonel H. Mostyn Owen, Captain C. L. Bates, and Major and Brevet Lieutenant Colonel W. H. Birkbeck. There were four DSOs, awarded to Major H. de C. Eastwood, Captain C. L. Bates, Lieutenant J. J. Brockbank, and Lieutenant Colonel W. J. Lockett who was attached to the 14th Hussars. Sergeant T.Gilton was awarded the Distinguished Conduct Medal. Twelve officers and four non-commissioned officers and one trooper of the KDG were mentioned in despatches, some of whom were serving attached to other units. Sergeant F. Webb of The Queen's Bays was awarded the Distinguished Conduct Medal:

> In the action at Leeuwkop, on the first kopje occupied by the Regiment, he continued firing from the eastern edge of the position, with a party of about four men, in the face of a Commando of Boers who were outflanking his party at about 50 yards, and who were repeatedly calling on our men

to surrender. Corporal Webb, when his Squadron was withdrawn, covered its retirement by continued firing, after the men round him were all hit. After keeping the Boers in check for several minutes, he was dangerously wounded, and unable to move. Shortly after this, the Boers, finding the close range fire had ceased, streamed over the position.

Colonel Fanshawe gained a fourth mention in despatches, while Captains H. P. Sykes, P. M. Sykes, J. Ward and Quartermaster Hopkins all received mentions, as did Sergeant Pope, Farrier Quartermaster Sergeant O. Preston and Trooper J. Roberts, the former for having commanded a troop in action following the wounding of his troop officer, and the two latter for rescuing Lieutenant Hill under a heavy fire. Captains Smith and Mullins were promoted to brevet-majorities, and Corporals C. Ginn and G. Buff were promoted sergeants for gallantry.[1, 2, 3]

The King's Dragoon Guards returned to England during 1903 and were quartered at Hounslow. Lieutenant Colonel S. Bogle Smith took over command from Lieutenant Colonel Mostyn Owen on 14 June 1902 and in turn handed over command to Lieutenant Colonel W. J. S. Fergusson on 14 June 1906. The Queen's Bays remained in South Africa, where at the end of 1902 Lieutenant Colonel Dewar was forced to give up command because of ill-health and was succeeded by Lieutenant Colonel Fanshawe. However, Fanshawe was promoted brigadier general in command of the 2nd Cavalry Brigade at Aldershot on 1 April 1907, the anniversary of Leeuwkop, and was succeeded by Major William Kirk. The Queen's Bays returned to England in January 1908, relieving The King's Dragoon Guards at Hounslow. On leaving South Africa the commander in chief wrote, 'The conduct of the 2nd Dragoon Guards (Queen's Bays) has been irreproachable in action, on trek and in camp.'[1, 2]

The King's Dragoon Guards, on arrival in England, were issued with the Broderick cap, and the much-prized Victorian pill-box forage cap with its gold lace rank distinctions for warrant officers was withdrawn. The Broderick cap was of blue cloth, round in shape with a projecting rim, but with no peak; it was worn with a chinstrap and the regimental cap badge, the Austrian double-headed eagle, in metal in the centre at the front. The Queen's Bays, on their return to England in 1908, were issued with a blue Broderick cap, which had a white patch carrying the regimental badge at the front, the patch being later changed to light buff.[4]

On 10 June 1904 the links between The King's Dragoon Guards and the Habsburgs were further strengthened, when the regiment paraded at Aldershot for inspection by the Archduke Frederick of Austria. The Emperor Franz Josef, as colonel, always took a close interest in his regiment, and there is an unconfirmed story that he promised that if ever, sadly, Britain and Austria were to find themselves on opposite sides during hostilities, any KDGs captured by the Austrians would be treated, not as prisoners of war, but as his personal guests.[5, 6]

In 1908 the Emperor Franz Josef celebrated his Diamond Jubilee and ordered that a number of silver and bronze medals be awarded to the officers and men of his British regiment. Lieutenant Colonel Fergusson wrote on 20 May 1908 giving the names of 23 officers and 93 men, all of whom would be serving with the regiment on 2 December 1908. In fact 36 silver and 100 bronze medals were despatched from Vienna to the regiment in Ambala, with a note to say that additional officers or NCOs might have arrived in the meantime. On 18 January 1909 Colonel Fergusson returned 13 silver and 5 bronze medals as surplus to requirements, which meant that one of the 23 officers recommended originally did not receive a medal, and two additional NCOs were awarded a bronze medal. The medals were presented at a special parade held at Ambala on 2 December 1908. Examples are to be seen in the Regimental Museum.[8, 9]

In 1907 the KDG moved from Aldershot to Hounslow. While at Hounslow the regiment was involved in a typical garrison duty, providing for a military funeral. On 27 September 1907 Colonel Donne, in charge of the Home Counties Regimental District Record Office, was buried with military honours. The King's Dragoon Guards was ordered to provide the gun carriage and Union Jack, and the band was to head the funeral procession; the funeral party was to consist of all available officers, NCOs and men, to be formed up in two ranks facing inward in front of the gun carriage. Dress was to be review order with officers wearing mourning bands on the left arm. In November 1907, having been relieved by The Queen's Bays on their return from South Africa, the KDG left for India, and on arrival were quartered at Ambala. [1, 2, 7]

On 14 June 1910 Lieutenant Colonel F. C. L. Hulton took over command of the KDG from Lieutenant Colonel Fergusson. During 1911 the King Emperor George V visited India, and a great Coronation Durbar was held at Delhi, in which The King's Dragoon Guards participated fully, providing mounted escorts for the Viceroy and a trumpeter dressed in state livery and black jockey cap for the King Emperor. At a Grand Review the regiment marched past the King at the walk, trot and gallop, and received much praise for its dressing and smartness. In 1914 the regiment moved to Lucknow, and on 14 June Lieutenant Colonel J. A. Bell-Smyth assumed command, taking over from Lieutenant Colonel Hulton.[1]

The *Cavalry Journal* of 1911 records an old custom of the officers of The King's Dragoon Guards, who were required to wear a regimental ring consisting of a plain hoop of gold set with an oblong signet of bloodstone, affixed to a swivel. On one side of the bloodstone was engraved the crest of the regiment, and on the other the wearer's personal coat of arms or initials. [10]

The Queen's Bays, on taking over the Hounslow Cavalry Barracks from The King's Dragoon Guards in 1907, were placed on the Home Establishment, comprising three sabre squadrons and a reserve squadron, with a strength of 26 officers, 2 warrant officers, 52 sergeants, 8 trumpeters, and 672 rank and file, with 50 chargers, 459 troop horses, and 6 draught horses. The Bays took over 150 horses from the KDG, received 150 remounts, and 10 trained horses from every cavalry regiment stationed in England and Ireland. By 1 April 1909 the regiment was up to strength, and all the horses were bays.[2]

On 27 May 1908 The Queen's Bays were on street duty in London for the visit of M. Fallières, the French President, and again on 19 November for the visit of the King and Queen of Sweden. On 16 February 1909 the regiment went to London for the State Opening of Parliament by the King. During 1909 the King approved the regimental motto of 'Pro Rege et Patria', and authorised the battle honours of Warburg and Willems, fought on 31 July 1760 and 24 June 1762, to be borne on the standard together with Queen Caroline's cypher within the Garter, in place of the Royal Cypher. The regiment was called upon to send drafts to India during the course of 1909. On 20 May 1910 The Queen's Bays took part in the funeral procession of King Edward VII from Westminster Hall to Paddington Station. During August and September the regiment participated in training and in the army manoeuvres on Salisbury Plain, and at their conclusion they marched into quarters at Aldershot. A War Office letter of the 26 May 1910 gave approval for non-commissioned officers to wear an arm badge on the tunic above the chevrons; the badge was to consist of the word 'Bays' in old English capitals within a laurel leaf surmounted by a crown, all in white metal. On 21 June a similar ornament was approved for the officers to wear on their pouch.[2, 11, 12]

Lieutenant Colonel W. Kirk completed his tenure of command and went on half-pay, handing over to Lieutenant Colonel H. W. Wilberforce. One of the first duties of the new colonel was to command a detachment of The Queen's Bays when King George V and Queen Mary paid a semi-official visit to Aldershot; the regimental band played on the 5th for the King's dinner in the Royal Pavilion, and His Majesty complimented Mr J. W. Faulkner, the bandmaster. The Bays also provided Sergeant J. Marks to carry the Royal Standard for the visit, together with a sergeant and two troopers as escort; for this duty Sergeant Marks was awarded the silver medal of the Royal Victorian Order. On 23 June the regiment lined the streets of London for the Coronation, and again on 23 and 29 June they provided street liners for the royal processions to the south of the Thames and to the Guildhall. The following month The Bays were sent again to London, this time in aid of the civil power, to help cope with the railway strike, but it proved unneccessary to call on their services.[2]

During 1911 the West Cavalry Depot at Seaforth was formed, and The Queen's Bays provided details for the staffing of the depot during October and November. As a result the regimental establishment was again changed, with the reserve squadron being absorbed into the three sabre squadrons, and the regimental strength altered to 23 officers, 2 warrant officers, 44 sergeants, 6 trumpeters, 621 rank and file, with 45 chargers, 542 troop horses and 14 draught horses. Changes were also made to the officers' mess dress.[2, 11]

On 30 September 1911, on the Ash Ranges at Aldershot, 'B' Squadron of The Queen's Bays won the Royal Cambridge Challenge Shield, with 'A' Squadron taking second place. The regiment was maintaining a tradition, for it had won this trophy in 1884, 1899, 1900 and 1901. In August 1912 Lieutenant Colonel R. L. Mullens was appointed to command the 4th Dragoon Guards. On leaving the regiment, he presented a cup to The Queen's Bays, to be known as the Inter-Troop Challenge Cup, for the most efficient troop in the regiment. The subjects to be examined were general efficiency in individual and troop training, musketry, skill at arms, and horsemanship. The Mullens Cup is still competed for annually and, to this day, is looked upon within 1st The Queen's Dragoon Guards as one of the highest awards that can be won.[2]

On 3 August 1914 Aldershot Command ordered mobilisation, two days ahead of the rest of the country. It was not unexpected; all leave had been stopped a week earlier and preliminary preparations had been put in hand. The Queen's Bays, along with the 5th Dragoon Guards and the 11th Hussars, formed part of the 1st Cavalry Brigade, commanded by Brigadier General C. J. Briggs, himself a KDG.[2]

The regiment, commanded by Lieutenant Colonel H. W. Wilberforce, was stationed at Willems Barracks in Aldershot. The reservists came in quickly, mostly within the first twenty- four hours, making up 36 per cent of the strength of the regiment. Many reservists were old NCOs or men who had fought in South Africa, providing a welcome fund of experience and steadiness for the young troopers who composed the bulk of the rank and file. Mobilisation was completed by 10 August, and the following day King George V and Queen Mary inspected the regiment, wishing all ranks the best of good fortune. The same day Captain H. W. Hall went ahead to France to arrange billeting in the forward concentration area. On the evening of the 14th The Queen's Bays marched to Farnborough Station and entrained for Southampton, embarking in the 13,000-ton *Minneapolis*. The transport sailed on the 16th, reaching Le Havre that evening and disembarking the troops the following morning in a deluge of rain. Fifteen French reservists and territorials were posted to the regiment to act as interpreters. On the evening of Monday, 17 August and during the 18th, the regiment entrained in the famous French box waggons, marked 'hommes 40, chevaux 8', for Maubeuge, ninety miles away. The journey was something of a triumphal procession, with much flag waving and cheering along the route, and gifts of flowers, tobacco and chocolate. The horses were fed and watered at Rouen, where the men were supplied with hot coffee. On arrival at Maubeuge, the Bays went into billets at Jeumont, Marpent and Solre-sur-Sambre, where they became the most advanced of all the British troops.[2]

The cavalry was ordered to move forward on the morning of 21 August, the 1st Brigade holding the crossings over the Aisne and Mons-Charleroi canal, with the Bays acting as advance guard. On the 22nd Lieutenant Kingstone recorded that 'Sergeant Tucker and I went out half a mile in front of the canal. We got off our horses and walked to the top of a little hill, and sat down with a map and field glasses. We had not been there more than ten minutes before we saw a German squadron move slowly across our front, not more than 400 or 500 yards off.' That evening the infantry took over the outpost line, and the Bays moved twenty miles to the left rear to the village of Audregnies, in order to provide flank cover. The 23 August was spent at Audregnies, resting and cleaning, although the sound of heavy gunfire reverberated throughout the day.[2] On the evening of the 23rd Sir John French, commanding the British Expeditionary Force, learnt that the French on his right were retreating, that the Germans were in much greater strength than previously reported, and were working round his left flank. The British had no option but to conform, but it took time to reverse all the supply and ammunition waggons which had closed up to the forward line. Orders reached the Bays at 10 p.m. on the 23rd, and they were saddled up and on the move for Quievrain, just south of Conde, by 11 p.m., where they were given the task of holding a stretch of the railway line from Mons to Valenciennes. Arriving at midnight the regiment found hundreds of miners and peasants assembled to help them dig a trench line, which was ready by dawn.

Daylight showed a dense mass of Germans, with bands playing and staff riding ahead, advancing about a mile to the right of the regiment's position. As the pressure increased the Bays were ordered to retire, conforming with the general movement but, so far, with no direct contact with the enemy. All that day the Bays moved hither and thither, coming under some shrapnel fire, but with no casualties, until after a confused and tiring day of marching and countermarching, they settled for the night in a field near St Waast.[2]

Throughout the 25th the Bays acted as rearguard, briefly contacting a patrol of Uhlans, one of whom was wounded and captured. The men eventually bivouacked in a field south of Le Cateau at 1.30 a.m., and were on the move again by 4 a.m. on the 26th, retreating to Reumont, and taking up a position around the village of Escaufort, where they were spectators of the Battle of Le Cateau to their right. As the day wore on the men came under shell fire, but suffered few casualties in their shallow trenches. By 3 p.m. the pressure was such that a further retirement was ordered, which continued throughout a wet and dark night, causing the squadrons to become separated. By the morning of the 27th the greater part of the regiment had come together again, and continued the retreat through St Quentin, Serancourt and Berlancourt, while a separated detachment under Major Ing fell back through Ham, collecting stragglers from other cavalry regiments as it went.

The Bays, marching early on the 28th, were in touch with patrols from a German cavalry brigade, when near Guiscard the enemy increased the pressure, bringing a battery into action and wounding three troopers and six horses. The Bays then broke contact, passing through the Allied outpost line covering Noyon, crossing the Oise and finally halting at Bailly, where they were at last rejoined by Major Ing with his detachment. On the 29th the whole regiment made a short march to the Aisne, and went into billets at Choisy-au-Bac north east of Compiègne. The retreat continued on 31 August, the cavalry covering the flank; the Aisne was crossed, Compiègne passed, and the Oise, and moving onto high ground the Bays halted near the village of Venitte. There the 1st Cavalry Brigade, with the 2nd on its left, extended the British line whilst maintaining contact with the French. That evening Brigadier Briggs decided to concentrate the brigade in the village of Néry, on the left bank of the Oise, where it would be in reserve behind the infantry of 3rd Corps.[2]

The Queen's Bays and 'L' Battery, Royal Horse Artillery, were the last to reach Néry. The 5th Dragoon Guards were billeted in the north end of the village, with their horses in the open; the 11th Hussars were in the centre with all men and horses under cover; and so the Bays of 'A' and 'B' Squadrons were billeted in some houses along the village street with their horses in the open, while 'L' Battery and 'C' Squadron were in a field to the south of the village. The brigade spent a quiet night, but awoke at 4.30 a.m. to a dense mist, which delayed any start, so that many of the men stood down and started to cook breakfast. At 5 a.m a patrol of the 11th Hussars came galloping in to report that they had been chased into Néry by German cavalry. Almost at once shells began to fall in the village, with the German gunners concentrating their fire on the horse lines of the Bays and on 'L' Battery, with such deadly effect that the survivors stampeded in terror. Brigadier Briggs at once ordered the 11th Hussars to take over the defence of the north end of the village, and gave Colonel Ansell of the 5th Dragoon Guards freedom to act against the right flank of the advancing enemy. [2]

In the meantime the officers and men of the Bays were improvising a firing line, while the Horse Gunners under Captain Bradbury, their second in command, struggled to manhandle three of their guns into action. Major Sclater-Booth, the battery commander, had been wounded and blinded by a shell burst as he ran to

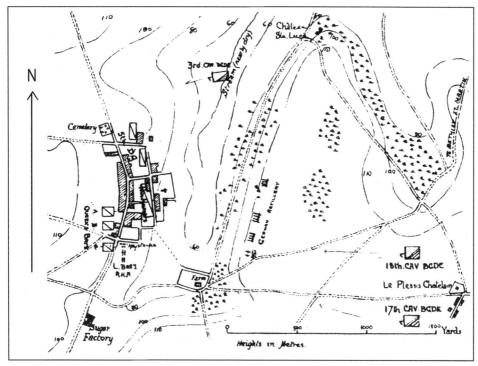

Néry

his guns. Lieutenant Lamb, the machine gun officer of the Bays, was able to gather together some of his men and, covered by a low brick wall, bring two machine guns into action, aiming at the flashes of the German guns, which could by now be seen only 800 yards away. Under cover of this fire 'L' Battery got its three guns into action against what proved to be eight guns of two German batteries, with a third battery shelling the village. [2]

Colonel Ansell with two squadronas of the 5th Dragoon Guards moved out of the north end of Néry in the mist to attack the German flank, only to run straight into the 9th Uhlans and 3rd Cuirassiers. In the following exchange of fire Ansell was mortally wounded, and the 5th Dragoon Guards fell back before this over-whelming force, fighting all the way. But the mist also covered the British weakness, and the German cavalry halted their advance.

The 1st Cavalry Brigade had been caught by surprise, at close quarters, by the whole of the 4th German Cavalry Division who were in greatly superior numbers; but the Germans were not sure what confronted them. This enabled individual officers and men to improvise a firing line, while the gunners served their 13-pounders in the open field, bringing up the ammunition from the limbers across twenty yards of fire-swept ground. Lamb, smoking a pipe, kept his two machine guns in action, concentrating on the German guns, and filling the belts by hand to keep the guns supplied while the water in the jackets boiled. The Germans now switched the fire of their third battery against 'L' Battery and the Bays. Two of the three guns were quickly knocked out, the remaining gun being served by Captain Bradbury, Battery Sergeant Major Dorrell and Sergeant Nelson, who had already been wounded. Bradbury had both his legs blown off, and as he was carried back, dying, he said to Colonel Wilberforce, 'Hallo, Colonel, they have been giving us a warm time, haven't they?' The lone remaining gun of 'L' Battery continued to fire until the last round had been expended.[2, 13]

As the mist cleared Wilberforce noticed that 'by degrees these eight [German] guns ceased firing, and we could see the guns distinctly every time anyone came near them; Vickers and all the rifles we had slated them with fire until the Germans gave up all attempt either to serve them or get them away.' Earlier in the day a small party of the Bays had managed to work forward to the right and occupy the sugar factory, from where they were able to check by rifle fire any German attempt to get around from that side. Eventually the Germans occupied some buildings to the east of the sugar factory, and Lieutenants De Crespigny and Misa launched a counter-attack with some fifteen men. They held the objective until De Crespigny was killed, and only three were left unwounded, Lieutenant Misa and two men.

Soon after 8 a.m. help arrived in the form of the 4th Cavalry Brigade, who coming upon the stampeded horses of the Bays feared the worst. As 'I' Battery's guns came into action, however, Lamb's machine guns prevented the Germans from bringing up their teams to withdraw their guns, and Major Ing took forward some of the Bays to capture the eight abandoned guns, the first to be taken by the BEF. By 9.45 a.m. the Germans had withdrawn.

The Bays lost Lieutenant Champion De Crespigny and four men killed, Majors G. H. A. Ing and A. E. W. Harman, Captains E. S. Chance and W. F. G. Renton, Lieutenants E. Walker and F. D. R. Milne, Second Lieutenants L. W. White and H. D. St G. Cardew, and thirty-five men wounded. About eighty horses were killed, and many of those stampeded were temporarily or permanently lost. Seventy-eight Germans were taken prisoner and their dead lay mainly around the captured guns – all cavalry or artillerymen.[2]

Captain Bradbury, Battery Sergeant Major Dorrell and Sergeant Nelson of 'L' Battery were all awarded the Victoria Cross. Lieutenant Lamb of The Queen's Bays was awarded the DSO and was mentioned in despatches, as were the seven Bays who served the machine guns, Lance Corporal Webb and Troopers Goodchild, Phillips, Fogg, Emmet, Ellicock and Horne. In addition Ellicock was given the DCM. [2, 13]

SOURCES

1   W. Clowes, *The King's Dragoon Guards, 1685-1920*, 1920.

2   F. Whyte and A. H. Atteridge, *The Queen's Bays, 1685- 1929*, 1930.

3   John Stirling, *Our Regiments in South Africa*, 1903.

4   *Journal of the Society for Army Historical Research*, vol. 60, 1982, p. 213.

5   *Regimental Journal of the 1st The Queen's Dragoon Guards*, vol. 1, no. 8, 1966, p. 667.

6   E. Belfield, *The Queen's Dragoon Guards*, 1978.

7   *Journal of the Society for Army Historical Research*, vol. 63, 1985, p. 123.

8   *The K. D. G.*, vol. 1, no. 2, 1930.

9   *Regimental Journal of 1st The Queen's Dragoon Guards*, vol. 4, no. 5, 1984, p. 1230.

10 *The Cavalry Journal*, vol. 6, 1911, p. 509.

11 Historical Records of The Queen's Bays, War Office folder 1/1, file no. 2, 1911.

12 *The Cavalry Journal*, vol. 10, 1920, p. 334.

13 *Official History of the War: Military Operations France and Belgium 1914*, 1926.

34

Retreat from Mons, Marne, Aisne, Messines, Ypres, Hooge
1914-1915

The action at Néry was over by 10 a.m., and the retreat resumed at 11 a.m. In the meantime the battlefield was cleared, the damaged guns removed, and Lieutenant Lamb replaced his losses in the machine gun and waggon teams with loose and captured horses. The weary way back continued in hot weather through Senlis, Moussy-le-Vieux and Gournay, until the Bays reached the northern outskirts of Paris on 5 September, spending the night at Aubepierre. The retreat from Mons was now over, and on 6 September the regiment received orders to advance, first to the east to Jouy-le-Chatel, and then north, to cover the advance of the British centre and left. The Bays were not seriously engaged, but managed to pick up a straggler from a German cuirassier regiment while clearing some farms. On bivouacking that evening the Bays had present 17 officers, 7 having been killed or wounded; 423 NCOs and men, with 114 dead, wounded and missing; but there were only 304 riding horses left of the original 527, and 48 draught horses out of 74.[1, 2, 5]

On 7 September the Bays advanced through Choisy, reaching the line of the Grand Morin river on the 8th, and pushing on to the Petit Morin at Sablonnières, where the crossings were held by the German cavalry and Jaegers of the Guard. The Bays machine guns 'did a lot of firing across the valley. Later, we galloped on down the road under rifle fire, and came into another position without losing any men. The Germans were eventually turned out of the village and their positions on the heights above it at about 12.30 p.m. after three hours' fighting; a lot of German wounded and dead lying about.'[1, 2, 5]

On 9 September the Bays moved off at 2 a.m acting as advance guard, reaching the River Marne at Saulchery, where they were to seize the crossings over the river. Captain Springfield, leading with 'A' Squadron, found the bridge intact but blocked with barbed wire and barricaded. As soon as the bridge was cleared the Bays crossed, and as they moved up the high ground on the other side they came under shell fire from one of their own side's batteries. 'Luckily no one was hit, although the shells burst unpleasantly close.' As the Bays moved forward Lieutenant Kingstone's troop was fired on from a small copse near the village of Le Thiolet. 'C' Squadron went to his assistance, resulting in a German officer and four men being killed and twenty-one prisoners taken. Lieutenant Barnard of 'A' Squadron killed another German and captured his companion. The regiment was billeted at Lucy le Bocage by 6 p.m.

The advance continued on the 10th and 11th through Bonnes to Breny on the River Ourcq, and on to Branges. On 12 September the Bays were ordered to clear the village of Braisne on the Vesle, and then to seize the heights above, which overlook the River Aisne. When 'C' Squadron under Captain Pickering came in sight of the village it seemed to be deserted, but as Lieutenant F. D. R. Milne's troop approached the outskirts they saw that the bridge was barricaded. Then fire was opened on them, causing several casualties and killing Milne's horse. The survivors galloped for cover, dismounted, and went into action. The rest of 'C' Squadron came up in support, being reinforced by 'A' Squadron, while 'B', under Captain Stone, dismounted and attacked from the left. 'Z' Battery, RHA, unlimbered and fired in support, checking some German reserves which tried to reinforce the village. The Germans fell back slowly as the Bays' attack enveloped them, and there was some stiff house to house fighting, as numerous snipers were cleared, one of them killing Captain Springfield and another wounding Captain Pickering. By 3 p.m. the village was cleared, with 200 prisoners taken, and one of the bridges over the river secured intact. The Bays were relieved by an infantry brigade, and they moved on to occupy the heights overlooking the River Aisne at Dhuizel. Sir John French particularly mentioned the good work of the Bays in clearing Braisnes in his third despatch.

On 13 September The Queen's Bays crossed the River Aisne, and moved onto high ground near Pargnan on the extreme right of the British line, while an attack was launched to capture the Chemin des Dames. On the 15th the Germans counterattacked but were beaten off, and on the 16th Sir John French, recognising that both sides had reached a position of stalemate, ordered the British to be 'strongly entrenched', and thus started the era of trench warfare. For the next few days the Bays were in and around Chavonne, coming at times under shellfire, as Lamb described: 'We sat about in the wood and off-saddled. The enemy now commenced to 'coal-box' the wood we were in. They dropped all over the place, and the detonation when they burst in the wood was terrific. However, they never got amongst us.' On the 20th the Bays moved into the trenches to relieve the 11th Hussars.

Our trench consists of a high bank with steps cut in it to fire from, and holes

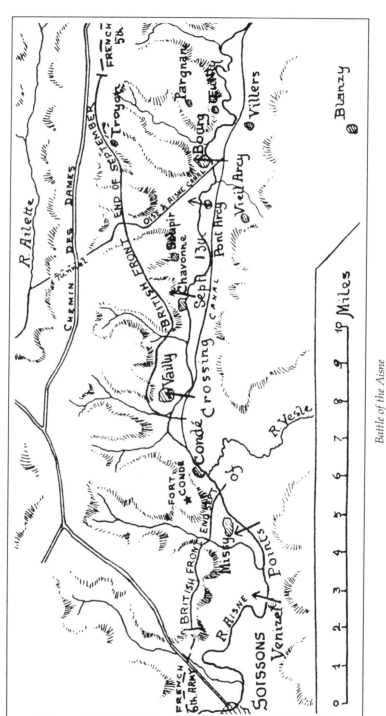

Battle of the Aisne

dug out all along, for the men to shelter in. During the day the Germans attacked the Wiltshire Regiment on our left. I got a fairly good target as they attacked. Whilst advancing the Germans made horrible noises, and seemed to be all talking at once. They were also encouraged by the blowing of cheap tinny- sounding trumpets.

During this attack Major Hall witnessed the Germans driving in front of them some 60 to 100 British prisoners, 'in order to mask our fire'.[1, 2, 5]

On 24 September the award was announced of the DSO to Lieutenant Lamb, and the DCM to Ellicock for their conduct at Néry. Some months later Trooper Goodchild was also awarded the DCM for his behaviour at Néry. On the 30th five reinforcement officers arrived, Lieutenant Single from Aldershot, and Kemmis (Inniskillings), Davidson (2nd Indian Lancers), Sartorius (6th Bengal Cavalry, and Robertson (Jacob's Horse). On 2 October news came through of French awards for the action at Néry: Colonel Wilberforce received the Légion d'Honneur, Croix d'Officier, and Lieutenants Heydeman and Lamb the Légion d'Honneur, Croix de Chevalier, whilst Corporal Short received the Médaille Militaire.[1, 2, 5]

The Battle of the Aisne had now been fought to a standstill, and both the Germans and the French were extending their flanks in Northern France and Belgium towards the sea, where a dangerous gap had opened up between the French left flank and the Belgian right. The British were now relieved on the Aisne, and transferred to cover this gap in the north. The Bays, in billets at Courcelles, were ordered to move on 3 October, their destination being kept secret; Lamb noted, 'Nobody knows yet what we are here for. We all believe that shortly we will be moved round to the extreme left flank.' The Bays marched via Violaine, Orrouy, where a short digression was made by some officers to Néry to photograph the graves, then on through Tricot, Montdidier, Amiens, Doullens, St Pol, to Béthune, which was reached on 11 October. On the 10th, 'A lot of new horses came to join the Regiment'.[1, 2, 5]

On the 11th 'One of our patrols, under Lieutenant Sartorius, had a brush with an Uhlan patrol, killing one and wounding another. Two men also of our own wounded.' On the 12th, on leaving Merville, 'B' Squadron bumped into some German cavalry and had one man wounded in the back. The same day Lieutenant Kingstone laid an ambush for some German cyclists and killed three of them. On the 13th, as the regiment moved forward, the situation remained very fluid, and as 'A', the leading squadron, moved north through Fletre towards Méteren, they came under shell and machine gun fire. Trumpeter Webb was killed and several men wounded. By midday the Bays were on the ridge of Mont des Cats running north from Méteren, and at 3 p.m. a German Jaeger battalion counter-attacked from the direction of St Jans-Cappel, but was driven off with some loss. The next day the regiment crossed the Belgian frontier, being billeted at Dranoutre.

On the evening of the 15th the Bays were ordered to capture Ploegsteert by means of a night march and attack. In the dark and rain, not knowing where the enemy were, 'after repeated long halts, whilst waiting for our advanced troops, who work dismounted, and creep along the ditches with fixed bayonets, we eventually got to Ploegsteert without opposition at 5 a.m.' Some days were spent in the vicinity, and on the 21st Colonel Wilberforce was invalided with muscular rheumatism, the command devolving upon Major J. A. Browning.[1, 2, 5]

> The Regiment was instructed to make and hold certain trenches just north of Messines on the night of the 30-31st. 'C' Squadron, under Major Terrot, were entrenched in front of 'B' and 'A' Squadrons, the latter being on the Ypres-Messines road. About 7 p.m. on October 30th, 'C' Squadron (temporarily under Captain Moncrieff) successfully withstood a strong attack by the enemy. At 5.45 a.m. on the 31st, the enemy strongly attacked against the left flank of 'C' Squadron, who had eventually to withdraw by successive troops and take up other positions in the rear. Up to now 'C' Squadron had about 30 men wounded, including some men from 'A' Squadron. Lieutenant Paul and three troop sergeants were killed, and Lieutenant Milne was wounded. At about 10 a.m. the Bays were ordered to retire from the line of the road. Just as they reached a hedge a hundred yards to the rear, Major Browning was killed, while standing out in the open getting the men to line the hedge and form a firing line. Major Matthew Lannowe assumed command, and led the regiment back to the road under heavy shellfire, where the regiment held the line all day.[1,5]

Captain Milne of 'B' Squadron recalled one incident:

> The calm of the evening was disturbed by a German band playing 'Deutschland Uber Alles', and then the charge sounded on the German bugles, and the Germans came through the hedge, advancing slowly in almost close formation, kettle-drummers beating their drums as they advanced. Rapid fire was opened on them from the trenches and houses, and great execution was done, the attack breaking up, some of the Bosch retiring and about 50 others taking cover in a barn, the door of which opened into the field. The only door on the roadside of the barn was about 12 feet from the ground, opening into the loft. Efforts were made to burn them out by forking hay, saturated with paraffin and lighted, through this door, but they did not succeed, and eventually Lieutenant Sartorius and Sergeant Wallace each crept round a side of the barn and emptied their

revolvers into the crowd. This caused the Bosch to shout 'Kamerad', and the whole lot surrendered. Lieutenant E. Walker has one of the Bosch side drums.

This side drum is now in the Regimental Museum. Twenty-four Germans were killed, eighteen wounded and thirty-two taken prisoner in this affair.

The Bays remained in defence of the Messines ridge and around Ypres until 24 November. On 12 November they were at the château of Hooge on the Menin road, remaining there until the 17th, and relieving the infantry as they came under intense pressure. It was bitterly cold, the area a sea of mud, and the men were subjected to continuous shellfire. The 18th to the 21st provided a short rest, and then the Bays relieved the Greys in the Kemmel trenches, where they stayed until the 24th, when they were withdrawn into Army Reserve and were billeted in farms around the village of Fletre, which lay between Bailleul and Hazebrouck. It was from here that the first home leaves were granted, and 'We spent our first Christmas in this place, and I think managed to be quite happy. People in England sent out all kinds of clothing, including some very nice things from the Queen and Princess Mary.' There they stayed until 22 February 1915. Of those who had come to France with the regiment, there remained only 6 officers and 299 men.[1, 5]

During the early part of 1914 The King's Dragoon Guards in India moved to Lucknow, and it was here on 31 August that the regiment received orders to mobilise, which it did by 11 September. On 8 October the KDG, under command of Lieutenant Colonel Bell-Smyth, entrained for Bombay with a strength of 16 officers, 532 men and 512 horses. On the 11th 'A' and 'D' Squadrons embarked on HT *Chilka* and the following day 'B' and 'C' on HT *Franz Ferdinand*. That day the regiment was brought up to full strength with 17 men and 42 horses from the Inniskilling Dragoons. On 12 and 17 November the KDG disembarked at Marseilles, moving a week later to La Source, Orléans to form part of the Lucknow Cavalry Brigade of the Indian Expeditionary Force. The other regiments of the brigade were the 29th Cavalry and the 36th Jacob's Horse with 'U' Battery RHA in support. At Orléans the KDG were joined by Major Wiiliams, who became second in command, Captain G. R. Cheape, Lieutenant Crossley, and 2nd Lieutenants Muir, Richardson, Gratton Holt and Wilson; in addition Captains Thompson, Scinde Horse, Johnson, 28th Cavalry, and Lieutenant Rimington, Indian Army Reserve, were posted to the regiment for duty. On 7 December the KDG moved to billets at Lillère, and on the 21st to Norrent-Fontes, moving further west again on Christmas Day to Lisbourg. During this period two KDG officers died: Lieutenant Hawkins was killed in action and Lieutenant White died of wounds, when they were serving away from the regiment.[3, 4]

On 9 January 1915 The King's Dragoon Guards, leaving their horses, travelled in buses to Béthune and then marched to Festubert, where they took over trenches from the 17th Lancers. 'A' and 'B' Squadrons occupied the forward trenches with the German front-line trenches between 70 and 150 yards away. 'C' and 'D' Squadrons stayed in reserve in the village of Festubert. Five men were wounded on moving in, and these were evacuated during the night. The following day the squadrons in reserve relieved the forward squadrons. Conditions were deteriorating as the water level in the trenches rose until it was above the waists of the men; by 10 January the men were up to their armpits in water, with the level still rising. On the 11th the regiment was ordered to abandon the flooded front line and take up position in the support trenches. A certain amount of sniping went on, causing two casualties, but the Germans made no attempt to occupy the abandoned trenches, and were, indeed, fully engaged in baling out their own. On the evening of the 11th the KDG, being relieved by the Inniskillings of the Mhow Brigade, returned to their billets at Lisbourg. Of the seven men wounded, one died from his wounds, but two officers and eighty men had to be evacuated sick, nearly all with frostbitten feet.[3, 4, 7]

During the remainder of January 1915 and throughout February, March, April and May, The King's Dragoon Guards were in billets at Lisbourg, Flechin, Oxelaere, and Rincq. At the end of January the regimental sergeant major, Mr A. Jacques, and SQMS Farthing were gazetted as 2nd Lieutenants. On 8 February 2nd Lieutenants Murray Johnson and Waggett joined the regiment, together with forty-five NCOs and men, and nineteen horses. On the 13th 2nd Lieutenants Burton and Rimington joined, to be followed on the 26th by a further reinforcement of twenty-seven NCOs and men. At the end of March Major Hunt and Captain Renton rejoined and took over command respectively of 'B' and 'C' squadrons. At the beginning of April strong parties af 200 men and 12 officers were engaged in digging third line trenches at Robecque. On 13 May Captain Denny was killed in action. During this long period mostly out of the line, it was decided in London that the Emperor Franz Josef of Austria could no longer remain as colonel in chief of the regiment, Britain being at war with Austria. At the same time the cap badge was ordered to be altered from the Austrian double-headed eagle, which the regiment had proudly worn since 1897, to the badge carried on the regimental buttons of the star and Garter surmounted by the crown, with the initials KDG in the centre.[4, 7]

On 31 May 1915 this long period behind the lines ended, and the KDG were ordered to take over the front line trenches from the 3rd Dragoon Guards at the château of Hooge near Ypres. The relief was carried out by 11 p.m. As the château was found to be free of Germans, it was occupied, together with the stables and

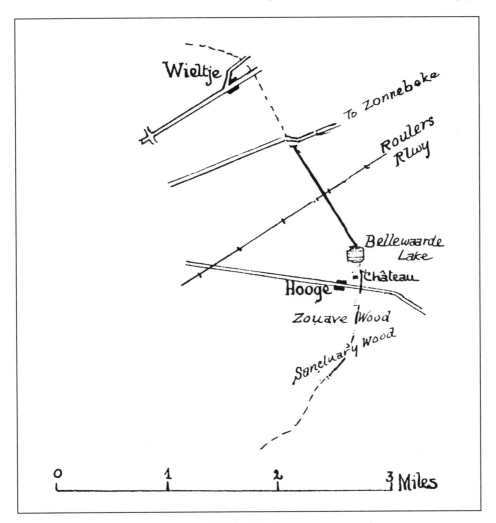

Battle of Hooge

annexe, which were sandbagged and loopholed. The trenches towards Belwarde Lake were wired and prepared for all-round defence, and communication trenches to connect with the 3rd Dragoon Guards' positions were prepared and occupied. Major Hunt was hit in Zouave Wood on the way in and Lieutenant Ward took over command of 'B' Squadron. Each night rations were brought up to Zouave Wood, where they had to be collected by the front line troops, invariably under shellfire. Throughout 1 and 2 June the Germans shelled the château heavily with high explosive, shrapnel and *minenwerfers*. During the bombardment the men took what cover they could in the cellars of the château and a nearby fort, dashing out during lulls to repair and improve the trenches. Lieutenant Colonel Bell Smyth was put in command of the sub-sector, Major Turner taking over command of the regiment.

At 5 p.m. on 2 June the shelling lifted and the Germans attacked from the north and east. On the right the attack from the east was driven back, but the thrust against the château succeeded. Of the two troops holding the position, the only survivors were Captain Cooper, who was wounded, and three men, two of whom were wounded. The Germans bombed out this small party and occupied the annexe, while the four survivors retreated through the remains of the stables to one of the trenches. The enemy then occupied the stables, but were counter-attacked by a party from 'A' Squadron on the left under Captain Cheape, and the stables were retaken and occupied.

On the left two squadrons occupied the fort and buildings of Hooge north of the Menin road. As the enemy attacked, the Maxim gun came into action against thirty to forty Germans at a range of 300 yards; it also drove in German working parties near the Belwarde Lake. KDG snipers posted in the roof of the fort and buildings found plenty of targets. Having taken part in the counter-attack on the stables, these two squadrons and the survivors at the stables were relieved by the Lincolnshire Regiment, the regiment withdrawing to Zouave Wood, then marching back through Ypres to huts at Vlamertinghe.

The KDG casualties at Hooge were Captain Renton killed, and Major Hunt, Captain Cooper, Lieutenants Alexander, Carleton Smith, Murray Johnson wounded; with 23 men killed and 44 wounded, of whom 4 later died of their wounds, whilst 6 men were missing, all of whom were later discovered to be killed. For this action Lieutenant Colonel Bell Smyth was awarded the CMG, Major Turner the DSO, Captain R. Cheape the Military Cross, Lance Corporal Carpentier the DCM, and Sergeant Davis and Corporal Waterman the French Croix de Guerre.[4, 7]

The Queen's Bays spent the whole of January 1915 in billets at Fletre, and on 8 February the regiment was inspected by King Albert of the Belgians. On 22 February there came sudden orders to leave Fletre; the regiment paraded dismounted and was carried in buses to Ypres, where after a night in billets it took over some trenches from the 16th Lancers near Zillebeke. The Germans had exploded a mine under the Lancers' trenches and had captured a part of an advanced trench, so that only some fifteen yards separated the two sides. During the relief there was shelling with trench mortars and *minnenwerfers*, which caused nine casualties among the men, whilst Major Ing was slightly wounded. On the 24th some bombing was carried out, and Trooper Rowan was killed. The 25th brought cold weather and snow, with Trooper Caine killed and another man wounded. On the 26th the Bays were sniped all day, and they carried out some retaliatory bombing of the German trenches. Two men were wounded. By the 27th the Bays had established a superiority over the German snipers, but two more men were wounded. The 28th was a quiet day and the Bays, with the exception of 'B' Squadron and the machine guns, were relieved by the 18th Hussars that evening, and returned to billets.

'B' Squadron, under Captain Sloane, had been left in the line to cooperate in a combined operation with the French on their immediate left, using their knowledge of the locality. The plan was for two mines, one French and one British, to be exploded simultaneousely under a crater which the Germans had rushed and seized on the 21st; the Bays would attack from the right, and the French from the left, recapturing the crater and joining hands in the centre. On a dark night in steady rain there was a delay because the French were not ready; then at 10.30 p.m. the British mine was exploded and the Bays stormed the crater to find no resistance. The advance was stopped by a steep eight-foot bank topped by sandbags, and voices were heard above it. Thinking these were the French, the interpreter was given a leg up and over, only to land in the midst of a somewhat shaken group of Germans. The interpreter managed to scramble back and it then became clear that the French had not started, as their mine was still not ready. The Bays now tried a bombing attack, but this failed because of the unreliability of the bombs. The venture was then abandoned and 'B' Squadron returned to the rest of the regiment, having had eight men wounded.

The Bays remained about Fletre for the whole of March 1915, then moved to billets around Thieusbroek until 23 April, when, following the first German gas attack, the Second Battle of Ypres started and the cavalry were needed to support the line against German pressure. The Bays were ordered to defend the village of Woesten and, on arrival, found it held by two battalions of French Territorials. They remained in and around the village, in support, until the 28th, when they returned to their billets. Colonel Wilberforce was transferred to staff duty, and on 1 May Lieutenant Colonel Lawson of the Royal Scots Greys assumed command of the Bays. The regiment remained in billets until 9 May, when it was ordered into reserve trenches south of Potijze, relieving the 9th Lancers. On the 12th the Bays moved into the front-line trenches, taking over from the 19th Hussars. These trenches were in a very bad condition, being shallow and straight, and as the men improved them at night they were digging through decomposing corpses. The noise, too, attracted some sniper fire, when the RSM, Mr Turner, was killed by a bullet through his neck.

On 13 May the Germans opened a heavy bombardment and the trenches of 'B' Squadron on the left were largely destroyed. 1 Troop on the extreme left had only

Sergeant Isles and one trooper left, Second Lieutenant Heron being killed and all other of his men killed or wounded, some of them buried in the wreckage of the trench. Even so a mouth organ concert was improvised during the continuing bombardment. At 10 a.m. the barrage lifted and the Germans could be seen advancing and were immediately engaged. On the right of the Bays the line was broken. Major Ing dashed out of the trench, rallied the retreating men and shepherded them back to reinforce the Bays line. The colonel sent the adjutant, Captain Kingstone, to brigade for reinforcements, which arrived in the form of the 10th Hussars, who were slowly able to re- establish the situation. The medical officer, Lieutenant Chapman, was killed, but on the evening of 14 May the Oxfordshire Yeomanry relieved the Bays, who returned to billets. The casualties had been severe, two officers killed and one wounded (Lieutenant Misa), 28 other ranks killed and 32 wounded, and this out of only 175 men in the line. Major Ing was awarded the DSO and Corporal Clarke the DCM.[1,6]

SOURCES

1   F. Whyte and A. H. Atteridge, *The Queen's Bays, 1685- 1929*, 1930.
2   War Diary, 1914, of Lieutenant Lamb, Queen's Bays, Regimental Museum, 1st The Queen's Dragoon Guards, Cardiff Castle.
3   War Diary, 1914, 1st King's Dragoon Guards, Regimental Museum, 1st The Queen's Dragoon Guards, Cardiff Castle.
4   W. Clowes, *The King's Dragoon Guards, 1685-1920*, 1920.
5   War Diary, 1914, 2nd Dragoon Guards, Regimental Museum, 1st The Queen's Dragoon Guards, Cardiff Castle.
6   War Diary, 1915, 2nd Dragoon Guards, Regimental Museum, 1st The Queen's Dragoon Guards, Cardiff Castle.
7   War Diary, 1915, 1st King's Dragoon Guards, Regimental Museum, 1st The Queen's Dragoon Guards, Cardiff Castle.

35

Somme, Arras, Fampoux, Morval, Cambrai
1915-1918

The Queen's Bays had two days' rest at Vlamertinghe, receiving reinforcements from base, including six young officers. On 16 May 1915 they were ordered back into the line at Hooge, taking up positions just below Belwarde lake and to the left of Hooge Château. On the 21st they were relieved by the 4th Dragoon Guards after a quiet time, with only one casualty, Sergeant Petty. They returned to Vlamertinghe, only to be called out again on the 24th by a sudden crisis arising from a German gas attack, which had caused heavy casualties, enabling the enemy to advance on either side of Belwarde lake. On setting out they were joined by 2nd Lieutenants Biddulph, Etherington and Mitcherin. The Bays strengthened the line south of the lake through Zouave and Sanctuary Wood, beating off one attack and losing one killed and three wounded from shellfire. The regiment was relieved during the early morning of the 26th, when the exhausted men returned to their billets. At the end of the month they were moved back eighteen miles west of Ypres to Hardifort, near Cassel, where they remained until the latter end of September 1915, waiting to be used as cavalry in a 'breakthrough' [1, 3]

This prolonged period of inaction was spent in training, refitting and inspections. On 12 and 19 June 2nd Lieutenants Macnaughton, Whitmore-Smith and Gray joined. On 14 July a move was made to billets at Rubrouck, while various digging parties were provided for reserve trenches from time to time. On 23 September 1915 the whole of the 1st Cavalry Division was concentrated further south near Mametz, the Bays being billeted at Ecques, in preparation for a 'breakthrough' during the Battle of Loos. Again, the 'gap' never materialised but false alarms moved the Bays to Hesdigneul on the 26th, to Cauchy à la Tour on the 29th, and to Enquin les Mines on 6 October. On 20 October the regiment went into winter billets around Lottinghem, where Lieutenant McGrath joined. There was a further move to new billets at Neuville near Montreuil on 11 November, where they remained until the end of March 1916. During the whole of 1915 The Queen's Bays had one officer killed and one wounded, 31 other ranks killed, 4 died of wounds, 57 wounded, and 2 missing; a total of 95 casualties. [1, 3, 6]

The King's Dragoon Guards, after their battle at Hooge in the early days of June 1915, remained in billets at Vlamertinghe, being shelled by long-range guns and receiving reinforcements of officers, Lieutenants Fox and Gladstone and 2nd Lieutenant Brown, and 36 NCOs and 13 horses. They heard that SSM Webb, Sergeant Stratford, Corporal Bramley and Troopers Habgood and Hill had all been awarded the Military Medal for their gallantry at Hooge Château. On 14 June the regiment moved back to their old billets at Rincq, where they stayed until 1 August 1915. During their time at Rincq, the KDG provided trench digging parties and were joined by Lieutenant Card and 43 reinforcements. On 1 August they were on the move again, Captain Cooper rejoining from England. They arrived at Halloy les Pernois on the 6th, where another 23 men and 38 horses came up from base. On 23 August a dismounted party of 13 officers and 300 men took over trenches from the Inniskilling Dragoons at Authuille, a sector which included MacMahon's Post and Mound Keep. Over the next few days much work was carried out improving the trenches, under a certain amount of intermittent shelling. The regiment remained in this quiet sector of the line until they were relieved by the 7th Dragoon Guards on 2 September. During this first spell in the Authuille trenches, from which they had an excellent view of Thiepval Château, Lance Corporal Barrett was the only casualty, being wounded in the arm by a stray bullet.[4, 5]

From 3 to 12 September the KDG were in billets at Halloy les Pernois, providing working parties. They were joined by 2nd Lieutenant Langford. On the 12th, 14 Officers and 240 men went back to the Authuille trenches, where over the next two days they were subjected to a certain amount of shelling, mortaring and rifle grenades, with one man, Trooper Grandison, being wounded in the leg. They were relieved by infantry on the 16th and returned to their billets, moving to Autheux on the 23rd, where the regiment remained until 13 October. They then moved to Bernaville, where they stayed until 22 October. During this period they were joined by Captain Alexander, who had to be evacuated two days later, having been accidentally shot. Lieutenant Colonel Bell Smyth was promoted, Major Williams taking over command of the regiment. Another move to Molliens Vidame took place on 23 October and Captain Longworth, Lieutenant Sprot, 2nd Lieutenants Adair, Moreton and Parker arrived from base. RSM Brewer was commissioned and posted to 'C' Squadron. November saw two more moves to new billets, on 18 November to Longpré, and on 16 December to Quesnoy Le Montant. The year ended with The King's Dragoon Guards having moved thirty-five times, and with a strength of 30 officers and and 592 other ranks, of whom 100 were dismounted reinforcements. Christmas Day saw Major Hunt rejoin the regiment as second in command.[4]

The KDG stayed at Quesnoy Le Montant until 26 March 1916, and were joined on 3 February by Captain Alexander and 2nd Lieutenants Peacock, Ward and Tiarks. They then moved to new billets at Gueschart, and whilst there the award of the Military Cross to Sergeant Major R. Holmes was announced. Another change of billets took place on 10 May 1916, when the regiment moved to Sericourt; where on 24 May Lieutenant Colonel Wickham assumed command. The *London Gazette* of 15 June announced that Colonel Hulton, Major Langton and 2nd Lieutenant Farthing had been mentioned in despatches for their services. The regiment moved again on 1 July, this time to Frohen le Grand, from where, on 19 July, a party of 8 officers and 300 other ranks went into support trenches at Neuville St Vaast, under Captains Wienholt and Alexander. On 30 July the detachment returned to the regiment, having had three men wounded.[2, 5]

A reinforced squadron of the KDG, under command of Major Spurrier, was inspected by King George V on 10 August 1916 at Wavrans. The remainder of the regiment moved to new billets at Humbercourt during August, and to Canchy in September, where various working and tunnelling parties, as well as sniping detachments, were provided for the front-line infantry. On 25 September, while at Morlancourt Camp, the KDG were ordered to be at forty-five minutes' notice for mounted action. On the 26th the regiment moved to the area of Mametz on the Somme battlefield and set up headquarters in Trones Wood, sending out patrols. One man was severely wounded and later died; his horse was also killed. The breakthrough never materialised, and after a number of marches and counter-marches the KDG returned to billets at Machy. At the end of September the regiment moved into winter billets at Miannay, again providing 8 officers and 270 other ranks for a Pioneer Battalion, which was formed from the Lucknow Brigade. During October 2nd Lieutenants Dick, Peacocke and Ward joined the regiment.

A number of KDG attached to other regiments distinguished themselveso: 2nd Lieutenant Percival, recently commissioned, was awarded the DCM, 2nd Lieutenant Alford the Military Cross, and Sergeants Ford and Darden the Meritorious Service Medal. Lieutenant Farthing was killed in action serving with the East Lancashire Regiment. On 1 December the Pioneer Battalion came under heavy shellfire, which killed two KDG and wounded two more. 2nd Lieutenants Parker and Bogle Smith joined at Miannay and three men of the Machine Gun Squadron were awarded the Military Medal, Sergeant Corley, and Troopers Vance and McIntosh. On 28 December another burst of shellfire killed one man and wounded two others. So ended another year of comparative inaction for the cavalry, but of twenty-five moves for the KDG.[2, 5]

The Queen's Bays, also in billets during the early part of the year, moved on 26 March 1916 from Neuville to Montcavrel, where they stayed until 23 June 1916.

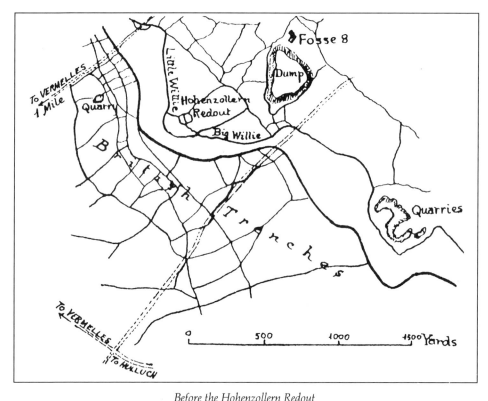

Before the Hohenzollern Redout

But during the winter dismounted detachments of the Bays did tours of trench duty on the old Loos battleground, opposite the Hohenzollern Redout. Captain Kingstone described the scene:

> The trench system was very complicated, and the whole of the area which we took over was overlooked by German observation posts on the big slag heap at Fosse 8. Nothing very startling happened during the first part of our time there, but there was always a good deal of bombing and patrol fighting going on at night. We lost a good number of men. When visiting in the summer of 1925, I found that there were at least twelve of our men buried there, together with four or five graves of unknown men of the Bays.[1, 6]

At the end of June 1916 the 1st Cavalry Division was moved up behind the Somme in the expectation that there would be a breakthrough, and after four days' marching the Bays arrived at Querrieu, where they were concentrated with the rest of the division in readiness for the opening of the Somme offensive on 1 July.

These hopes were to be dashed, and the Bays, together with the rest of the cavalry, moved in and around Querrieu for the whole of July, until on 9 August they were withdrawn to the north west of Amiens, still waiting hopefully for the breakthrough. On 6 September they were moved forward to the Carnoy valley, bivouacking and seeing tanks for the first time. On 15 September the Bays were ordered forward, and 'A' Squadron under Major Pinching, MC, advanced into a valley south west of Leuze Wood. The ground, after the battle, was a wilderness, and two patrols under 2nd Lieutenants Yeatherd and Macnaughten were sent on to maintain contact with the infantry. Yeatherd's patrol came under heavy fire as it advanced dismounted, every member being wounded, and Yeatherd himself missing. One of the wounded died the next day and Yeatherd's body was discovered a week later. Major Pinching was also wounded during this action.

The Bays continued to wait around Carnoy until on 17 September they were withdrawn into billets near Daours, moving again on 23 September to Blangy sur Turnoise. October saw two more moves to Auchy les Hesdin and then to Bonieville, where it rained continually, reducing the area to a sea of mud, with the horses left out in the open. On 9 November a welcome move took place back to winter billets in the area of Montcavrel, where the Bays remained until the beginning of April 1917. The regiment's casualties during 1916 had been 2 officers killed and 2 wounded, 17 other ranks killed, 35 wounded and 3 missing.[1,6]

In January 1917 a Pioneer Battalion was formed from the 1st Cavalry Brigade, with the Bays contributing 6 officers and 225 other ranks, with Major Pinching commanding the battalion. The Bays company, under Captain Kingstone, had the task of improving the railway line from St Pol to Arras. On a foggy January morning they suddenly saw a German aeroplane flying very low, and then landing in a field not far away. While the pilot remained in his cockpit, the observer climbed out and walked to examine a signpost on the nearby road. The Bays were unarmed, but dashed with their picks and shovels towards the now-running observer. The pilot opened fire with his machine gun — the shots went high — as the observer reached the plane and started to climb in. At that moment the pilot accelerated, throwing off his observer, who was duly captured, while the plane roared off into the fog. Captain Kingstone commented: 'The observer was not a bad little chap.' During February Lieutenant Colonel Lawson fell sick and Major Pinching returned from the Pioneer Battalion to assume command of the Bays, also hearing that he had been awarded the Croix de Guerre, and Sergeant Parker the Médaille Militaire.

Lieutenant Colonel Lawson rejoined on 7 April, as the regiment was concentrating with the rest of the 1st Cavalry Division in an open field at Agnières near Arras, before moving round on 10 April to Fampoux, to be ready to exploit a successful attack made by the 4th Infantry Division. 'This was the worst night that we ever spent during the war. There was a foot of snow on the ground, and we had no cover or shelter of any description.' On 11 April the Bays were ordered to advance through Fampoux to seize Greenland Hill beyond the village, and then to exploit north eastwards. As the leading troop reached Fampoux, the German shellfire increased, especially around the crossroads in the centre of the village through which they had to pass. Lieutenant Grant and a trumpeter were wounded, and Captain Kingstone ordered his, the leading squadron, to take cover while a patrol of 2nd Lieutenant Quested and six men galloped forward to find out what was happening. As the patrol advanced they came under heavy shell and machine gun fire which killed or wounded all seven horses, with the men, now dismounted, taking what cover they could and eventually making their way back at dusk under cover of a snowstorm. It now transpired that the infantry had been pinned down, and the attack which the Bays were meant to exploit had never taken place. Kingstone now had to extract his squadron:

> Two troops had been able to get into a big sandpit, and so had escaped much of the shell fire. The leading troop had suffered very heavily, particularly from a direct hit from a large shell, which had brought a wall down on top of them. It took a considerable time to extract such men as were alive from the debris. The result was that they had lost 17 men killed and wounded, including Sergeant Chelmsford, their troop-sergeant, and Corporal Smith.

The shelling gradually diminished and by 5.30 p.m. the regiment was withdrawn, returning on the 12th to its billets at Agnières. The casualties had been heavy: Lieutenant Grant and 2nd Lieutenants Quested, Ascoli and Beddington had been wounded, 4 other ranks had been killed and 18 wounded; of the horses 14 were killed, 6 were wounded and 3 died of exposure. On 13 April 2nd Lieutenant Walker and two other ranks were wounded while on a digging party on Vimy Ridge.

The Queens Bays were moved on 16 April from the wretched area of Agnières to better billets at Hesdin, where on 20 April they heard the sad news that Major Pinching had died from the effects of an operation in London. The Bays remained at Hesdin until 14 May, when they were moved to new billets at Marles les Mines to the west of Lens, staying there until 4 June 1917. During June there were further moves to billets at Nedon, Nedonchelle and Busnettes, when on 9 July the Bays left Busnettes for billets at Le Sart, where they stayed until 29 July 1917. [1, 7]

The King's Dragoon Guards started 1917 in billets at Miannay, where during January they heard that Temporary Lieutenant Colonel Wickham had been appointed a brevet lieutenant colonel and had been mentioned in despatches, as had RSM Burdett. SSM Webb, transferred to the Machine Gun Corps, had been awarded the Military Cross, and Sergeant Stratford the Military Medal. At the end of February a move was made to new billets at Talmas, and then on 17 March the KDG moved up as the Germans retired from the Thiepval-Gommecourt position to the Hindenburg Line, one KDG being captured during this operation. Over the next few days they followed up the German withdrawal in operations around Mory, eventually taking up positions facing the Hindenburg Line. While maintaining contact with the enemy two men were killed and six wounded, one of whom later died of his wounds; thirteen horses were killed or died of exposure in the very severe weather. On 28 March 1917 the regiment was relieved by the 17th Lancers and went into billets at Aveluy.[2, 8]

The KDG stayed at Aveluy until 16 May 1917 when there followed a series of moves, until on 3 June they took over the trenches at Hargicourt. After a quiet start the Germans, from 6 June, shelled and mortared heavily each day, with the result that over the next four days one man was killed and nine were wounded, of whom three died of their wounds. Active patrolling took place, and on 12 and 14 June the Germans attacked the 29th Lancers on the left of the KDG. Both attacks, which came from the direction of Cologne Farm, were driven off with the help of supporting fire from the KDG. On 16 June the regiment was relieved by the Inniskilling Dragoons, but not before another man was wounded by mortar fire. The KDG then rested for the 16th at Jeancourt, moving back into trenches at Hargicourt that night, when they had the Canadian Dragoons on their right and the 36th Jacob's Horse on their left. During the next four days the trenches were shelled and subjected to machine gun fire, and a small German bombing party was driven off. The regiment was relieved by the Jodhpur Lancers, moving to take over the front-line posts of the Inniskillings. This spell of duty was subject to the usual shelling and machine gun fire, and one man was killed. The KDG supported the 2nd Lancers and Central India Horse in bombing raids on Cologne Farm before being relieved on 29 June, marching to billets at Hamelet and moving again on 5 July to Le Mesnil.

On 7 August the KDG were back in the line, taking over trenches from the Lancashire Fusiliers at Epehy. On 9 August heavy shelling resulted in 'B' Squadron losing a sergeant killed and seven men wounded. The KDG patrolled actively, and during one such patrol Sergeant Woolsey was wounded. On 18 and 19 August Bird Post was mortared and shelled. Some sixty gas shells fell around the position, but to no effect, as the wind was in the wrong direction. The regiment was relieved by the Manchester Regiment on the 19th, marching back to Le Mesnil and then taking over trenches from the Northumberland Fusiliers at Le Vergier on the 20th and 22nd. A fighting patrol on the night of the 23rd, commanded by Lieutenant Muir, encountered and attacked a strong enemy patrol, capturing three Germans and routing the rest, but losing one man killed and one wounded. For this action Lieutenant Muir was awarded the Military Cross. On the 27th a KDG standing patrol at Ascension Wood was attacked with bombs by an enemy patrol, which was driven off, and three nights later a German surrendered to this post. Lieutenant Moreton proved a very successful patrol leader, being awarded the Military Cross for his leadership during this period. The KDG were relieved by the 17th Lancers on 3 September.

The regiment was back in the trenches at Vadencourt on 13 September 1917, taking over from the 2nd Lancers. Lieutenant Brown led a reconnaissance patrol of thirty-nine men on 20 September and encountered an enemy patrol of the same strength near the German front-line wire. When the Germans were ten yards off, Brown led a charge which captured three of the enemy, wounded eight others, and put the remainder to flight. The KDG suffered two men slightly wounded. Lieutenant Brown was awarded the Military Cross for this engagement. On 30 September the Rifle Brigade relieved the regiment, who returned to Le Mesnil.

On 1 October 1917 The King's Dragoon Guards received orders to be prepared to move back to India. During their time in France through 1917, they had moved twenty-two times, with nine spells of duty in the trenches, each of which averaged ten days. On 7 October the regiment entrained at Péronne with 31 officers and 566 other ranks and 585 horses, arriving at Marseilles on the 10th, and embarking on the M.V. *Minnetonka* and *Bohemian* on 15 and 26 October 1917. The King's Dragoon Guards disembarked at Bombay on 20 November 1917 and were stationed at Meerut.[2, 8]

At the end of July 1917 The Queen's Bays moved up to Dickebusch for another great offensive, which was expected to turn the German flank and clear the Belgian coast. Once again their waiting was to end in disappointment, and on 2 August the cavalry were withdrawn back into billets at Munc Nierlet, then to Isques and Audisque south of Boulogne, where they stayed until 5 October. By 12 October they were back in their old billets at Hesdigneul until on 11 November the cavalry was brought forward in readiness for the Battle of Cambrai. Five days' marching found the Bays in a camp at Buire, and on the 21st they were bivouacked ready to exploit the advance of the tanks and infantry. By 11 a.m. the Bays were at Bois des Neuf as the tanks attacked and captured Cantainge, whereupon 'A' and 'C' Squadrons moved across the open ground clearing Cantaigne and taking 100 prisoners. 'B' Squadron secured the left flank, forcing 300 of the enemy to the west of Cantaing to retire. All three squadrons then consolidated in front of Cantaing. The Bays were relieved on the 22nd, returning to billets at Metz, having had 3 officers wounded, 1 man killed and 20 wounded. Lieutenant Barnard received the Military Cross for his conduct at Cambrai, and Sergeants Ballard and Evans, Corporals Tennant and Allanson, and Trooper Magner all received the Military Medal.

Whilst at Metz the Bays supplied a company of men for a 1st Cavalry Brigade infantry battalion, and in very cold and wet weather held a section of the line on the Bapaume-Cambrai road, having ten men wounded. On 26 November the Bays moved to billets at Cappy. From 2 to 4 December the 'infantry' company did another spell in the trenches near Vaucellette Farm, having three men wounded. The regiment remained at Cappy until 22 December, moving to Buire, where on the 23rd they were bombed by German aircraft, with the loss of 11 horses killed and 22 wounded. During 1917 The Queen's Bays had had 8 officers wounded, 4 other ranks killed, 58 wounded and 1 missing.[1, 7]

On 15 January 1918 the Bays were back in the trenches, holding a line of posts in front of Vadencourt, where they remained until the 26th. There was a certain amount of shelling, with the Bays patrolling actively, but the front was quiet and no casualties were suffered. The regiment remained at Buire from 26 January to 15 February 1918, when the 'infantry' company went back into the trenches at Vadencourt. They patrolled regularly without encountering the enemy. There was little shelling, but much aerial activity. The Bays were relieved on 5 March and went into reserve at Vendelles, where they prepared for a raid on the German front line to secure prisoners before handing over the sector to an infantry division.[1, 9]

The raidng party was commanded by Major Richardson, Queen's Bays, and was divided into three parties, made up from the 5th Dragoon Guards and the 11th Hussars as well as the Bays, each regiment providing about eighty picked men. The first party under Lieutenant Miles of the 5th Dragoon Guards attacked a post known as 'Eleven Trees', which the staff thought to be unoccupied but the front-line troops knew to be held. 'As had been expected "Eleven Trees" was occupied, and the German inhabitants were captured without any difficulty and without casualties on our part.' There were ten Germans in the post under an NCO. Two were taken prisoner and the rest killed. The main party then advanced up to the enemy's wire, 'covered by a barrage. Two Bangalore torpedoes were fired, but failed to do any serious damage to the wire...It was impossible to advance any further, our party threw bombs at the enemy, and after waiting eleven minutes under intense machine gun fire, were ordered to withdraw.' The raiding party suffered thirty-six casualties, all in the vicinity of the wire, of which three Bays were killed and seventeen wounded.[1, 9]

SOURCES

1 F. Whyte and A. H. Atteridge, *The Queen's Bays, 1685-1929*, 1930.

2 W. Clowes, *The King's Dragoon Guards, 1685-1920*, 1920.

3 War Diary, 1915, 2nd Dragoon Guards, Regimental Museum, 1st The Queen's Dragoon Guards, Cardiff Castle.

4 War Diary, 1915, 1st King's Dragoon Guards, Regimental Museum, 1st The Queen's Dragoon Guards, Cardiff Castle.

5 War Diary, 1916, 1st King's Dragoon Guards, Regimental Museum. 1st The Queen's Dragoon Guards, Cardiff Castle.

6 War Diary, 1916, 2nd Dragoon Guards, Regimental Museum, 1st The Queen's Dragoon Guards, Cardiff Castle.

7 War Diary, 1917, 2nd Dragoon Guards, Regimental Museum, 1st The Queen's Dragoon Guards, Cardiff Castle.

8 War Diary, 1917, 1st King's Dragoon Guards, Regimental Museum, 1st The Queen's Dragoon Guards, Cardiff Castle.

9 War Diary, 1918, 2nd Dragoon Guards, Regimental Museum, 1st The Queen's Dragoon Guards, Cardiff Castle.

36

German March Offensive, Advance to Mons, Afghanistan, Dakka
1918-1919

After the raid on Eleven Trees, The Queen's Bays remained at Buire and Brie until the commencement of the German offensive, which opened with tremendous force against General Gough's weak 5th Army early on the morning of the 21 March 1918. The night of the 20th/21st was very quiet; a dense fog that covered the whole front persisted until the afternoon, which greatly assisted the new German tactic of infiltrating storm troop units to push on as far as they were able, leaving strong points behind to be mopped up later. The German bombardment opened at 4.30 a.m., drenching the battle zone with shellfire and with long-range artillery searching out the advanced zone to a depth of twenty miles behind the front line. The infantry assault was not launched until 8 a.m., but by 1 p.m. on the 21st the Bays were moved up to a position of readiness at Bernes. A dismounted party, under Captain Barnard, MC, was sent to the assistance of the 24th Division, taking up a position east of Vendelles, where it came under heavy and continuous shellfire. On the morning of the 22nd the dismounted party became engaged with the German advance, until at 12 p.m. all the cavalry dismounted parties were ordered to rejoin their regiments, but not before Lieutenant Waddell had been killed, Lieutenants Gordon and Mitton wounded, and four other ranks killed, thirty wounded and one missing.

Meanwhile the main body of the regiment had been ordered back to Estrée-en-Chaussée and then to Athies. At 11 p.m. they were ordered to cover the crossings over the Somme between Brie and St Christ. The Bays were again withdrawn throughout the 23rd and 24th, when at night German aeroplanes bombed the horse lines, killing one man and wounding six. At 4.30 a.m. on the 25th a dismounted party under Lieutenant Todd was moved up to Montauban in support of a rapidly changing and fluid defence. Major Rome of the 11th Hussars took over command of the 1st Dismounted Brigade from Lieutenant Colonel Lawson of the Bays, who assumed command of the 1st Dismounted Division. Throughout the 25th the 1st Dismounted Division was heavily attacked, at the same time being

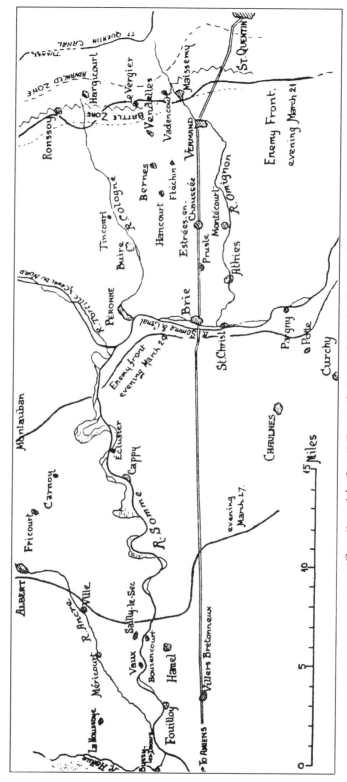

*Operations of the Queen's Bays in the Retreat before the German offensive
(21–30 March 1918)*

subjected to intense shelling and machine gun fire. Every attack was beaten off, but Lieutenant Todd was wounded, six other ranks were killed and twenty-four wounded. German pressure meant that part of the line of the Somme had to be given up, and at midnight the Bays formed the rearguard, withdrawing for seven miles on the 26th to Ville sur Ancre.

Early on the morning of 27 March the Bays took up a position on the Ancre at Mericourt, filling a gap in the line. They came under more shelling and had three men wounded. Late on the 27th the regiment was moved to Sailly le Sec, crossing the Somme to fill another gap in the line at Bouzencourt. By now the rapid German advance was threatening Albert, and to the north and south the Ancre had been crossed. The main southern thrust was directed towards Amiens, with the object of dividing the French and British armies. General Gough, commanding the hard-pressed 5th Army, organised a scratch force of reinforcements, made up of strag-glers, the staff of a machine gun school, clerks, labour corps and tunnelling com-panies, plus some Canadian and American engineers, all under the command of Brigadier General Carey and known as Carey's Force. On 28 March the Bays' positions at Bouzencourt were strengthened by the addition of members of the motley Carey's Force, whom the Bays proceeded to organise, while enemy artil-lery and snipers remained active all day. Throughout the 29th the Bays held the Bouzencourt position, sending a squadron under Captain Sutherland to the east of Hamel village. Lieutenant Colonel Lawson, Bays, resumed command of the 1st Dismounted Brigade, establishing his headquarters at Hamel. During the 29th desultory shelling wounded Lieutenant Benham and three other ranks.

On 30 March the Germans made a determined effort to break the improvised line, cobbled together to defend Amiens. At 10.30 a.m. they began a fierce bombardment of the whole of the front held by the 1st Cavalry Brigade, concentrating on the village of Hamel. At 12.30 p.m. the infantry assault com-menced, with the German storm troops attacking in waves. Each successive assault was repulsed and in front of Captain Single's squadron alone sixty-seven dead Germans were counted. Captain Single, MC, was commanding 'B' Squadron, together with detachments of Carey's Force, and as the German attack closed in on the position, some of the latter began to show signs of wavering. Single jumped up onto the parapet, encouraging them, successfully, to stand fast; but at the same time he fell, mortally wounded. At once he called out to Captain Sackville-West, who was near him, 'For God's sake, don't let these men go back.' He then propped himself up and went on exhorting the waverers. West wanted to bring assistance, but Single told him, 'Old boy, you are being attacked now, and I won't have myself on your hands. Now you must get off and stop them.' A prisoner captured during this attack revealed that the enemy had been ordered to capture Hamel at all costs, but by 4 p.m. the assault had failed all along the line of the 1st Cavalry Brigade. Single was evacuated to a casualty clearing station, but died of his wounds during the night. That day, in addition to Single, Lieutenant Paul was wounded, and seven other ranks were killed with twenty-three wounded, and two men posted as missing.[1, 2]

Easter Sunday, 31 March, was a comparatively quiet day, but snipers wounded five men. Colonel Lawson was now holding the line with the three dismounted cavalry regiments of the 1st Cavalry Brigade, reinforced by some 300 odds and ends from Carey's Force. The next day, 1 April, was also quiet, with only slight shelling during the afternoon; but, even so, five men were wounded and German snipers were again very active, killing Lieutenant Barclay. That night a reconnaissance patrol of 'C' Squadron, led by Captain Sutherland, surprised and rushed a German post, killing the garrison and bringing back a prisoner of the 228th Infantry and three light machine guns. On 2 April the 300 men of Carey's Force were withdrawn when the 18th and 19th Hussars took over their positions. At dusk on 3 April a raiding party of two Bays officers and twenty men set out to surprise a German machine gun post, but an enemy listening post gave the alarm and the raiders had to return under heavy machine gun fire. That night the Bays were relieved by infantry, and after a march of eight miles bivouacked at Bussy les Daours. The regiment had been fourteen days and nights countering the German offensive, and had lost over 25% of its effective strength: 2 officers had been killed and 6 wounded, while 20 other ranks had lost their lives, 111 had been wounded, and 3 had been posted as missing. The Queen's Bays were not to see action again until the final victorious advance began on 8 August 1918. On 17 April Lieutenant Colonel Lawson was promoted to command the 2nd Cavalry Brigade; he had commanded the Bays in the field since 1 April 1915, and Lieutenant Colonel Ing, DSO, took over from him. During June the adjutant, Acting Captain Whitmore Smith, Lieutenants Hamelin and Schnier received the Military Cross and RQMS Bird the DCM. During May the regiment was in billets at Estrée Blanche and Ray sur Authie, where it stayed until 17 July, when it moved to Authieule, setting out from there on 5 August to move up to the front.

During 5, 6 and 7 August the Bays marched to their rendezvous just behind the front line, ready to follow up the offensive to be launched by seven British, Canadian and Australian infantry divisions supported by 400 tanks. The artillery barrage opened at 4.20 a.m. on 8 August 1918 with the Bays moving forward from Longueau just outside Amiens and sheltering by 5.20 a.m. in some woods to the west of Villers Brettonneux. The regiment was scheduled to follow up the advance of a Canadian and an Australian infantry division and to keep in touch with the advancing troops. Mounted patrols were sent forward, two from 'A' Squadron

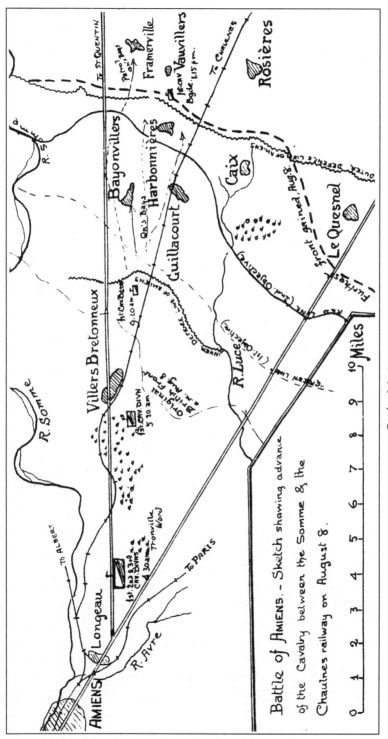

Battle of Amiens

Battle of AMIENS. – Sketch shewing advance of the Cavalry between the Somme & the Chaulnes railway on August 8.

under Lieutenant Solaini and Sergeant Mann, and two from 'B' Squadron under Lieutenant Cockrill and Sergeant Ford. By 6.20 a.m. the Canadians were across the St Quentin road and at 7.40 a.m. 'B' Squadron on the left under Captain Magnay and 'A' Squadron on the right under Captain Barnard advanced along the line of the Amiens-Péronne- St Quentin road.[1, 2]

'A' Squadron with 'C' in support followed up the attack on Harbonnières, capturing two trench mortars and some thirty prisoners. On coming over a ridge the squadron was fired upon, whereupon two troops charged the German position, killing twenty of the enemy and capturing twenty-six prisoners and two machine guns. The squadron rallied in the valley south of Harbonnières, coming upon a party of the enemy and cutting down seven and capturing two, as well as a machine gun and a Lewis gun. At 9.30 a.m. 'A' Squadron advanced again, only to be met by heavy machine gun fire. It was forced to retire back to the cover of the valley, but Lieutenant Solaini's troop on the right caught another German party, killing ten of them and capturing nine others, together with a waggon and eight horses. A second attempt to advance behind Whippet tanks was equally unsuccessful, with Captain Barnard and Lieutenants Solaini and Thomlinson being wounded, and again the squadron sought the cover of the valley.

Lieutenant Hannan now took over command of 'A' Squadron, holding the line of the railway and engaging the retreating Germans with rifle and Hotchkiss fire. Two waggons coming up the road were captured along with an officer, twelve of the enemy with the waggons being killed. 'C' Squadron now came forward in support, galloping over the crest south of Harbonnières and taking up a dismounted position on the flank of the advance, where they captured a lorry full of officers' kit trying to escape from Harbonnières. By 10 a.m. the Australian infantry had cleared the Germans out of Harbonnières, but as the enemy seemed to be massing for a counter-attack, the Bays remained dismounted in support of the Australians. The counter-attack did not develop but any attempt to advance was met by heavy fire, until the Germans eventually fell back to the east of their old trench line. The Bays then occupied the abandoned trenches in support of the Australians until 3.30 p.m. when they were withdrawn to Guillaucourt.[1, 2]

'B' Squadron on the left also had an exciting day. Lieutenant Cockrill's patrol somehow managed to get through the German rearguard and entered Bayonvillers, where they found the village full of startled Germans. As the patrol came into action a British armoured car entered the village from the Amiens road, and both mistook each other for enemies. Cockrill extricated his men from the confusion and pushed forwards towards Framerville. The rest of the squadron had been delayed by the number of field telephone wires lying hidden in the grass, but Captain Magnay pushed on through the infantry until the squadron came under heavy machine gun and rifle fire on reaching the line of the railway, suffering several casualties. The squadron dismounted and as they engaged the enemy, who were falling back, Lieutenant Carabine dashed out under heavy fire to bring in a wounded man. Once the Australians had cleared Bayonvillers, the squadron rejoined the rest of the regiment at Guillaucourt. Meanwhile Lieutenant Cockrill had come across a supply train hit by shells and captured its crew of three, one of whom was naked and badly burned. Lieutenant Cockrill's patrol then linked up with the other patrol under Sergeant Ford, and they both entered Framerville, only to be fired on by two armoured cars (almost certainly British) which killed a horse and wounded one man. The combined patrols then withdrew, under heavy fire, rejoining 'B' Squadron at 12.15 p.m., having taken two officers and seventy-seven Germans prisoner in their foray. Ludendorff was to write in his memoirs after the war, 'The 8th August, 1918 was a black day for the German arms.'

The night of 8/9 August at Guillaucourt was spent under long-range shellfire, but without any casualties. The next few days, until the battle was broken off on 12 August, were spent moving up behind the infantry waiting for the chance to exploit any success. On the 9th the Bays were bombed by German aircraft, sustaining some casualties, and on the 10th the regiment was strafed on the march, suffering nine men and some horses wounded. On 11 August the Bays were withdrawn to Longueau, near Amiens, where they remained until the 19th, when the 1st Cavalry Division was moved around to support a new offensive by General Byng's Third Army north west of Bapaume. The new offensive opened on 21 August with the Bays in reserve. They suffered some casualties from shellfire while watering their horses in Foncquevillers, then moved back to billets at Authieule. Over the next few days the regiment remained in support, waiting for a chance to operate and having one casualty by shellfire. The Bays stayed at Authieule, and then Bebreuve, until 31 August. On the 28th they heard that Lieutenant Colonel Ing had been awarded a bar to his DSO and that Captain Magnay and Lieutenants Cockrill and Solaini had won the Military Cross. Sergeants Spain and Ford and Corporal Bordman received the DCM, and Sergeant Elliott had a bar to his MM. Fifteen other Military Medals were also awarded. On 11 September the regiment heard that Lieutenant Carabine had won the Military Cross.

Throughout September 1918 The Queen's Bays were kept in reserve, until in October they came forward to follow up the advance. On 8 October the 1st Cavalry Brigade was operating in support of an American corps advancing over the Le Cateau heights, when 'C' Squadron was sent forward to keep in touch with the 11th Hussars. As it reached the outskirts of Prémont, it came under heavy shellfire, having one man killed and five wounded. The following day the regiment found itself near Le Cateau, when Captain Misa was wounded while on liaison duty. The advance continued on 10 October, but from the 11th the Bays remained at Maretz until the 25th, when they moved to Lesdain until 7 November. From the 7th to the 11th they marched through Cambrai, Auchy, Peruwelz to Basècles, fourteen miles from Mons. At 7 a.m. on 11 November 1918 The Queen's Bays advanced through a line of Allied infantry north west of Herchies. Shortly after 10 a.m. a patrol from 'C' Squadron came in contact with an Uhlan patrol two miles east of Montigny le Lens, and another patrol from 'B' Squadron encountered some more Uhlans at Masnuy St Pierre. Twelve prisoners were taken, without any Bays casualties. At 11 a.m. the order came to stand fast, and the sound of firing coming from the direction of Mons ceased. The war was over.

On 1 December The Queen's Bays crossed the border into Germany as the leading regiment of the 1st Cavalry Brigade, to be a part of the outpost line on the Rhine. On 17 December they moved into winter quarters at Elsdorf, some twenty miles west of Cologne. There they started to demobilise, and by the end of March 1919 they could only form a skeleton cadre.[1, 2]

The King's Dragoon Guards remained in India at Meerut until the end of October 1918, when they moved to Risalpur, changing stations with the 21st (Empress of India's) Lancers. On 2 May 1919 the Afghan army had occupied some wells across the border of India. The Afghan Amir Amanullah was warned to withdraw, but his answer was to send more regular troops to reinforce those at the wells and to move other regular Afghan units to various points on the frontier. On 6 May the KDG were ordered to mobilise; 17 officers and 360 other ranks marched throughout the night of the 7th/8th up the Grand Trunk Road, arriving in Peshawar at 6.45 a.m. The police had discovered that the Afghan postmaster at Peshawar was organising a rising in the city in conjunction with the invasion of India by the Afghans. At 1.30 p.m. on 8 May the KDG seized all sixteen gates of the city, establishing machine gun posts at each, and provided mounted patrols to cover the walls in between the gates. One man tried to escape over the walls, but a KDG soon caught him at lancepoint. For the next three days the regiment maintained its hold on Peshawar, while some of the leading Afghans and agitators were arrested.[3, 4, 5, 9]

On 11 May the 1st Cavalry Brigade, of which the KDG formed a part, received sudden orders to march to Jamrud Fort on the frontier. On the 13th they moved on to Landi Khana and then to Dakka, a village in Afghan territory, north west of the Khyber Pass. They arrived at 1.30 p.m. to find Dakka deserted. The Afghans had established a camp and advanced base at Dakka, which had been bombed on the 11th by the RAF, causing the Afghans to abandon everything and flee. On the15th 'B' Squadron under Captain Cooper went out to deal with some snipers and was

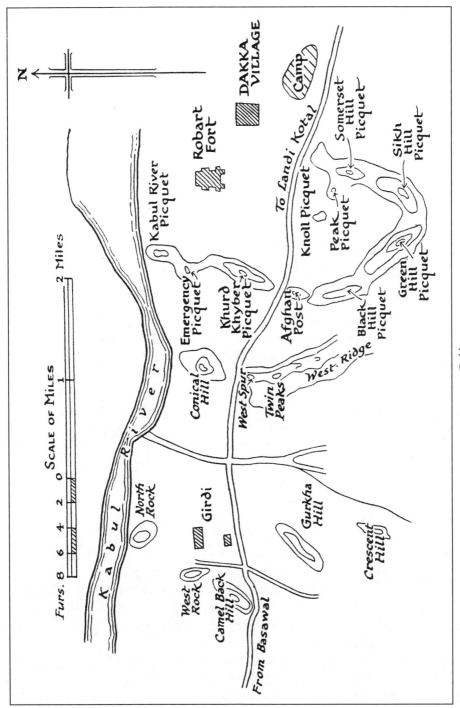

Dakka

fired upon. Lieutenant Card was slightly wounded but the accompanying guns of the RHA soon dispersed the tribesmen.[4, 5, 6, 9]

At 5 a.m. on 16 May 'B', 'C' and 'D' Squadrons moved out to reconnoitre towards Basawal. Two troops under Lieutenants Card and Jacques examined Robart Fort and Sherabad Cantonment and found them clear of the enemy. The Khurd Khyber Pass was cleared by the infantry and, with the KDG acting as advance guard, moved into the plain beyond. Two troops under Lieutenants Ward and Bogle Smith occupied a prominent conical hill as about a hundred of the enemy retired. While the rest of the regiment closed up, these two leading troops gave chase, galloping for about 1,000 yards but unable to come to grips because of the bad going. They therefore dismounted and dispersed the enemy with rifle fire. As the Afghans took refuge in Girdi village, the two troops remained as left flank guard while 'D' Squadron under Captain Wilson secured Girdi. The 15th Sikhs then advanced to clear the hills beyond, but it at once became clear that the Afghans had assembled in large numbers and resistance was stiffening. Supported by machine guns and artillery, a strong Afghan force started to work around the British left flank. A withdrawal to camp was ordered, and immediately the Afghans came on with great boldness and rapidity.[4, 5, 6, 7, 9]

As the 15th Sikhs and KDG retired by bounds, the Afghan rifle fire, augmented by two guns, began to cause casualties. This, in the need to take back the wounded, slowed down the retirement. Many of the Sikhs were evacuated on the KDG horses, and all the wounded were got back safely. As the force retired through the Khurd Khyber Pass, with 'B' and 'C' Squadrons KDG covering the withdrawal, the Sikh company picketing the heights was pulled in. 'D' Squadron then held the pass, retiring through 'B' and 'C' who had leapfrogged back to cover the exit to the pass. As the squadrons debouched onto the plain by Robart Fort, they came under heavy fire from the so-called 'friendly' Mohmand village of Lalpura, and this caused a number of casualties. While two troops under Lieutenant Parker, and another under Lieutenant Barrett, took up positions at Sherabad Cantonment, to cover the guns of the RHA, the remainder of the squadrons formed up south of Robart Fort under Captain Cooper. The enemy was now advancing in heavy numbers across the open plain. Captain Cooper, forming 'B' and 'C' Squadrons into line, gave the order to charge. The two squadrons rode through the Afghans, scattering them and doing severe execution. Captain Cooper, in the lead, was badly wounded in the shoulder. To the rear, Captain Wilson quickly rallied 'D' Squadron, ready to reinforce Cooper, and then covered his withdrawal. The charge had halted the enemy and gained the time needed for the infantry to pull back into camp. [3, 4, 5, 6, 7, 9]

During the early part of the action Lieutenant Ward, having moved his troop, had reported to Captain Cooper. On leaving, he failed to rejoin his troop and was next seen being brought in, severely wounded, by his orderly, Lance Corporal Sheppard, who had placed him on his own horse, as Ward's had been killed under him. Ward died of his wounds, and Sheppard was himself so badly hurt that he also succumbed. It is presumed that they were caught on their way back to their troop.[4, 6, 7]

It was not until the troops left in the camp at Dakka saw the KDG charge that they realised the proximity of the enemy. 'A' Squadron under Captain Hadfield immediately saddled up, taking up position outside the camp, until by 1 p.m. the whole force was back within the perimeter. The Afghans then occupied the high ground overlooking the camp and opened up with rifles and artillery. At 9 p.m. that night an attack was launched from the south west in an attempt to storm the camp, but it was beaten off by the troops manning the perimeter. The following day the battle recommenced, with the KDG remaining in reserve, their horses being held in a nullah to give some protection against rifles and shellfire, which caused some casualties. On the arrival of five more battalions of infantry, together with a battery of howitzers, from Landi Khana, the heights were stormed and cleared and the Afghans retreated. During 16 and 17 May the KDG had one officer killed and one wounded; three other ranks were killed, five died of wounds and twenty-five were wounded. Twenty-two horses had been killed and twenty-eight wounded. For this action Captain Cooper was awarded the DSO and Lieutenants Card and Waggett the MC, with Sergeant Browning and Lance Corporal Sheppard receiving the DCM, and Corporal Bell the Military Medal.[3, 4, 5, 6, 7, 9]

From 18 May, and onwards throughout June and July, active patrolling was carried out on a daily basis by the regiment, with contact being constantly established with the enemy. There was constant sniping and from time to time patrols were fired at and returned the fire; but the regiment was not engaged in any major action, although the hills around Dakka were teeming with the enemy. On 19 June a squadron of the KDG helped to lay an ambush along with a company of the 1/9th Gurkha Rifles. The KDG rode towards the enemy and then feigned a retirement. As the cavalry pulled back, the enemy at once started to follow up, only to be caught by two Gurkha platoons in hiding, who killed ten to fifteen tribesmen before they were able to get back into cover. The Third Afghan War had become a typical frontier operation, requiring constant vigilance against a crafty enemy, who would take immediate advantage of any carelessness. On the night of the 23/24 June, which was particularly dark, a party of tribesmen crawled into the camp and quietly cut the ropes holding a large KDG tent, which was used as a coffee shop and had another tent of similar size attached to it. As they rolled up the first tent, they realised they had not been spotted, so they proceeded to remove the second tent as well, working so quietly that the sentry did not notice until they were dragging their booty away. He then raised the alarm, but not before the tribesmen had managed to disappear with their prizes.

On 8 August 1919 the peace treaty with Afghanistan was officially signed, and the following day, in true frontier style, 'A' Squadron, out on patrol, had a brush with fifty of the enemy and had one trooper wounded. On 12 August three squadrons turned out to chase some raiders who had cut and stolen telegraph wire. Three men carrying wire were seen in the Khyber Pass; one was killed, the other two got away. On 25 August 1919 the KDG left Dakka and marched by easy stages back to their station at Risalpur, arriving on 28 August.[3, 4, 5, 9]

SOURCES

1 F. Whyte and A. H. Atteridge, *The Queen's Bays, 1685-1929*, 1930.

2 War Diary, 1918, 2nd Dragoon Guards, Regimental Museum, 1st The Queen's Dragoon Guards, Cardiff Castle.

3 W. Clowes, *The King's Dragoon Guards, 1685-1920*, 1920.

4 War Diary, 1919, 1st King's Dragoon Guards, Regimental Museum, 1st The Queen's Dragoon Guards, Cardiff Castle.

5 *The Third Afghan War, 1919.* Official Account, Government of India, 1926.

6 *The K. D. G.*, vol. 1, p. 54.

7 'Action at Dakka, May 16 & 17, 1919', *The K. D. G.*, vol. 4, no. 8, p. 341.

8 E. Belfield, *The Queen's Dragoon Guards*, 1978.

9 *The Cavalry Journal*, vol. 10, 1920, p. 446.

37

Moplahs, Iraq, Germany, England, Egypt, India
1919-1939

The Queen's Bays remained at Elsdorf on the Rhine until March 1919, when the regiment returned home. One of that number relates, 'At Antwerp, late in the evening of the 29th March, 1919, there embarked aboard the ex-Russian steamer *Mogileff*, bound for Southampton, which was reached at 8 a.m. on the 31st, a bare handful of cavalrymen in full war equipment – the remnants of those who had departed in 1914.' The Bays were some sixty strong. Despite the hour of their eventual arrival at York, Colonel G. H. Ing, CMG, DSO, who had preceded the regiment from Elsdorf, was awaiting them at the railway station. The weeks slipped by rapidly, and the re-forming of the Bays was going on apace. The strength, by the middle of June, had increased to 19 officers and 499 other ranks. Then came a bolt from the blue. 'The Regiment will embark at Liverpool for Palestine on 24th June 1919.'

The Bays landed at Port Said on 6 July, moving to the base at El Kantara, where they were remounted from some of the 50,000 horses then at the Remount Depot. A short journey in two steamers took them to Beirut, from whence they marched to the railhead at Rayak, then went by train to their camp at Aleppo, which they reached on 3 August. They stayed at Aleppo until, with the French assuming control of Syria, they began an epic march of 500 miles, starting out on 12 November and arriving at Sarona near Jaffa on 13 December. 'The Bays arrived at dusk to find that their cosy camp consisted of one wretched marquee standing on the side of a slope under pouring rain.' Whilst at Sarona the wives and children joined the regiment from England. 'The jackals howled around at night, and those of the wives and children who experienced the summer of 1920 in Sarona have vivid recollections of their sleepless nights there.' One squadron was detached for a month to provide an escort for the High Commissioner when he paid a visit to Es Salt in Jordan. On 7 November 1919 the Bays entrained for Port Suez, embarking for India, and arriving in Bangalore in April 1921 with a strength of 29 officers and 562 other ranks.[1, 2]

During 1921 there was a rebellion of the Moplahs on the Malabar coast. The

Moplahs were the descendants of the Arabs, who had for centuries traded along the Malabar coast and then settled, retaining their Arab customs and Muslim religion. A number of Moplah officers and men from a disbanded Moplah light infantry regiment stirred up trouble and aggravated grievances which always lay just under the surface of a Muslim minority living in a largely Hindu area. The disaffection started with rioting, which developed into guerrilla warfare in the hilly area adjoining the coast. Then the trouble suddenly flared up at Calicut, after a British police officer had been killed, and the town was only saved by the cool action of its garrison, a company of the Leinster Regiment.[1] A movable column was formed consisting of the 2nd Dorsets and a detachment of the Bays with some supporting arms. The Bays detachment, commanded by Major Stone, was made up of one troop of 'A' Squadron and three troops and the Hotchkiss troop from 'B'. On 20 August the column set out from Bangalore by train, getting as far as Shoranur Station, where it was found that the line to Calicut had been torn up by the Moplahs. The troop from 'A' Squadron was sent on to Calicut, while the rest of the Bays reconnoitred the surrounding area, patrolled the railway line and provided escorts for convoys. The rebels were constantly seen in small and large bands, but they never attacked and disappeared into the jungle when chased. The Bays were never in action, but on two occasions, when going to the relief of a besieged village and later to the assistance of a hard-pressed company of the Dorsets, the mere sight of the cavalry led to the retreat of the rebels. Most of the area was dense jungle, however, and impracticable for cavalry, and in September the Bays detachment returned to Bangalore. Three NCOs were left behind to help with convoys and pack mules, and did not rejoin the Bays until 1922; one of them, Corporal Archdale, was mentioned in despatches for his conduct in rallying native drivers who had panicked when fired upon by the Moplahs.

In January 1922 the Bays provided an escort for the Prince of Wales when he visited Bangalore. In July Lieutenant General Sir Hew Dalrymple Fanshawe was appointed as colonel of the regiment to succeed General Sir William Seymour, who had died. The Bays left Bangalore in February 1924 to relieve the 3rd/6th Dragoon Guards at Sialkot. Lieutenant Colonel Osborne took over command in September 1924 from Lieutenant Colonel Rome. While at Sialkot 7 officers and 110 other ranks were awarded the India General Service Medal with the clasp 'Malabar'. During 1924 The Queen's Bays were awarded the following battle honours for their services in the 1914-18 war: 'Mons; Le Cateau; Retreat from Mons; Marne, 1914; Aisne, 1914; Messines, 1914; Armentières, 1914; Ypres, 1914-15; Frezenberg; Bellewaarde; Somme, 1916-18; Flers-Courcellette; Arras, 1917; Scarpe, 1917; Cambrai, 1917-18; St Quentin; Bapaume, 1918; Rosiers; Amiens; Hindenburg Line; St Quentin Canal; Beaurevoir; Pursuit to Mons; France and Flanders, 1914-18.' The

Bays left Sialkot on 2 December 1926, embarking at Bombay for England in the *City of Marseilles* on 1 January 1927. On arrival they were stationed at Colchester, where they remained until 31 March 1928.[1]

While The King's Dragoon Guards were at Risalpur, Lieutenant Colonel Williams resumed command in October 1919 during a period when many KDG were being demobilised, their release having been deferred owing to the Afghan War. On 14 December 1919 a draft of seventy-four other ranks joined from England, but even so, when on 10 January 1920 a much reduced regiment left Risalpur for Karachi, embarking on the SS *Coconada*, it had a strength of only 9 officers and 136 other ranks. Arriving on 26 January at Basra in Iraq, or Mesopotamia as it was then called, the main body was met by the advance party of two officers and four other ranks, and went into camp at Makina. On 30 January the KDG embarked on a river steamer and were carried up the Tigris to Kut, from where they continued by train, reaching their tented camp at Hinaidi, Baghdad, on 4 February. Four days later another draft of fifty-eight men arrived from England, and by the end of May the regiment's strength had been built up to 13 officers and 283 other ranks. On 21 April the KDG were taken by train into Persia to a summer camp at Karind, but on 12 July an order came to return at once to Baghdad as an Arab rebellion had broken out. [3, 4, 5]

The KDG arrived back in Baghdad on 16 July, having dropped off an officer and the Hotchkiss machine guns at Quaritu. Throughout the rest of July and until mid-October strong patrols of squadron strength constantly operated around Baghdad. On 10 August three officers and fifty-one men entrained for Baqubah to carry out a punitive expedition against the rebel areas. A village sixteen miles from Baqubah was destroyed as punishment, when a rearguard action had to be fought as the force returned to Hinaidi, the only casualties being four horses wounded. On 5 November 1920 the KDG entrained for Kut and Basra, embarking on the transport *Rio-Negro* and arriving at Southampton on the 28 December with a strength of 16 officers and 146 other ranks. They proceeded to Redford Cavalry Barracks in Edinburgh. By the end of January 1921 the regiment had built up its numbers to 24 officers and 490 other ranks.[3, 4, 5, 6]

The spring of 1921 in Britain was marked by industrial unrest. When on 4 April general mobilisation was ordered to help deal with the coal strike, 231 reservists rejoined and an armoured car section was brought under command. For six weeks the KDG had a strength of 31 officers with another 16 attached, and 670 other ranks with 36 more attached. However, the only duty for which the regiment was called upon was for the regimental musical ride and vaulting team to perform in front of the Crown Prince of Japan, and to provide the royal visitor with mounted escorts. Demobilisation took place on 1 June, although the KDG were called upon to send three officers to Ireland to assist in the administration of martial law during that unhappy period. On 14 June 1922 Lieutenant Colonel Little from the 20th Hussars took over command. In September the Maidstone coroner commented on the gallant conduct of Corporal Clifford, KDG, who, whilst on leave, rescued a child and put out a fire at Eccles in Kent. The regiment remained in Edinburgh until October 1923 when it moved to the British Army on the Rhine and was stationed in Deutz Cavalry Barracks at Cologne.[6, 7]

From 1921 to 1929 the KDG musical ride was much admired and in demand at horse shows in Scotland, Germany and the south of England. The responsibility for excellence rested with the rough riding sergeant major, a position held by two famous men during this period, SSMs Page and Crotty. The ride consisted of either twenty-four or thirty-two men, supported by the mounted kettledrummer and four mounted trumpeters, with the dismounted band providing the music. The ride was always performed in the 1914 full dress, entering the arena at the walk, working up into a trot, then canter, until the final charge at the gallop. [8]

During 1923 the KDG were awarded their battle honours for the 1914-18 war: 'Somme, 1916; Morval; France and Flanders, 1914-17; Afghanistan, 1919.' In May 1924 a move was made to Kronprinz Barracks, Kalk, Cologne. Command of the regiment had been taken over on 13 January 1925 by Lieutenant Colonel Chappell, and on the 20th the regiment moved to Wiesbaden. On 20 August 1926 Colonel Chappell accepted a new standard from Lieutenant General J. P. DuCane at an impressive mounted parade at Dotzheim, Wiesbaden. In December the KDG left Germany for Aldershot, taking over Beaumont Barracks from the Royal Dragoons. During 1927 the organisation of the regiment was changed from three sabre squadrons to two sabre squadrons and one machine gun squadron, the latter being carried in motor lorries. In July 1927 the Corps of Dragoons, Hussars and Lancers were designated 'Cavalry of the Line'. King George V visited Aldershot during May 1928 and the KDG provided a mounted standard bearer and two orderlies for the visit. His Majesty inspected the regiment at drill on the Long Valley. In January 1929 Lieutenant Colonel Howes assumed command, a move to Tidworth being made at the end of September. The regiment was 'met by the mounted band of The Queen's Bays, who played them into Assaye Barracks, their home for the next two years. This was not the only effort made by The Queen's Bays to give us a hearty welcome to Tidworth, as they provided parties of men to off-saddle and groom our horses on arrival, which was extremely friendly and an act worthy of the Bays.'[7, 8, 9]

In April 1931 the KDG monogram was taken into use as the arm badge to be worn by NCOs above their chevron stripes of rank, and this replaced the star and Garter badge, which had been worn since 1915 when the Austrian double-headed eagle had been removed. During the course of 1930 the Canadian Cavalry Regiment, the Mississauga Horse, based at Toronto, was officially affiliated to The King's Dragoon Guards. To mark the occasion the Canadians ordered one of their officers to shoot a moose, and its head, duly mounted, was presented in 1931 to the officers' mess. At the same time the KDG were affiliated to the Australian 1st/21st Light Horse in Sydney. After nearly three very happy years at Tidworth the KDG moved to Hounslow on 3 October 1931, and four days after their arrival the sad news came of the death of Captain W. H. Muir as the result of a riding accident, at the age of thirty-six. Captain Muir had achieved fame as a show jumper on his celebrated horse Sea Count, and had carried off numerous trophies, including the Aga Khan Cup, the King's Cup and the Coronation Cup. For five years he had been a member of the British team at Olympia and of the KDG team that in 1926 won the International Team Jumping at Strasburg, as well as winning many other trophies. On 1 October 1932 the regiment embarked on the *Dorsetshire* for passage to Egypt, arriving at Abbassia, Cairo, at the end of the month and relieving the 17/21st Lancers. [10, 11, 12]

Early in 1928 The Queen's Bays at Colchester learnt that the 7th Australian Light Horse had been affiliated; the regiment was already affiliated to the Canadian Militia Cavalry Regiment, the Governor General's Bodyguard of Toronto. On 24 September 1928 Lieutenant Colonel Heydeman assumed command and the Bays moved to Tidworth on 22 October. At the end of its first year at Tidworth the general's inspection report read: 'The Regiment has done very well and it is very well commanded and officered. It is particularly quick and good in the field and smart in barracks. The horses are of a very good stamp and in very good condition. A Regiment which it is a pleasure to have under one's command.' With the reorganisation of the cavalry, Major Evelyn Fanshawe on returning from an equitation course at Weedon was given command of the mechanised machine gun squadron, and Captain Draffen, on his return from an advanced mechanical course at Woolwich, was made equitation officer. Lieutenant General Sir Hew Fanshawe handed over as colonel of the Bays, a post that he had held from 1921, and was succeeded by Lieutenant General Sir Wentworth Harman.

On 29 September 1931 The Queen's Bays moved to Shorncliffe, and on 24 September the following year Lieutenant Colonel Heydeman handed over command to Lieutenant Colonel Wootten. During 1931, for the first time in the regiment's history, the Bays won the inter-regimental polo tournament, with a team composed of Captains Draffen, Barclay, G. H. Fanshawe and E. D. Fanshawe. George Fanshawe was also Champion at Arms in the Military Tournament that year, and Evelyn Fanshawe was first in two of the individual skill at arms championships. These successes were to continue, for in the five years from 1932 to

1936 the Bays won the Champion at Arms each year, and carried off no less than sixteen first prizes. Lance Corporal Palandri was Champion at Arms Mounted in 1933, 1934, 1935, 1936 and 1937. The regimental diary records: 'The following telegram was received from the 1st King's Dragoon Guards [In Egypt], in connection with the successes obtained by the Regiment at the Royal Tournament, Olympia in May, 1933:— "Hearty congratulations Olympia successes K. D. Gds." '

At the end of 1933 the regiment moved to join the 1st Cavalry Brigade at Aldershot, going into Willems Barracks. During the Silver Jubilee celebrations of 1935, The Queen's Bays took part in the review and march past before King George V and celebrated the 250th anniversary of the raising of the regiment. On 3 October Lieutenant Colonel Evelyn Fanshawe took over command of the regiment and led it out on brigade training, followed by army manoeuvres for the last time as a horsed cavalry regiment. The Inspector General of Cavalry had told all cavalry colonels that he could no longer see any future for horsed cavalry as such, and so he was asking that the bulk be converted to armour. In October 1936 The Queen's Bays started the sad process of conversion to a light tank cavalry regiment, after 252 years as horse soldiers. Some of the Bays horses were transferred to other regiments, but a large number were destroyed or sold. By March 1937 the regiment was reduced to 40 chargers and 53 troop horses, and had received in exchange 10 light tanks, 20 tractors, 6 motorcycles and 15 15 cwt Morris trucks. It was a slow process: cavalrymen had to learn how to become fitters, drivers and gunners. By June 1937, however, sufficient progress had been made, in spite of the shortage of equipment, for a troop to demonstrate an attack in cooperation with the Royal Horse Artillery before Prince Chichibu of Japan.

On 12 May 1937 seven officers and 110 other ranks represented the regiment at the Coronation of King George VI, including a horsed troop under Second Lieutenant Weld, which took part in the procession. On 14 May the King was graciously pleased to appoint Her Majesty Queen Elizabeth to be the colonel in chief of The Queen's Bays (2nd Dragoon Guards). On 24 July the regiment was visited by its new colonel in chief, who expressed the same charm and interest that Her Majesty has shown ever since. The Bays moved to Tidworth on 8 October 1937, joining the 2nd Light Armoured Brigade, together with the 9th Lancers and the 10th Hussars. Seven years later an officer wrote, 'The three regiments were destined to fight side by side in many battles throughout the years ahead, and never has there been greater comradeship between regiments. The 2nd Armoured Brigade was from first to last a happy family, and as a result its war record was second to none.'

On 2 January 1939 Lieutenant Colonel Fanshawe handed over command to Lieutenant Colonel Beddington, and by the spring of 1939 enough equipment had arrived for three fully equipped and manned squadrons to start collective training. The Bays were warned that in the trooping season of 1939-40 they would be going to Palestine; then in the mounting tension in Europe reservists started to be called up in July for training. On the 29th the colonel in chief visited her regiment to present a new standard. The old standard was laid up in St Michael's Church at Tidworth on 25 August, as the last few days of peace faded away.[13, 14]

The King's Dragoon Guards, having arrived in Egypt in 1932, were called upon to provide a number of ceremonial parades as part of the cavalry brigade, and for the King's Birthday and Accession Day at the Gezira Sporting Club, and these became annual events. In January 1933 Lieutenant Colonel Sprot took over command of the regiment, and an exercise was carried out in the desert near Mena, when an early experiment in Army/Air Force cooperation was carried out: a landing strip was cleared and RAF transport planes flew in water for the horses and men. The colonel of the regiment, Lieutenant General Sir Charles Briggs, visited Abbassia in February 1935 and four days of inspections and exercises were arranged. Garrison life provided its round of soldiering, sport and amusement. One local employee who had been dismissed wrote to his erstwhile employer:

> Kind Sir,
>   In opening this epistle you will behold the work of a dejobbed person and a very bewildered and much childrenised Gentleman who was violently dejobbed in a twinkling by your good self. For Heaven's sake, Sir, consider the catastrophe as falling on your own head remind yourself as walking home at the Moon's end of five savage wives and sixteen voracious children with your pockets filled with non-existent £.s.d. not a solitary sixpence. As to reason given by yourself, Esquire, for my dejobment, incrimination was laziness. No, Sir, it was impossible that it was myself whom has pitched sixteen infant children into this world can have a lazy atom in his mortal frame, and the sudden departure of £11 monthly has left me on the verge of the abyss of destitution and despair. I hope this vision of horror will enrich your dreams this night and the good Angel will meet and pulverise your heart of nether milestones so that you will awaken and with alacrity hasten to rejobulate your servant.
>   So mote it be. Amen. Yours despairfully, ANANIAS SOLOMON[11,15]

In 1935, the 250th anniversary of the raising of The King's Dragoon Guards, there was a move to India. The journal recorded, 'Despite wars and rumours of war, the Regiment left Abbassia at 9.30 a.m. on 2nd October 1935. We should have left on the previous Sunday, but as a result of the war scare in the Mediterranean, HT *Nevasa* was delayed two days at Gibraltar.' Mussolini, the Italian dictator, had invaded Abyssinia, a crisis that was soon to be followed by crisis after crisis in rapid succession. On arrival at Trimulgherry in the Deccan on 14 October, the KDG were reorganised back to the Indian establishment, which consisted of regimental headquarters and three sabre squadrons, each squadron being of four troops, bringing the regiment up to 27 officers and 564 other ranks, with 54 officers' chargers, 532 riding horses and 41 pack horses. On 1 April 1936 news came through from Headquarters, India that The King's Dragoon Guards were to return home at the end of 1937 and would be converted to a motor cavalry regiment. Lieutenant Colonel Sprot retired during the year and was succeeded by Lieutenant Colonel Tiarks. On arrival in India the regiment applied to the War Office for permission to restore to the KDG the Habsburg double-headed eagle, in place of the star and Garter badge, which the regiment had been ordered to wear since 1915 when Britain had been at war with Austria. Permission was granted on 18 September 1937 and the officers, RSM and bandmaster wore the double-headed eagle as from 2 February 1938, while the rest of the regiment put up the new badge as soon as supplies became available. On 23 October 1937 a sad event took place when The King's Dragoon Guards held their last mounted parade at Secunderabad for inspection by Major General Nicholson, commander of the Deccan District. The KDG embarked at Bombay in the *Somersetshire* on 13 November 1937, bound for England, and arrived at Southampton on 6 December. The regiment had been ordered to leave behind in India over 100 men for cross-posting to other cavalry regiments, and many of those who accompanied the regiment home were 'time expired', so that on reaching Aldershot the strength had been reduced to 25 officers and 249 other ranks. The regimental diary records, 'Training for mechanisation commenced, greatly handicapped by the shortage of men and vehicles.'[11, 16, 17, 18]

December 1938 saw the establishment of the KDG as a light tank regiment with fifty-eight light tanks. However, for the whole of 1938 the regiment was the proud possessor of four obsolete Mark II light tanks and one contemporary Mark VI, but by July 1938 the situation had been transformed and the KDG had fifty-three Mark VI light tanks. During 1936 the Canadian Militia was being reorganised, and the Mississauga Horse and the Governor General's Bodyguard, as the two Toronto cavalry regiments, started to work closely together. On 13 December 1936 the two regiments were amalgamated to become the Governor General's Horse Guards. The new regiment retained its alliance with both The King's Dragoon Guards and with The Queen's Bays, and this close connection continued with the amalgamation of our two regiments in 1959 until the present day. In July 1938 the KDG received the first sixty 'Militiamen', together with an increasing number of regular reservists recalled to the colours.[11, 18, 19]

Both The King's Dragoon Guards and The Queen's Bays lost their ranking as cavalry of the line in 1939, and became the two senior regiments of the newly formed Royal Armoured Corps. The other two regiments which normally took precedence over the KDG and the Bays were the Life Guards and the Royal Horse Guards, but at this time the two latter were still horsed cavalry. The King's Dragoon Guards, when first embodied, had been ranked as the second regiment of horse in the English service, and in numbers, rates of pay at that time, and all other particulars, it was placed upon an equality with the Royal Regiment of Horse Guards. Both the KDG and the Bays during 1939 in England played hard at sport, but also prepared hard for the war that seemed to many to be inevitable.[20]

SOURCES

1 F. Whyte and A. H. Atteridge, *The Queen's Bays, 1685-1929*, 1930.

2 War Diary, 1919, 2nd Dragoon Guards, Regimental Museum, 1st The Queen's Dragoon Guards, Cardiff Castle.

3 W. Clowes, *The King's Dragoon Guards, 1685-1920*, 1920.

4 War Diary, 1919, 1st King's Dragoon Guards, Regimental Museum, 1st The Queen's Dragoon Guards, Cardiff Castle.

5 War Diary, 1920, 1st King's Dragoon Guards, Regimental Museum, 1st The Queen's Dragoon Guards, Cardiff Castle.

6 War Diary, 1921, 1st King's Dragoon Guards, Regimental Museum, 1st The Queen's Dragoon Guards, Cardiff Castle.

7 War Diary, 1923-1929, 1st King's Dragoon Guards, Regimental Museum, 1st The Queen's Dragoon Guards, Cardiff Castle.

8 *The K. D. G.*, vol. 3, no. 7, 1947, p. 9.

9 *The K. D. G.*, vol. 1, no. 1, 1930, p. 3.

10 *The K. D. G.*, vol. 1, no. 5, 1932, p. 168.

11 War Diary, 1930-1938, 1st King's Dragoon Guards, Regimental Museum, 1st The Queen's Dragoon Guards, Cardiff Castle.

12 *The K. D. G.*, vol. 1, no. 2, 1930, p. 42.

13 War Diary, 1927-1932, 2nd Dragoon Guards, Regimental Museum, 1st The Queen's Dragoon Guards, Cardiff Castle.

14 W. Beddington, *The Queen's Bays, 1929-1945*, 1954.

15 *The K. D. G.*, vol. 1, no. 7, 1933, p. 243.

16 *The K. D. G.*, vol. 2, no. 5, 1936.

17 *The K. D. G.*, vol. 2, no. 6, 1936.

18 *The K. D. G.*, vol. 2, no. 8, 1937.

19 *The K. D. G.*, vol. 3, no. 2, 1938, p. 6.
20 *The K. D. G.*, vol. 3, no. 3, 1939, pp. 45, 59.

38

England, France, Somme, Egypt, Beda Fomm
1940-1941

The King's Dragoon Guards and The Queen's Bays found themselves at the receiving end of hundreds of cavalry reservists as the orders for general mobilisation, issued on 1 September 1939, brought men from many cavalry regiments back to Aldershot and Tidworth. At the same time the two regiments provided soldiers for various headquarters and training regiments. With the outbreak of war on 3 September, both regiments spent the first months settling down, sorting out, and engaging in individual training of the reservists and recruits.[1, 2, 3, 4]

On 11 September the KDG moved to billets in the Wimborne area of Dorset. Lieutenant Colonel Tiarks left to become second in command of 7th Armoured Brigade in Egypt, his place being taken by Lieutenant Colonel Paton. At the beginning of November the regiment moved to the vicinity of York, and was billeted around Stamford Bridge, as part of 1st Light Armoured Brigade of 2nd Armoured Division. At the end of October a parachute scare had brought orders to the Bays to move to the East Coast. On 2 November Lieutenant Colonel Beddington was posted to a staff appointment and Lieutenant Colonel George Fanshawe took over command. The regiment moved on 3 October to three villages near Cambridge as a part of 2nd Light Armoured Brigade of the 1st Armoured Division. Both regiments trained during a very cold and hard winter, the Bays having received the new Mark VIC light tank, which arrived without its armament of a 7.92 mm and a 15 mm Besa, so that a sheet of 3-ply wood covered the mounting where the guns should have been. In December the K.D.G. heard that Major Crossley had been awarded the OBE for his services in Palestine when on secondment to the Transjordan Frontier Force. But these winter months of 1939-40 were the period of the 'Phoney War', when the ser..e of urgency that was to come later was sadly lacking.[1, 2, 3, 4]

It had been decided to equip the 2nd Armoured Brigade, to which the Bays belonged, with cruiser tanks, as and when they became available, and to abolish the Mark VI light tanks. The first A.9 cruiser tanks arrived in April, armed with a 2-pounder gun and a .303 co-axially mounted Vickers machine gun. In the meantime the KDG had been ordered to hand over to the 3rd Hussars the best of their Mark VI light tanks, as the 3rd were under orders to go to Norway. Lieutenant Lord Clifton, KDG, became the first officer chosen to be attached for battle experience to a cavalry regiment serving in France; unhappily, he was captured during the retreat to Dunkirk, when commanding a troop of the 15/19th Hussars. On 3 May 1940 The Queen's Bays received orders to proceed overseas. Two different types of cruiser tank started to arrive, the A.10 and the A.13, but neither in sufficient numbers to make up the regiment's strength, and they came without any tools or spare parts. Nine Mark VICs still had to be taken, their Besa guns arriving as the Bays were about to sail – still nailed down in their packing cases, and the first sighting the gunners had so far had of them. Before sailing on 14 May the Bays were inspected, first by King George VI, and then immediately afterwards by their colonel in chief, Queen Elizabeth, a double event that must be unique in the annals of the British Army.[1, 2, 3, 4]

The Bays embarked at Southampton on 19 May 1940. By then the German offensive had changed the situation. Instead of going to a training area in France, the 1st Armoured Division was ordered to concentrate west of Brussels. Before the division could arrive, the Germans were reported at Cambrai, a few miles east of the division's disembarkation port at Le Havre. It was also found out that Le Havre had been mined, so the convoy was diverted to Cherbourg, where the Bays arrived, disembarking on 20 May. The tanks left by rail on 21 May, with the crews mounting and assembling the guns en route, arriving at the small station of Brevai forty miles south of Rouen on the 22nd. Unloading had to be carried out with ramps improvised by the ingenuity of the crews. The tanks moved with difficulty through the mass of refugees crowding the roads to Aigleville, eight miles north, where they were joined by the wheeled transport which had come by road.[1, 3]

By now the enemy had taken Amiens and Abbeville, and had leading elements across the Somme, approaching Rouen from the east. The Bays were ordered to move to the Forêt de Lyons east of Rouen to destroy any enemy encountered and to deny the approaches to the Seine. Having reached their positions, the Bays, with the 4th Border Regiment carried in lorries, received urgent orders to seize crossings over the Somme, cut off the advanced German elements, and so relieve the pressure against the right of the retreating BEF. Captain Barclay was sent ahead to reconnoitre the Somme crossings, which were said to be held by the French. On arrival there were no French troops to be seen, but plenty of German motorised patrols; a patrol on foot revealed that the whole of the north bank of the river was held by the Germans. Captain Keyworth, who had become separated from the regiment in the dark, joined a small group of scout cars from brigade and ran into a German road block by the Somme. Keyworth and his other leading tank charged

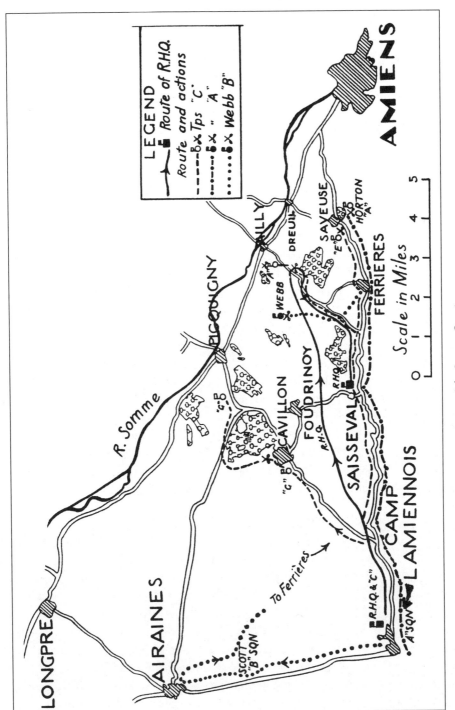

Action of the Somme Crossings

the block, both being blown up on mines; Keyworth and his gunner got out and were taken prisoner.

On 24 May Major Asquith, commanding 'C' Squadron, was ordered to seize the bridges at Dreuil, Ailly and Picquigny, using three troops, with the 4th Borders and 'B' Squadron in support. On the right Lieutenant Viscount Erleigh of 2 Troop, with only two Mark VIC tanks, advanced on Dreuil, and when a mile from the objective, his leading tank was fired at by an anti-tank gun. Returning the fire, the tank spun round and withdrew. A machine gun then opened up on Erleigh's tank from a water tower. It was silenced for a time with fire from the Besa, but it soon reopened on the Border infantry, who suffered casualties. Small arms fire then started from various directions, with a fire fight ensuing, but no progress could be made against the dug in Germans. Erleigh's two tanks then attempted an outflanking movement. Sergeant Bunn on the left met no opposition; the Border infantry in the centre were involved in a mêlée with bayonets, rifles and machine guns; Erleigh on the right was blocked by a wood. The two tanks withdrew when they heard that a cruiser troop was coming up, but it failed to arrive and at 1 p.m. 2 Troop, out of ammunition and petrol, was ordered back into reserve. For this action Viscount Erleigh was awarded the Military Cross.

In the centre, 3 Troop with TSM Ayling found the bridge at Ailly destroyed, but two platoons of the Borders managed to cross and, covered by fire from the tanks, engage the enemy. Eventually they had to withdraw under increasing pressure, but gave the tanks more targets as they came back. Ayling was awarded the Military Medal for his leadership.

On the left, Lieutenant Gavin's 4 Troop made for the bridge at Picquigny, meeting a French officer who told them that the Germans were in great strength. The Border company who had taken another route ran into an ambush in a wood, taking many casualties and having their lorries set on fire. Gavin went to their aid, engaging parties of enemy infantry, driving them out of the wood and coming under heavy anti-tank and machine gun fire from Picquigny. He took up a position to cover the Border survivors, as about fifteen wounded straggled back during the day, to be evacuated in French ambulances. At 8 p.m. Gavin withdrew, taking thirty more Border survivors back on his tanks, with nine others making their way on foot.

Major Asquith with squadron headquarters had followed Erleigh's and Gavin's troops, engaging some German infantry near Picquigny. Lieutenant Nicholson's Cruiser Troop from 'B' Squadron, with only two runners, was under command, and Asquith sent him to Erleigh's support. When Nicholson arrived, one tank having broken down en route, Erleigh had withdrawn, so Nicholson also retired, hitting a mine and coming under anti-tank fire, but without sustaining any serious damage.

'A' Squadron, under Captain Horton, now down to four light tanks, was ordered to the assistance of the Border Regiment at Dreuil, which they reached at 3 p.m. and engaged the Germans. The tanks fired broadside and made two runs as they traversed the position; on the third run Horton's tank was knocked out by an anti-tank gun. The gunner, Corporal Brown, was killed and the driver, Trooper Hunt, wounded, but in spite of his wounds he brought the tank out of action. Another tank was disabled, and the remainder retired. Hunt was awarded the Military Medal for his gallantry.

By early afternoon 'B' Squadron was the only one still capable of action, and was sent to the assistance of 'A'. As Lieutenant Webb's troop advanced it managed to shoot down a low-flying enemy aircraft, but then encountered anti-tank and machine gun fire, while three German tanks closed and opened fire. TSM Snoswell's tank was hit twice and knocked out. TSM Merrin then moved his tank in front, as Webb's tank received two direct hits, killing him and his driver. Merrin managed to rescue the wounded gunner, Corporal Parsons, and then retired. For his bravery Merrin was awarded the DCM. In this first action the Bays had lost one officer killed, two officers wounded and one captured, one troop sergeant major and three other ranks killed, and two wounded, with six missing.[1, 3]

The next day, 25 May, was spent in rest and maintenance, with the crews catching up on sleep. On the night of the 25th and through the 26th the brigade moved to an area twelve miles south of Abbeville to cooperate with the French in an attack on 27 May on the German bridgehead at Abbeville. At 6 a.m. the Bays attacked on the right, with the 10th Hussars on their left, 'C' Squadron leading with a composite squadron of 'B/A' in support. As 'C' passed over the crest of the Bailleul ridge through a lightly held line of French riflemen, they were met by a hail of heavy machine gun and anti-tank fire from strongly held enemy positions in a wood to the north of Limeux. Four of the leading tanks were hit at once, and as the remainder started to withdraw, giving each other supporting fire, Lieutenant Aitken was killed, with TSM Ayling and Sergeants Barnard and Blythe wounded. Lord Erleigh's troop was then sent round to the right, but could not get forward due to the intensity of the enemy's fire, Erleigh's tank being knocked out. Contact had been lost with the 10th Hussars and Lieutenant Dance's troop was sent to reconnoitre the Huppy area, where the 10th were thought to be. On meeting a French officer Dance learnt that the Germans were holding Huppy in strength, and that the 10th Hussars had had heavy casualties. By this time the enemy at Limeux were quiet, and it was decided to test the position again with a troop of the composite squadron, 'B/A'. Lieutenant Behrens with Sergeants Norton and Burnside moved through the French infantry on the ridge. As soon as the three tanks were well clear, the Germans opened up. Norton's and Burnside's tanks were destroyed,

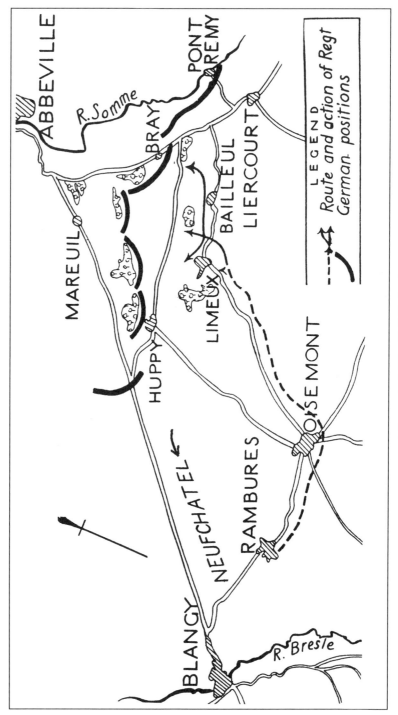

Action against Abbeville Bridgehead

Norton being killed and Burnside mortally wounded and his crew killed. Norton's driver managed to extricate and save his wounded gunner. Only Behrens's tank managed to return safely. That evening the Bays withdrew, leaving the 9th Lancers in support of the French infantry. That day Lieutenant Aitken, two sergeants and six troopers were killed, one warrant officer, one sergeant and five other ranks were wounded, and twelve tanks were lost.[1, 3]

During the next four days the remaining tanks of the brigade reorganised into a composite regiment of two squadrons of the 9th Lancers and one each of the 10th Hussars and Bays, the latter under Major Asquith. On 31 May Colonel Fanshawe was injured in a motor accident and was evacuated to England, Major Sykes taking over command. The composite regiment remained at St Leger until 5 June, when the Bays squadron was ordered to advance to the River Bresle to contact the 51st Highland Division and to counter-attack a German breakthrough near Abbeville. The Bays were bombed as they made their way forward to the infantry around Fressenville. During the afternoon the squadron succeeded in driving back three German attempts to infiltrate the Gordon Highlanders, and at dusk it was withdrawn south of the Bresle. On the two following days the squadron held crossings over the Bresle without coming into contact with the enemy, but on 8 June a move of sixty miles had to be made to the vicinity of Rouen to counter a German penetration. Contact was made with a German armoured division at about 3.30 p.m. and fire was exchanged.

In the meantime Major Sykes reorganised the remainder of the regiment: 'A' Squadron became lorried infantry, armed with rifles and light machine guns under Captain Barclay; the remnant of the tanks from all three squadrons was organised into 'B' Squadron under Major Scott; and the 'B' Echelon, having given up many of its lorries to 'A' Squadron, was reorganised under Captain Blackett. The Bays were now to operate in three parts.

On the evening of 7 June, Major Sykes and the dismounted 'A' Squadron were ordered to hold the line of the Andelle to block another German breakthrough north of Poix towards Rouen. Major Scott's tanks, together with a squadron of the 10th Hussars, were to deny the crossings over the Epte, but on arrival found the enemy in possession and so took up positions forward of the Andelle. Early on the 8th Major Sykes was ordered to fall back across the Seine at Pont de l'Arche, and on the 9th he was told to deny the crossings, the bridges having been blown. He was given fifteen miles of river to cover with his dismounted squadron. The French, who were meant to be holding the line of the Seine, had nearly all retreated south. That evening 'A' Squadron was withdrawn, leaving Captain Hibbert and his troop to patrol up to the Pont de l'Arche.

Major Scott's 'B' Squadron of tanks, covering the Andelle, was attacked by a

German Panzer division at about 11 a.m. on 8 June and was gradually pushed back to the line of the Rouen- Beauvais road. Late that night the squadron was withdrawn, crossing the Seine by the Pont St Pierre and at Les Andelys, but having a narrow escape from being cut off at the Andelle, only avoided by the daring of TSM Merrin, who discovered – and avoided – a force of German tanks that had crossed ahead of them. Scott's squadron with the 9th Lancers and brigade head-quarters, fifteen tanks in all, remained to support and encourage the French, and withdrew alongside as they retreated. On 14 June, with the Germans driving the French before them south of Rouen, Scott rejoined the regiment near Le Mans.

On 10 June Major Asquith's 'C' Squadron was told to hold the crossings over the Eure, south of Louviers, but on the 11th the squadron was withdrawn and made its way to La Ferrière, reaching Le Mans on 13 June. On the 14th the Bays were reorganised back into the original four squadrons, and on the 15th news came that the French had asked for an armistice. The regiment left for Brest at once, reaching the port that evening, with the main party embarking on HMT *Isle of Man*. They remained on board throughout the 16th, while the harbour was cleared of mines, and eventually reached Plymouth early on 17 June. The vehicle party was left to load the vehicles at Le Mans, and the fifteen remaining tanks were put onto railway flats, but never reached the coast. The drivers made their way in the lorries to Cherbourg, and so to England. The Bays then assembled under canvas at Longbridge Deverill in Wiltshire, moving to Warminster on 1 July 1940.[1, 3]

On 18 June 1940 The King's Dragoon Guards moved from Yorkshire to Hasel-bech, near Northampton, where it was still short of nearly half its tanks. On 17 July Lieutenant Colonel McCorquodale took over command of the regiment, and on the 24th the Prime Minister, Mr Winston Churchill, inspected the KDG. On 18 August the regiment was moved to Stetchworth, south of Newmarket, to be ready for any possible invasion, while the Battle of Britain was being fought over-head. His Majesty the King inspected the KDG on 23 August, and the Duke of Gloucester on 5 September. On 6 October the regiment was ordered back to Northampton, when it became clear, in spite of the strictest secrecy, that a move overseas was in the offing: the issue of tropical kit was otherwise hard to explain away. The KDG embarked on the old Cunard liner SS *Scythia* at Liverpool on 16 November, with a strength of 38 officers, and 482 other ranks. The convoy reached Durban, where the regiment was transferred to the troopship *Dunera* and where two days were spent as guests of the hospitable South African people. Port Tewfik was reached on 23 December and the main body disembarked at Port Said on the 30th, proceeding by train to an empty stretch of desert at El Quassassin, where they went under canvas at Tahag Camp. On 1 January 1941 the colonel was told that the KDG was to convert to an armoured car regiment and to be equipped with the South African-built Marmon Harringtons.[2, 4]

The Queen's Bays on their return from France spent the summer and autumn of 1940 at Warminster as part of 1st Armoured Division in the Canadian Corps, forming the GHQ central reserve for counter-attack against invasion south of the Thames. By October, when a move was made to billets near Farnham in Surrey, the regiment had been brought up to strength with a full complement of cruiser tanks. On 7 October Lieutenant Colonel Draffen assumed command from Lieutenant Colonel Fanshawe. The winter was spent in training and exercises, and on 14 February 1941 the Prime Minister, Mr Winston Churchill, accompanied by the Polish General Sikorski and General de Gaulle, the leader of the Free French, visited the regiment. On 2 June 1941 the Bays moved to Marlborough, where they stayed for the rest of their time in England. On 8 August 1st Armoured Division was told to prepare for embarkation to the Middle East, and the Bays were equipped with thirty-six British Crusader and sixteen American Honey tanks. The Queen paid a farewell visit to the regiment on 18 September and the Bays embarked on the *Empire Pride* at Gourock on the Clyde on 24 September.[1, 3]

The King's Dragoon Guards at Tahag Camp started to receive their armoured cars, and to hand over their light tanks, during January, when they were suddenly asked to send a squadron up to the desert at once to relieve the 11th Hussars. 'B' Squadron, under Major Crossley, left Tahag on 27 January 1941 and reached the regimental headquarters of the 11th Hussars at Mechili in Libya on 2 February, having suffered their first casualty when Trooper Lamour kicked and exploded one of the small red Italian hand grenades which littered the desert. After two days spent in maintenance after the long drive, 'B' Squadron became the advance guard to a thrust by 7th Armoured Division across the desert to cut off the Italian army retreating from Benghazi along the coast road. On 4 February Msus was reached and on the 5th 'B' Squadron resumed the advance, reaching Gotsdes, fifty miles to the south west as the advance guard to a composite force, commanded by Lieutenant Colonel Combe with the 11th Hussars and including the 1st Rifle Brigade and the 4th RHA.[2, 4]

Reports were then received that enemy traffic was moving north and south along the main coast road. 'B' Squadron was ordered to cut the road and hold it until the Rifle Brigade were able to get into position. Lieutenant Chrystal's and Lieutenant Taylor's troops were soon established astride the coast road with Beda Fomm slightly to the east. Lieutenant Delmege's troop was sent south west to give warning of any enemy approach from that direction. As soon as Chrystal and Taylor had taken up their positions, they came under artillery fire from an Italian column trying to retreat down the coast road. Major Crossley had been given two

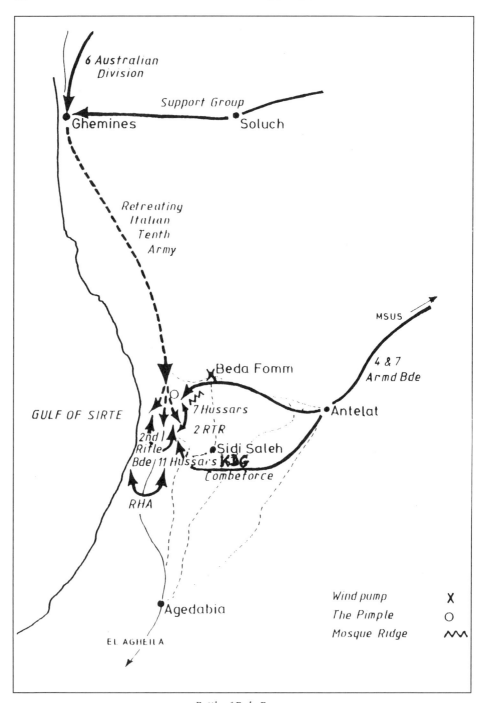

Battle of Beda Fomm

25-pounders and two anti-tank guns, which he sent to help stiffen the defence. Lieutenant Cubitt's troop moved to support Chrystal and Taylor, followed by Major Crossley, who brought the rest of the RHA into action on the north-east flank. Moving about a mile further north, Major Crossley positioned some anti-tank guns on the Italian flank, where they were able to shoot up a long line of halted lorries. Thus the KDG fired the opening shots in the Battle of Beda Fomm.

As the whole squadron fired on the enemy columns, white flags began to appear; individual armoured cars dashed across the desert to round up prisoners, but some of the more determined Italians managed to inflict casualties on both 'B' Squadron and the artillerymen. A hundred and fifty Italians surrendered to Corporal Ashbrooke, armed only with a jammed Bren gun, then forty officers surrendered to Lieutenant Taylor. By this time some 350 prisoners had been collected and driven away in lorries, as the Rifle Brigade began to arrive. An Italian anti-tank gun scored a direct hit on Lieutenant Cubitt's armoured car, killing him and all his crew.

As the Rifle Brigade dug in across the coast road, 'B' Squadron moved to the west to stop enemy columns from escaping around the slenderly held position. The Italian Breda 20 mm guns, firing high explosive and armour-piercing bullets, gave both Chrystal and Taylor a lot of trouble, but the armoured cars certainly drew that fire away from the infantry. The Italians now counter-attacked, reoccupying some of their former positions, so both Chrystal and Taylor stayed with the Rifle Brigade while Sergeant Watkins was sent to round up another Italian column trying to get away to the west. As night came on, 'B' Squadron moved five miles south, blocking the main road, and the night passed without incident. SSM Buckley had been sent to meet and escort the supply echelon, and on his way surprised an escaping Italian column, which he attacked single-handed, capturing 100 prisoners and handing them over to the Rifle Brigade before rejoining the squadron.[2, 4]

On the following day, 6 February, KDG patrols moved south to round up escaping Italians and take up observation positions. SSM Buckley came upon a fort where he captured another 150 Italians, and in the evening the squadron captured the walled town of Zuetina. This could only be approached by a causeway, but the garrison of 2,000 surrendered, so rendering a very difficult operation unnecessary. That night 'B' Squadron leaguered at the fort, guarding prisoners numbering fifteen times their own strength and firing at Italian columns trying to escape. As the greater part of the Italian army in Cyrenaica had now been captured by 7th Armoured Division at Beda Fomm, 'B' Squadron moved south on Agedabia, capturing the airfield and several aircraft by a charge in line, and on 8 February moved on seventy miles to El Agheila, sending out patrols into Tripolitania.[2, 4]

On 14 February 'B' Squadron was attacked by fourteen Messerschmitt 110s, when four KDG were wounded, but two of the aircraft were brought down by the combined fire of nearby Bofors gunners, together with the Vickers guns on the armoured cars. This was the first appearance of German aircraft in the desert. Later that day the 11th Hussars lost a car by a direct hit from a Junkers 87, the dreaded Stuka. The next day the attacks from the air continued, but that evening Lieutenant Colonel McCorquodale appeared, followed shortly afterwards by the rest of the regiment. For his fine example at Beda Fomm SSM Buckley was awarded the DCM.[2, 4]

SOURCES

1   W. Beddington, *The Queen's Bays, 1929-1945*, 1954.
2   D. McCorquodale, B. L. B. Hutchings and A. D. Woozley, *History of The King's Dragoon Guards, 1938-1945*, published by the regiment.
3   War Diary, 1939-1941, The Queen's Bays, Regimental Museum, 1st The Queen's Dragoon Guards, Cardiff Castle.
4   War Diary, 1939-1943, 1st King's Dragoon Guards, Regimental Museum, 1st The Queen's Dragoon Guards, Cardiff Castle.

39

El Agheila, Msus, Tobruk, Bir Gubi
1941-1942

By 18 February 1941 The King's Dragoon Guards, with some anti-aircraft gunners and four RAF armoured cars, were the only British troops west of Agedabia, the 7th Armoured Division having retired to Cairo for a well-deserved rest and refit. The KDG came under command of 1st Australian Corps when on the 20th the 17th Australian Infantry Brigade arrived at Agedabia, much to the relief of the KDG. 'C' Squadron was forward at El Agheila, and on the 20th Lieutenant Williams engaged three enemy armoured cars, a motorcycle combination and a truck with an anti-tank gun. Williams had been told that Lieutenant Weaver's troop was coming to his assistance, so when he was returning along the coast road and saw a stationary armoured car blocking the road, he merely sounded his horn. The car moved off to one side and as they drew near, Williams saw that it had eight wheels and a German cross on the side of its turret. In passing there was a sharp exchange of fire at point blank range, a bedroll from Williams's car was knocked off, but no other damage was recorded by either side. This was the first British contact with the newly arrived German Afrika Korps.[1, 2]

Over the following days the Messerschmitt 110s stepped up their low-level attacks. On 22 February Lieutenant Kellie was wounded, later dying, while Major Lindsay, Sergeant Foster and Trooper Findlay were all slightly wounded. On the 24th the enemy occupied Agheila fort, knocking out an armoured car and capturing Lance Corporal Allen and Trooper White. Trooper Close, who was badly wounded, pretended to be dead. The enemy party, which included two heavy and five light tanks as well as armoured cars, towed away the damaged Marmon Harrington, and then Close was picked up by his troop leader as the enemy retired. On the 27th 'C' Squadron was heavily attacked by the Messerschmitts, losing three armoured cars and three trucks, with Captain Delmege and nine other men being wounded. On 1 March Lieutenant Howard and Corporal Short of 'A' Squadron were surprised and captured by an enemy patrol when they dismounted and climbed a hill to get better observation. Throughout the rest of February and March patrol activity, brushes with the enemy on the ground and in the air, became a daily routine. On 29 March one of Lieutenant Taylor's armoured cars was hit in an action in the Wadi Faregh with two German eight-wheelers. The crew evacuated the damaged car and ran towards Taylor, with the Germans chasing and only 200 yards away. Taylor turned back and picked up the crew, but Trooper Roberts was hit by a shell as he was pulled aboard and another shell blew off Lance Corporal Hullah. Taylor was awarded an MC for his brave action. On 31 March Lieutenants Budden and Whetherly reported a concentration of 200 enemy tanks moving eastwards, and on 1 April the KDG were ordered to withdraw as it was clear that an enemy offensive was being launched against the weakly held British positions.

During the next few days the KDG withdrew through Agedabia to Antelat and then to Sceleidema, where the regiment found considerable panic and confusion in what had been the rear areas. Msus was reached on 4 April and Mechili on the 5th, then Derna, arriving in Tobruk on the night of 7/8 April. During the retreat Corporal Cedras died of wounds received from a Messerschmitt attack, when nine other men were wounded. In Tobruk the KDG came under command of 9th Australian Division and were concentrated on the Lysander aerodrome. By 10 April Rommel, who had outrun his supplies, was closing up on the town and the Stuka raids began in earnest. There was such a shortage of infantry within the perimeter that the KDG formed a dismounted 'B' Squadron of 150 men under Captain Selby, who held a section of the line near the Fig Tree. Lieutenant Williams was appointed liaison officer to the Australian Division, and on reporting to divisional headquarters in the caves on the escarpment, saw two men sitting at a table without any badges of rank. Saluting, Williams said, 'L. O. from the K. D. G, sir.' A very Australian voice came back, 'K. D. G, what mob are they?' Williams, glaring over his spectacles, replied, 'The King's Dragoon Guards, sir, formed in 1685, many years before Captain Cook set sail for Australia!' This story only came to light when Colonel Lloyd, one of the two men and GSO1, told other KDG officers months later, and revealed that the other man had been General Morshead, the Australian general commanding the garrison.

On 30 April and 1 May Rommel launched his last major attack on the perimeter, making a salient two miles deep and two miles wide. The attack was eventually held, mainly due to the accurate shooting of the gunners. 'B' Squadron in the perimeter suffered casualties, but held their line. Rommel now turned his attention to Egypt, and the Tobruk garrison settled down to the siege. A reinforced 'C' Squadron, with all the armoured cars, the dismounted 'B' Squadron and two troops from 'A' manning light tanks, all under Major Drabble, remained in Tobruk, while the remainder with RHQ were withdrawn by sea to the Delta for re-equipment.

The two light tank troops, under Lieutenants Budden and Fraser, were just west of Fort Pilastrino, and the dismounted squadron was moved north of the escarpment and astride the road to Derna, where it stayed. 'C' Squadron remained on the Lysander Aerodrome in an anti-parachute role, with troops on the Gubi and Satellite Aerodromes. In May some captured German and Italian tanks were dug into the perimeter and 'C' Squadron was called upon to man six of them. There were two daily patrols, to the Derna road and to the El Adem road in the south, where the Italians had built some wooden observation towers, in one of which Lieutenant Weaver was killed in October.

The KDG got on extremely well with the offhand and generous Australians. Sergeant Hall, reporting to an Australian battalion HQ, asked a private soldier for his colonel. The Australian, putting his hand to his mouth, yelled, 'Oi, Bill!' and a head and shoulders with a crown and a pip immediately appeared. Sergeant Hall was shaken. The men lived in dugouts in the sand, under constant bombing and shellfire, with all supplies very scarce, except for food always generously shared out by the Australians. A regular bathe in the sea was appreciated.

The dismounted squadron, holding some 3,000 yards of perimeter, was so stretched that every man had to be on guard every night, with the only reliefs coming from the echelon. At the end of May Captain Palmer relieved Selby, giving the latter a rest, and during this time a successful raid was carried out on the enemy, killing a number and creating the impression that the perimeter was more strongly held than it was. Palmer was awarded the Military Cross and Sergeant Berryman the Military Medal for their part in this raid.[1,2]

In June 1941 Operation Battleaxe was launched from Egypt to relieve Tobruk, starting on the 15th. The besieged garrison waited expectantly for news, but by the 19th it was clear that Rommel had been too strong and the attack had been a failure. Disappointment was inevitable but the two daily patrols and the manning of the dug-in tanks continued. The dismounted squadron was evacuated back to Egypt in July and at the same time the 9th Australian Division was relieved by the British 70th Division. By the beginning of November preparations were being made for a break-out by the garrison to coincide with an offensive from Egypt. 'C' Squadron, carrying Sappers to lift the mines, was to lead the 4th Royal Tanks and the infantry in an assault on the strongly held German positions to the south east of the perimeter, and then exploit towards El Duda ridge. Just before the breakout, the KDG held a church parade and were treated to a pacifist sermon from the padre. As soon as the last hymn had been sung, Major Lindsay roared, 'King's Dragoon Guards ... Attention! In spite of the sermon you have just heard, you will go out against the enemy and smite him hip and thigh!'

At dawn on 18 November 1941 the advance across the frontier began, and before first light on the 21st 'C' Squadron led the break out of Tobruk through the

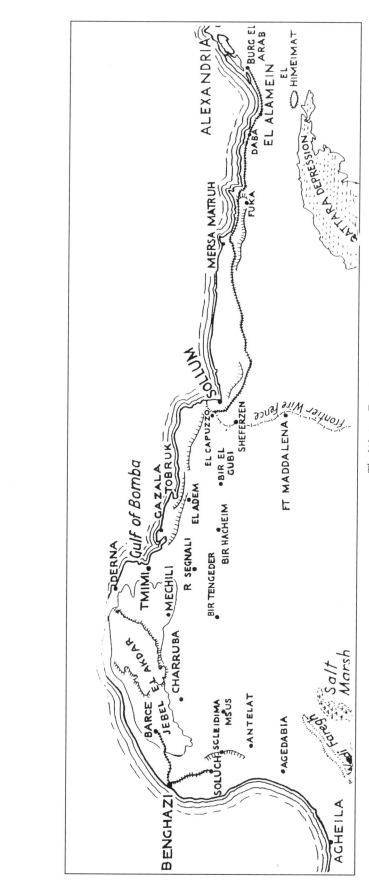

The Western Desert

perimeter under cover of a heavy artillery barrage, the noise and smoke concealing their movement for a time, before 'all hell broke loose'. The Sappers jumped off the Marmon Harringtons and started lifting booby-trapped teller and S mines as well as Italian box mines; 4th Royal Tanks went through the gaps and the Black Watch charged the dug-in enemy. 1,100 German prisoners were captured; the KDG lost thirteen out of fifteen armoured cars, but with only one fatal casualty, Trooper Dean. Lindsay led the squadron in a Dingo, receiving a direct hit on the side where he was sitting and wounding him severely in the ear. Several other KDG were wounded, and QMS Swinburne and his driver, going out to recover some of the damaged cars, were killed by a German Spandau post which suddenly came to life after it had been passed. 'C' Squadron was then withdrawn to the Lysander Aerodrome, where the fitters managed to recover and repair seven of the knocked out and mined armoured cars, so that by the afternoon of the 22nd, when the squadron was called forward again, there were five active troops available. The Eighth Army was having a difficult time. Auchinleck had taken over personal command and the Tobruk garrison was ordered to make redoubled efforts to capture the El Duda ridge. On 24 November the KDG sent out probing patrols, when Lieutenant Franks, with another car commanded by Corporal Muir, disappeared and were presumed captured. Shortly afterwards Lieutenant Beames was severely wounded and his driver and operator killed. Lieutenant Gardener of the RTR saw what was happening, brought up his tank, dismounted and rescued Beames, who, sadly, died of his wounds before he could be got to medical help. Gardener was awarded the Victoria Cross for his gallantry.[1, 2]

On 27 November, after very heavy fighting by the tanks and infantry, contact was established with the New Zealanders of Eighth Army on Belhamed Ridge, thus forming a corridor into Tobruk. The KDG was given the task of policing this corridor and bringing into Tobruk the whole of the New Zealand Division and HQ 13 Corps, through a series of minefields. During this operation Lieutenant Windle was slightly wounded, but was drowned when the ship evacuating him was bombed. Over the next few days there was a lull in the battle, and on 1 December 'C' Squadron contacted the 11th Hussars and 12th Lancers of 7th Armoured Division, while Major Lindsay was able to speak on the wireless to RHQ and Colonel McCorquodale. Lieutenant Franks was found in the German hospital at El Adem, but he died two days later. The rest of his troop had been taken prisoner. On 15 December 'C' Squadron was withdrawn to Cairo for leave and re-equipment. On arrival they were greeted with the news that Lieutenant Windle had been awarded the MC, and Sergeants John, Lodge and Battersby the MM.

The rest of the King's Dragoon Guards, on being evacuated from Tobruk, arrived at Abbassia in May 1941, where they remained without equipment for three months, until on 1 September they were ordered back to the desert and took over forty-five armoured cars of the 11th Hussars. By 15 September patrols were being sent out from Bir Kenayis along the frontier wire, and on the 30th contact was made with a German patrol of five armoured cars, two of which were Marmon Harringtons, presumably captured during the retreat. Major Drabble and Lieutenant Chrystal were wounded when they were examining a gap in the frontier wire on foot, and their armoured car, following, exploded a mine.

The 11th Hussars relieved the KDG on 6 November. The regiment withdrew to Bir Diqnash, where it was re-equipped with thirty new Marmon Harringtons, in time for the offensive which started on 18 November. The KDG led the 4th Armoured Brigade, comprising the 8th Hussars and the 3rd and 5th Royal Tanks. By 22 November the brigade reached Sidi Rezegh Aerodrome and was heavily engaged with the German panzer divisions, suffering serious casualties both of tanks and men. The KDG protected the rear and gave exact information about enemy movements. On 24 November Rommel gathered together his panzers and, moving south from Sidi Rezegh, made a dash for the Egyptian frontier and the British lines of communication. While the infantry was left to hold on to Sidi Rezegh, 7th Armoured Division was sent after Rommel, the armoured cars providing a screen behind which the tanks could gather. By the 26th the threat was countered and contact was re-established with the New Zealanders on Belhamed. By the 27th Rommel was withdrawing, the KDG constantly providing information about the movements of the 15th and 21st Panzer Divisions and the Italian Ariete Armoured Division. Over this period Lieutenant Lillingston, Lance Corporal Hughes, and Troopers Smith and Hamilton were killed. On 30 November 'A' Squadron was attacked by Stuka dive bombers, who inflicted some damage and wounded Trooper King.

During the first days of December Rommel concentrated his forces between Sidi Rezegh and Bir Gubi. On the afternoon of the 3rd Sergeant Ashbrooke of 'B' Squadron, with Corporal Redfearn in a second armoured car, captured two Italian M13 tanks and one light tank. On 4 December the infantry attacked Bir Gubi, while the KDG patrolled on the south and north flanks. 'B' Squadron under Major Rydon captured an Italian tank with its crew, and 'A' Squadron under Major Hellyer took thirty-five Italians prisoner and then discovered a large petrol dump containing 50,000 gallons. They managed to destroy 10,000 gallons before being chased off by eleven enemy tanks. To complete the day, Lance Corporal Fining of 'A' Squadron shot down a Messerschmitt 110 with a captured Italian Breda mounted on his armoured car. The following day 'A' Squadron managed to destroy another 12,000 gallons of enemy petrol, but Captain Heywood and three of his men from the REME Light Aid Detachment were captured when trying to recover

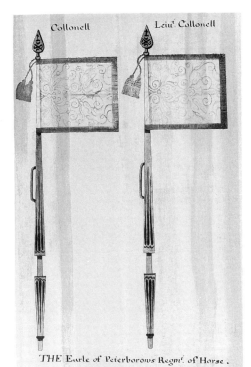

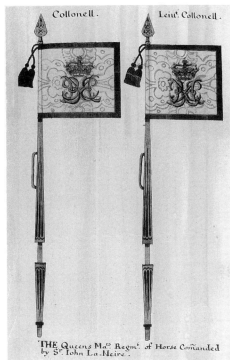

ABOVE: Original standards of The Queen's Regiment of Horse (KDG), commanded by Sir John Lanier, and the Earl of Peterborough's Horse (Bays). From a Manuscript Book in the Royal Library, Wimdsor Castle, dated between October 1685 and May 1686. (By gracious permission of Her Majesty The Queen.) BELOW, LEFT: Trooper, The Queen's Regiment of Horse (KDG), 1704. Watercolour by Pierre Turner, property of the late Captain R. G. Hollies Smith. RIGHT: Trooper, The Queen's Bays, 1763. Watercolour by R. Simkin, Regimental Museum, 1st The Queen's Dragoon Guards.

David Garrick, the actor, cornet in The Queen's Own Regiment of Horse (Bays), 1730.

Barnastre Tarleton, cornet in The King's Dragoon Guards, 1775. He later achieved fame as a dashing cavalry leader in America. Portrait in the National Portrait Gallery.

Officer, 1st King's Dragoon Guards, circa 1780. Watercolour by R. Wymer, Regimental Museum, 1st The Queen's Dragoon Guards.

General Le Marchant, The Queen's Bays, 1794, founder of the Royal Military College. From an engraving by J. D. Harding.

1st King's Dragoon Guards at Waterloo, 1815. Painting by R. Caton Woodville, property of the author.

ABOVE, LEFT: Officer of The Queen's Bays in levée dress, 1830. Lithograph by L. Mansion and L. Eschauzier. RIGHT: Trooper, The Queen's Bays, 1835. Watercolour by R. Simkin. BELOW: 1st King's Dragoon Guards moving quarters, Newcastle-upon-Tyne, 1824. Painting, Regimental Museum, 1st The Queen's Dragoon Guards.

ABOVE: 1st King's Dragoon Guards, 1846. Watercolour by Henry Martens, property of the author. BELOW: 1st King's Dragoon Guards on Laffans Plain, Aldershot, circa 1873. Watercolour by Orlando Norie, Regimental Museum, 1st The Queen's Dragoon Guards.

ABOVE: The Queen's Bays, circa 1875. Watercolour by Orlando Norie, Regimental Museum, 1st The Queen's Dragoon Guards. BELOW: Major Marter, 1st King's Dragoon Guards, capturing Cetewayo, the Zulu king, in the Ngombe forest, 1879. Watercolour, Regimental Museum, 1st The Queen's Dragoon Guards.

Officers, 1st King's Dragoon Guards, Norwich, 1894. The group contains eight colonels or commanding officers. Standing: Lt Marter, Major Benbow, Capt Bates, Lt Wildes, Lt Graham, Lt Quicke, unknown. Seated: Capt Fergusson, Lt Eastwood, Col Hulton, Gen Sayer, Major Lawrence, Capt Matthews (Riding Master), Capt Levita. On ground: Col Bell Smyth, Lt Langton, Lt Berners, Col Bogle Smith. Photograph, Regimental Museum, 1st The Queen's Dragoon Guards.

The Queen's Bays crossing a drift, South Africa, 1901. Photograph, Regimental Museum, 1st The Queen's Dragoon Guards.

ABOVE LEFT: The Queen's Bays searching a Boer farm, South Africa, 1901. Photograph, Regimental Museum, 1st The Queen's Dragoon Guards. RIGHT: Major N. M. Smyth, VC, The Queen's Bays, in South Africa, 1901. Photograph, Regimental Museum, 1st The Queen's Dragoon Guards. BELOW, LEFT: Drum horse, 1st King's Dragoon Guards, 1911. Watercolour by R. Simkin. RIGHT: Drum horse, The Queen's Bays, 1906. Postcard, E.F.A. Military Series.

ABOVE: The Queen's Bays in Le Carnoy valley awaiting the breakthrough, Somme, 1916. Photograph, Imperial War Museum. BELOW: Lt P. Moreton, 1st King's Dragoon Guards, on the Somme, 1916. Waiting for the breakthrough, beside Tank C5, in which Capt Inglis was to win the DSO a few days later, when tanks made their first appearance at Pozières. Photograph property of Major J. A. Moreton.

ABOVE: 1st King's Dragoon Guards, Dakka, Afghanistan, 1919. Photograph, Regimental Museum, 1st The Queen's Dragoon Guards. BELOW: Corporal Ayling, The Queen's Bays, in marching order, 1935. Photograph property of Brigadier G. Powell, QDG.

ABOVE: The last mounted parade of 1st King's Dragoon Guards, Secunderabad, 1937. Water-colour by Snaffles. Officers' Mess, 1st The Queen's Dragoon Guards. BELOW: The Queen's Bays returning from France, 1940. Photograph, Imperial War Museum.

ABOVE: Marmon Herrington armoured cars of the 1st King's Dragoon Guards in Benghazi, 1941. Lieutenant Colonel D. McCorquodale and Captain M. Arkwright on the right. BELOW: The breakout from Tobruk, 1942. The Black Watch moving up on El Doda, with a wounded man being helped to the KDG Marmon Herrington armoured car on the ridge. Photographs, Imperial War Museum.

a damaged Marmon Harrington. Lieutenant Maxwell managed to free 700 British and New Zealand prisoners of war, capturing their escort and thereby winning a Military Cross. On the 7th 'A' Squadron captured thirty Italians and sixteen Germans.

With Bir Gubi secured, the regiment extended its patrols from Bir Gubi to Bir Hacheim, and on 9 December Lance Corporal Robinson of 'B' Squadron towed a grounded Tomahawk fighter with its pilot fifty miles across the desert back to the echelon. On the 9th the Royal Dragoons arrived from Syria to relieve the KDG, which was drawn into reserve under the 22nd Guards Brigade, but continued to patrol with one squadron west of Bir Hacheim. On the 13th Lieutenant Collie's troop of 'B' Squadron was heavily engaged, losing two armoured cars, while the third car raced forward picking up most of the crews. Collie and Trooper Carr were captured.

By 18 December KDG patrols were south of Mechili and on the 20th an advance of sixty miles was made to Charruba, where 200 Italians were captured. RHQ was heavily dive-bombed by forty Stukas, when Captain Collie, Lance Corporal Butler and Trooper Theobald were wounded. On the 23rd the KDG were ordered to take El Abiar and Benghazi, and on the 24th 'A' Squadron entered Benghazi from the east, just before a patrol from the Royals entered from the south. The KDG occupied and patrolled the town until relieved by 4th Indian Division on the 26th, when the regiment moved to Ghemines on the road between Benghazi and Agedabia. An order to move back to the Delta was cancelled owing to lack of petrol and enemy activity, and the regiment moved west of Msus, with 'A' and 'B' Squadrons driving twenty-five miles to the east of Agedabia. The first days of 1942 were fairly quiet, until on 9 January Lieutenant Maxwell destroyed two enemy lorries, capturing three prisoners. On the same day a supply column from the echelon ran into a minefield in the Wadi Faregh, where it was attacked by a German armoured car; Lieutenant Hall was severely wounded and the lorry crews taken prisoner. However, the Germans left two men with Hall, who died of his wounds. Lieutenant Ward and eleven men were taken prisoner. On the 17th 'A' Squadron destroyed a German wireless lorry, killing one man and taking three prisoner. On 21 January the regiment learnt it was to return to Cairo, but no sooner had the order been issued than it was countermanded and the KDG found themselves escorting a convoy from Msus to Benghazi. Rommel was now on the move again, and the KDG were reduced, by sheer wear and tear, to one composite squadron of nineteen Marmon Harringtons and three Morris armoured cars.

Having moved back to Tobruk, the composite squadron went on 28 January to Giovanni Berta, coming under 30 Corps and the 4th Indian Division. During this period some Libyan Arabs came under command and one of their officers, a White

Russian, Major Peniakoff, insisted on attaching himself to the KDG. He was later to achieve fame as the leader of 'Popski's Private Army'. Lieutenant Budden and four men were captured out on reconnaissance by four German armoured cars, but were later freed when the Germans were put to flight by three armoured cars of the 12th Lancers. At 1700 hours on 6 February Eighth Army ordered the KDG to return to the Delta for rest and refitting, and by 1730 hours the regiment was on the move, arriving at Abbassia on the 10th, where 'C' Squadron was reunited with the rest of the regiment.[1, 2]

The Queen's Bays had sailed from the Clyde, arriving at Cape Town on 30 October 1941, where they enjoyed South African hospitality until 4 November, when the convoy sailed. Reaching Suez on the 25th, the Bays disembarked and moved to a bleak camp at Amriya, ten miles west of Alexandria, where they prepared their twenty-five Crusaders, seventeen Honeys and six close support tanks for the desert. By 20 December 1st Armoured Division, including the Bays, had arrived at the frontier wire and were grouped for the next ten days south of Tobruk. On 30 December the division moved forward to Bir Hacheim, staying for four days, then advancing to cut off Rommel at Agedabia. Before the Bays could arrive the Germans had retired to El Agheila.[3, 4].

Rommel now showed his extraordinary capacity for recovery,. On 21 January 1942 he struck at the forward troops at El Agheila, consisting of the 201 Guards Brigade and the 1st Support Group. The 1st Armoured Division and the 4th Indian Division were too far back to give effective support, because of difficulties of supply. Michael Halsted in his diary remembered 22 January: 'Woke up at 7.25 to hear "Anyone awake? We've got to move at 8.30!" My word, we didn't half shift. The Germans have broken out with two columns of tanks, 30 and 40, and our Support Group have had to withdraw. Now we are halted south of Saunnu wait-ing. This game is full of surprises.' That night the 'B' Echelon was attacked, but the enemy were beaten off and thirty Italians made prisoner. Two of 'A' Squadron's tanks were hit when milling about the Saunnu-Antelat-Msus area on 23 January. Halsted's diary gives the flavour of the day:

> Shelling on our right and a lot of smoke. Coming up we could see Jerry on the far ridge. Can see one Mark VI blazing and another pulling out. Hope chaps OK. Many lorries burning near us. Suddenly we have to move. George's tank won't start. Sgt Smith goes up in his tank and Cpl Work gets George's tank out OK. Then Cpl Smith's won't go, and rather under the nose of Jerry we have to tow him back.[3, 4, 5]

On the 24th the Bays took up a blocking position across the Antelat-Msus track, with 'C' Squadron on the right, 'A' on the left and 'B' in reserve. 'C' was soon in action and had one tank hit, but destroyed an anti-tank gun and took some German prisoners. 'B' meanwhile knocked out a German Mark IV tank. Halsted noted: 'I spot a truck and some men on the far ridge. I suspect a trap so go up very carefully. They all tumble out with their hands up. I shout "Italiano?" One replied "Nein, Deutsches." We get them back OK. Luckily Bradley remembers loot. We each get a Parabellum automatic. He gets binoculars and a short bayonet. We had forgotten to disarm them!' The 25th saw a small column in trouble, and 'C' Squadron under Major Streeter was sent to its assistance, coming into action and destroying several anti-tank guns and vehicles. Lieutenant Frankau was badly wounded, however, and Trooper Rowney was later awarded the Military Medal for pulling him out of his tank under heavy fire. Corporal Minks and two men were killed, and Captain Patchett was reported missing, later confirmed as having been captured. While 'B' and 'C' Squadrons withdrew, 'A' remained as rearguard but found itself under heavy pressure when a German anti-tank gun got round to the rear of the squadron and a column of German tanks attacked from the front. Major Barclay's tank was hit and set on fire, Sergeant Harvey had to abandon his tank because of an electrical failure, and Lieutenant Glynn's tank was also hit, Glynn himself being killed. The survivors from all three tanks hid in some camel scrub, waiting for the battle to pass them by, and after two days behind the enemy lines were able to make their way to 4th Indian Division at Benghazi.

The Bays fought a confused rearguard action all the 25th, described by Colonel Draffen as 'disorganised into three columns', until they fell back to the track running east from Msus to Mechili. Halsted described the confusion:

> Horrible jumble of guns and transport streaming away with us. No orders so kept on trying to shoot whatever appeared. Some guns went down for a few minutes and then we saw the German tanks coming on and on at us from two sides. Horrible as their shelling kept getting on to us, and our little guns just couldn't reach them at all.

At 11 a.m. 'B' Squadron counter-attacked near 'The Bog', checking the German advance and inflicting casualties, but losing Major Blackett and Lieutenant Fleming as prisoners, and Sergeant Rutherford and a number of men killed. The brigade was then ordered to retire on Charruba. 'Finally our dear old tank [Halsted's] blew up the ghost and had to be abandoned. Piled onto Stephen's tank with most of my kit. In getting into Charruba, Stephen's tank with ten chaps of us on board fell over a sharp drop onto the Colonel's tank below. No one was hurt, but we had to leave his tank too.' [3, 4, 5]

The 'Msus Stakes', as the scramble back became known, finished at Charruba, where the brigade was reorganised into a single composite regiment, with the 9th Lancers commanding and finding two squadrons. The third composite squadron, made up of the remnants of the Bays and 10th Hussars, consisted of a Bays squadron headquarters under Major Streeter with two Bays troops totalling nine tanks, seven of which were Honeys. Major Sykes was crushed between a tank and scout car and had to be evacuated, but during the 26th some more Honeys arrived, and a third troop was formed under Lieutenant Halsted. Sergeant Cockwill and Trooper Harris also rejoined, being rescued by some South African armoured cars capturing their escort. Lieutenant Colonel Draffen formed a tank salvaging party, which by 1 February was at Mechili, having recovered six cruiser tanks and blown up two cruisers and six Honeys. The regiment's losses had been eight tanks knocked out, twenty-nine lost through various failures. The 'Left Out of Battle' party under Major Dance reached Mechili on 29 January, where it was joined by Major Barclay, Sergeant Harvey and six men, after their time behind the enemy lines. By 2 February the Bays were near El Adem, being re-equipped with tanks from 22nd Armoured Brigade. By 5 February the regiment was ready for action again.[3, 4]

SOURCES

1 D. McCorquodale, B. L. B. Hutchings and A. D. Woozley, *History of The King's Dragoon Guards, 1938-1945*, published by the regiment.

2 War Diary, 1939-1943, 1st King's Dragoon Guards, Regimental Museum, 1st The Queen's Dragoon Guards, Cardiff Castle.

3 W. Beddington, *The Queen's Bays, 1929-1945*, 1954.

4 War Diary, 1940-1943, The Queen's Bays, Regimental Museum, 1st The Queen's Dragoon Guards, Cardiff Castle.

5 M. Halsted, *Shots in the Sand*, 1990.

40

The Cauldron, Knightsbridge, Bir Hacheim, Gazala, El Alamein Line, Alam el Halfa
1942

The Queen's Bays now had one squadron equipped with Crusaders and two with Honeys, all old tanks with holes plugged from former battles. On 6 February 1942 the Bays moved astride the Trigh Capuzzo at Dahar El Aslagh, staying for ten days, then moving a few miles west to Sidi Muftah to guard the minefield between Alem Hamza and Bir Hacheim. There was little enemy ground activity, but SQMS Malcolmson was killed in a combined Stuka and ME 109 attack on the 'B' Echelon. The Bays were relieved on 21 April by 44th RTR, moving to the south of Tobruk, where they were inspected by the Duke of Gloucester on 1 May. New tanks began to arrive, including some Grants and reconditioned Crusaders. On 24 May the 2nd Armoured Brigade moved forward to a position on the Trigh Capuzzo five miles east of Knightsbridge, as Rommel was expected to attack. The regiment now had twenty-eight Crusaders, twelve Grants and six Honeys.[1, 2]

On 25 and 26 May Rommel moved south around Bir Hacheim, and on the 27th the 2nd Armoured Brigade was ordered to take the enemy in flank. 'B' Squadron was soon in action, the three tanks of Lieutenant Sherbrooke's troop being knocked out and several men killed, so Captain Baker, under cover of smoke, went in a scout car to rescue the survivors. Halsted remembered, 'Then − the enemy in sight, four to five thousand yards to our right front. They appeared as a black mass of moving and stationary vehicles. Apparently they were two united columns being joined by some more.' Colonel Draffen ordered 'C' Squadron with its Grants to engage frontally. They opened fire at 3,000 yards, but soon closed up to 1,500 yards, shooting to great effect; then the two Crusader squadrons charged in from the right flank, led by 'A' Squadron under Viscount Knebworth. Colonel Draffen and the adjutant were standing on the back of their Crusader, going here, there and everywhere, encouraging everyone. As the tanks swept over the German positions, infantry and gunners came out with their hands up, and the Rifle Brigade following collected up 250 prisoners of the 90th Light Division. They found fifteen guns destroyed by the Bays.

Halsted described his part:

> Jerry opens up on us now, and then anti-tank shells whip past us with sharp
> cracks. They keep so low that their flight can be seen by the swirl of sand
> just below the projectile. Very uncomfortable feeling that one may come
> inside at anytime. A very satisfying target in front of us. Gun firing well and
> the line rolling slowly on to the Jerry line. Then – a terrific crash and a cloud
> of smoke inside the tank and driver Wilson reels sideways. An anti-tank shell
> has come through the front. I nipped out of the top. Loader Mounsey came
> out of the side door, then a shell landed and made a nasty mess of him. I
> turned for morphia for Bradley who was out behind the tank but that was
> the last thing that I did. I felt a great blow on the face, and fell to the
> ground.[5]

For this action Colonel Draffen was awarded the DSO, Major Streeter and
Lieutenant Cummings the MC, and Sergeant Bunn the DCM. 'C' Squadron had lost
two Grants, with Sergeant Norfolk, Corporal Gristock and Trooper Mounsey killed,
and Lieutenants Radice and Halsted wounded. The two other squadrons lost three
Crusaders, with Corporals Ellis and Emery killed, and Lieutenant Anthony, Sergeant
McGuiness and Trooper Rowney wounded.[1, 2]

On 28 May the Bays moved to Aslagh, where at 5.30 a.m. on the 29th they
woke to find a concentration of enemy vehicles within 1,500 yards of their leaguer.
'C' Squadron attacked frontally with 'A' coming in from the flank, and within an
hour three anti-tank guns, several lorries and thirty-five prisoners from the 90th
Light had been taken. Almost at once 2nd Armoured Brigade was attacked by
more than a hundred German tanks. Viscount Knebworth's tank was knocked out
and all morning the Germans came on, first from one direction, then from another.
The battle continued until early evening, when a sandstorm reduced visibility. The
Bays were left still holding their position, having eliminated ten German tanks for
the loss of five of their own. Captain Baker, commanding 'B' Squadron, was killed
and Corporal Waterhouse badly wounded, together with other casualties. The
30th saw the Bays in reserve, although Sergeant Fox was killed by a stray anti-tank
shell and the echelon was dive-bombed by Stukas, with the loss of some lorries.

The Afrika Korps was now pulling back, leaving a strong rearguard on Bir El
Aslagh. The Bays took over the surviving tanks of the 9th Lancers and 10th
Hussars and became the only armoured unit left in the brigade. The regiment
remained quietly in position until 2 June, but the general situation had taken a turn
for the worse, with the 150th Brigade Box at Sidi Muftah being overrun and the
1st Army Tank Brigade destroyed. The Bays moved to Eluet El Tamar to cover the
gap, where they stayed until 5 June. They were then ordered to stop a German

advance from Bir El Harmat, and at once two Crusaders were hit, with Trooper Foster killed and others wounded. On 6 June a squadron of the 8th Hussars came under command and 'A' and 'B' Squadrons were merged under Major Weld, while Lord Knebworth went back to organise another squadron from recovered tanks. The Bays were told to go to the rescue of 22nd Armoured Brigade at the Knightsbridge Box, but they were unable to get round owing to enemy armour hull down on top of the escarpment, which was too steep to climb. They remained hull down engaging the enemy as best they could, and taking some prisoners who walked into their overnight leaguer. But the Germans had that day overwhelmed the 10th Indian Infantry Brigade as well as four regiments of artillery, and the Battle of the Cauldron had been lost.[1, 2]

During 7, 8 and 9 June the Germans intensified their pressure against the Knightsbridge Box, but were beaten off by shellfire. The Bays were now left with a strength of eleven Grants, ten Crusaders and one Honey, while the attached 8th Hussars had six Grants and one Honey. On 10 June they moved south, being joined by Lord Knebworth with a re-formed squadron of Crusaders. That night the gallant Free French had evacuated the Bir Hacheim Box, and on the 11th Colonel Draffen was ordered to take the Bays north, ending the day at Nadurat El Ghescheuasc, halted, facing the German armour. At first light on 12 June the Bays were in action against twenty-five tanks and more anti-tank guns to the south, with another forty to fifty tanks moving around their left flank. Major Streeter's tank was knocked out and Lieutenant Dean was wounded. By 3.30 p.m. heavy attacks were launched against the Bays and against 4th Armoured Brigade to their right, resulting in a fighting withdrawal under constant attack at close quarters.

On the 13th the Bays were relieved by Valentines of the 32nd Army Tank Brigade and withdrew for food and replenishment; but by 2 p.m. they were off again to the succour of the Scots Guards Box on the Maabus ridge, north of Knightsbridge. As they came into position Sergeant Gramson's tank was hit, killing him and Corporal Work. As evening drew on the forward companies of the Scots Guards were overrun, Lord Knebworth's tank was hit and his operator wounded. At 3.30 a.m. on 14 June the eleven remaining tanks, two Grants and nine Crusaders, started for the Acroma Gap, the Knightsbridge Box having been evacuated. Here the Bays awaited the inevitable German attack, and it was not long in coming. At 10 a.m. forty-one panzers attacked but were driven off, the Bays losing one Crusader, leaving them with ten tanks. At 3 p.m. the Germans came on again, driving in the infantry and Valentines posted on either side, so that the Bays found themselves faced by seventy enemy tanks and with their flanks wide open. Supported by the 11th RHA and the 1st Rifle Brigade this small force held firm until evening, when it was forced to withdraw slightly, losing Major

Weld's and two other tanks in the process. Here the remaining seven tanks and seven surviving guns of the 11th RHA held on until dark. That day the echelon was dive-bombed by Stukas, killing Lieutenant Pollock, Sergeant Clare and Corporal Kinsella. At midnight the exhausted Bays withdrew through the Acroma Gap, Colonel Draffen and Major Barclay being seen sitting, sound asleep, on the back of Barclay's Crusader, the colonel's tank having broken down. The Battle of Gazala was over.

Nineteen days of fighting resulted in the award of the Military Cross to Captains Crosbie Dawson and Tatham Warter and the medical officer, Captain Lewis. Sergeant Harris, Troopers Davidson and Emery, as well as Sergeant Edwards of the attached 9th Lancers, were awarded the Military Medal.[1, 2]

The King's Dragoon Guards had remained at rest in the Delta during January to March 1942. On 19 March 'C' Squadron departed once more for the Western Desert, to be followed nine days later by the rest of the regiment. The KDG came under command of 7th Armoured Division and moved to the south east of Bir El Gubi. The armoured car screen at this time was posted on the line Gazala-Rotonda Segnali-Garet El Asida, while the infantry positions stretched along the minefields between Gazala and Bir Hacheim. On 1 May 1942 the KDG took over from the 12th Lancers, patrolling extensively in hot and unpleasant conditions against an alert and better-equipped German adversary. They began to receive some armoured cars with their turrets removed and replaced by a captured German or Italian gun, but even this subterfuge made them no match for the faster and better-armed Germans. Throughout May the armoured car patrols played hide and seek with the enemy, relying on their desert knowledge and skill to outwit them. The Luftwaffe continued its attentions, twelve Messerschmitt 110s strafing 'A' Squadron for forty minutes without a break, destroying ten vehicles, and mortally wounding Corporal Stocks and SSM Acres. Two planes were destroyed, and a third came so low that it bent the wireless aerial on Captain Hellyer's armoured car.[3, 4]

Rommel opened the Battle of Gazala on 25 and 26 May, moving south of Bir Hacheim and overrunning the 7th Motor Brigade. The KDG were ordered to fall back on Bir El Gubi, being attacked several times en route by their own aircraft and suffering casualties as a result. Armoured car regiments operated far in advance of the main body, making identification difficult, and so were frequently attacked by friendly as well as enemy aircraft. As the Battle of Gazala unfolded, the KDG patrolled between Bir El Gubi and Bir Hacheim, having frequent clashes with the enemy and his supply columns. On 30 May 'C' Squadron, commanded by Major Palmer, was detached to 4th Armoured Brigade north of Knightsbridge. On arrival Major Palmer was ordered to move south of Bir Hacheim and engage enemy

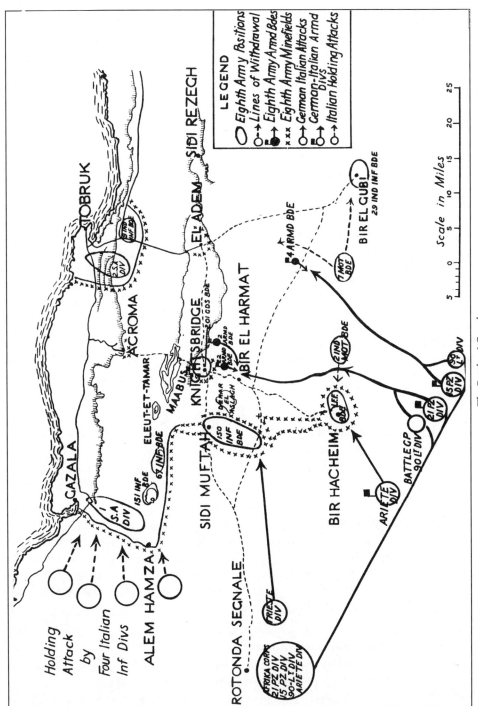

The Battle of Gazala

LEGEND

Eighth Army Positions
Lines of Withdrawal
Eighth Army Armd Bdes
Eighth Army Minefields
German Italian Attacks
German-Italian Armd Divs
Italian Holding Attacks

Scale in Miles

Holding
Attack
by
Four Italian
Inf Divs

TOBRUK
GAZALA
ACROMA
ALEM HAMZA
ROTONDA SEGNALE
SIDI MUFTAH
MAABUS
KNIGHTSBRIDGE
EL ADEM
SIDI REZEGH
BIR EL HARMAT
BIR HACHEIM
BIR EL GUBI
ELEUT-ET-TAMAR

S.A DIV
151 INF BDE
69 IND BDE
2 S.A DIV
3/10 INF BDE
101 GDS BDE
150 INF BDE
DEIR EL HAMATH
22 ARMD BDE
2 ARMD BDE
201 GDS BDE
4 ARMD BDE
7 MOT BDE
2 IND MOT BDE
29 IND INF BDE
3 IND MOT BDE
5 PZ DIV
21 PZ DIV
32 LT DIV
90 LT DIV
BATTLE GP
ARIETE DIV
TRIESTE DIV
AFRIKA CORPS
21 PZ DIV
15 PZ DIV
90 LT DIV
ARIETE DIV

transport. On 1 June a number of German trucks and some infantry were captured. Lieutenant Richardson's troop then came on some German tanks with their crews sunning themselves. Sergeant Fletcher's car, mounted with a captured 47 mm gun, scored a direct hit, setting one tank on fire. The remainder quickly withdrew a short distance, barring the way to any further excursions. On 3 June 'C' Squadron rejoined 4th Armoured Brigade, encountering on their way two German tanks which knocked out Sergeant Battersby's armoured car, killing Trooper Britten.

Rommel had now turned his attention to the Free French at Bir Hacheim, which held out until 10 June. Following its capture the Afrika Korps advanced on 12 June towards Acroma and El Adem, pushing back the British armour and inflicting heavy casualties. On the night of 15/16 June the Allied armour wended its way through the minefields to Tobruk, led by Major Palmer and 'C' Squadron, then on through the perimeter wire to the Bardia road. On the 16th 4th Armoured Brigade, which still had sixty tanks, moved towards Sidi Rezegh and Belhamed, again led by 'C' Squadron. On the 17th Lieutenant Batt's troop encountered the leading tanks of 21st Panzer Division, which soon overran Sidi Rezegh, engaging the tanks of 4th Armoured Brigade while 'C' Squadron patrolled both flanks of the battle. Sergeant Battersby's troop, becoming mixed up with the German tanks, calmly moved parallel with them, reporting all the time. It was a day of retreat, only lightened by the squadron coming across the Eighth Army's abandoned ration dump at Belhamed and heaping every car and vehicle with crates of tinned milk, bacon and fruit.

The KDG spent 18 to 21 June patrolling between the Egyptian frontier and Bir El Gubi, while Rommel launched his assault on Tobruk. 'C' Squadron then withdrew with the rest of 4th Armoured Brigade to the desert south of Mersa Matruh, where on 24 June it was ordered back to Alexandria to re-equip with new armoured cars. The Tank Delivery Regiment was eventually found near Amriya outside Alexandria, where new cars were drawn, and 'C' Squadron was joined by the rest of the regiment on 30 June.[3, 4]

At the end of May 1942 the rest of The King's Dragoon Guards were under command of 7th Motor Brigade, patrolling to the south of the Free French at Bir Hacheim. On 1 June the regiment moved to the west, where Lieutenant R. Whetherly captured nine German lorries with fifteen prisoners, while 'A' Squadron in an attack by Italian aircraft lost three lorries. On the 2nd 'A' lost three Marmon Harrington armoured cars on a new German minefield near Rotonda Segnali; Lieutenant J. Delmege was thrown from his turret and had to be evacuated, and Sergeant Leslie received injuries from which he later died. While the battle raged to the north, the KDG had a quiet time – apart from the attentions of the Luftwaffe. 'A' Squadron lost two lorries and an armoured car on 12 June. By the 13th the regiment was back in the Bir El Gubi area and was able to rescue odd parties of the Free French escaping after the fall of Bir Hacheim, including a tall Senegalese who consumed immediately a complete 7lb tin of marmalade. Sergeant Cranfield of 'A' captured two half-tracks and a staff car, with six Germans.

From 15 June to the 23rd the KDG continued to patrol in support of 7th Motor Brigade, generally covering the southern flank. On the 17th regimental head-quarters was heavily attacked from the air, at a cost of seven trucks, a staff car and an armoured car, with another armoured car badly damaged. Fortunately no one was hurt. 'A' and 'B' Squadrons were also strafed, losing vehicles. On the 18th Lieutenant Colonel McCorquodale moved to a new job and Lieutenant Colonel Kidston of the 12th Lancers arrived to take over command; at the same time Major Luck returned to the Delta, Captain A. Delmege taking over 'B' Squadron. The 19th saw more air attacks, RHQ losing two more trucks and 'A' losing an armoured car to an anti-tank gun. However, the technical adjutant arrived from the Delta with eight Daimler armoured cars, which proved to be a great improvement on the Marmon Harringtons. On the 21st RHQ was again visited by the Luftwaffe, losing the ambulance, the MO's staff car and a water truck, and Colonel Kidston was slightly wounded. News now came through of the fall of Tobruk and the regiment moved north to pick up any survivors, but without success. Throughout the 23rd to the 26th the KDG retired back through the frontier wire to a position to the south of the New Zealanders holding Mersa Matruh. On 29 June the regiment was ordered back to the Delta to refit, leaving all its remaining armoured cars with 'A' Squadron, who came under command of the Royals. They, in turn, had retired back by 30 June to the Alamein line. On 5 July 'A' rejoined the rest of the regiment.[3, 4]

The Queen's Bays had made their way back to Mersa Matruh by 16 June, where they rested for a few days before going on to Amriya. They were expecting a period of rest, but on the 24th they were told to return to Fuka to be re-equipped for immediate action, a squadron at a time, as the situation was so serious. By the evening of the 25th the Bays were at Fuka, 120 miles to the west, and on the 26th half of 'B' Squadron was given Honeys and put with a squadron of the 4th Hussars under Major Lord Knebworth, while the other half and one troop of 'A' was similarly equipped under Major Weld. RHQ and 'C', which was to be attached to the 3rd County of London Yeomanry, were given scout cars and Grants; the balance of 'A' was sent back to the Delta under Captain Harman. On the 27th the Bays' Honeys moved forward and at last light briefly engaged an enemy column at Abu Batta, losing two tanks with Lieutenant Taylor and Sergeant Mordue killed. There was confused fighting on 28 and 29 June, with more withdrawals as the Germans took Matruh. On the 30th the Bays made a successful attack on an enemy column of 3,000 vehicles, until a covering force of twenty-three Panzer IIIs and IVs came forward, knocking out Lieutenant Sherbrooke's tank and capturing him and his crew. By now the Bays were far to the west, and the night was spent breaking through the enemy leaguers, without a shot being fired by either side, until at 8.30 a.m. on 1 July a point a few miles west of El Alamein was reached after having covered fifty-four miles.

'C' Squadron, detached with the CLY, were breakfasting near El Daba on 30 June when four Italian motorcyclists, followed by a staff car and a general's caravan, then a long column of tanks neatly dressed two by two, appeared 1,500 yards away. Within minutes 'C' had knocked out at least seven tanks and the CLY had accounted for another six, by which time the Italians had fled. On 1 July 'C' was sent forward onto the Alamein position to engage a column of enemy tanks attacking an Indian infantry brigade, when they knocked out two tanks without loss to themselves. On 2, 3 and 4 July the squadron was in action on the Ruweisat ridge, stopping German attacks against the El Alamein defences. On 5 July the squadron handed over its three remaining Grants and two Honeys, having during the week destroyed twenty enemy tanks, as well as several guns and lorries, for the loss of three unrecovered Grants.

During this period Knebworth's composite squadron of Bays and 4th Hussars was on the Miteiriya ridge, also stopping the German attacks. On 2 and 3 July four Honeys had been lost, and by the 5th the squadron was down to one Crusader and three Honeys, three tanks commanded by Bays and one by a 4th Hussar. Sergeant Barker described the day:

> Away we went, Lord Knebworth leading, just his tank and our troop. As we approached the wadi, we kept our machine guns in action. The German infantry was bewildered at our attack and gave themselves up in their tens; my tank had around it 60 to 100 prisoners with their hands in the air. Then in front of me I saw my officer's tank burst into flames and the crew abandon; then Lord Knebworth reversed towards it to give them protection; his tank was also knocked out and became a raging furnace as the crew abandoned it.

Sergeant Barker then charged forward with the sole remaining Honey and silenced the anti-tank gun which had caused the damage, giving covering fire to the two dismounted crews. Sadly, Knebworth was mortally wounded by a rifle bullet on his way back. Sergeant Barker continues:

> Near a stone cairn, we dug his grave. I emptied his pockets. He had everything a soldier should carry and generally doesn't. Then, wrapping him in a blanket, we laid him to rest. There was no clergyman or officer,

but just four of his men who respected him. We commended him to God; then gave him his last salute. He was a splendid soldier and a good man.

On 16 July 'C' Squadron of the Bays was back in the line with reconditioned Grants, and on the 24th regimental headquarters and 'A' Squadron joined 'C' south of the Ruweisat ridge, while 'B' stayed in the Delta at Khatatba. The Bays covered the withdrawal from an abortive attack by the Australians on the Miteiriya ridge on 27 July. 'C' Squadron returned to the Delta at the beginning of August, RHQ and 'A' Squadron took over positions at the south of the Alamein position near Himeimat. At this time Lieutenant General Montgomery took over command of the Eighth Army and made it clear to all ranks that there would be no further withdrawals. On 14 August RHQ returned to Khatatba, where 169 reinforcements had arrived, and at the end of August Colonel Draffen departed on promotion and Major Barclay took over command. In the meantime Major Dance, commanding the detached 'A' Squadron, had gone sick and had been replaced by Captain Tatham Warter.[1,2]

On 31 August Rommel made his move to the south of the Alamein position against the Alam el Halfa ridge, where Montgomery had placed 16 medium, 240 field and 200 anti-tank guns, as well as 400 tanks, and 100 infantry anti-tank guns. 'A' Squadron had taken up a forward concealed position east of the Ragil depression, awaiting the German advance. As the leading Mark IIIs appeared, followed by Mark IVs, the Bays opened fire at a range of 900 yards, many of their 2-pounder shells being seen bouncing harmlessly off the German armour. But they did halt the advance, giving themselves time to fall back behind Alam el Halfa, drawing the Germans on to the main defences while the Bays retired to the reverse slope. During 1 September the battle raged in front, and by the 3rd, 'We went to inspect the damage. In front of the British positions was a graveyard of Nazi tanks. It was an encouraging sight.' 'A' Squadron, after the battle, joined the rest of the regiment at Khatatba, where on 10 September the first of the new Sherman tanks arrived. By the end of September 'B' and 'C' Squadrons were equipped with Shermans, and 'A' had Crusaders mounted with 6-pounder guns. On 20 October Colonel Barclay attended a lecture at Amriya, where General Montgomery explained his battle plan, which on his return that evening the Colonel passed on to all the officers. After dark on 21 October the Bays moved forward to their battle positions.[1,2]

RHQ and 'B' Squadron of The King's Dragoon Guards at Amriya in the Delta were re-equipped by 12 July. Lieutenant Colonel Kidston had been appointed to command his own regiment, the 12th Lancers, and Lieutenant Colonel Hermon of the Royals took over command of the KDG, Majors Drabble and Whetherly going

El Alamein and Alam el Halfa

to staff appointments. Major Crossley became second in command and Captain Hellyer took over 'A' Squadron on its return. RHQ and 'B' Squadron moved back on 13 July to cover an area in front of the Alamein position along the Deir Alinda, to be joined on 21 July by 'A' and 'C' Squadrons, and extending their patrols to cover the Himeimat area for a time. Patrolling continued through July and August under very trying conditions of heat, dust and flies, until Rommel made his bid against the Alam el Halfa position. The KDG, like the Bays, had sensed a marked change in the atmosphere with the arrival of Montgomery, but, to start with, as 'old desert hands' they were not enthusiastic about being told how to do things by a newcomer from England. Alam el Halfa and Alamein were to change that. As the Afrika Korps advanced on 31 August, 7th Armoured Division fell back towards Alam el Halfa, with the KDG observing the German movements from the flanks. Rommel did as Montgomery wanted him to do and attacked the Alam el Halfa ridge, while the KDG watched the RAF pound the enemy columns around Deir Ragil. By 3 September Rommel was withdrawing, defeated, leaving much of his armour wrecked on the battlefield. The KDG followed up the German retreat, and by the evening were in contact with the enemy, who had halted in the Deir Munassib, leaving behind them the wrecks of thirty-nine tanks and numerous other vehicles. On 10 September 1942 the KDG were withdrawn to the Delta for rest.[3, 4]

SOURCES

1 W. Beddington, *The Queen's Bays, 1929-1945*, 1954.
2 War Diary, 1940-1943, The Queen's Bays, Regimental Museum, 1st The Queen's Dragoon Guards, Cardiff Castle.
3 D. McCorquodale, B. L. B. Hutchings and A. D. Woozley, *History of The King's Dragoon Guards, 1938-1945*, published by the regiment.
4 War Diary, 1939-1943, 1st King's Dragoon Guards, Regimental Museum, 1st The Queen's Dragoon Guards, Cardiff Castle.
5 M. Halsted, *Shots in the Sand*, 1990.

41

El Alamein, El Agheila, Tripoli, Mareth, El Hamma
1942-1943

The Queen's Bays now prepared for what General Montgomery had told all the men of the Eighth Army was to be the decisive battle. The main British attack at El Alamein would be in the north on a front of about eight miles from the coast road to the Miteiriya ridge, with diversionary attacks on the Ruweisat ridge and holding attacks at the south of the forty-mile front at Himiemat. Four infantry divisions would assault on the night of 23 October 1942, behind a barrage of 800 guns, to clear a northern and southern corridor; 2nd Armoured Brigade, with the Bays and 10th Hussars in the lead, would exploit the northern corridor.[1, 2]

The regiment, under Colonel Barclay, moved off at 9.30 p.m. After a refuelling halt by Alamein station at midnight, it advanced through the Allied minefields, reaching and crossing the first enemy minefield at 4 a.m. on the 24th. A second minefield blew off the tracks of three Shermans, and then news came of a third minefield ahead, with the infantry held up 3,000 yards short of their objective. At dawn an enemy anti-tank gun opened up, knocking out a scout car, but was dealt with by a troop of 'B' Squadron, which deployed, as did 'C' to their left, losing two of their Shermans on mines. The two squadrons had an Australian battalion to their north and the Gordon Highlanders to the south, and at once they engaged the enemy anti-tank screen, which was supported by tanks. By 8.30 a.m. ten out of twenty enemy tanks facing them had been knocked out, and 'A' Squadron moved up between 'B' and 'C'. By the afternoon the Sappers had cleared passages through the minefields, and the Bays advanced on their final objective, Point 33. Immediately the two leading squadrons were met by a storm of 88 mm fire, which, within a minute, immobilised four tanks of 'B' Squadron and two of 'A'. Major Weld and Lieutenants Radice and Gay were wounded.

The eighteen remaining Shermans of 'B' and 'C' were then counter-attacked by a large number of German and Italian tanks, while the Bays linked up with the 9th Lancers to the south under cover of a smoke screen. The two regiments then fought side by side until dark, when the enemy withdrew defeated, leaving behind twenty-six tanks. Surrounded by mines the two regiments went into close leaguer, staying throughout the night under machine gun and sniper fire. The fitters had managed to repair most of the damaged Shermans, and these with the ammunition lorries were brought into the leaguer by Lieutenant Parker; Sergeant Godwin and Corporal Fiddler of the fitters were awarded the MM for their work under fire that day. There were now only twelve of the original twenty-nine Shermans left, so that a composite squadron was formed under Major Manger and Captain Crosbie Dawson.

The next day, 25 October, the Bays were again ordered to take Point 33. As the composite Sherman squadron advanced, supported by the Crusaders of 'A' Squadron, the two leading troops were met by 88 mm fire at point blank range, from guns which the Germans had brought up during the night and stationed behind their wrecked tanks. Five Shermans were knocked out in quick succession and Lieutenants Christie-Miller and Barnado killed. The Bays withdrew to their former position and, hull down, engaged the guns. News came that the medical officer, Captain Lewis, had been killed. During the afternoon the Bays, and the Australian infantry on their right, were attacked by a strong force of enemy tanks, but between them they destroyed eighteen and the attack was not pressed. At 5.30 p.m. the 10th Hussars attempted to take Point 33, but lost eight Shermans to the 88s.[1, 2]

The 26th was spent preparing for a night attack by the 7th Motor Brigade on Point 33. Eleven replacement Shermans and five Crusaders arrived, giving the Bays a strength of eleven Shermans and sixteen Crusaders. During the night 2nd Rifle Brigade of 7th Motor Brigade established themselves on a feature south of Point 33, called Snipe, where they were attacked by tanks but knocked out fifty-two of them before they ran out of ammunition. Lieutenant Parker of the Reconnaissance Troop managed to work his way onto Point 33, followed by a troop of 'A' Squadron. They were joined by Major Tatham Warter and Lieutenant Dallas but compelled to withdraw by anti-tank fire. A second attack in the afternoon petered out, and the Bays leaguered for the night with the 9th Lancers, while preparations were made for a night attack by 133rd Infantry Brigade, which was successful. The Bays remained in reserve during 28 October, coming up at 9 a.m. to meet an enemy tank attack, which was dealt with by the artillery. An Italian tank was knocked out with the first shot at a range of 3,500 yards. By 8 p.m. the Bays were relieved by the Staffordshire Yeomanry of 8th Armoured Brigade, and withdrew to Alamein station for refitting.

Spare crews and fifteen diesel Shermans arrived, and 'B' Squadron was re-formed under Captain Crosbie Dawson. On 2 November the Bays moved up again as part of a force consisting of 1st and 7th Armoured Divisions, which was to exploit a exploit a breakthrough to be made by 151st and 152nd Brigades supported by 9th Armoured Brigade. As the regiment cleared the minefields it came under heavy 88 mm fire from along the Rahman track, 'B' losing two tanks and 'A' one. As the early morning mist cleared the Bays and 9th Lancers engaged the guns from hull down positions, silencing them after three quarters of an hour. Major Tatham Warter was killed and Captain Harman took over command of 'A' Squadron. Throughout the day there was a succession of tank battles as 21st Panzer Division sought to restore the situation. By nightfall every tank had used all its ammunition, but 1st Armoured Division was still in place east of the Rahman track, and sixty-six enemy tanks had been destroyed, eleven of them by the Bays.

Throughout 3 November the Germans held Tel El Aqqaqir and the Rahman track, but a night attack by 5th Indian Infantry Brigade broke through, and the Bays with twelve Shermans and twelve Crusaders advanced eight miles before encountering, and overcoming, a rearguard. Four miles further on, 'C' Squadron was halted at Tel El Mansira by an anti-tank gun which knocked out Sergeant Dumbleton's tank, killing him and his turret crew. At dawn on the 5th Tel El Mansira was found abandoned and the chase continued for fourteen miles, when another rearguard was encountered, which withdrew after a short fight. The Bays, now four miles south of Daba, dealt with one 88 mm, but were held up by two more skilfully sited guns. By 7.45 a.m. on the 6th the exhausted crews had covered another forty miles, but petrol and diesel was short. The seven remaining diesel Shermans were sent to the 9th Lancers, while Captain Harman took the six Crusaders to form a composite squadron with the remnants of the 10th Hussars. Harman reached high ground twenty-five miles north, near Charing Cross, where the enemy could be seen streaming away, but there was no petrol to pursue them. Heavy rain fell during the night of the 6th, and on the 8th the coast road was cut; but the retreating enemy, saved by the rain, had slipped through.

The Queen's Bays as part of 1st Armoured Division remained near Matruh for three days, collecting new tanks, and then moved on to El Adem, where they heard that they were to undergo training and re-equipment at Tmimi. Having handed over most of their Shermans to the 22nd Armoured Brigade, the regiment went into camp in the desert near Tmimi, where they stayed for the next three and a half months. During this period ten new officers joined, Major Weld returned from hospital, and a number of decorations were announced: Lieutenant Colonel Barclay, Captain Nicholson, Lieutenants Tomkin, Ward and Parker, together with the medical officer, Captain Conway, were all awarded the Military Cross; the Military Medal went to Corporal Cockroft, and Corporal Andrews won a bar to his MM. The Bays left Tmimi for Tripoli on 1 March, 1943, reaching Ben Gardane near the Mareth Line on the 13th.[1, 2]

The King's Dragoon Guards arrived at Tahag Camp, Qassassin on 14 September 1942 – the very same patch of desert they had occupied on their arrival in 1940. On the 21st they were warned for duty in Cyprus, but in fact found themselves at Mena on the outskirts of Cairo, where they stayed for six weeks, being subjected to order and counter-order. On 12 November the regiment, re-equipped on the basis of one Daimler and two Marmon Harrington armoured cars per troop, started back for the desert, having sent 'B' Squadron on ahead for attachment to 8th Armoured Division. On arrival at Acroma on the 16th 'B' rejoined, and the regiment moved to Taormina until the 24th, where many of the Marmon Harringtons had their turrets removed and replaced by captured German 20 mm and 37 mm guns, scrounged from Tobruk along with other useful stores. The KDG arrived at Agedabia on 26 November and remained in reserve to 7th Armoured Division until 1 December.[3, 4]

On 2 December Major Palmer arrived back from Cairo, taking over command of 'C' Squadron, Captain Chrystal returning to 'B' as second in command to Major Delmege. Palmer and Chrystal were summoned to the corps commander, Sir Oliver Leese, because of their intimate knowledge of the ground around El Agheila. On arrival they found themselves questioned by General Montgomery, and their plan for a reconnaissance bypassing El Agheila to the south was adopted. Lieutenant Richardson with three jeeps, each with two men, set out with a backup party of a petrol lorry and six armoured cars under Captain Chrystal. The jeeps left Chrystal's party on the frontier between Cyrenaica and Tripolitania, and making a wide sweep to the south and then north west they reconnoitred a way to the coast road near Marble Arch, some eighty miles west of El Agheila. The jeeps were spotted by an enemy aircraft but not attacked. Soon afterwards, however, three ME 109s spent twenty minutes unsuccessfully searching the area where the jeeps had been hurriedly hidden among inadequately small shrubs. Richardson realised that they would be seen as soon it became light the following day, so he turned the jeeps round and the tired drivers drove back over the rough and unknown ground throughout the night. Shortly after daybreak they rejoined Captain Chrystal, and the whole party was back with the regiment on 6 December, with a detailed report of the way to outflank the German position at El Agheila. For this outstanding patrol Lieutenant Richardson was awarded a bar to his Military Cross.

On 7 December a second patrol of four Marmon Harringtons and three jeeps found that the oasis at Marada was not held by the enemy, and captured two Germans who had driven from El Agheila to Marada. On 10 December the KDG led the New Zealand Division and the 4th Light Armoured Brigade on a left hook around the El Agheila position over the ground reconnoitred by Richardson. On the 13th they were rejoined by the Marada patrol, whose jeeps had run into a minefield, losing two, with Sergeant Manning killed and Corporal Reeve and Trooper Edward wounded; the remaining jeep drove 150 miles with the five survivors, two of whom were wounded. The main part of the regiment crossed the frontier wire, cutting a fifty-yard gap and pressing on to leaguer ten miles south of the coast road, pausing to give the New Zealanders time to catch up. On the 16th the KDG made contact with enemy patrols along the coast road. The New Zealanders and 4th Light Armoured Brigade closed the road, hoping to cut off the enemy. By chance the retreating remnants of the three German divisions hit a gap between the New Zealanders and the Armoured Brigade, enabling the majority to escape, although the tanks of the Scots Greys, the New Zealand artillery and the RHA inflicted heavy losses on the moving mass of German armour and lorries. The discipline shown by the Afrika Korps in what was for them a very tight spot earned the admiration of their adversaries.

On the 17th the regiment advanced towards Nofilia, which was found to be held by a strong rearguard, but on the 18th the advance continued to Sirte. The KDG were relieved by the Royal Dragoons on 21 December and spent until the 25th resting. During the afternoon of Christmas Day the regiment moved to the west of Sirte, coming up against the enemy, dug in around Buerat, on the 28th. There they remained until 15 January 1943. Over this period they took a number of prisoners, knocked out a German armoured car and some trucks, and captured a staff car. While at Buerat Lieutenant Batt's troop was being chased by German tanks, when a Stuka managed to knock out one of his armoured cars, killing Trooper Stoker and wounding Sergeant Phillips. Batt managed with great skill to extricate his troop, and was awarded the Military Cross for his gallantry and cool leadership. The regiment rescued several RAF pilots whose planes had been shot down, on one occasion winning a race with a German armoured car. Several prisoners from the German 3rd and 33rd Reconnaissance Battalions were captured, one of whom remarked on seeing the double-headed eagle, 'Ah, the KDG – I captured your officer, Collie, last year!' The KDG, under command of the New Zealand Division, formed the strongest of friendships and developed the greatest admiration for the New Zealanders.

On 15 January 1943 the advance resumed, with The King's Dragoon Guards acting as flank guard to another 'left hook' past Tripoli to cut off the retreating enemy. The rough nature of the ground made slow going, but a German rearguard was encountered near Beni Yulid, where the regiment took over the lead, and three Stukas killed four men of 'C' Squadron, wounding seven more. On the 23rd Tripoli fell, largely due to the pressure exerted by the southern outflanking move. By 24 January the bad going over 200 miles of appalling mountainous country had reduced the strength of the regiment to twenty-seven Marmon Harringtons and three Daimlers, and so two composite squadrons were formed, until Captain Birks came up from the Delta on the 28th with sixteen AEC armoured cars. On the 24th 'B' Squadron's rapid advance surprised the ground staff of a German aerodrome at Bir El Ghnem, capturing twenty prisoners and four lorries. Major Lindsay rejoined on the 25th, taking over 'C' Squadron, with Major Palmer moving to 'A'. The regiment now concentrated for rest and maintenance south west of Tripoli until 3 February, when it resumed the advance on the southern flank.

A small party under the commanding officer, Colonel Hermon, went to Tripoli on 3 and 4 February for a Victory Parade of 7th Armoured Division and the 51st Highland Division before the Prime Minister, Mr Winston Churchill. On the 5th Lieutenant Eggleton knocked out a German eight-wheeled armoured car with his first shot and damaged another, and Lieutenant Hempson disabled and captured a four-wheeler. 'C' Squadron's water cart, an essential piece of equipment in the desert, went to Dehibat where there was a good supply of water; but there also turned out to be four German armoured cars. Troopers Price and Fenn hid in the house of a Frenchman, who supplied them with Arab clothes. They then walked past the Germans and their captured water cart, and after a twenty-five-mile hike rejoined the regiment the following day. The four German four-wheelers then drove up a track where they met Sergeant Fletcher's troop, which knocked out the command car and captured the German officer and his crew. When Price and Fenn arrived back after their tramp, the German officer exclaimed, 'So that's how you got past me!' On 6 February 1943 the KDG crossed the frontier into Tunisia, leaving behind the desert with all its particular folklore.

From 7 February until the 20th there was constant patrol activity, with clashes between the regiment's patrols and those of the enemy, and with the Luftwaffe active, strafing and bombing. A German four-wheeler was knocked out on the 8th; on the 9th Sergeant Smith's car was hit, with Trooper Ventress, Corporal Emons and Smith all wounded, and Lieutenant Macartney's car had a shell through the turret which took away the wireless set and hurt no one; on the 10th Lieutenant Phillips's armoured car was hit, Corporal Winnard being killed with Phillips and Trooper McAleavey wounded; on the 13th RHQ was bombed and strafed; Sergeant Berryman's AEC was captured on the 14th, and a 'B' Squadron Humber was disabled on the 19th. By the 20th the advance had got under way, Medinine was reported clear and Foum Tathouine isolated, as the Eighth Army slowly closed up to the Mareth Line. From 20 February to 1 March 1943 the KDG patrolled up to the Mareth Line, which proved to be dangerous work as the enemy came out at night laying mines on tracks that had previously been cleared. The CO's AEC was blown up, then 'A' Squadron lost two AECs on mines and a Scammell which was trying to recover them, another AEC was lost on the 26th. As the armoured cars became held up in the enclosed country of Tunisia, individual troop leaders started to work forward on foot, Lieutenant Cassels carrying out a particularly bold patrol, ending in a skirmish with some German infantry, for which he was awarded the MC. On 1 March the Royal Dragoons relieved the KDG, who returned to Ben Gardane to be re-equipped with Humber and AEC armoured cars, saying goodbye to the South African Marmon Harringtons which had been their mainstay since they arrived in the Middle East.

The Queen's Bays arriving at Ben Gardane on 13 March 1943 also found themselves in a new atmosphere – very different from their three long years in the desert, as they took up concealed positions in a wadi near Medinine in preparation for the assault on the Mareth Line. The infantry attacked under cover of a heavy bombardment on the night of the 20th, but by the 22nd German counter-attacks made Montgomery shift the emphasis to his left flank. The King's Dragoon Guards, having been re-equipped with brand new Humber armoured cars and enough AECs for one per troop, moved on 11 March, leading the 8th Armoured Brigade and the New Zealand Division in the projected left hook to El Hamma, so outflanking the Mareth Line. The advance started at dawn on the 20th, but the going was so bad, with rearguard opposition near Bir Soltane, that with RHQ in the lead the hills overlooking El Hamma were not reached until the evening. It then transpired that the gap was only lightly defended by an Italian Sahara division. The difficult country had also held up the New Zealand Division with 8th and 2nd Armoured Brigades, who did not arrive until nightfall. During that night the enemy reinforced the gap with 21st Panzer and the German 164th Infantry Divisions. On the 21st the New Zealand infantry probed forward, while the KDG protected the left flank and tried to find a way through the Tebaga mountains. Colonel Hermon was wounded in the leg (just as the news came through that he had been awarded the DSO), so Major Lindsay took over command. Trooper Joy was killed in low-level ME 109 attacks and Corporal Winstanley and Trooper Jones were wounded, but a total of 1,500 Italians were captured together with twelve guns. From 22 March until the main attack started on the 26th the KDG continued to probe, 'C' Squadron capturing sixty-three Italians and three guns on the 23rd, and rescuing their twenty-fourth RAF pilot.

On the 26th and 27th the KDG observed the battle, collecting many prisoners in numerous skirmishes. Lieutenant Macartney of 'B' Squadron captured two guns, killing twelve enemy and driving a further 200, whom he had cut off, to be picked up by the infantry – a feat for which he was awarded the MC. 'C' Squadron captured over a hundred German prisoners. On the 28th 'A' Squadron opened up with everything they had, eventually being ordered to lead the advance, at which point they handed over two complete Italian battalions, which they had captured, consisting of 32 officers and over 700 men of the Spezia Division – a record bag for an armoured car regiment. By the 29th the enemy were withdrawing and the KDG were ordered to dash for Gabes, which was entered by 'B' Squadron before midday. As 'A' and 'B' Squadrons pushed forward until they met German rear-guards, it became apparent that the enemy intended to make a stand on the Wadi Akarit.[3, 4]

On 26 March The Queen's Bays, after the terrible approach march, were in position to support the New Zealand Division's drive to El Hamma. At 4 p.m. the New Zealand infantry and 8th Armoured Brigade moved forward with tremendous artillery and air support, and an hour and a half later they had reached their objectives. The Bays on the left of the El Hamma road, with the 9th Lancers on the right, moved up in support, and then pushed forward on their own. They soon came under fire from anti-tank guns, tanks and infantry, and found themselves amidst the burning tanks of 8th Armoured Brigade. The intense fire held up the advance and by dusk the main objective had not been reached, so Colonel Barclay ordered them to press on in the dark. The German gunners, unable to aim, were taken completely by surprise; many gun crews surrendered and 250 prisoners were taken. The Bays had lost a number of tanks but had suffered surprisingly few casualties. There was a two-hour pause for the moon to come up, and then the advance continued. The enemy was utterly disorganised, with parties of infantry, lorries, guns and staff cars appearing indiscriminately, some bumping into the advancing tanks, and all being shot up or captured, while lines of red tracer from the tanks' machine guns pierced the moonlight. Lieutenant Gill destroyed a huge 210 mm gun; two 88 mm, a 75 mm, five lorries, two tractors and a Volkswagen were knocked out, and many more were overrun. At 4.30 a.m. the regiment halted.[1, 2]

At dawn on the 27th El Hamma was in sight. The 2nd Armoured Brigade had broken right through the 21st Panzer Division, who were now trapped between the brigade and the New Zealanders. The Bays had lost two Shermans knocked out, two with their tracks damaged, and Sergeant Smith's Crusader trapped in a deep wadi. Smith and his crew hid in a slit trench as German tanks approached, firing at the top of their trench, luckily with armour-piercing shots. Soon a very smart Italian officer walked up and asked, 'Are you my prisoner, or am I yours?' Smith immediately claimed him, whereupon the officer led Smith round a bend in the wadi and surrendered an 88 mm in perfect order, together with twelve officers and thirty-eight soldiers.

The Bays were now ordered to capture El Hamma. As they advanced they encountered a screen of 88 mm guns from the 15th Panzer and 164th German Infantry Divisions, who had been rushed up to hold open the road. Four tanks were lost, with Corporals Nolan and Evans killed, and Captain Widdrington and

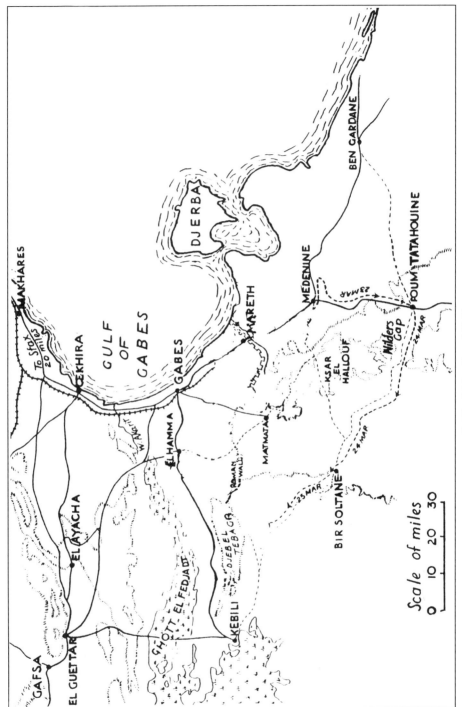

Battle of Mareth and El Hamma

four men wounded. The Bays withdrew to positions overlooking El Hamma, having captured twenty-three prisoners. On the 28th shelling killed Troopers Doxon and Every in the echelon, and wounded five others. Dawn on the 29th revealed that the Germans had retreated. The Bays advancing through El Hamma, until some seven miles further on they were halted by a strong enemy rearguard. The Germans held their position on the 30th and 31st, then on 1 April the Bays were moved back into reserve. For their part in the Mareth battle Major Weld and Lieutenant Jackson were awarded the Military Cross, and Sergeant Smith and Cockwill the Military Medal.[1, 2]

The battle of the Gabes Gap to force the German position at the Wadi Akarit was carried through by the infantry of the 51st Highland Division, 50th Division and the 4th Indian Division. The KDG waited to lead the New Zealand Division through the Gap.

SOURCES

1 W. Beddington, *The Queen's Bays, 1929-1945*, 1954.
2 War Diary, 1940-1943, The Queen's Bays, Regimental Museum, 1st The Queen's Dragoon Guards, Cardiff Castle.
3 D. McCorquodale, B. L. B. Hutchings and A. D. Woozley, *History of The King's Dragoon Guards*, published by the regiment.
4 War Diary, 1939-1943, 1st King's Dragoon Guards, Regimental Museum, 1st The Queen's Dragoon Guards, Cardiff Castle.

42

Wadi Akarit, Gabes, Tebaga, Tunisia, Tripoli, Algeria

1943

The King's Dragoon Guards made their way through one of the two gaps in the minefields, cleared by the Royal Engineers, between the Wadi Akarit and the Tebaga Fatnassa mountains. As they drove, they realised the achievement of the infantry of 4th Indian Division in storming the almost perpendicular heights of the mountains through which they were passing, and in clearing such a position in a day and a half. The regiment then passed through the plain beyond, encountering disorganised resistance some thirty miles north of Akarit. 'B' Squadron shot up numbers of stray lorries, taking both German and Italian prisoners, until the leading troops came to some high ground held by a rearguard of German infantry supported by Tiger tanks. On 8 and 9 April the squadrons probed forward, destroying an 88 mm and several lorries, as well as capturing prisoners from 10th Panzer Division around Mezzouna, which was still held by the enemy. 'B' Squadron had a particularly good day on the 9th, reaching Triaga airfield, catching the Germans by surprise and destroying nine lorries, two staff cars and a Volkswagen, which did a complete somersault when hit by fire from Lieutenant Cassels.[1, 2]

The Queen's Bays moved through the Gabes Gap on 7 April, 'A' Squadron, under Major Harmon, fanning out to the south in a sweep which netted twenty prisoners, while 'C' Squadron knocked out a Panzer Mark IV Special and a Mark III Special with only two shots. The crews were taken prisoner. On the 8th the regiment closed up to Mezzouna, which on the 9th was found to be clear, and so the advance continued for twenty miles until a force of twenty to thirty enemy tanks, including Tigers, was encountered. The German tanks steadily withdrew and the Bays followed up carefully, moving from one hull down position to another. On the 10th the Bays withdrew to Bouthada, near Sfax, resting for five days.[3, 4]

The KDG were ordered to cut the main coast road north of Sfax on 10 April. Moving across, they made contact with the 11th Hussars, who had just taken Sfax, and occupied the village of Djebeniana. On the afternoon of the 11th the KDG led the New Zealand Division forward, having lost contact with the enemy, until Lieutenant Weinholt rounded a corner to be knocked out by a German anti-tank gun at fifteen yards' range. Lance Corporal Saunders was killed and Trooper Barton wounded. Early on the 12th 'B' Squadron entered Sousse, to be met by enthusiastic crowds. Lieutenant Cassels's troop pushed on north, surprising and capturing a German machine gun post, but coming up against the 90th Light Division on the hills three miles beyond Sousse. Here the KDG had the doubtful honour of being the first British troops to come under fire from a new German weapon, the multi-barrelled Nebelwerfer mortar, which became known as 'Moaning Minnie' from the noise its bombs made as they descended. Early on the 13th the KDG, with the New Zealand Divisional Cavalry, made a sweep to the west. 'A' Squadron soon captured a German major and two privates from 15th Panzer Division, complete with a large 18-ton half- track and two lorries, shortly followed by another truck and sixty Italian Coastguards. By evening the regiment had closed up towards Enfidaville, where it was clear that the enemy was making a stand. On 14 April the KDG moved to an area west of Kairouan, resting until the 19th and then moving round to join the First Army for the final battle in North Africa.[1,2]

The Queen's Bays also moved round to join the First Army when, on 14 April, the tanks, their sand-coloured look now changed to dark green, were loaded onto transporters. By the 23rd they were concentrated three miles south of Goubellat. Moving out with the 9th Lancers, the Bays made slow progress, advancing five and a half miles by dusk, delayed principally by mines which had been laid in standing corn and were very difficult to detect. Sergeant Brookfield lifted a large number, but six Crusaders and two Shermans were damaged. Next day the advance continued; crossing another minefield with the help of the Royal Engineers, the Bays reached the Goubellat plain, which turned out to be crossed by a succession of deep wadis. The final wadi was passable at only one place and as 'A' Squadron emerged in single file, the tanks came under anti-tank fire from a wood at 100 yards' range. Lieutenant Cottier's, Sergeant C. W. Smith's and Sergeant Grayson's Crusaders were knocked out, the two sergeants being killed, together with Corporals Hunt and Garforth, and Troopers Newbold and Terry. A further two Crusaders were lost on mines. Because of this opposition the line of advance was changed to a thrust to the east and north east.[3,4]

The Bays set out at 7 a. m. on 25 April and, after an advance of three and a half miles, engaged some German tanks and anti-tank guns near Borj Baj Hamba. A Tiger, firing from a flank, knocked out six of 'B' Squadron's tanks, including Major Crosbie Dawson, Lieutenant McVail, Sergeants Rowe and Southam, and Corporals Avis and Thompson, with Troopers Parkes and James killed. However, three tanks were soon recovered and one German tank was destroyed. RHQ was then attacked from the air and the regiment came under heavy shellfire. The whole area was dominated by the rocky peak of Kournine, which was unsuccessfully attacked by the infantry on the nights of 25, 28 and 29 April. On the 30th the Bays attempted to press on to the north east, but were unable to make progress as the terrain was completely covered by German tanks firing from hull down positions. 'B' and 'C' Squadrons each lost a tank, with Sergeant Webber and two others wounded. On 1 May a 'C' Squadron tank was knocked out by an 88 mm, with three men wounded, but it was recovered after dark. The momentum of this nine days' battle was now petering out, and on 2 May the Bays concentrated at Bou Arada, where news came through that Lieutenant Colonel A. H. Barclay had been awarded the DSO, Major Weld and Lieutenant Jackson the MC, and Sergeants Cockwill and L. W. Smith and Trooper Haggard the MM.[3, 4]

The King's Dragoon Guards rested at Kairouan until 19 April, when the regiment came under command of the French XIX Corps. The KDG experience of First Army was not happy: order and counter-order caused the commanding officer to seek an interview with General Montgomery. 'Sir, we are at present the right-hand unit of First Army under French XIX Corps. May we remain where we are and become the left-hand unit of Eighth Army, and then be able to draw some rations?' As a result the regiment returned to Eighth Army and was revictualled! On 22 April a jeep patrol from 'B' Squadron met some Arabs. While talking to them, the patrol was confronted by more Arabs and fifteen Germans and withdrew. On the regiment's left was a mixed French regiment of Goums and Spahis, who operated in unorthodox ways, most of which were extremely unpleasant for their enemies. On the 26th Corporal Erratt was killed on a mine while bathing in a nearby river, and Trooper Wood was wounded. On the 27th an advance was made with a combined force of the KDG, 3rd RHA, and a company of the 2nd KRRC, all under command of Colonel Lindsay, which by the 29th had pushed through one range of hills, establishing contact with the Goums, who had been making a similar advance on their left but lost three AEC armoured cars on mines.[1, 2]

On 30 April orders came to move 100 miles to Bou Arada, where the KDG were given a sector to cover from Pont du Fahs to Kournine, bringing them into the same area as The Queen's Bays. The first few days were spent patrolling and on 7 May 'B' and 'C' Squadrons advanced, engaging in numerous small skirmishes, knocking out an enemy armoured car, a large lorry and a motor cyle combination, and capturing another combination, which resulted in thirteen dead Germans and three prisoners. The lorry turned out to be the concert party of 21st Panzer Division, complete with musical instruments and a piano which had lost two keys when The Queen's Bays inadvertently opened fire on the captured lorry before mutual recognition was achieved. On the 8th an AEC of 'C' Squadron with

Lieutenant Curtis became bogged down when under fire from anti-tank guns. As the crew were getting the car free, it was strafed by Spitfires and then two German tanks with infantry came on the scene, capturing Curtis, Corporal Darley and Trooper Embleton. The Spitfires then attacked Squadron HQ, wounding Major Cairns in the hand. Major Eden of 4th RHA asked a driver at Squadron HQ where the colonel could be found and was told 'Oh, he's much further forward, sir, this is only Squadron HQ.' On 9 May Captain Chrystal and Trooper Tarrant were wounded by shellfire, and on the 11th the regiment moved to Grombalia, when the news came that Tunis and Bizerta had fallen.[1, 2]

As the KDG moved to Grombalia, enemy equipment littered the roadside together with columns of dejected Germans marching into captivity. The prisoner of war cage at Grombalia already contained 15,000 Germans, with more pouring in. On 12 May the KDG advanced in line across the Cap Bon Peninsula collecting prisoners. During the day Corporal Darley and Trooper Embleton arrived back, having been briefly captured by the 21st Panzer Division. On the previous night the Germans had given them two rifles and at first light these two KDG, followed by the remains of 21st Panzer Division, 1,500 men, marched off until they met the Derbyshire Yeomanry, to whom the two KDG handed over their erstwhile captors, now their prisoners. The Germans had been terrified of falling into the hands of the Goums and presented Darley with a staff car in gratitude for their safe conduct to captivity. Lieutenant Curtis returned the following day, having been handed over by the Germans to the Italian military police for evacuation to Italy on a hospital ship. While he was waiting to be embarked, some German soldiers tried to rush the ship and the Italians opened fire on them. Then some Bostons bombed the area. Curtis persuaded the Italians that the harbour was too dangerous a place and so managed to escape.[1, 2]

The Queen's Bays advanced on 8 May to Oudna, 'A' Squadron moving up to support a KDG squadron and then taking over the advance as far as a T-junction, where a Crusader was hit by German tanks and anti-tank guns, who halted further progress. 'C' Squadron moved round on the right flank, destroying a half-track and capturing some lorries and eighty prisoners. On 9 May the Bays continued to lead the advance to cut off the Cap Bon Peninsula, dealing with pockets of isolated resistance and capturing Creteville by 11 a. m. against 88 mm and artillery fire from 90th Light Division. 'C' Squadron then took on entrenched German infantry and anti-tank guns, methodically eliminating them and knocking out a German Tiger, when Major Weld was wounded and Captain Widdrington took over command of the squadron. Before daylight 'C' Squadron advanced again, destroying two 88 mm guns. At midday two German officers appeared carrying a white flag, saying that their unit had run out of ammunition and was willing to surrender.

Tripolitania and Tunisia

Three quarters of an hour later two more German officers arrived, folowed at 4 p. m. by 12 officers and 400 men of the Hermann Goering Division. Later more Germans from the 19th Flak Division appeared, followed by some Italians from the Superga Division, making the bag for the day more than 600 officers and men. Captain Widdrington was awarded the Military Cross for his leadership of 'C' Squadron that day.[3, 4]

The Queen's Bays were ordered to push on to Grombalia, which had been taken by the 9th Lancers, but German tanks were still giving the 10th Hussars trouble in the hills behind, and so the reserve squadron at the rear of the column, which happened to be 'C', was sent to their aid. A German envoy soon arrived to say that their tanks had run out of ammunition and that the crews, all from 10th Panzer Division, had destroyed their vehicles and were on their way to surrender. For the next five days the Bays helped to round up Germans and Italians, all driving themselves into captivity by the thousand. Guns and equipment lay abandoned everywhere. The Allies took 250,000 prisoners and less than 700 Germans managed to escape by sea. The official day of victory was declared as 12 May 1943, and the Bays were that day issued with khaki drill.[3, 4]

The King's Dragoon Guards settled down on the west coast of the Cap Bon Peninsula, while The Queen's Bays were at Soliman. Officers and men of both regiments visited each other. On 27 May 1943 1st Armoured Division, together with the KDG, moved from Tunisia to Tripoli. After a tiring march of four days both regiments found themselves allotted barren areas of desert to the south of Tripoli near Souani Ben Adem. The KDG area was slightly relieved by an Italian fort, while both regiments were able to send parties to bathe in the sea each day. On arrival the KDG heard that Major Llewellen Palmer had been awarded the DSO, Captain Chrystal a bar to his MC, Lieutenants Batt, Macartney, Clarke and Cassels the MC, while Sergeant Robinson received the DCM and Sergeants Norton, Plumb, Mercer and Corporal Erratt were awarded the MM.[1, 2, 3, 4]

During June the Bays converted their last Crusader squadron to Shermans, and on the 14th Lieutenant Colonel Hermon returned to the KDG from hospital, resuming command. On 21 June both regiments paraded for inspection by a mysterious General Lyon, who turned out to be His Majesty The King. At the end of June the Bays, after an intervention by General Montgomery, moved to a camp site by the sea twenty miles west of Tripoli at Olivettia, where they heard that Captains Harman and Widdrington had been awarded the Military Cross, and SSM Strutt, Sergeant Brookfield and Corporal Merewood the Military Medal. The Bays were now told that they would have to hand over all their tanks and move to Algiers, where in three months' time they would be re-equipped. The KDG spent the summer at Tripoli training, being frequently inspected, sending parties to the

beach, organising sports facilities of various kinds, and on 19 July hearing that
Major Lindsay had been awarded the DSO, Lieutenants Hempson and Curtis the
MC, Sergeant Plumb a bar to his MM, and Sergeants Redfearn, Poynton and
Addison and Trooper Lamour the MM. On 28 July the KDG were allowed to
move to an area by the sea, where troop training went on apace. On 19 August an
AEC armoured car overturned, killing Lieutenant Lofthouse and Trooper Brothers.
By the beginning of September new equipment, including six American half-track
vehicles mounted with a French 75 mm gun, began to arrive. The regiment was
now fully equipped with Humber armoured cars, which were sent off in batches for
waterproofing. On 5 September the KDG moved to a transit camp in preparation
for the invasion of Italy, and on the 9th came news of the Allied landings at
Salerno.[1, 2, 3, 4]

The King's Dragoon Guards now had three sabre squadrons, each of five troops,
with a squadron headquarters, a headquarter squadron, and a gun battery of six 75
mm guns mounted on American half-tracks, capable of direct or indirect fire. The
wait to embark was long and tedious, for things had not been going well with the
landings and the German resistance had been fierce. Allied hopes had been raised
by the surrender of Italy, but the Germans had reacted quickly and the British and
American divisions faced only first-class German troops. On 20 September 1943
the KDG embarked regimental headquarters together with 'A' and 'B' Squadrons in
two American LSTs, 70 vehicles and 200 men being packed into each ship. 'C'
Squadron, under Major Cairns, and the echelon, under Major Delmege, remained
briefly at Tripoli, embarking on 28 September.[1, 2]

The Queen's Bays left Olivettia on 1 August 1943, arriving north east of Enfidaville on the 4th, where they took over the guarding of 7,000 Italians in a
prisoner of war camp at Medjez el Bab. The Bays stayed at Medjez el Bab until 1
October, when they entrained for a three-and-a-half-day journey, arriving finally at
Boufariq, fifteen miles south of Algiers. They were quartered in three farms, and
very soon the whole regiment was under the cover of a roof, for the first time in
three and a half years. During the winter of 1943/44 very little training took place,
mainly because of the incessant wind and rain, but also because of the limitations of track mileage placed on the few Shermans with which the regiment was
equipped. However, none of this prevented the officers and men getting on very
friendly terms with the local inhabitants, and the Algerian atmosphere with its
strong French influence appealed to all ranks.[3, 4]

In December Major Dance happened to meet an officer of The Governor
General's Horse Guards of Canada, which was the Canadian regiment affiliated to
the Bays. The GGHG was the armoured reconnaissance regiment for the Canadian
5th Armoured Division, then on its way to Italy. A reunion dinner was arranged at

Boufariq, which twenty officers of The Governor General's Horse Guards attended, to be followed three days later by another dinner when the Bays sergeants' mess entertained twenty-three members of the GGHG sergeants' mess. Later that winter Major Dance attended a ceremony at Sfax, when the 2ième Dragons of the French Army celebrated the escape and return to their regiment of eight officers and thirty-eight men who had escaped from Occupied France.[3, 4]

In January 1944 Lieutenant Colonel Barclay gave up command and left to attend a course in the USA, being succeeded by Lieutenant Colonel Asquith, who for the past three years had been engaged on special duties. Over this period when the Bays were not in action there were a number of changes in the regiment. Major Streeter left for the United Kingdom and was succeeded as second in command by Major Manger. In February 1944 Majors Weld and Hibbert rejoined, taking over 'B' and 'C' Squadrons; Major Dance took over 'A' Squadron from Major Harman, who returned to the United Kingdom, and Major Rich commanded Headquarter Squadron.[3, 4]

During January more Shermans arrived, and in February the Bays moved to Sidi Aissa on the high plateau for regimental training, which was much needed, for many of the reinforcements arriving were not sufficiently trained to take their places in the tank crews. The months of March and April 1944 were spent back at Boufariq. In April orders came that 1st Armoured Division would shortly be moving to Italy, and the Bays were at once made up to strength with Shermans and other vehicles, including some Stuart tanks with their turrets removed for the reconnaissance troop. In May the whole regiment was sent on a week's leave, most of the men going to the Surcouf rest camp, and on their return the tanks and vehicles were loaded at Algiers on six ships, the *Empire Clarence*, *Fort Gaspereau*, *Samsteel*, *Samewash*, *Fort Marin* and *Samoa*. The main body of the Bays embarked on the *Durban Castle* on 24 May 1944 and sailed in convoy for Italy the following day.[3, 4]

SOURCES

1 D. McCorquodale, B. L. B. Hutchings and A. D. Woozley, *History of The King's Dragoon Guards, 1938-1945*, published by the regiment.
2 War Diary, 1943-45, 1st King's Dragoon Guards, Regimental Museum, 1st The Queen's Dragoon Guards, Cardiff Castle.
3 W. Beddington, *The Queen's Bays, 1929-1945*, 1954.
4 War Diary, 1943-45, The Queen's Bays, Regimental Museum, 1st The Queen's Dragoon Guards, Cardiff Castle.

43

Salerno, Naples, Garigliano, Abruzzi,
Advance to Perugia
1943-1944

Early on 24 September 1943 The King's Dragoon Guards in their two LSTs un-shackled their vehicles as the craft ran in towards the beach. As they arrived, the great bow doors opened and the KDG drove out at speed. They had arrived at Salerno, to be the armoured car regiment for the British 10 Corps, part of the Fifth Army, coming under immediate command of 23rd Armoured Brigade. On the 26th the KDG moved out of the assembly area, driving through Salerno and Maiori to Ravello, then on the 28th advancing north across the mountains from Maiori. Lieutenant Phillips of 'A' Squadron found a bridge across the River Sarno intact, although strongly held by the enemy. He managed to get onto it but was driven off by heavy fire. He then kept it under observation until the arrival of 131 Brigade, who captured the bridge intact; and he himself finished the day at Scafati. Another 'A' Squadron patrol encountered stiff opposition at a second bridge, with heavy fighting taking place, but the Germans managed to blow the main bridge over the Sarno before retiring. 'B' Squadron advanced on the left nearer the sea, reporting Castel-mare and Gragano clear of the enemy, but they reached the line of the Sarno to find all the bridges blown and the Germans in strength on the far bank of the river. 'B' Squadron lost Trooper Rowell, badly wounded and later dying, to a sniper.

On the 29th the right hand troop of 'B' Squadron spotted a German eight-wheeled armoured car. They opened fire on it, but the trigger mechanism jammed on the leading car and the German hit the Humber, killing Corporals Crowther and Tullock and wounding Corporal Vince. Lieutenant Knapp of 'B' Squadron even-tually managed to find a way through many demolitions and considerable German resistance, and by evening was established with the Scots Greys to the west of Torre Annunziata. The 30th saw little progress in the built up country which, combined with widespread demolitions, made it impossible for the armoured cars to get off the roads. The squadron support troops, consisting of sections of KDG acting as infantry and carried in an armoured White scout car, proved their value,

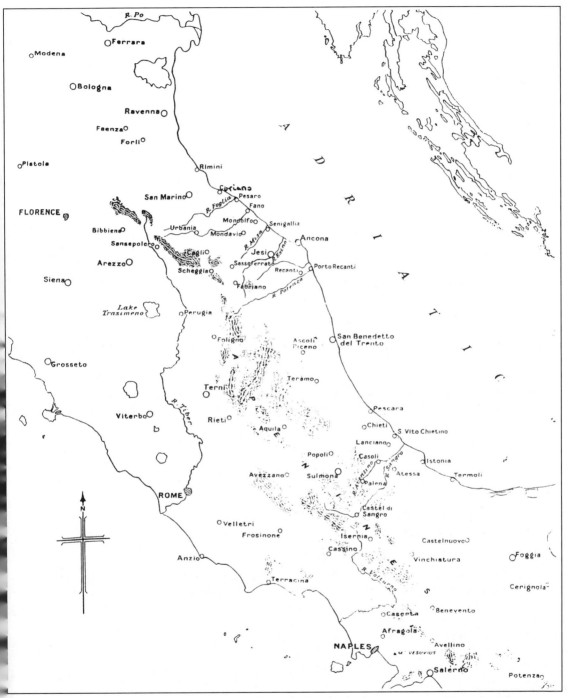

Central Italy

getting forward where the cars could not go and providing protection for the vehicles in the close country.

On 1 October the Scots Greys with some American infantry found that the Germans had retired. Two troops of KDG 'A' Squadron immediately moved forward, advancing rapidly through Torre Del Greco, Resina and Portici, and by 9.30 a.m. were in the centre of Naples. From Torre Annunziata to Naples the rest of the regiment, following behind, was fêted by cheering crowds, who showered the vehicles with fruit and flowers. While 'A' Squadron enjoyed the honour of being the first into Naples, 'B' moved east and north of the city, capturing a German staff car and its officer occupants, and hitting a lorry. Lieutenant Cassels, after being held up by trees felled across the road, which were cleared by the RE, managed to work his way forward to La Rotunda on the northern outskirts of the city, where he was stopped by enemy tanks and infantry. On the following days progress was slowed by the extensive demolitions, and the KDG soon learned that any advance in Italy was dependent on the speed of the Engineers in laying Bailey bridges, clearing obstacles, and lifting mines.

'C' Squadron and the echelon landed at Salerno on 3 October, joining the regiment on the 5th. Progress was severely hampered by the bad weather and the repeated obstacles of blown bridges and roads, with the American infantry advancing in front of the armoured cars. By the 6th the line of the River Volturno was reached and found to be strongly held by the enemy. Lieutenant Smith of 'C' Squadron was sent to reconnoitre the mouth of the Volturno as a possible crossing point. Having carried out a successful patrol, Smith, while turning in a Dingo scout car, ran over a mine and was killed. Troopers Mason and Murfitt were also wounded, Mason's wound proving fatal. On 8 October the KDG were withdrawn in foul weather to the outskirts of Naples, where the regiment installed itself under cover in the local lunatic asylum. Over the next few days they cleaned up and explored Naples, while Major Lindsay searched for horses to form a mounted troop. By 26 October, when the KDG left the 'loony bin' and Naples, the nucleus of a mounted troop under Lieutenant Richardson had been formed with official approval.[1, 2]

On 26 October the KDG moved to the east of Capua, the Volturno having been forced earlier, and over the next few days pushed slowly forward through difficult country. Trooper Leach was wounded on a mine on the 27th, and on the 31st Lieutenant Richardson took out the first of many mounted horse patrols through the mountains. That day news came through that Major Delmege had been awarded the Military Cross. By the first week in November patrols were slowly moving forward to the line of the River Garigliano and RHQ was situated at Roccamonfina, but the deteriorating weather and constant demolitions made it an infantry war. In terrible conditions and against the most determined opposition,

201 Guards Brigade seized Monte Camino, a bare mountain in front of Cassino and a part of the German winter line. On 10 November, the War Diary records 'We are now to be responsible for a part of the line'; and so started a miserable winter of infantry work for the KDG.[1, 2]

Captain Maxwell took the first party of fifty dismounted men into the line, patrolling forward through the mine-infested ground as far as the Garigliano river. One patrol found itself in a minefield, when Trooper Gillow was killed on an 'S' mine, with Sergeant Mercer and Corporal Jeeps wounded. The rain never seemed to let up day or night. The regiment was then ordered to take over the Daimler armoured cars of the 11th Hussars, who were returning to England, and this was eventually accomplished after 'C' Squadron had extracted itself from mud which held the squadron in its grip for two days. 'C' Squadron then relieved 'A', losing Trooper Stevens on a mine, and had a number of contacts with German parties, who, knowing the location of their mines, patrolled aggressively at night. Lieutenant Mann remembered:

> As night fell we manned the slit trenches, and I heard the sound of much rustling below the bank of the road. The noise was too far away to fire at with any chance of hitting anything, and as it could only have been a cow, we should have given away our positions. At midnight we went to contact our neighbours, the London Irish, in a farm about 100 yards to our flank; the position was deserted, with blankets and kit strewn around.

A local farmer informed the KDG that a party of thirty Germans had occupied the farm and had stolen most of his cattle, as well as capturing the entire London Irish post.[1, 2, 3]

On 22 November 'B' Squadron took over a new part of the line near the mouth of the Garigliano opposite Minturno, patrolling down to the blown bridges over the river. On the night of 3 December Corporal Ingleby of 'A' Squadron had an encounter with a German patrol, capturing four Germans of the 94th Infantry Division. There were further skirmishes on the 11th and 14th, and Sergeants Walsh and Bevan were wounded when one of own own mines exploded, Walsh later dying. The regiment was relieved by the 6th Grenadier Guards on the night of the 17th, and withdrew to rest in the village of Zuni. There Christmas was spent, with a regimental church service and the men wearing for the first time their recently issued medal ribbons of the Africa Star. News came through that Major Robin Whetherly, KDG, on attachment to the partisans in Yugoslavia, had been killed. On 4 January 1944 a move was made to Albanova, south of the Volturno, where Major Crossley rejoined the regiment as second in command, Major Lindsay taking over 'C' Squadron, and Major Cairns the PRI.[1, 2]

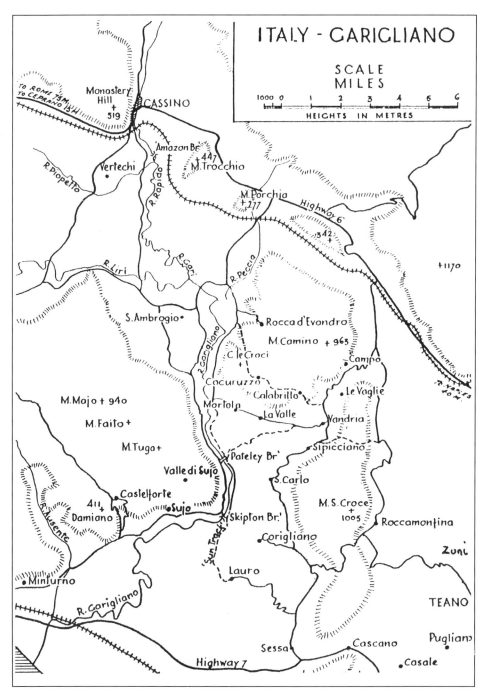

Garigliano

On 7 January 'A' Squadron went back into the line around San Carlo on the Garigliano, followed by 'C'. One evening Lieutenant Mann

> posted all the guards and returned to SHQ to study the maps. We sat round a table eating our bully beef and biscuits and drinking huge mugs of sweet tea. Mick [Lindsay] suddenly looked up, 'What size boots do you take, Mickey?' I admitted to size 12. 'What monsters! I shall send you to look for the mines tomorrow morning, your feet will find them!' [During the course of that patrol] I carefully prodded the ground, and I had not gone five yards before my stick struck something hard, and, sure enough, I uncovered a mine. We had walked through a minefield without stepping on one! We cleared a path, marking it with boulders. There was not a man who could not tell of such hairbreadth escapes.

On the night of the 17th 'C' Squadron and the guns carried out a 'Chinese attack', to create a diversion as 46th and 56th Divisions crossed the Garigliano. 'A' and 'B' Squadrons were then put into the line to plug a gap of six kilometres south of Monte Camino, where they, in turn, put in a 'Chinese attack' on the 28th, bringing upon themselves a barrage of heavy mortar fire. Captain Howes, an attached South African officer, and Lieutenant Armstrong were killed by a booby-trapped Teller mine.

On 1 February a strong party of 'B' Squadron crossed the Garigliano, taking over from a company of the KOYLI who had been ambushed. Lieutenant Cassels found the positions very exposed, both to the weather and the enemy. A patrol under Lieutenant Mann captured two prisoners and a German patrol attempted to snatch a prisoner from 'A' Squadron on the other side of the river. The German officer in command stabbed to death an Italian in cold blood and was himself shot at three yards' range by Corporal Smith and captured. The slit trenches were filled with water; Mann wrote, 'We lay soaked to the skin, our wet clothes clinging to us, and we could not stop our limbs shaking. By now I was completely lousy and flea-ridden, and the itching and cold discomfort added to the gloom.' The cold was so intense that at dawn a number of men had to be carried down from the aptly named Monte Purgatorio. On the 3rd Lieutenant Bloomfield of 'A' crossed the river to find that the enemy had retreated some 1,500 yards. He contacted a patrol from 'B' and then reinforced them until the Leicesters came up to relieve the KDG.[1, 2, 3]

The regiment was then moved to cover the area in front of a ridge, San Ambrogio, where the Rivers Gari and Liri converged to the west of Cassino. The men were situated in small isolated posts around ruined houses, only approachable by night. The troops were so thin on the ground that every man was on duty in slit trenches every night, often up to his waist in water in teeming rain, and having to keep under cover during daylight. All the posts were mortared and shelled continuously, and the German patrols were exceptionally daring and active – one night killing or capturing an entire post of the neighbouring Leicesters. On 8 February Sergeant Macdonald was killed by a mortar bomb which fell at his feet. The 15th saw the bombing of the monastery, as the New Zealanders and 4th Indian Division attacked Cassino town, with the men having a grandstand view. On the 20th Trooper Richards was killed by a mine, and 'B' brought in a prisoner wounded on one of their own mines on the 23rd. The most dangerous post covered 'The Two Bridges', a stone bridge and a Bailey, on the lateral road used to supply the troops to the flank. The Germans constantly tried to destroy these bridges, and on the 24th 'B' Squadron had a spirited fight with a German patrol that had got onto them at last light. On the 26th an enemy patrol was seen off by 'C' Squadron, but as a result the 'C' squadron post was so heavily shelled that it had to be evacuated. The following night another enemy patrol stumbled into the British mines, leaving plentiful signs of blood and numerous weapons to be found next morning, but no bodies. On the 28th Lieutenant Clay drove off yet another German party. Captain Maxwell was severely wounded in the legs on an 'S' mine on 5 March. At last, on 13 March, the KDG were relieved by the 44th Recce Regiment, after nearly ten weeks continuously in the line as infantry. The adjutant's comment in the War Diary summed up the regiment's feelings: 'Apart from the exceedingly hard work which our present infantry work imposes on officers and men, they are getting very fed up with it, doing a job for which they aren't trained, and which completely disorganises us as an armoured car regiment. So far from being a temporary stopgap, we seem to be more permanent than the infantry themselves.'[1, 2]

The King's Dragoon Guards were now required to form a fourth squadron, 'D', to bring themselves into line with a new establishment. Officers and senior NCOs were taken from the existing squadrons, with Major Crossley commanding until the return of Major Delmege from a course. The adjutant, Captain Hutchings, was to be second in command and his place was taken by Captain Chrystal; the troop leaders were Lieutenants Batt, Macartney, Napper, Woozley and Mann, sixty-three reinforcements forming the bulk of the troopers. The new squadron trained at Sarno, six miles east of Vesuvius, arriving on 23 March in the middle of an eruption which showered the area with lava dust to a depth of feet. A month later the regiment was on the move again, complete with its fourth squadron, taking over positions from the Polish 12th Podolski Lancers in the Abruzzi mountains in the area of the Sangro on 24 March. The squadrons were widely separated, holding a long front; with 'A' at Montazzoli, 'B' at Pescopennataro, 'C' at Colle di Mezzo, and 'D' at Castiglione, with RHQ and the echelon at Agnone.

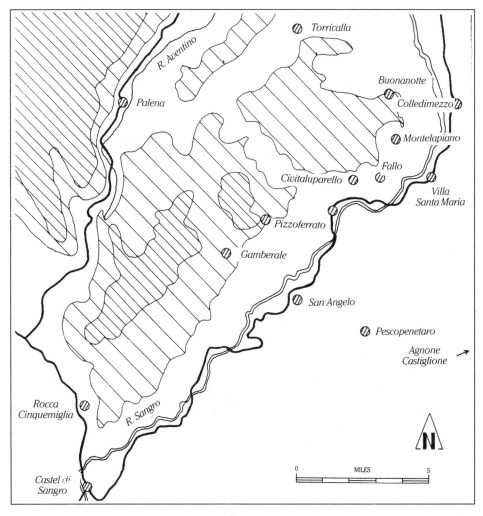

Sangro

In this mountainous country what few roads existed were mostly obstructed either by blown bridges or demolitions or booby-trapped trees. The regiment held an extended front against the entire German 114th Jager Division while Allied strength was being built up around Cassino. During this period the KDG by aggressive patrolling led the Germans to think they were opposed by the entire 18th British Infantry Division (which was not in Italy). 'B' Squadron at Pescopen-nataro sent forward patrols to the Sangro, and across it by foot and mule to the village of Pizzoferato, where the Italian guerrillas, under the leadership of Beniamino, had held out against four German efforts to capture the village. 'B' then patrolled forward two miles to Gamberale, which was still held by the enemy. On 3 May Sergeant Baldwin of 'A' Squadron killed a German in a patrol skirmish, and on the 7th two partisans were wounded operating west of Gamberale. It was then decided to capture Gamberale, but the Germans forestalled the attempt by withdrawing to the line of mountains behind Palena Station. Gamberale was occupied by the partisans, with 'C' Squadron patrolling as far as Palena Station. On 13 May Major Crossley, with Sergeant Vince and Troopers Ware and Wilkinson, arrived simultaneously at the top of a mountain with a German patrol. Both sides decided that discretion was the better part of valour and beat a hasty retreat, but Ware and Wilkinson, off to a flank and unaware of this, calmly occupied the mountain OP and remained in observation until the next day.

The squadrons now engaged in long-distance patrols, supported by mules carrying rations and equipment and remaining out for several days at a time. On 17 May Colonel Hermon heard that he had been promoted, and Major Palmer went to the North Irish Horse. Command of the regiment was taken over by Lieutenant Colonel Crossley. Many Polish deserters came in over this period to 'D' Squadron. On the 22nd Lieutenant Weinholt lost a foot when a wandering cat set off a trip-wired mine. Over May and June the horse troop provided each squadron with useful mounted assistance. The Germans heavily shelled the regiment's positions on 25 May, killing Trooper Owen and wounding Sergeant Robinson and Corporal Dunham. Lieutenant Mann ambushed a patrol of the Household Cavalry Regiment which had gone 'spare', luckily with no damage done, and it was sent on its way 'in words that were unambiguous if hardly Parliamentary'. Lieutenant George cut the wire around an enemy position, killing some of the sunbathing occupants before withdrawing; and on the 29th Sergeant Poynton killed three Germans, Lieutenant Moffatt killed one in another patrol encounter, and the regiment lost its first prisoner when Corporal Soames was wounded on a Teller mine.

Throughout the early part of June the patrolling continued, prisoners being taken and casualties inflicted on the enemy. Sergeant Forbes took a prisoner and killed a German at Castel di Sangro. On 3 June, however, Lieutenant Miller's entire patrol was ambushed, with Lieutenant Musselwhite, Troopers Aubrey and Ward being wounded and captured, together with seven others. Only Corporals McDonald and Fieldhouse got away. On the 6th a patrol ran into a minefield at Rocca Cinquemiglia, Sergeant Robinson and three men being killed; and on the 9th Trooper Pargeter stepped on an 'S' mine, which wounded him and Sergeant Baldwin and killed Sergeant Baker. This run of misfortune ended on 10 June with the news of the Allied progress at Cassino, and the regiment resumed its armoured car role, starting off to catch up with the advancing Fifth Army and arriving ten miles west of Rome on 12 June.[1, 2]

The Queen's Bays disembarked at Naples on 27 May 1943, moving to a transit camp at Afragola and thence to Matera, about thirty-five miles from Bari, where they remained training, as a part of 1st Armoured Division, until the beginning of August 1943.[4, 5]

On 14 May 1943 The King's Dragoon Guards reached Narni, moving on twenty-seven miles on the 16th to Todi. Throughout the move forward, from the utter destruction around Cassino to the outskirts of Rome and then as they advanced rapidly north, the men had been greatly encouraged to see the vast quantity of German equipment destroyed by the RAF and in the battles of the Liri valley. There were vehicles by the hundred, with Tiger and Panther tanks strewn alongside the roads. On 17 May the regiment was in action again: Lieutenant Bloomfield of 'A' Squadron surprised a party of German engineers about to blow up a bridge, and then had a battle with some tanks at Migliano. 'C' Squadron on the main road to Perugia met a self-propelled gun which hit the leading Dingo, killing the driver and wounding the commander. The troop leader then reversed his Daimler armoured car into a ditch, where it had to be abandoned. The Germans later towed the car away and set it on fire.[1, 2]

By the 18th the enemy had stabilised his line and the regiment advanced only two miles. The 19th saw 'D' Squadron in the lead; Lieutenant Mitchell on the left made some progress, but stirred up a hornets' nest of opposition; on the right, Lieutenants Woozley and Thompson advanced along parallel roads to within a mile of Perugia. Thompson, faced with the last two miles along a dead straight road with no cover, had almost reached a T-junction when a machine gun opened up on the leading Dingo. This reversed behind Thompson's armoured car, and as he traversed his turret to engage the gun, the doors of a barn opened and a 75 mm anti-tank gun inside hit the Daimler, killing the driver and setting the car on fire. Thompson and his gunner both baled out badly burned.[1, 2]

On 20 June Lieutenant Napper of 'D' Squadron entered Perugia, the Germans having abandoned it the previous night, to an ecstatic welcome from the inhabitants. Lieutenant Woozley advanced to the Tiber, where he found all the bridges blown, but picked up six Panzer Grenadiers and an Austrian deserter from the Luftwaffe as prisoners. 'B' Squadron with support sections from 'A' and 'D' occupied Monte Giogo overlooking the Tiber three miles north of Perugia, and here Colonel Crossley was severely wounded by a burst of Spandau fire while visiting the position. Major Lindsay took over command of the regiment. The Germans continually shelled this exposed position, wounding Captain Richardson and Lieutenants Fargher and Cundall (Fargher died of his wounds), and killing Sergeant Page, Corporals Jickells and Abbots, and Trooper Riddelough, and wounding four other men.

'D' Squadron on the right continued active patrolling, but was unable to advance further as the enemy held Monte Croce in strength, dominating the area. 'A' Squadron on the left had more success, reaching Agello twelve miles to the west of Perugia and on the 24th capturing a 75 mm gun intact at Monte Melino, together with seven prisoners, but losing Trooper Bailey killed. 'A' made such good progress that the whole regiment was moved around to operate to the west of Perugia. On the 29th Lieutenant Munro of 'C' entered Magione, shooting up the German defenders, who scattered in all directions, two being run down. Sergeant Innes of 'A', entering a house, burst into a room and surprised a small party of the enemy who were demanding wine and women, killing the lot. Another patrol captured five Germans and a 105 mm gun. By 30 June the northern end of Lake Trasimene had been reached, and Magione was found to be clear.

Lieutenant Munro again excelled himself, charging a Spandau post, killing four Germans and capturing nine Panzer Grenadiers; for his bravery he received an immediate award of the Military Cross. Lieutenant Curtis also captured five prisoners and made good progress. On 2 July the enemy withdrew and the KDG advanced slowly, with great help from the Engineers, who cleared extensive demolitions and mines. On 3 July 'D' Squadron, seven miles to the north of Magione, had a hard battle around the village of Preggio. Lieutenant Mann came to a massive crater where the road had been blown down the mountainside, and caught and killed the two men of a Spandau team, but came under heavy fire from Monte Muro. A battalion of Gurkhas put in a flanking attack under cover of fire from the armoured cars and cleared Muro. With the road repaired by the Engineers, Mann supported by Lieutenant Woozley advanced towards the River Niccone on the way to Umbertide, with Woozley occupying Preggio. Both troops came under intense artillery, mortar and machine gun fire, and Mann found his troop in the middle of a German company position. Both troops fired at numerous targets, Lieutenant Courage came up, giving excellent support from the gun battery, but little progress could be made owing to mines and the strength of the opposition. A 105 mm shell landed in front of Woozley's Daimler, shredding every tyre and riddling the storage bins and bedding. One prisoner and three deserters were taken, and next day eleven dead Germans were found around the position. On the following day foot patrols reached the Niccone, after finding countless mines and demolitions.

The regiment went into rest on 5 July at Castel Rigone. Major Chrystal, while clearing a path for his men to bathe in Lake Trasimene, trod on an 'S' mine and was killed; Montgomery had said that he was the finest troop leader in the British Army. 'D' Squadron was back in action on the 10th, and on the 13th a dismounted patrol found itself trapped in the middle of a German company position. The patrol managed to withdraw, but Corporal Myers was wounded and taken prisoner. The remainder of the Regiment remained in rest until 16 July. By this time, and until the regiment left Italy in December, 1944, The King's Dragoon Guards had been longer in continuous action during the war than any other unit in the British Army, and since landing at Salerno had sustained heavier casualties than any other unit in the Royal Armoured Corps.[1, 2]

SOURCES

1 D. McCorquodale, B. L. B. Hutchings and A. D. Woozley, *History of The King's Dragoon Guards, 1938-1945*, published by the regiment.
2 War Diary, 1943-45, 1st King's Dragoon Guards, Regimental Museum, 1st The Queen's Dragoon Guards, Cardiff Castle.
3 Major Mann's diary, in possession of the author.
4 W. Beddington, *The Queen's Bays, 1929-1945*, 1954.
5 War Diary, 1944-45, The Queen's Bays, Regimental Museum, 1st The Queen's Dragoon Guards, Cardiff Castle.

44

Arezzo, Florence, Sansepolcro, Gothic Line, Coriano, Croce, Ceriano

1944

By the middle of July the Germans were conducting an orderly and slow withdrawal in their own time, fighting rearguard actions and making skilled use of the mountains, extensive demolitions and minefields. The King's Dragoon Guards were back in the line at Arezzo on 16 July. Patrolling commenced in an area where the roads were blocked by fallen trees, blown bridges and many mines, and there were constant brushes with the enemy and casualties inflicted. The German artillery was particularly active, shelling any movement – when Lieutenant Bloomfield ditched his Daimler, 121 shells landed around it within an hour. 'B' Squadron entered the village of San Polo on the morning of the 18th to find that the day before the Germans withdrew they had taken all the men of the village, except the parish priest, to a nearby olive grove, made them dig three large graves, and had then bayoneted the men into the graves and placed a number of explosive charges among the dead and dying. This was said to be punishment for partisan activity.

Sergeant Beale of 'C' Squadron led a patrol as far as Monte Veriano, to find the area held in strength by the Germans, but with no anti-tank guns. He then allotted each vehicle a sector and drove right in amongst the enemy. Corporal Kay in the leading Dingo leapt out hurling grenades into each slit trench and firing his tommy-gun until it was shot out of his hand; he then went back for a Bren and started again. The Germans managed to knock out a supporting M10 with an Ofenrohr, killing the driver. With only a single casualty Sergeant Beale captured the position, taking two officers and twelve men prisoner and counting thirty dead Germans, twenty of whom had been accounted for by Corporal Kay alone. Both Sergeant Beale and Corporal Kay were given an immediate award of the Distinguished Conduct Medal.

King George VI paid a visit to the KDG at Arezzo on 25 July, the same day that a Daimler received a direct hit from a 105 mm shell, wounding the crew slightly. On the 27th Lieutenant Napper was killed by a shell when out on an observation post. By 3 August 'A' Squadron had advanced to Monte Ferrato, where a patrol under Lieutenant Howes attacked a strongly defended German company position, killing two Germans and capturing eight more. The patrol was then heavily shelled: Corporal Biddell was killed, and Sergeant Barrett and Corporals Marks and Long were wounded, Long later dying of his wounds. Corporal Marks was awarded the MM for his leadership. That day Trooper Cook of 'B' Squadron was wounded by a shell. On 5 August the KDG went into reserve until the 11th, when the regiment moved to the area around Florence.

At first the regiment's role was to act as a mobile reserve to the infantry units closing up to Florence, but that soon changed when 'D' Squadron was ordered into the city to support the 8th Indian Infantry Brigade and the Italian partisans. The Germans had blown all the bridges over the River Arno with the exception of the Ponte Vecchio, where they had blown down the houses on each side of the streets approaching the bridge. 'D' Squadron managed to get Lieutenant Mann and his troop across the Arno via a weir, made just passable by the Engineers. The other squadrons patrolled actively both on foot and, where possible, in cars, and on 14 August Trooper Bailey was killed and Corporal Taggart and Troopers Naisbitt and Gard were wounded when caught by a mortar barrage. Within Florence Mann set up his headquarters over a bar in the Piazza Vittorio Emmanuele, where he was to support the Indians and partisans who were clearing the city. When the Engineers building a Bailey bridge across the Arno at the Ponte Trinita were being harassed by sniper fire from Fascist elements, Mann's troop was called out and Corporal Ingleby with a snap shot killed the sniper, who was at an upstairs window. After a few days Lieutenant Woozley's troop was brought into the city to reinforce Mann. By day both sides strove to maintain the illusion that the city was not to be fought over, and civilians thronged the streets, but by night the German parachute troops and their Fascist allies infiltrated back and Mann was twice called out to repel attacks. On 17 August Woozley led out a strong force of armoured cars and infantry, which successfully cleared the enemy from the north-eastern part of the city. The following day Woozley led an advance with infantry to San Domenico on the road to Fiesole, but ran into stiff opposition. After a fierce fight he managed to withdraw without casualties.

On 20 August the KDG were drawn back into reserve by Lake Trasimene, where 'D' Squadron received its fifth squadron leader, Major Phillips, since it had been formed just five months previously. On the 26th the regiment was back in action, taking over from the 12th Lancers in the area of Sansepolcro. The Allies were now edging up to the Gothic Line. On 27 August a patrol of 'C' Squadron was ambushed, losing a Dingo with Lieutenant Hethey and Corporal Price killed and Trooper Burton a prisoner. The following day 'C' lost two more Dingos on mines, with Corporal Eckersley and Trooper Shakespeare wounded and Trooper Chambers dying of his wounds. Another of their patrols drove off a party of fifteen Germans, killing two of them. The 31st saw Major Hellyer and Captain Batt badly wounded when the White scout car in which they were travelling went over a mine and the armoured engine cover was blown off to land on their heads. Another 'C' Squadron patrol had a brush with the enemy in which one German was killed and Trooper Moynihan was wounded. Often the mines had been laid so deeply that several vehicles could go over them before enough pressure was built up to set them off. This made them impossible to detect.

The KDG used Sansepolcro as a base, patrolling well forward in the mountainous country over a large area of no man's land, with the enemy operating in much the same way. Again the KDG horse troop came into its own. 'D' Squadron sent a strong patrol comprising Lieutenant Woozley's and Mann's troops to establish a patrol base two miles further forward. The patrol soon encountered and cleared a dug-in German position after a short fire fight, in which one German was severely wounded and taken prisoner. A base was then established, from which Mann took a patrol forward to the Alpe della Luna to observe the German positions, then brought artillery fire down on them. On 1 September another patrol of five men ran into a German post, capturing one, but losing Sergeant Bourhill killed and Corporal Cameron taken prisoner. On 5 September Sergeant Redfearn of 'B' stalked and captured an enemy observation post, and the following day a patrol ran into a minefield, which severely wounded Sergeants Mercer and Mitchell and slightly wounded Corporal Miller and four others. When not on patrol all ranks enjoyed the comfort of Sansepolcro. As the Engineers slowly mended bridges and cleared roads the regiment pushed forward, and by the 13th the Passo di Viamaggio on the road from Sansepolcro to Rimini had been reached. On the 14th a foot patrol pressing on ran into an ambush, losing Trooper Whitlock killed and Sergeant Gutteridge and Corporal Waters captured. Patrolling involving a number of brushes with the enemy continued until the 24th, and then on 25 September the KDG were withdrawn into reserve.[1, 2]

The Queen's Bays moved out of the camp in Southern Italy on 6 August 1944 and moved north with the rest of 1st Armoured Division, concentrating around Civita Nova on the Adriatic coast, about twenty miles south of Ancona. Here they were issued with twelve of the new 76 mm Shermans. The infantry attack to pierce the Gothic Line started well on 27 August, with the Germans being pushed back from the Metauro to the River Foglia; by 1 September they were back on the line of the Conca. As the pressure on the enemy continued, Lieutenant Colonel Asquith was ordered to secure crossings over the Marano, the next river beyond the Conca, by first light on 5 September. A company of the 1st KRRC with the 10th

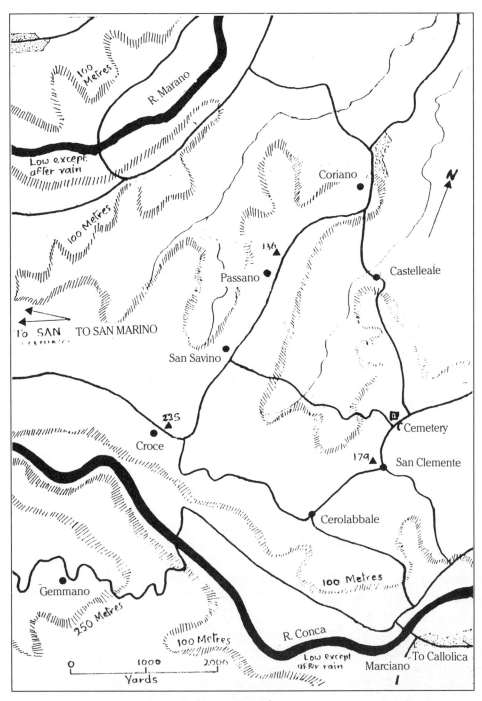

Gothic Line, September 1944

Hussars moved onto the San Clemente ridge between the Conca and the Marano and reported that, contrary to the information given to the Bays, the ground to the west, including Coriano, San Savino and Croce, was still strongly held by the enemy.

With 'A' Squadron on the right and 'C' on the left, the Bays started their attack at 3 p.m., but traffic congestion on the approach roads delayed 'C', who had no time to gather themselves before they were launched. Four tanks were immediately hit by anti-tank fire from Gemmano on the left and from Croce to the front. Before the start line was even reached, the rear-link tank had been knocked out, cutting off communications with RHQ; then three more tanks were lost, but the two leading troops managed to move up just short of the start line, where they took up positions under the squadron second in command, Captain McVail. Major Hibbert, the squadron leader, was to the north with two more tanks. Here 'C' Squadron was attacked by bazookas and came under shell fire; Sergeant Jones and Trooper Edwards were killed and their tank was knocked out. With no cover, and now in the lead, Major Hibbert did the only possible thing, taking squadron headquarters forward into the centre of the German position. One German tank was knocked out before the tanks of squadron headquarters suffered the same fate. Major Hibbert and Captain McVail with their crews managed to take refuge in a German dug-out, where after a series of hair-raising adventures they made their way back to their own lines, arriving two days later. Captain Conway, the medical officer, went forward to the assistance of 'C' Squadron and was wounded. The colonel then managed to make contact with Lieutenant Ward of 'C', and as a result ordered 'A' Squadron to go to their assistance. In trying to do so 'A' lost a number of Shermans through the tracks coming off on the steep and slippery hillside.

As night came on, the Bays drew back into leaguer, reduced to only eighteen tanks, although many were later recovered. During the night shelling killed one man and wounded four others. 'A' Squadron had lost three men wounded that day and 'C' had Major Hibbert, Lieutenant Read and six men wounded, Corporal Turner later dying of his wounds, and five men missing, as well as those killed. As a result of the tank losses the rest of the regiment was formed into a composite squadron under Major Weld. Prisoners that had been taken came from the 26th Panzer Division and the 29th Panzer Grenadier Division, and the Germans had also brought the 1st Parachute Division to help hold the Coriano position.

Just before daylight on 5 September the composite squadron moved back to positions below the crest line, where they remained throughout the day, at nightfall leaguering in the divisional gun lines near San Clemente. Even though the regiment was not called into action for some days, Lieutenant Butcher, Corporal Jones, the CO's driver, and Trooper Peacock were killed by shellfire, and Sergeant

Foster with Troopers Taylor and Maylin were wounded. 'C' Squadron was re-formed with new tanks on the 7th, but German resistance was so fierce along the Coriano ridge that the infantry made only little headway. The Bays were not back in action until the 13th, when they supported the 43rd Gurkha Lorried Infantry Brigade in an assault on the San Savino – Passano ridge. It was a difficult night approach march, with the Shermans towing the Gurkhas anti-tank guns and the Gurkha crews riding on the back of the tanks.

The initial objectives were gained by the 8th and 10th Gurkhas with 'B' and 'C' Squadrons in support. As 'B' moved forward into Castelleale it was greeted by a barrage of mortar shells and the discovery that the engineer bridging tank, by which the squadron was to cross the Marano, was not in place; as a result the squadron had to turn round and follow 'C' Squadron at its crossing point. The leading tank was hit, Lieutenant Gay being wounded, with the Gurkhas on its back all killed and the towed gun destroyed. Then Major Weld's tank turned over, blocking the way. Eventually Sergeant Jordan guided the tanks on foot and the Marano was crossed just before first light. But 'B' Squadron then had to drive as fast as possible to cross over to their position to the left of 'C', and in so doing Sergeant Birch's Honey and Sergeant Southam's Sherman were blown up. On their arrival on the San Savino – Passano ridge an enemy counter-attack was dealt with by Sergeant Bruton's Browning machine gun, and the squadrons put their tanks into good hull-down positions. As the hot day wore on, 'A' Squadron and RHQ moved up in support and all squadrons were subjected to mortar bombardment and shellfire; Lieutenants Yates and Williams, Sergeant Spence, Corporal Andrew and Troopers Swain and Chapman were all wounded. The 9th Lancers then carried out the second phase of the attack by supporting the 18th Infantry Brigade in its advance to the Ripabianca feature, with 'A' Squadron moving up to give the advancing 9th Lancer tanks fire cover, and in so doing brewing up two of the enemy's armour. The day's fighting had resulted in the capture of a very strong position and nearly a thousand prisoners from some of the best German divisions.

From 14 to 18 September The Queen's Bays rested, taking in reinforcements and new tanks. On the 19th 'A' Squadron supported the Yorkshire Dragoons in seizing a small hill across the River Ausa, with the rest of the regiment moving up two miles to the east. That afternoon the Bays moved forward with the 1st KRRC to attack the high ground across the Ausa between San Marino and Ceriano, and were rejoined by 'A' Squadron, who were low on petrol. The going across deep wadis was very slow and by nightfall Colonel Asquith decided to close up and halt the regiment until the morning.

At 3 a.m. on 20 September The Queen's Bays moved off to attack the Ceriano ridge, and before first light a Honey of the reconnaissance troop was knocked out by a bazooka close to the start line. As dawn broke, Ceriano to the left rear and Point 153 above San Martino i Venti to the right could be seen overlooking the advance. It soon became clear that the high ground around Ceriano was still held by the enemy, and some passing infantry confirmed that they themselves had been driven off it during the night. A Troop of 'B' Squadron moved up the Ceriano feature with guns blazing, demolishing each building as they advanced, but they came under anti-tank fire from Point 153 on their right and all three tanks were knocked out, with only Sergeant Burley's tank reaching the top of the hill. The crews baled out, making their way back to the squadron. 'C' Squadron then moved forward onto the crest to engage the anti-tank guns, but every time a tank moved from its hull down position it came under accurate sniper and machine gun fire from close quarters. At the same time the whole regiment was subjected to heavy shellfire, which killed Lieutenant Bunn. With 'A' Squadron short of ammunition and petrol, and the casualties already suffered, Colonel Asquith suggested that an immediate attack against such opposition was unlikely to succeed. The brigadier insisted that the attack must go in at once.

With 'C' Squadron leading, followed by the four remaining tanks of 'B', the Bays came under a storm of 75 mm and 88 mm anti-tank fire from a series of different enemy positions as soon as they emerged onto the forward slope. Tanks were hit at once. There was no cover and when 'C' asked for permission to withdraw, it was told to hold on. It then became clear that the tanks were sitting ducks on a forward slope that was devoid of any cover, and smoke was put down to enable them to withdraw. Only three tanks of 'B' Squadron returned. As each tank was knocked out, the crew baled out, but those who tried to get back came under sniper and machine gun fire at very short range from houses still occupied by the Germans. There were many casualties; only those who lay low during daylight had a chance of survival.

The 9th Lancers came up to renew the attack, but when the situation was made clear by Colonel Asquith, the advance was called off, the remnants of the Bays withdrawing to the area of the previous night's leaguer. The morning's fighting had cost the Bays Captains McVail and Parker, Lieutenants Ward and Franklin killed, with Captain Rowland and Lieutenant Gay creeping back after dark. Five officers and eighteen other ranks had been killed, thirty-nine other ranks were wounded, and two men were missing — a total of sixty-four trained tank crewmen as casualties, of which thirty-nine came from 'C' Squadron. Lieutenant Gay was awarded the Military Cross and Sergeant Burley the Distinguished Conduct Medal. The one consolation for this miserable day was that the Bays' attack had enabled the Canadians on their right to take San Fortunato, and as a result Rimini was entered that night, with the Gurkhas occupying Point 153. Then the heavens

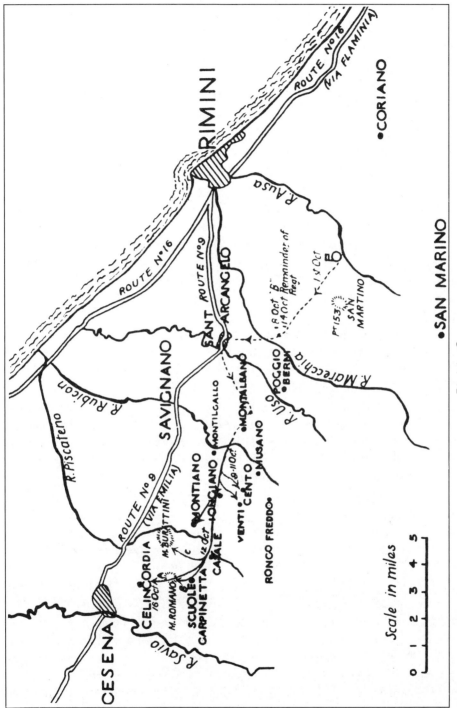

Rimini – Cesena

opened and the rain continued, bringing the movement of vehicles to a halt; but even so the Gurkha brigade on the night of the 22nd advanced 8,000 yards, fighting their way across the Marecchia and capturing Santarcangelo. The forcing of the Gothic Line had cost the Eighth Army 14,000 casualties, and there were no longer the reinforcements to replace such losses. As a result 1st Armoured Division was to be broken up, and 2nd Armoured Brigade would become an independent brigade.

The Queen's Bays were given ten days to absorb reinforcements, most of whom came from the disbanded Wiltshire Yeomanry, who were excellent material, being trained and battle-experienced soldiers. During the Gothic Line battles the Bays had lost five officers, three sergeants and fifteen other ranks killed; one sergeant and four other ranks missing; six officers, four sergeants and fifty other ranks wounded. Three new officers and twelve other ranks arrived on 18 September, followed by sixty-three more reinforcements a few days later. New tanks arrived on the basis of one 76 mm and two 75 mm Shermans per troop, so that by 1 October 1944 the regiment was again ready for action, moving up to support 138th Infantry Brigade of 46th Division.

On 1 October the Italian weather again intervened and torrential storms rendered the ground and rivers impassable to armour. By the 8th 'B' Squadron managed to cross the Marecchia and Uso, and then the Fiumicino, placing a troop in support of each battalion of 138 Brigade, with a fourth troop moving over to Cento to give support to 10th Indian Division. By the 11th Longiano was captured, and Captain MacCallan with two troops supported the 6th York and Lancasters in a successful, but fierce, battle for the Montiano ridge. On the night of the 12th 'B' Squadron, now concentrated at Montiano, supported the 2/4th KOYLI as they crossed the Rubicon, capturing Casale. On the 13th the 2/4th KOYLI and the 6th Lincolns captured Carpinetta under covering fire from 'B' Squadron. The following day 'B' crossed the Rubicon to support the advancing infantry, but were unable to get beyond Casale owing to demolitions and craters. On the evening of the 14th 'A' Squadron under Major Harman moved up to reach Casale, followed by 'C'.

By the morning of 15 October the Engineers had cleared the way sufficiently for 'B' Squadron to support the 6th Lincolns as they attacked Scuole. However, Lieutenant Cassidy's tank ditched on the way forward, so he changed with Sergeant Glastonbury, the two remaining Shermans of his troop reaching the start line at Casa Almerici. Sergeant Southam's tank led with the infantry, Cassidy giving supporting fire. Scuole was cleared house by house, and the ridge beyond occupied. The next objective was Monte Romano, which the infantry gained with supporting fire from the Shermans. Cassidy's two tanks then moved forward to deal with some houses beyond, where the Germans were collecting for a counter-attack. As the tanks approached enemy soldiers emerged from slit trenches surrendering, but by now the infantry was fully occupied with other pockets of resistance, and Cassidy moved back to help. In the afternoon Lieutenant Hough came up in support, but his troop ran into heavy shellfire which wounded him and Sergeant Shaylor. Lieutenant Brooks's troop was then sent to relieve Cassidy, and on the way up he was able to help another company of the Lincolns forward. On the right 'C' Squadron assisted the Lincolns in capturing Monte Buratini, but to the left 10th Indian Division was held up in the steep hills east of the Savio, from whence came most of the defensive fire that still swept Monte Romano, to which the enemy were still clinging, the last obstacle before Cesena.

The divisional commander now ordered Cassidy to advance to Celincordia without the infantry, but supported by Brooks. Neither of the Browning machine guns on Cassidy's two tanks was working properly and there were only four rounds of high explosive remaining. Sergeant Southam was leading when, rounding a corner, his tank slipped off the road down a bank. Trooper Webster managed to crawl back after dark, but Southam and the rest of the crew were taken prisoner. Cassidy was then ordered to withdraw but, while reversing, his tank was hit by high explosive and armour-piercing shells. The crew baled out and got back in spite of intense machine gun fire. Then the tank was hit again, bursting into flames. The Germans now counter-attacked, gaining ground on both flanks of Monte Romano, so that Brooks's troop, reinforced by Sergeant Bruton's tank, came under fire from three sides. The Bays and the Lincolns held onto their isolated spur until nightfall, when the 2/5th Leicesters of 139 Brigade relieved the Lincolns and 'A' Squadron took over from 'B'.

On 16 October Major Harman arranged for 'A' Squadron to support the Leicesters in their advance on Celincordia, but they, while clearing a part of Monte Romano, had been driven back by a counter-attack. As Lieutenant Munro advanced in support, the Germans put an armour-piercing shot into the remains of Cassidy's tank just twenty yards in front of him, and so the attack was put off until after dark. During the night the Leicesters cleared Monte Romano, Munro's tanks helping with the last pockets of resistance. On the 17th 'A' Squadron gave supporting fire as the Leicesters pushed forward, 'C' Squadron helping the 5th Sherwood Foresters around Buratini. On the 18th 'A' Squadron leapfrogged forward troop by troop as the Leicesters advanced, and Celincordia was captured by 4 p.m, Corporal Harvey playing a prominent part in the final assault. On the right 'C' Squadron helped the Foresters forward to Cesena, where the Germans still held out stubbornly in the outskirts of the town. The Bays were now relieved by the 10th Hussars and withdrew to Montalbano for a short rest. An entry in the

War Diary of 46th Division recorded: 'A satisfactory feature of the present battle has been the effectiveness of tank and infantry co-operation.' Another satisfactory feature from the Bays' point of view had been the comparative scarcity of casualties.

The Bays remained at Montalbano until 7 November, the whole regiment being under cover for the first time in Italy — even if the houses were dirty, leaking and overcrowded. By the time the Bays moved back into the line, again in support of 46th Division, the advance had moved some ten miles further forward and the fighting was now in the foothills between the Rabbi and Ronco rivers.[3, 4]

SOURCES

1 D. McCorquodale, B. L. B. Hutchings and A. D. Woozley, *History of the King's Dragoon Guards, 1938-1945*, published by the regiment.
2 War Diary, 1943-45, 1st King's Dragoon Guards, Regimental Museum, 1st The Queen's Dragoon Guards, Cardiff Castle.
3 W. Beddington, *The Queen's Bays, 1929-1945*, 1954.
4 War Diary, 1944-45, The Queen's Bays, Regimental Museum, 1st The Queen's Dragoon Guards, Cardiff Castle.

45

Gothic Line, Monte Poggiolo, Lamone, Senio, Argenta Gap

1944-1945

The King's Dragoon Guards remained in reserve until 1 October 1944, when Lindforce was formed, under command of Lieutenant Colonel Lindsay, and consisting of the KDG, with 1st RHA and 73rd Medium Regiment R.A. as artillery, 1/4th Essex, Lovat Scouts, Nabha Akal Indian Infantry, and 11th LAA all acting as infantry. In addition 2 AGRA, two RE companies, and the 12th Indian Pack Transport Company were in support. RHQ, KDG was expected to administer this considerable force with only the additional help of two staff officers. Lindforce was to take over the advance on 10 Corps' main axis along Route 71, which wandered north east over the mountains and formed a main part of the Gothic Line.

As only one squadron was able to operate on this narrow and mountainous axis, 'A' Squadron established two troops on Poggio di Sapigno. While the infantry cleared the mountains on either side of the road, the armoured cars had to wait for the Engineers to erect Bailey bridges. On the 7th the road was open, the Germans withdrew, and Sarsina and Sorbano were occupied. Beyond Sorbano there were many demolitions, although Lieutenant Bloomfield with a foot patrol reached Mercato Saraceno, to find the enemy still in position. On 11 October the 18th Infantry Brigade took over and Lindforce was disbanded; the KDG concentrated in the area of Poppi, squadrons arriving as roads and weather permitted. On 12 October the regiment moved further back into rest around Perugia, where it remained until 13 November, then moving across to the Adriatic coast to take over a sector of the Po valley north of Cesena on the line of the River Ronco.

The regiment came under command of Porterforce, commanded by Colonel Horsburgh Porter of the 27th Lancers, and made up of a variety of British and Canadian units. The Germans held the Fosso Ghiaia and had observation over the very flat and open countryside, which was waterlogged after the heavy autumnal rains. 'A', 'C', and 'D' Squadrons were in the line, at once establishing contact with the Germans, and by aggressive patrolling soon gained the ascendancy. By 16 November 'D' Squadron had cleared its area up to the Ronco and 'A' Squadron had established a post 1,000 yards beyond the Fosso Ghiaia on the edge of Ravenna airfield, having captured seven men of a German patrol. 'C' Squadron also crossed the Fosso on foot, and with the help of a troop of armoured cars from 'D', which had lifted sixty mines in order to come round from their sector, a joint attack on the village of Molinaccio was made, driving out the Germans and capturing two prisoners. By that evening the KDG had pushed forward 2,000 yards and were patrolling another 1,000 yards beyond that. This enabled the Royal Canadian Engineers to bridge the Fosso Ghiaia, and the Germans lost their first waterline. By the evening of the 17th the second line had gone as well, when 'C' Squadron in its armoured cars again stormed Molinaccio, to which the enemy had returned during the night, and the Canadian Westminster Regiment was able to establish a company in the village with a platoon beyond the canal.

Over this period there were two unhappy incidents: the first when a 'C' Squadron sentry fired on and killed his relief, and the second when Sergeant Barrett of 'A' was wounded, returning from a patrol, by fire from another 'A' Squadron post. Corporal George with a patrol of four men fought a German patrol of ten, killing two of them without any KDG loss. On the 18th Sergeant Senior found a bridge intact across the canal near Molinaccio, guarded by eight Germans, whom he captured; then, crossing the canal, the troop worked along the main road lifting mines for 2,000 yards. The following day a joint attack by Lieutenant Munro's and Sergeant Little's troops resulted in the capture of an entire enemy platoon of nineteen men. The 20th saw 'A' Squadron bag four prisoners, and a patrol led by Sergeant Redfearn took two more and killed another four of the enemy. Redfearn was awarded the MM for his gallant conduct on this patrol. 'C' Squadron had two men wounded by mortar fire, and a 105 mm shell hit a Daimler armoured car on the 22nd, destroying it but not causing any casualties. As the KDG advanced, so the front narrowed, until by the 20th only 'D' Squadron had room to operate.

The KDG then received a series of orders and counter-orders until on 1 December 'D' Squadron found itself relieving 2/10th Gurkhas along the line of the Montone. Apart from patrolling to pick up any spare Germans, 'D' had a quiet time, while the Canadians mounted an assault across the Montone and Ravenna was captured. On 12 December The King's Dragoon Guards were ordered to drive to Taranto at the southernmost point of Italy, and there to embark in LSTs for the Piraeus in Greece, where fighting had broken out between the Communist ELAS guerrillas and the small British occupying force.[1, 2]

The Queen's Bays, having moved up from their brief rest at Montalbano, found that the continuous rain had so swollen the rivers that it was impossible to find a crossing for the tanks. By 9 November 1944 the infantry had cleared up to the line of the Montone and had occupied Forli. By 12 November the water levels had fallen sufficiently for the Engineers to prepare a crossing of the Montone at Terra del Sole, but they came under heavy shell and machine gun fire. By 2 a.m. on the 13th the first tank of 'B' Squadron crossed, followed by 'C' under Major Rowland, who was across by daybreak. 'C' Squadron was to support the 6th York and Lancasters in capturing Monte Poggiolo, but between them and their objective stood the village of Ciola on top of an intermediate hill. Lieutenants Read and Beak deployed their troops in front of the infantry and charged Ciola with guns blazing. The enemy immediately withdrew 500 yards, and the York and Lancasters, following, picked up numbers of prisoners. After a further bombardment of the ground beyond, the Germans withdrew into the medieval castle on Poggiolo with its thirty-foot-thick walls. That night an infantry patrol surprised the garrison of sixty men, capturing all of them. 'A' Squadron on the 14th supported the KOYLI in taking Villa Grappa, while 'B' with the Lincolns advanced alongside.

'C' Squadron carried out a hazardous outflanking move on Poggiolo on the 15th, carrying infantry on the back of their tanks and successfully pushing forward along terrible mountain tracks. Lieutenant Read was wounded and Trooper Heeler killed by a chance shell. The other squadrons supported their infantry as they worked their way forward to the next river position, the Cosina, to which the main body of the enemy had withdrawn. Major Rowland was awarded the Military Cross for his fine leadership of 'C' Squadron on Monte Poggiolo. The weather had reduced the roads and Bailey bridges to a sorry state, and it was not until the 21st that 'A' and 'B' Squadrons were able to form a fire group, using their high-velocity guns to destroy the houses which sheltered the Germans defending the Cosina. Resistance was so stubborn that it took two days to drive the enemy back to the river line, and a night assault to secure a bridgehead, with the infantry then working forward.

By midday on 24 November 'A' Squadron, supporting the 1st KRRC, had advanced seven miles beyond the Cosina and were across the Marzeno a mile south of Faenza, closely followed by 'B'. German reaction was so fierce that the bridgehead was only expanded slowly, 'B' Squadron attacking Belvedere, where Lieutenant Brooks and Trooper Marsh were killed by shellfire on the 25th. The next major obstacle was the River Lamone, and time was needed to bring up ammunition and supplies over the deteriorating roads, not helped by the continuous rain.It was 3 December before the infantry made a successful assault across the Lamone, securing the high ground beyond; and it was late on the 4th before two troops of 'B' Squadron under Captain Fleming were able to cross and support the KOYLI beyond Errano in attacking a church and some buildings 1,000 yards away at the top of a bare slope. Sergeant Bovill in the leading tank lost a bogie on a mine about three quarters of the way up, completely blocking the track. With an hour of daylight left, and under heavy shellfire, the infantry consolidated and 'B' Squadron retired to Errano.

The tank blocking the track was moved by the Engineers on the 6th, and Lieutenant Clough was ordered to reach the church to support the KOYLI and drive out the enemy from the buildings they still held. Sergeant Thompson, leading, received a direct hit, but his tank was towed off the track and the rest of the troop joined the infantry. To the left 2 Troop, advancing to support the Lincolns, lost its leading tank on a mine, which blocked that track. By nightfall 'B' Squadron had five tanks up with the infantry and four more in reserve in the valley. On the right 'A' Squadron had Lieutenant Levett's troop among houses to the right of Castel Rameiro, though Levett himself was then wounded by a shell.

On 7 December the infantry attacked with the Bays in support. Lieutenant Munro's troop moved up to bolster 'A' Squadron, and by mid-morning the Lincolns with 'B' in close support had reached the line Magnana to Rinaldo. One of the tanks became ditched and Lieutenant Walker was wounded trying to tow it out, leaving Sergeant Shaylor as the lone tank in that troop. Lieutenant Clarke took his troop down a track to Rinaldo to assist the Lincolns, and an additional troop from 'A' Squadron climbed the ridge above the valley to give support to the York and Lancasters, who were about to attack Celle. This attack was postponed owing to the tiredness of the infantry, and when it did go in met with no success. Little progress was made during the 8th, with the tanks having to endure heavy shelling and the supply situation worsening as the roads continued to crumble.

The 9th opened with an intense barrage of enemy shellfire onto the infantry and tank positions, and reports started to come in of German attempts to infiltrate on the left flank of the KOYLI. This was followed by a massed enemy infantry and tank assault along the roads leading from Celle, mainly against the Lincolns in front of Magnana. By 10 a.m. the Lincolns' forward positions had been overrun and both tanks of Lieutenant Munro's troop knocked out by armour-piercing shot, but not before they had inflicted considerable damage on the enemy. The crews managed to make their way back to Sergeant Shaylor's tank on foot. Lieutenant Munro was awarded the Military Cross for his exceptional gallantry. The Germans continued to come on, surging around the farmhouses forming the infantry strong points; in the centre the Lincolns were forced to yield more ground, but despite heavy losses the positions were held.

After four hours of intense fighting, accompanied by a terrific artillery duel, the German attack began to lose impetus. By midday the position was stabilised, with no ground lost except in the centre. It was clear that the 90th Panzer Grenadier Division, who were the assaulting troops, had lost severely. Sergeant Shaylor, on his own, had had some excellent shooting as the Germans swarmed over the ridge to his front, while on the low ground to the right a troop of 'A' Squadron, commanded by Corporal Price, was engaged in the fiercest fighting. Throughout the night the enemy tried to stalk Price's tanks, one bazooka man being killed when only six yards away; then one of the tanks broke down, blocking the way so that the other two could not move. The next morning more than eighty dead Panzer Grenadiers were found around the wrecked farmhouse which the troop had been holding. Price was awarded the Military Medal for his gallantry.

'C' Squadron in reserve at Petrigone was to relieve the exhausted tank crews of 'A' and 'B', but difficulty of access to the forward positions meant that the relieving crews had to be brought up in infantry Bren gun carriers, dismount, and trudge with all their kit and spare ammunition up the mountainside in the dark to take over seventeen tanks, including two of the 11th RHA. The infantry was being relieved at the same time by the New Zealand Division, both infantry and tank crews being taken back in the carriers. 'C' Squadron lost three men wounded by a sudden burst of shelling on the 11th, and that evening a German counter-attack with tanks was beaten off. One tank was so near the enemy, and under such continuous shellfire, that its crew members were unable to dismount for a hot meal or a brew for three days; the crew was then relieved, but the commander, Sergeant Beauchamp, insisted on staying with his tank. On their way back the other four members of the crew had a grenade thrown at them from a New Zealand picket, wounding three of them. The new crew was then attacked by an enemy bazooka team, Trooper Crisp spotting and killing one German and wounding another, while Sergeant Beauchamp killed a third with a grenade. Shortly afterwards a New Zealand post dealt with another bazooka party coming up from a different direction.

Late on 14 December the New Zealand Division and 10th Indian Division attacked, and by the 15th the New Zealand armour had passed through 'C' Squadron. 'C' remained in action that day, supporting the advance, and engaging targets as they appeared, until that evening it was withdrawn to Errano for a rest. By the evening of 16 December Faenza had been taken by the New Zealanders. The Queen's Bays then moved back to Pesaro, arriving on the 20th, and joined up there with the rest of 2nd Armoured Brigade.

Early in January 1945 some Shermans arrived which were mounted with 17-pounders. The Bays remained in rest until 16 January, when 'C' Squadron moved back into the line on the Senio, near Faenza, to relieve the 10th Hussars, with the other two squadrons in reserve. Each troop was billeted in a farmhouse and regular shoots on enemy posts took place, Sergeant Beauchamp destroying six Spandau nests in one operation. Otherwise life was fairly quiet, with nightly tank patrols to eliminate enemy outposts and to support the infantry in capturing prisoners.

By the end of January the front held by the Bays was widened, 'C' Squadron being relieved, with 'A' and 'B' now in the line. On 2 and 3 February the regiment moved to another sector of the line in front of Ravenna at Villa Nova; each squadron had three dismounted troops in the line acting as infantry, with the fourth troop supporting in its tanks. The remainder of the tanks were parked fifteen miles back in Ravenna under the care of a maintenance party. Life was quiet, apart from sporadic shelling, but it was exhausting with too few men carrying out too many tasks in the winter weather. On 2 and 3 March the Bays were relieved, re-mounted in their tanks, and withdrawn to a training area at Massa near Ravenna, where they joined the rest of 2nd Armoured Brigade. Each armoured regiment of the brigade was attached to one of the infantry brigades of 78th Division, to prepare and train for the spring offensive. The Bays were with 38th Brigade.

On 10 March 36th and 38th Brigades were ordered back into the line, with 'A' Squadron of the Bays accompanying 38th Brigade to their old area in the Lugo sector. The same tactics of shelling the enemy by day and patrolling by night helped to establish good relations with the men of the 38th Irish Brigade. The remainder of the regiment trained hard for the coming operations, until 'B' Squadron relieved 'A' on 27 March. On 8 April the Bays moved up to Borgo fra Giovanni in preparation for the forthcoming offensive, with 'B' Squadron and 11th Brigade covering a gap of 5,000 yards on the Senio opposite Cotignolo. A troop of 105 mm tanks from 'A' and 'C' Squadrons moved up to take their place among the 1,500 guns massed for the main attack.

At midday on 9 April 800 heavy bombers saturated the area between the Senio and the Santerno with fragmentation bombs, so as to avoid cratering the ground in front of the advance, while 1,000 medium and fighter bombers attacked gun positions, headquarters and any target that moved. At 4 p.m. the air bombardment ceased and the artillery took over for three hours, when at 7 p.m. the infantry attacked. 11th Infantry Brigade had little opposition, with the Germans evacuating their positions as well as Cotignolo. But the 2nd Lancashire Fusiliers on the right got caught in a minefield, with Lieutenant Lyle of 'C' Squadron going to their rescue and losing Trooper Boyter killed and Sergeant Glastonbury wounded when their tank blew up on a mine. 'B' Squadron had a direct hit on one tank, but with no casualties.

The offensive was going well, and on the 10th and 11th the whole regiment concentrated in support of 38th Brigade north of Lugo. The bridgehead across the Santerno was not secured until the 12th when the Bays crossed over, joining 38th Brigade at Mondaniga by last light. 'B' Squadron leaguered with the 2nd Inniskilling Fusiliers on the right, and 'A' with the 1st Royal Irish Fusiliers on the left. At

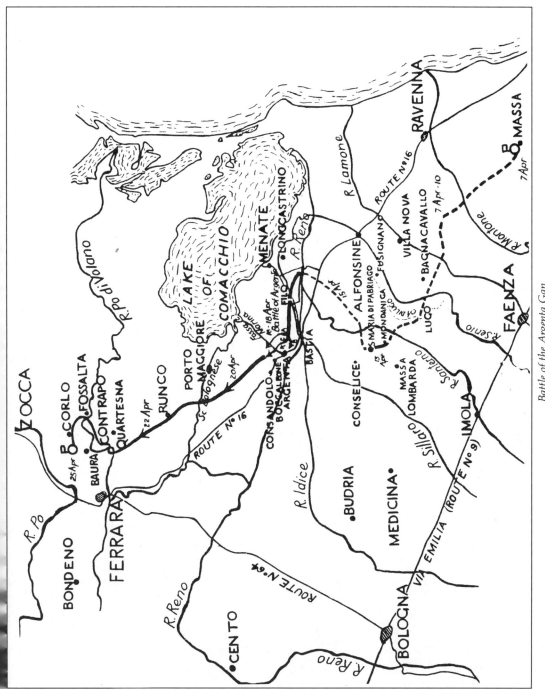

Battle of the Argenta Gap

dusk 'A' was straddled by a concentration of mortar bombs, which wounded two men. Before first light the infantry advanced, followed by the tanks, and by dawn the church at San Maria di Fabriago had been reached. As the infantry and tanks pushed forward, mortar and small arms fire came from around the various houses. Each was methodically attacked, resulting in a flow of prisoners. The leading tank in each troop moved slightly ahead of the infantry, covered by the fire of the other two tanks. When opposition was met, the tanks would silence it, and then the infantry would go in to clear up. The first serious resistance was met at Scola Fossotone after the tanks and infantry had advanced three miles. 'A' Squadron and the Royal Irish Fusiliers cleared a series of buildings on both sides of the canal, and the Engineers got an Ark into position, enabling 'C' Squadron to cross, covered by smoke from 'A', and move forward to catch up with the infantry. In the meantime 'B' Squadron had been making steady progress with the Inniskillings on the right.

By midday 7,000 yards had been gained and the 9th Lancers, with the London Irish carried in Kangaroos, passed through to attempt to seize the bridge over the Reno at Bastia. Six hundred prisoners had been taken by the brigade, and the Bays had lost one NCO killed and one wounded. The regiment consolidated in the positions which had been reached. The way was now open to assault the Argenta Gap, a narrow two-mile defile between Lake Comacchio and the Reno, which ran for five miles to beyond Argenta. The attack by the 9th Lancers and London Irish met with strong opposition, and 'C' Squadron of the Bays was sent to support the 11th Infantry Brigade six miles east of Bastia, where the Engineers were putting a Bailey bridge across the Reno. There 'C' Squadron was joined by the rest of the regiment late on 15 April.

On the 16th the advance started again with 'C' Squadron on the right supporting the 1st East Surreys, and 'B' Squadron on the left with the 5th Northamptons. 'B' made a certain amount of progress until halted about a mile south of Argenta by stiffening resistance, when Lieutenant Clarke was mortally wounded by a sniper. 'C' fought its way around the outskirts of Argenta until it encountered an extensive minefield south of the Fossa Marina and flail tanks were called in to clear a way. The flails worked very slowly against opposition from elements of three German divisions, indicating that the Fossa Marina was the main line of resistance. At this point Lieutenant Perkins was wounded.

That night the 2nd Lancashire Fusiliers with 'A' Squadron in support forced a bridgehead across the Fossa Marina. An Ark was placed in position, and by first light one troop was across in support, losing one tank soon after to an enemy tank, followed at 8 a.m. by another. Then a German counter-atttack with tanks developed and two more troops were ordered across, one tank being lost on a mine but the rest fanning out and helping the Fusiliers to regain lost ground. Two troops of 'C' Squadron were also pushed across onto the lateral road 1,000 yards beyond the canal, where one tank was knocked out by an enemy gun. Next 'B' crossed with the 2nd Inniskilling Fusiliers, swinging left and behind Argenta, where they had to fight their way forward house by house, losing a tank to a Tiger but capturing many prisoners. By nightfall they were facing the Germans across the no man's land of Route 16. On the right 'C' with the Irish Fusiliers met determined resistance. Lieutenant Read's tank was hit and Read wounded. By midnight the Fusiliers had reached their objective at the Scolo Cardinolo, another canal.

By first light on the 18th the Northamptons had taken Argenta, but 'B' with the Inniskillings had a stiff fight to capture the village of San Antonio to the north west, taking fifty prisoners from 29th Panzer Grenadier Division. Turning along Route 16 there was another battle for Casa Tomba, which required not only tanks and infantry, but flamethrowers, and took until evening. 'C' made progress with the Irish Fusiliers, who captured two Mark IVs intact with the crews asleep inside. By evening it was clear that the battle for the Argenta Gap had been won. The morning of 19 April saw the 9th Lancers and the London Irish passing in pursuit to the north, while 'B' and 'C' Squadrons helped the infantry clear up remaining pockets of Germans around Argenta.

For their part in the battle Major Rich and Lieutenants Saunders and Clough were awarded the Military Cross, Sergeant Bruten the DCM, and Sergeants Battles, Bovill and Foster the Military Medal.

On 20 April 'A' Squadron with the Lancashire Fusiliers, and 'B' with the East Surreys, crossed the Scolo Bolognese, making progress against constant scattered opposition and establishing a bridgehead by dark. A counter-attack knocked out two of 'B''s tanks with bazookas, but the position was soon re-established. The 21st saw 'B' Squadron helping the East Surreys clear the Ronco, then moving on to the Po di Volano canal, while by dusk 'C' Squadron had moved up alongside with the Northamptons. On the 22nd 'A' Squadron with the Lancashire Fusiliers took Contrapo, and on the 23rd a bridgehead was established across the Po di Volano by the Northamptons with 'C' in support, coming up against fierce resistance at Fossalta, which was taken with many prisoners. The 24th saw 'C' Squadron reaching Corlo, where Lieutenant Beak was wounded before 'B' took over and captured the village. On 25 April 'B' Squadron helped the Lancashire Fusiliers and the East Surreys clear the area south of the Po and westwards; the whole regiment was then concentrated in the village of Boara, three miles north east of Ferrara. In the meantime the Eighth Army was chasing the defeated enemy northwards. On the 27th the weather broke, and on the 28th the Americans captured Piacenza, followed on 2 May by the surrender of all German forces in Italy. The Queen's Bays had reverted to command of 2nd Armoured Brigade on 30 April.[3,4]

SOURCES

1 D. McCorquodale, B. L. B. Hutchings and A. D. Woozley, *History of The King's Dragoon Guards, 1938-1945*, published by the regiment.
2 War Diary, 1943-45, 1st King's Dragoon Guards, Regimental Museum, 1st The Queen's Dragoon Guards, Cardiff Castle.
3 W. Beddington, *The Queen's Bays, 1929-1945*, 1954.
4 War Diary, 1944-45, The Queen's Bays, Regimental Museum, 1st The Queen's Dragoon Guards, Cardiff Castle.

46

Greece, Syria and Lebanon, Palestine, Germany
1945-1956

The King's Dragoon Guards' advance party arrived in an LST at the Piraeus in Greece on 19 December 1944. As the ship entered the harbour a Greek tug flying the royal flag preceded it, only to be shelled by a Communist ELAS gun. The tug fled back to sea, whereupon a Greek submarine entered the harbour and shelled the gun. The LST then berthed and disgorged its load of troops, without any interference from the ELAS they were arriving to suppress. The staff had allotted the KDG an open field, but Colonel Lindsay was not prepared to accept such treatment and moved the regiment into some empty bomb-damaged villas around the edge of Kalamaki airfield. The next day the KDG moved to billets in Glifadha two miles to the south, and near to Phaleron.

The War Diary described the local situation:

> The whole military set up is very queer, and the politics even queerer. The opposite army is ELAS, who are in fact a fairly orderly Radical party, now being run and dictated to by KKE, which is allegedly a thoroughgoing anarchist party. As they do not wear any uniform, and rely chiefly on furtive sniping, one cannot tell who is who, or where they are. We have a fair body of troops in Athens, but there is no sort of front line, as ELAS infiltrate all round and behind them. The main road to Athens from the sea is only safe to armoured cars by day, but by night convoys of supply vehicles go through unmolested.

Many British military installations had been sited around Athens in what had seemed to be the most appropriate places, without any thought that they might find themselves in the midst of a civil war and in hostile territory. On 21 and 22 December 'B' Squadron rescued REME workshops sited in a foundry in the Imittos district of south-east Athens. A blown bridge and roadblocks were dealt with, resulting in three ELAS being killed and three being taken prisoner, but 240 men and 130 vehicles were brought to safety. Over the following days a number of patrols were carried out: one by Lieutenant Dorell to Voula resulted in three

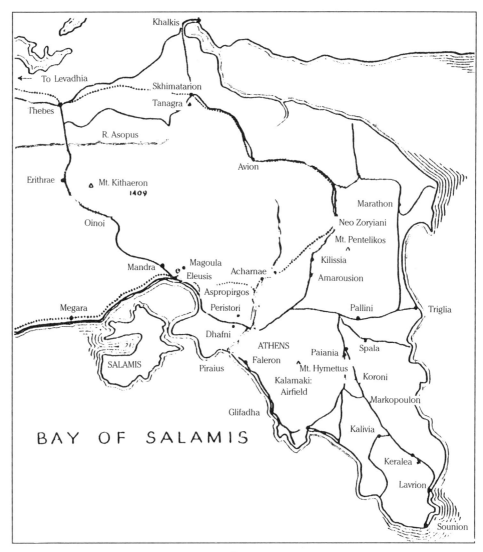

Greece

more guerrillas killed and three captured, as well as one of their headquarters being destroyed; other patrols escorted Mr Churchill, Mr Eden and Field Marshal Alexander to Athens.[1, 2]

On Boxing Day ELAS attacked Kalamaki airfield, slightly wounding two KDG sentries and blowing up an armoured car on prepared charges, which wounded all three crew members. Patrols to the guerrilla strongholds of Vari and Koropi resulted in a Daimler armoured car being blown up on the 26th, and two Daimlers and a Dingo on the 29th, all luckily without any casualties. On 31 December another rescue operation was successfully mounted to bring to safety 120 men of the 50th RTR, stationed ten miles east of Athens at Pallini wireless station, which provided the only direct link between Athens and London. Several roadblocks covered by sniper fire had to be cleared, resulting in fourteen ELAS killed and three prisoners being taken.

The hostilities had completely disrupted food supplies to the starving civilian population, and every squadron set up soup kitchens, with 1,000 meals being served each day. On 4 January 1945 a major raid was carried out on the ELAS stronghold of Peristeri: 'A' Squadron cleared roadblocks and captured a bridge intact, while Spitfires gave close support in strafing the enemy. 'C' seized another bridge, and then 'C' and 'D' dashed into Peristeri as the ELAS soldiers, making themselves easy targets for the Spitfires, fled to the hills. Six lorries, forty-four rifles, 500 lb dynamite and boxes of ammunition were captured, and at least fifty casualties inflicted on the enemy.

By early in January ELAS had been expelled from Athens. At dawn on 6 January the KDG, with a squadron of the 40th RTR, the 2nd Beds and Herts, and supporting troops under command, set out for Thebes, to clear the whole of Attica of ELAS. 'D' Squadron, leading, had an easy run until reaching the Bay of Elevsis, where the road had been blown along a causeway and the point was covered by mortar and small arms fire. Eventually two troops found a way round by difficult tracks, taking the defenders in the rear, killing some and capturing eight prisoners. The two troops then moved back to the road to cover the REs repairing the damage, before 'D' Squadron moved on to Aspropirgus, which was full of the enemy, who were promptly chased out into the hills. The rest of the regiment had entered Elevsis, where there was opposition from every direction. A sweep was made by the armoured cars and tanks, which collected a number of prisoners, but systematic clearing by the infantry was left to the morrow. 'B' Squadron was sent on to clear the villages of Mandra and Magoula and encountered resistance. A Sherman from 40th Tanks blew up on a mine, blocking the road behind two of 'B' Squadron's troops, which were only extricated after dark, when another tank came up to tow away the disabled Sherman. Colonel Lindsay decided to leaguer for the night on Elevsis aerodrome outside the town; as the War Diary relates, it was 'much to the disgust of the Americans there', who tried to keep aloof from the operations in Greece.

On 7 January 'C' Squadron took the lead, 3rd Greek Mountain Brigade having arrived to clear Elevsis, and met no opposition driving through hilly country for the first fourteen miles. At the village of Oinoi the leading troop surprised a number of ELAS, while three support troops cleared the high ground. Both activities inflicted casualties on the enemy. Half a mile further on, at the entrance to the pass of Dryoschephalae, the road was blown at the entrance to a gorge, and the holdup was covered by fire from positions around the old fort of Eleutherae which wounded Sergeant Wait and a Sapper. In spite of a deterioration in the weather, RAF Spitfires gave close support, and with all the armoured cars and tanks shooting up the Communist positions, the ELAS soldiers incurred many casualties but hung on most bravely. When the pass was eventually forced, there was little daylight left and Colonel Lindsay decided to retire to Elevsis for the night, taking back thirty prisoners.

On the morning of the 8th 'A' Squadron found the pass undefended. The Greek Mountain Brigade came up to picket the hills on either side of the road, leaving the KDG to enter Thebes at 2 p.m. to a tumultuous welcome. 'A' and 'B' Squadrons spread out over the plain of Asopus, meeting some opposition and shooting up numerous retreating bands of guerrillas, but mainly collecting the now dispirited Andarte soldiers of ELAS – the bag for the day was 250. Civilian hostages were freed in Thebes and huge quatities of ammunition and explosives were captured. The following day 'B' and 'D' Squadrons were despatched to chase the enemy; 'B' reached Levadhia, and 'D' Khalkis, with both inflicting more damage on the fleeing enemy. 'B' picked up M. Partsalides, and 'D' M. Zevgos, both ELAS delegates sent south to ask for an armistice, and both were immediately escorted back to Athens. The KDG were then relieved by the 4th Reconnaissance Regiment, commanded by Colonel Delmege, himself a KDG.

On 9 January the KDG were ordered to return to Athens, making a sweep to the east en route. 'C' Squadron, leading, shot up 200 ELAS in a village but the regiment made slow progress over very bad roads and several 'blows'. Marathon Dam was visited to ensure that it was still intact, and at Nea Zoryiani more arms and ammunition were discovered. The regiment arrived back at Glifadha on the 10th, having captured 470 prisoners, killed at least 222 ELAS, and taken 9 vehicles, 8 guns, 10 machine guns, 150 rifles, 75 tons of ammunition and explosives, and a bomb dump containing 195 bombs, varying in size from 50 to 1,000 kilos.

On 13 January The King's Dragoon Guards moved from Glifadha to excellent billets at the resort of Kifissia, ten miles north east of Athens, each squadron having a complete hotel to itself. On the 14th Major Macartney, returning from Athens, failed to hear the challenge of a sentry and was shot, dying from his wound. A few days later Trooper Pickles was shot dead by a Greek sentry. On 21 January 1945 The King's Dragoon Guards held a parade church service on the Areopagus, by the Acropolis, on the spot where St Paul preached to the men of Athens. The band of the 4th Hussars played, and the Revd Macmanaway officiated. Although at the time the regiment could not know it, its wartime service had come to an end, and the Areopagus provided a fitting tribute to those KDG who had died or been wounded, and a thanksgiving for those who had survived.

The King's Dragoon Guards remained at Kifissia for the next three months, during which time 350 KDG who had served overseas for more than five years returned to England under the Python scheme. Python meant that the regiment was to lose ten of its senior officers and all its NCOs, with the exception of five sergeants and twenty corporals. On 18 April 1945 the KDG embarked at the Piraeus in the troopship *Neuralia*, bound for the Middle East, arriving at Beni Yusef camp outside Cairo on the 23rd. The regiment was re-equipped with Staghound armoured cars. Throughout May 1945 250 reinforcements were absorbed, replacing the men going home, and intensive training continued. On 23 May The King's Dragoon Guards left Egypt, driving north to Syria and the Lebanon, where trouble had broken out between the French and Arabs.[1, 2]

Syria and the Lebanon had been French Protectorates, but the Arab inhabitants, looking for independence, resented French attempts to restore colonial rule. As a result the French had reacted with force. The War Diary described the role of the KDG as 'to restore order in the event of trouble'. Over the next few months the regiment was spread over an huge area, with squadrons or troops at one time or another at Beirut, Damascus, Tripoli, Latakia, Tartous, Tel Kalakh, Soueida, Baalbek, Palmyra and Raqqa.[2]

By 10 June 1945 'D' Squadron was patrolling Beirut and on the 12th 'A' Squadron went to Baalbek in the Bekaa valley. On 2 July 'B' Squadron was sent to Tel Kalakh near Tripoli to succour a French garrison which had been cut off, and where the situation remained extremely tense throughout the month. On the 3rd two troops of 'A' Squadron encamped on Damascus race course, their task being to escort high-ranking French officers who were otherwise unable to move about the town safely. Two troops of 'B' Squadron, known as Mannforce, went on the 6th to Latakia where the French had fired at a crowd, killing nineteen. The KDG then provided an escort for the victims' funerals in order to prevent further trouble. Mannforce, together with the 2nd Sherwood Foresters, was called to Banias on 10 July when the French opened fire on the town with mortars and machine guns; later Lieutenant Mann took a party to the Turkish frontier to bring back the horses and French officers of a French Tchekass Cavalry unit, whose men had deserted; on the 14th another detached troop went to Tartous where more trouble was expected. Throughout July there was trouble at Latakia, Banias and Tripoli, when numbers of the locally recruited Troupes Speciales constantly deserted, causing the French to fire on them and the Syrians to return the fire, with the KDG holding the ring in between.[2]

Early in August 'B' Squadron went to relieve the 3rd Hussars at Raqqa, in the

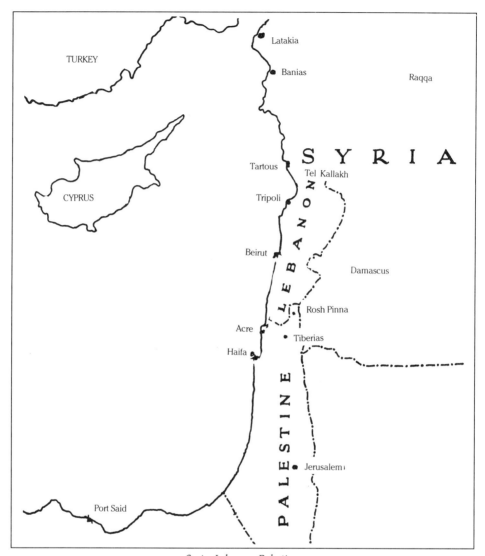

Syria, Lebanon, Palestine

northern Syrian desert east of Aleppo and on the River Euphrates. On 1 September 'C' Squadron had to intervene at Tartous when some Syrians opened fire on a party of Frenchmen, killing an officer and wounding two sergeants. At the beginning of October the KDG squadrons came together and moved to Palestine. They were first stationed at Lajjun before being distributed over northern Palestine, RHQ and 'B' Squadron at Rosh Pinna, seventeen miles north of Tiberias, 'A' at Tiberias, 'C' at Haifa and 'D' at Acre. All squadrons were fully engaged in an internal security role, which included establishing cordons around settlements, road blocks, searches and escort duties. By November squadron situations had changed: RHQ was still at Rosh Pinna and 'A' at Tiberias, but 'B' had moved to Camp 260 near Nahariya with the 3rd Grenadier Guards, and with two troops stationed in Jerusalem as the High Commissioner's escort; 'C' remained at Haifa, while 'D' were now at Rosh Pinna with two troops at Metulla.[2]

With the surrender of all the German forces in Italy, The Queen's Bays moved on 15 May 1945 from south of the River Po to Monselice north of the river, and then on the 20th to Pegi on the River Isonzo to be a part of the occupying force taking over Venezia Gulia from the Jugoslavs. During August another move was made – to Sagiletto, south of Palmanova, where the 2nd Armoured Brigade was concentrating before sending home time-expired men. On 26 August 225 time-expired men of the Bays left for England. Young reinforcements arrived to make good this loss of experienced and battle-trained soldiers, and the time was spent in bringing the newcomers up to regimental standards. A threat from Communist Italian partisans caused the Bays to move to Valdagno, north west of Vicenza, in September, where they remained until 13 June 1946. On 6 March 1946 Lieutenant Colonel Sykes took over command of the regiment from Lieutenant Colonel Asquith.[3,4]

On 13 June 1946 the Bays moved to the Italian barracks in Palmanova, taking over from the 7th Hussars. Their role was to support the Cameron Highlanders in any show of force that might be needed in the face of Jugoslav designs on Venezia Gulia. This move was meant to be temporary but, as so often with Army life, became rather more extended. The regiment remained at Palmanova for a full year, sailing for Egypt on 13 June 1947. On arrival at Port Said the Bays went into a tented camp at Fanara, and were told that they were to be reduced to a cadre of 33 officers and 146 other ranks. A parade on 13 September 1947 saw the disbandmenmt of 2nd Armoured Brigade. The following day Lieutenant Colonel Savill arrived to take over command of The Queen's Bays from Lieutenant Colonel Sykes. The tanks were handed in on 29 September and drafts of men were sent to The King's Dragoon Guards, the 4/7th Royal Dragoon Guards, the 3rd Hussars and the 15/19th Hussars. On 30 November 1947 15 officers and 187 other ranks embarked at Port Said in the SS *Samaria*, disembarking at Liverpool after being played in to the quayside by the band.

The Bays in their reduced state reassembled at the Dale Barracks, Chester, on 9 January 1948 with the role of helping the Territorial Army, and were equipped with a small number of Comet tanks. From 7 June to 8 September the greater part of the regiment ran a summer camp for the Territorials at Castlemartin ranges. During October official approval was given for The Queen's Bays to be affiliated with the 1st Special Service Battalion of the Union Defence Forces of South Africa. During 1949 the regiment again ran a summer camp for the Territorial Army at Merrion Camp, Pembroke. On 19 October 1949 The Queen's Bays were honoured with a visit by the colonel in chief, Her Majesty Queen Elizabeth, when the regimental war memorial was unveiled. This great day was followed on 1 November 1949 by the regiment moving to the British Army on the Rhine.[3, 4]

The years immediately after the war were a difficult period for regular regiments. *The K.D.G.* in its editorial for 1946-47 commented:

> Troops and whole squadrons no longer wear the same faces for months and years on end. They are lucky if they do so for a week. Whereas a reinforcement used to be a pleasant windfall – useful, unusual and inexperienced – nowadays, within a year of his arrival, he will be able to count himself among the older inhabitants. The Sergeants' Mess is the backbone of the Regiment – an old and very true cry – but within six months half of our sergeants will have left us for good.[5]

The King's Dragoon Guards found themselves in Palestine in 1946 engaged in the routine of internal security. Early in January 'B' Squadron was called out to intercept an illegal landing of Jewish immigrants. Later in the month, at a demonstration, Captain Hollies Smith was wounded when a mortar bomb hit the branches of a tree above him and a fragment of shrapnel lodged in his thigh. Roadblocks and searches were established at Samakh, and the Jewish colonies at Ein Zittin and Hadera were cordoned and searched. Each month involved a similar round of duties: in February Safad police station was attacked, resulting in more searches and roadblocks, while at the end of the month a large arms cache was discovered at the Jewish Birya settlement, which continued to be a source of trouble. During July 1946 the regiment was concentrated at Muqeibila for training, ostensibly free of internal security duties, but that did not prevent it being called upon to engage in a massive search of Tel Aviv. At the end of 1946 Lieutenant Colonel Lindsay handed over command to Lieutenant Colonel Radford.[2]

In January 1946 403 Independent Tank Squadron was raised to form the demonstration squadron at the Middle East School of Infantry at Acre. It was largely staffed and trained by the KDG; the squadron leader was Major Mann, with Captain Gaskin as second in command, and SSM Dilworth and four KDG sergeants providing the backbone of the NCOs.[5]

In April 1947 The King's Dragoon Guards handed over its operational role to the 17/21st Lancers and moved to Libya, with RHQ and two squadrons stationed in Benghazi, one squadron in Barce and another in Homs. The regiment's role was again internal security, but the country was peaceable and operational duties were mainly concerned with preventing pilferage. The squadron at Homs was moved to Misurata in Tripolitania, and the regiment's training suffered from this wide dispersal across 600 miles. For the first time since the outbreak of war, the regiment's families joined their menfolk, together with the regimental band. There was a great deal of sport, including polo and the KDG pack of hounds, football, cricket, hockey, bathing and yachting. In February 1948 The King's Dragoon Guards embarked 33 officers and 220 other ranks in HMT *Scythia*, landing at Liverpool on 11 March and going to Watton Camp near Driffield in Yorkshire. [2, 6]

Lieutenant Colonel Luck took over command of the KDG from Lieutenant Colonel Radford in June 1948. The regiment was under strength because of a shortage of recruits, and became a 'Territorial Army assistance unit', spending the year in training three Yeomanry regiments. The KDG also provided the permanent staff for the North Irish Horse. Lack of men meant that the regiment was reduced to a headquarter and two sabre squadrons only. During the year acquaintance was renewed with the affiliated Canadian and Australian regiments, the Governor General's Horse Guards and the Royal New South Wales Lancers. In November 1948 recruits, mainly National Service men, began to arrive, and early in 1949 the KDG moved from Yorkshire to Ulster, being stationed at Lisanelly Camp at Omagh. 'A' Squadron was re-formed, and throughout 1949, 1950 and 1951 the National Servicemen were trained to a high standard in preparation for their continuing service with the Territorial Army on their release from the Regular Army. During the summer months the regiment ran training camps for the North Irish Horse and the Fife and Forfar Yeomanry. During 1951 Lieutenant Colonel Lindsay took over command of the regiment from Lieutenant Colonel Luck. During the Korean War Lieutenant Lidsey, KDG, served on attachment with the 8th Royal Irish Hussars, distinguishing himself in helping to hold off the Chinese advance during the latter stages of the Battle of the Imjin River, for which he was mentioned in despatches. On 13 January 1952 The King's Dragoon Guards arrived at Hamburg in Germany to become the armoured car regiment of the 6th Armoured Division. [7, 8, 9, 10, 11, 12]

The Queen's Bays left Chester for Wessex Barracks, Fallingbostel, the main party arriving on 2 November 1949. On 1 March 1950 Lieutenant Colonel Weld took over command of the regiment. With the approval of both the colonel-in-chief, Her Majesty Queen Elizabeth, and the colonel, Brigadier Kingstone, the Bays celebrated Gazala Day for the first time on 27 May 1950. The years 1950 to 1955 were spent in training in Germany. The regimental polo team won the inter-regimental polo tournament in 1950, beating the 7th Hussars, and in 1951 the hockey team lost only three times in fifty-four matches. On 16 November 1952 Lieutenant Colonel Manger assumed command. During this period warm relationships were established with the French Deuxième Régiment de Dragons (Condé Dragons), and in November 1952 a representative party of The Queen's Bays paraded with their French sister regiment, when the standards of the two regiments were paraded side by side, an event which is believed to be unique. During the Gazala Day celebrations in 1954 the French Condé Dragons and the American 2nd Armoured Cavalry Regiment both sent representative contingents. In September that year the Queen's Bays left Fallingbostel for Assaye Barracks at Tidworth. [4, 13, 14, 15, 16]

The Queen's Bays spent only an eventful three months at Tidworth. On 16 October 1955 the colonel-in-chief, Her Majesty Queen Elizabeth, The Queen Mother, inspected her regiment and spent the day with the officers and men. On 10 December the main body of the regiment embarked in HMT *Asturias* for service in the Middle East, landing at Aqaba in Jordan on 19 December. The regiment had one squadron stationed at Maan on the Hejaz railway, with RHQ and two squadrons at Aqaba, with an observation post on the Jordan-Israel border opposite Eilath. From January to March 1955 'B' Squadron went to Habbaniya in Iraq, and in July 'A' Squadron was moved to Zerka to train with the Arab Legion. Then, in October 1955, a squadron was sent to Sabratha in Libya to take over from the 14/20th King's Hussars, leaving the regiment widely scattered over the Middle East. On 9 November Lieutenant Colonel Armitage took over command.[4, 16, 17]

The King's Dragoon Guards spent the years 1952 to 1956 in Germany as a part of the British Army on the Rhine. In addition to training, the regiment provided patrols along the border with East Germany. In April 1953 the regiment left Hamburg, moving to McLeod Barracks, Neuminster. During 1954 Lieutenant Colonel Cairns assumed command, taking over from Lieutenant Colonel Lindsay. At Neuminster the KDG worked with the Danes, taking part in their manoeuvres in September 1954 and establishing very close relationships with the Danish contingent in NATO throughout their time in Germany. To mark this close connection, Brigadier Storch, commanding the Danish Force in Germany, presented the regiment with two lances formerly carried by the Royal Danish Lancers from their formation in 1791 until they were disbanded in 1842. On 7 March 1956 The King's Dragoon Guards left Germany for Piddlehinton in Dorset, where they stayed until May, then embarking in HMT *Dunera* for Malaya and disembarking at Singapore on 30 May 1956.[18, 19, 20, 21, 22]

SOURCES

1 D. McCorquodale, B. L. B. Hutchings and A. D. Woozley, *History of The King's*

Dragoon Guards, 1938-1945, published by the regiment.

2 War Diary, 1945-48, 1st King's Dragoon Guards, Regimental Museum, 1st The Queen's Dragoon Guards, Cardiff Castle.

3 W. Beddington, *The Queen's Bays, 1929-1945*, 1954.

4 War Diary, 1945-59, The Queen's Bays, Regimental Museum, 1st The Queen's Dragoon Guards, Cardiff Castle.

5 *The K. D. G.*, vol. 3, no. 5, 1946-47, p. 2.

6 *The K. D. G.*, vol. 3, no. 7, 1947-48, p. 2

7 *The K. D. G.*, vol. 4, no. 2, 1949.

8 *The K. D. G.*, vol. 4, no. 3, 1949-50.

9 *The K. D. G.*, vol. 4, no. 4, 1950.

10 J. M. Strawson, *The Irish Hussar*, published by the regiment.

11 *The K. D. G.*, vol. 4, no. 5, 1950.

12 *The K. D. G.*, vol. 4, no. 6, 1951.

13 *Regimental Journal of The Queen's Bays*, vol. 1, no. 2, 1950.

14 *Regimental Journal of The Queen's Bays*, vol. 1, no. 3, 1951.

15 *Regimental Journal of The Queen's Bays*, vol. 1, no. 5, 1953.

16 *Regimental Journal of The Queen's Bays*, vol. 1, no. 6, 1954.

17 *Regimental Journal of The Queen's Bays*, vol. 1, no. 7, 1955.

18 *The K. D. G.*, vol. 4, no. 7, 1952.

19 *The K. D. G.*, vol. 4, no. 8, 1953.

20 *The K. D. G.*, vol. 5, no. 1, 1954.

21 *The K. D. G.*, vol. 5, no. 2, 1955.

22 *The K. D. G.*, vol. 5, no. 3, 1956.

47
Aqaba, Libya, Malaya, Amalgamation
1954-1959

The role of The Queen's Bays in Jordan was to deter the Israelis from attacking Aqaba. Known as 'O' Force, the Bays had a company of the Foot Guards attached from the Guards Brigade in the Canal Zone. This company changed every four months, and 'O' Force was commanded by the commanding officer of the Bays. The half-squadron at Habbaniya in Iraq trained the Iraqi Army how to operate and maintain Centurion tanks, with which the Iraqis were about to be equipped.

Lieutenant Ormrod, commanding the reconnaissance troop in Headquarter Squadron of The Queen's Bays, remembered his time in Jordan with affection:

> Lieutenant Colonel Manger was the only member of the Regiment with experience of Jordan. He loved the country and was determined that we all should see as much of it as possible. Our first military exercise was a trip to Azrak in January 1955. Soon our vehicles broke down. Vehicle unreliability and lack of spares were a major problem. The Land Rovers were a great success, though uncomfortable when driven at any speed across the desert, and they were underpowered. I finished the year with only two serviceable scout cars. They got stuck in soft sand, and the suspension could not take the hammering it received from the rocky desert. Exercises in North Jordan included the Arab Legion, who were very friendly . . . I have fond memories of Jordan, the freedom of military manoeuvre, the Aqaba beach, the coral and tropical fish unsurpassed anywhere, Petra, Jerusalem and the holy places, the opportunity to visit Damascus, Beirut and Cairo.[1]

During its stay in Jordan the regiment was visited three times by His Majesty King Hussein. In February 1956 The Queen's Bays were relieved at Aqaba by the 10th Royal Hussars; the squadrons still in Jordan moved to Tripoli in Libya, with a strength of 20 officers and 351 other ranks, and there joined the detached squadron at the Cavalry Barracks, Sabratha.[2]

'A brave attempt was made to re-establish polo. Twenty-three officers played polo in the next two years. The credit must go to Charles Armitage for his single-minded ambition to recreate polo in the regiment.' Limited training on a regimental basis continued, but with the other units of the brigade situated 130 and 900 miles away, brigade training became impossible. In August 1956 an officer from divisional headquarters reached the regiment, while it was out on exercise in the desert, to tell the Bays to return to barracks immediately and prepare for war. They at once brought their equipment up to battle efficiency, absorbing 191 reservists who had been recalled to the colours, and with a strength of 38 officers and 776 other ranks got ready to take part in the Suez Operations. The Army had planned that the 10th Armoured Division, of which the Bays were a part, should be ferried to Alexandria, but the Foreign Office then pointed out that the treaty with Libya precluded troops stationed in that country from being used in operations against other friendly Arab states. In view of the age of the vehicles, it was perhaps just as well that such an operation was never attempted, but the result for the regiment was a considerable anti-climax.

In the meantime Arab sensitivities had been so inflamed by the British and French action at Suez that the Bays remained in Libya carrying out internal security duties. During October and November 1956 there were a number of attempts at insurrection in Tripolitania. The regimental families had to be evacuated to safety, some being flown back to England, guards were strengthened and vehicles moved in sensitive areas with due caution. Patrols were mounted in Zavia, and by a combination of tact and good humour, law and order was maintained. By Christmas the reservists had returned home and the families had rejoined their menfolk. But the atmosphere in Libya had changed and was never to regain its former friendliness.[2, 3]

During 1957 The Queen's Bays were awarded the following battle honours for their service in the 1939-45 war. (The battle honours in capitals are those which are emblazoned on the regimental standard.) **North West Europe,** 1940: THE SOMME 1940; Withdrawal to the Seine. **North Africa,** 1941-43: Msus; GAZALA; Bir El Aslagh; The Cauldron; Knightsbridge; Via Balbia; Mersa Matruh; EL ALAMEIN; Tebaga Gap; EL HAMMA; El Kourzia; Djebel Kournine; TUNIS; Creteville Pass. **Italy,** 1944-45: CORIANO; Carpineta; LAMONE CROSSING; Defence of Lamone Bridgehead; RIMINI LINE; Ceriano Ridge; Cesina; ARGENTA GAP.[2]

On 24 July 1957 it was announced in Parliament that The Queen's Bays were to amalgamate with The 1st King's Dragoon Guards. 'It came as a bitter and unexpected blow to all of us. If it has got to be, we would rather amalgamate with the King's Dragoon Guards than with any other regiment; we are happy and proud to do so. We hope all, both serving soldiers and Old Comrades, will support the new regiment to the utmost, so that the spirit and traditions of The Queen's Bays may live on.' The Bays also learnt that they were to cease to be a tank regiment and that they were to convert to an armoured car role.[2, 3]

Throughout 1957, after a very wet winter, the Bays trained in Libya, then prepared to move back to England, embarking in HMT *Dilwara* at Tripoli on 13 December 1957. Sabratha, situated on the shores of the Mediterranean, had provided adequate regimental training areas and ranges, while both officers and men had been able to indulge in a variety of sporting activities. The regiment's stay at Sabratha was described as 'A quiet station, with little scope for riotous living. The usual after duty habit was to rush to the beach in fleets of three-tonners. Storms wash piles of ugly seaweed onto the beach, but all the same the attraction of long hours in the water surmounts such bothers.'[2, 3, 5]

The King's Dragoon Guards, on arrival in Malaya, found that they were one of two armoured car regiments, equipped with Daimler armoured cars and Ferret scout cars, operating in the peninsula. By June 1956 General Templer's policy of fortified villages and curfews was starting to take effect. Chin Peng's Communist guerrillas were finding it harder to operate, and although they had a plentiful stock of both British and Japanese weapons, supplies were becoming more difficult to obtain as rural villages became more secure. There was, however, plenty of fight left in the guerrillas and they still retained the element of surprise and choice of target.

The KDG relieved their old friends the 11th Hussars, taking over operational control on 26 June 1956. Regimental headquarters with 'B' Squadron was stationed at Seremban in Negri Sembilan, with 'C' Squadron forty miles to the north in Kuala Lumpur, while 'A' was 200 miles to the south in Johore Bahru. The regiment worked in support of the Gurkha Division, with each squadron operating as an independent unit, under its own brigade. In addition the regiment provided a training squadron on Singapore Island, which also had to be available for internal security duties and was heavily involved in containing serious rioting which broke out in Singapore city on 25 October.

The KDG were used to keep the main roads open, escorting convoys and VIPs, as well as setting ambushes, patrolling and carrying out village perimeter checks. 'A' Squadron had a contact in Johore, and at the notorious village of Kulai made 47 arrests of suspected terrorists out of a total of 174 made during a search operation. The squadron also went into Singapore city with the training squadron during the rioting, keeping the main roads and the back streets clear. 'C' Squadron at Kuala Lumpur was even more dispersed, with troops detached at Kajang and Nee Soon, but the whole squadron had to pack up suddenly and drive to Singapore during the riots, when it was stationed at Kallang Airport. It made twenty-seven arrests before returning to Kuala Lumpur. During the six months of 1956 that the

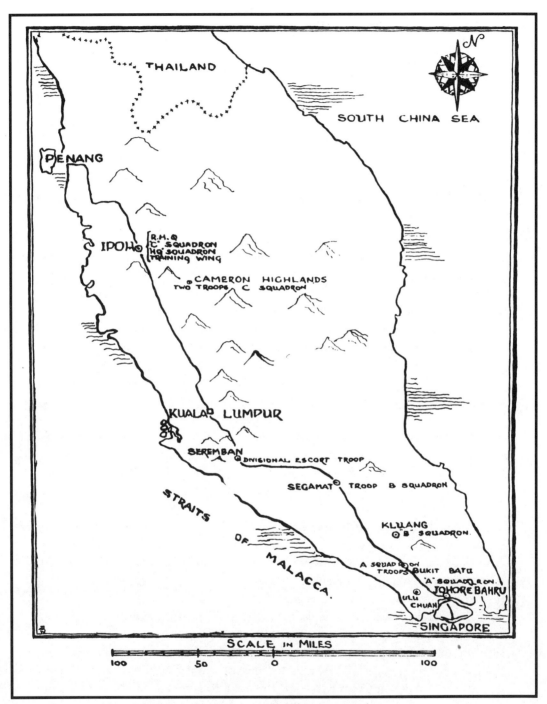

Sketch map showing regimental deployment in Malaya, 1959

regiment was in Malaya, its vehicles covered 600,000 miles. Back in Britain, Major Selby, KDG, scored two notable sporting successes, winning the Prince of Wales Cup and the King's Cup at the Royal Tournament at Earl's Court.[5, 6]

The King's Dragoon Guards took the news of amalgamation in the same spirit as The Queen's Bays:

> The big talking point in the regiment in 1957 was the shattering news of the amalgamation of the regiment with our sister regiment, The Queen's Bays. After 273 historic years the separate identities of two famous regiments will cease on 1st January, 1959, and the phoenix of a new regiment will rise from the ashes of the old. It is a sad blow to both regiments. We, on our part, are determined to make a success of the amalgamation.[7]

By May 1957 the situation in Malaya had been brought sufficiently under control for the number of armoured car regiments to be reduced, and when the 15th/19th Hussars went home in May, The King's Dragoon Guards took over responsibility for road security for the whole of Malaya, except for Selangor and Negri Sembilan. These two states were covered by a newly raised Malaya Armoured Car Regiment to which a number of KDG officers and NCOs were seconded. Regimental headquarters and Headquarter Squadron moved to Ipoh, where they were joined by 'C' Squadron; 'B' moved into Kuala Lumpur until 'Merdeka', or self-government, which was granted on 31 August 1957, when it moved to Kluang in North Johore. Because of manpower problems the training squadron joined RHQ, and in spite of the regiment's dispersal, the general carrying out the yearly administrative inspection commented, 'I was most impressed with all I saw of the K.D.G.' Yet another detachment of ten Ferret scout cars was formed later to act as the Gurkha Division Escort Troop, staying at Seremban. So the KDG were scattered in packets over an area the size of England, which provided a nightmare for the RSM, but a paradise for the troop and squadron leader.[5, 7]

On the night of 6/7 March 1957 a party of 'B' Squadron, while still at Seremban under command of Corporal Derench, opened fire on three Communist terrorists who walked into their position. The next day the body of Cheong Fatt, a section leader of the so-called 3rd Independent Platoon of the Malayan Republic Liberation Army, was found; it later transpired that a second terrorist, Tak Lan, had been wounded, but had escaped. 'C' Squadron, supporting the Commonwealth Brigade, had a detachment 6,000 feet up in the Cameron Highlands, and was responsible for the security of a particularly hilly and winding road that climbed from the plains to the highlands. Towards the end of the regiment's time in Malaya demoralised Communist terrorists used this road to surrender, and the KDG, by the end of March 1958 had accepted the surrender of a total of forty terrorists along it. On 27 September 1957 Lieutenant Colonel Selby took over command from Lieutenant Colonel Cairns. Sergeant Carter was awarded the Meritorious Service Medal, and Captain Lidsey and Lieutenant Gibson were mentioned in despatches. The regiment had by the end of 1957 travelled 1,851,094 miles since it arrived in Malaya.[5, 7, 8]

During 1957 The King's Dragoon Guards were awarded the following battle honours for their service during the 1939-45 war. (The honours shown in capitals are those emblazoned on the regimental standard.) **North Africa**, 1941-43: BEDA FOMM; DEFENCE OF TOBRUK: (a) Tobruk, 1941; (b) Tobruk Sortie, 1941; (c) Relief of Tobruk; Gazala; Bir Hacheim; DEFENCE OF ALAMEIN LINE; Alam el Halfa; El Agheila; ADVANCE ON TRIPOLI; TEBAGA GAP; Point 201 (Roman Wall); El Hamma; Akarit; TUNIS. **Italy**, 1943-44: Capture of Naples; Scafati Bridge; MONTE CAMINO; Garigliano Crossing; Capture of Perugia; Arezzo; GOTHIC LINE. **Greece**, 1944-45: Athens.[7]

Owing to manpower shortages 'B' Squadron was disbanded in February 1958 and was used to reinforce 'A' and 'C' Squadrons, only to be re-formed at the end of August before the regiment returned home. During October the KDG handed over to the 13th/18th Hussars and on 20 October the regiment embarked on HMT *Empire Fowey*, arriving at Southampton on 11 November, where it was greeted by a fanfare from the trumpeters of The Queen's Bays and then the Bays band played the KDG home. Lieutenant General Sir Richard Hull, the Commander in Chief Far East sent the KDG this message on their leaving Malaya.

> During the last two and a half years your tasks were to keep the roads open throughout the Federation for the safe passage of Security Forces and civilians and to give aid to the Civil Power in Singapore in time of need. Your vehicles travelled some three million miles on patrol and escort duties and as the Communist Terrorist threat outside the jungle decreased you went on foot into the jungle in search of the enemy. You have earned the respect and admiration of all units of the Security Forces and of the Federation and Singapore Police Forces. You go home now to amalgamate and I know that you will take with you to 1st The Queen's Dragoon Guards the same efficiency, high spirit and enthusiasm which you have shown here. Good luck to you all.[5, 7, 8]

The Queen's Bays landed at Southampton on 19 December 1957 and went straight to Perham Down, Tidworth. On 15 March 1958 the Bays handed in their tanks for the last time. With amalgamation pending, the flow of recruits dried up during the spring and in August 'B' Squadron ceased to exist. 'C' Squadron trained and converted to the new regimental role in armoured cars, while 'A' in April went to the Castlemartin camp to train and help the Territorial Army with their summer camps, returning to Tidworth in September.[4]

On 1 November 1958 The Queen's Bays paraded before their colonel in chief, Her Majesty Queen Elizabeth, The Queen Mother, for the last time. The parade took place on the Tattoo Ground at Tidworth, not far away from where, on 21 November 1688, King James II had reviewed the regiment, and on the same ground where on 25 July 1939 the colonel in chief had presented the regiment with the standard that was now being paraded for the last time. The Bays were in their new armoured cars, and the standard party, in full dress, was mounted on bay horses.[4, 5]

In addressing the regiment Queen Elizabeth said:

> At the beginning of next year The Queen's Bays will be amalgamated with The King's Dragoon Guards, and I know well the sadness which you all must feel. The merging of two proud regiments, each with its own history and traditions, will call for a high degree of tolerance, patience and understanding. That the amalgamation will be successful I have no doubt — for its success will depend on your loyalty, and on the loyalty and support of the Old Comrades, who have always proved themselves such a pillar of strength in days gone by. That I am to be colonel in chief of that regiment, the 1st The Queen's Dragoon Guards, is to me a source of very special pride.[4]

On 11 November 1958 the band of The Queen's Bays was at Southampton on the quayside to play home The King's Dragoon Guards, returning from Malaya. On landing the KDG went straight to Perham Down to prepare for their amalgamation with the Bays on 1 January 1959. They held their own final Parade at Perham Down on 20 November 1958. The official Amalgamation Parade took place at Perham Down on 20 February 1959. On 2 March the colonel in chief presented 1st The Queen's Dragoon Guards with its new standard in London in the grounds of Clarence House, amidst the flowering crocuses. The new regiment was commanded by Lieutenant Colonel Selby.[4, 5, 8, 9]

The standard of 1st The Queen's Dragoon Guards has emblazoned on its silken folds the following battle honours, showing the combined service and gallantry of the two senior cavalry regiments of the line over the 274 years of their separate existence: Blenheim, Ramillies, Oudenarde, Malplaquet, Dettingen, Warburg, Beaumont, Willems, Waterloo, Sevastopol, Lucknow, Taku Forts, Pekin 1860, South Africa 1879, South Africa 1901-1902, Mons, Le Cateau, Marne 1914, Messines 1914, Ypres 1914-15, Somme 1916-18, Morval, Scarpe 1917, Cambrai 1917-18, Amiens, Pursuit to Mons, France and Flanders 1914-18, Afghanistan

1919, Somme 1940, Beda Fomm, Defence of Tobruk, Gazala, Defence of Alamein Line, El Alamein, Advance on Tripoli, Tebaga Gap, El Hamma, Tunis, North Africa 1941-43, Monte Camino, Gothic Line, Coriano, Lamone Crossing, Rimini Line, Argenta Gap, Italy 1943- 45.[5]

SOURCES

1 Lieutenant J. J. Ormrod, The Queen's Bays, letter to author.
2 War Diary, 1955-58, The Queen's Bays, Regimental Museum, 1st The Queen's Dragoon Guards, Cardiff Castle.
3 *The Regimental Journal of The Queen's Bays*, vol. 2, no. 1, 1957.
4 *The Regimental Journal of The Queen's Bays*, vol. 2, no. 2, 1958.
5 E. Belfield, *The Queen's Dragoon Guards*, 1978.
6 *The K. D. G.*, vol. 5, no. 3, 1956.
7 *The K. D. G.*, vol. 5, no. 4, 1957.
8 *The K. D. G.*, vol. 5, no. 5, l958.
9 *The Regimental Journal of 1st The Queen's Dragoon Guards*, vol. 1, no. 1, 1959.

48

Wolfenbuttel, Omagh, Borneo, Aden, Detmold,
1959-1970

The King's Dragoon Guards and The Queen's Bays, having amalgamated at Perham Down on 1 January 1959 with the role of an armoured reconnaissance regiment, equipped with Alvis Saladin armoured cars, did not remain in England for long. By 10 March 1st The Queen's Dragoon Guards took over from the 12th Royal Lancers in Wolfenbuttel in Germany. On 13 June a combined Waterloo/Gazala Day was celebrated. In 1959 and in 1960 the new regiment won the BAOR inter-regimental polo cup. In December 1959 Lieutenant Colonel Selby gave up command and in January 1960 Lieutenant Colonel Harman took over.[1, 2]

The QDG was given the areas of Shropshire, Hereford, Monmouth and South Wales from which to recruit; during 1960 the last of the National Servicemen had been called up, and recruiting became a high priority. During 1959 the regiment took over the permanent staff role for the Shropshire Yeomanry, relieving the 4th Hussars. Lieutenant Colonel Knapp was commanding officer, with Captain Dent as adjutant and Captain Leckie as quartermaster. The regiment's allied regiments were The Governor General's Horse Guards of Canada, the 1st Reconnaissance Regiment of Ceylon, the 1/15 Royal New South Wales Lancers, and the 1st Special Service Battalion, Union Defence Forces of South Africa. In 1964 the 11th Cavalry (Frontier Force) of Pakistan was also allied to the QDG. [3, 4, 7]

The next years at Wolfenbuttel were to pass 'in the familiar pattern known to all who have served in Germany — inspections, field firing at Hohne, individual and collective training, autumn manoeuvres and sport'. In April 1962 The Queen's Dragoon Guards received an augmentation to its establishment of its own QDG helicopter flight, the 22nd Recce Flight, manned and flown by QDG, commanded by Captain Chamberlain, with Captain Railton and Lieutenant Stenhouse as pilots, and comprising four Skeeter helicopters, thus starting the regiment's long association with Army flying. In June 1962 Lieutenant Colonel Harman handed over command to Lieutenant Colonel Body. On 15 May 1963 the last of the National Servicemen were dined out by the commanding officer, squadron leaders and the

SSMs, and after twenty-five years the regiment was again all regular. In November 1964 it left Wolfenbuttel for Omagh in Northern Ireland.[4, 5, 6, 7]

Lieutenant Colonel Muir assumed command of the regiment on its arrival at Lisanelly Camp, Omagh. Shortly afterwards QDG was ordered to send an air-portable squadron to Cyprus in ten days' time. Tradition has it that on Day 9 a telephone call from the Ministry of Defence said, 'Sorry, a terrible mistake. We meant an armoured car squadron, and we meant Borneo and not Cyprus.' Within six days all the squadron's vehicles, together with a flight of QDG helicopters, were loaded onto the MV *Myrmidon* at Belfast, and 'B' Squadron flew out to Singapore three weeks later, moving on to take up station in the west of Sarawak. The Indonesians, led by their President Sukarno, were attempting to annex a defenceless Sabah (formerly Sarawak and North Borneo), conducting a guerrilla war and committing numerous border violations. The Commonwealth forces were considerably strengthened and infantry company bases were established half a mile to two miles behind the 1,000-mile border, with the difficult task of trying to dominate by patrolling and ambushes. The area consisted of rugged mountains covered by tropical forest, intersected by rivers; there was a single track metalled road, with appalling dirt side tracks. The QDG squadron, commanded by Major Bull, first supported the 1/10th Gurkha Rifles, then the 1st Royal New Zealand Infantry Regiment and later the 2nd Royal Malay Regiment.[1, 8, 9]

Squadron headquarters, with two and a half troops, was at Engkillili, on the banks of the swiftly flowing Batang Lupar river, 100 miles east of Kuching. The air troop, consisting at first of two Austers and later two Sioux helicopters, was at Simman-gang, while two troops were at Sungei Tengang, forty-five miles east of Engkillili, with half a troop at Batu Lintang on the border. The squadron's tasks consisted of an escort by one troop of the daily convoy of twenty to thirty three-tonners which plied between Simmangang and Kuching; secondly, to dominate the border area by patrolling with the help of Iban trackers, and laying of ambushes; thirdly, to gain the support of the local tribesmen and encourage them to assist the security forces; and fourthly for the Saladin armoured cars to give fire support with their 76 mm guns and machine guns to the infantry patrols operating in the border area.[1, 8, 9]

At the end of May Major Johnston took over command of the squadron whose base was situated on a small knoll around a police post, with sandbagged *bashas* for everyone, weapon pits and bunkers. The post was surrounded by barbed wire and *panjis* (sharpened slivers of bamboo). One night an Indonesian patrol of three men reached the perimeter fence, but were driven off. The rainfall in Borneo measured nearly 200 inches a year, falling mainly in a daily deluge around dusk, giving way to clear skies in the morning. Foot patrols visited

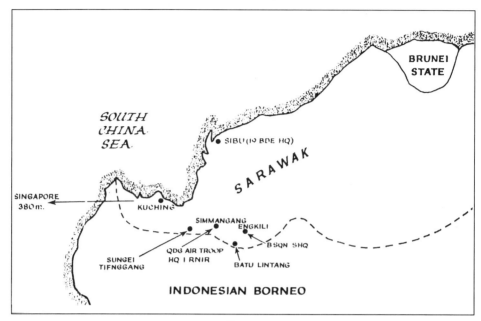

Borneo

at least one longhouse en route, and at these villages on stilts were always welcomed. Corporal Hathaway, our 'doctor', was immediately surrounded by anxious mothers – paludrine and asprin were cures for everything! It was fascinating to see in these huge dwellings, where all the villagers and animals sleep under one roof, pictures of the Royal Family, Cary Grant, Liz Taylor, and of course, The Beatles.

Other patrols used a long boat fitted with an outboard motor to cruise up and down the rivers.

The two Saladins at Batu Lintang were dug in and helped to repel an attack on the 1/10th Gurkhas soon after their arrival. The Air Troop meanwhile flew so successfully on border surveillance and support that Captain Chamberlain was mentioned in despatches. After a six-month tour 'C' Squadron relieved 'B', which returned to Omagh. On 6 November a 'C' Squadron patrol with the KOSB engaged an Indonesian incursion and the squadron base was moved from Engkillili to Wong Padong, halfway to Simmangang. In February 1966 'C' Squadron returned to the regiment in Omagh.

The rest of the regiment was based at Lisanelly Barracks, Omagh, with the air troop, now promoted to a squadron, eighty miles away at Aldergrove. 'A' Squadron, between April and June 1965, was engaged in training the Yeomanry in

England; 'C' Squadron went to Wales to recruit. On 7 July 1966 Her Majesty Queen Elizabeth, The Queen Mother, visited her regiment and, having inspected and taken the salute, spent time with the families and the Old Comrades, visiting the sergeants' and officers' messes. At the end of July 'A' Squadron left for Sharjah in the Persian Gulf, and at the end of September the rest of the regiment flew to Aden to take over from the 4/7th Royal Dragoon Guards, with regimental head-quarters based at Mareth Lines, Falaise Camp, in Little Aden.[8, 9]

The Queen's Dragoon Guards was widely dispersed throughout the Middle East: RHQ and Headquarter Squadron remained at Little Aden, 'A' Squadron spent a quiet time in the Persian Gulf at Sharjah, 'B' Squadron operated on the plain of Lahej, based on Little Aden, while 'C' Squadron was up in the Rad-fan based at Habilayn, with detached troops at Dhala and Musaymir. Another detachment was at Beihan on the edge of the Empty Quarter. The QDG Air Squadron was based at Falaise Airfield in Little Aden. Squadrons were moved around every three months or so, partly to give the hard-worked operational squadrons in Aden a change and a rest, and also to give everyone varied ex-perience.

The squadron in Sharjah operated over a huge area, sending patrols to Muscat and Oman, where Nizwa was visited, and then in October 1966 the squadron set the pattern by undertaking a 900-mile drive around the main towns of Muscat to show the flag; first going to the Buraimi Oasis and Ibri, and then returning to Nizwa, Bid Bid and Muscat, and back to Sharjah via Sohar. The squadron gave constant support to the Trucial Oman Scouts.[1, 10, 12]

The squadron in Little Aden was mainly concerned with internal security, involv-ing guard duties and patrols. A detachment in the native town of Sheikh Othman supported the resident infantry battalion, and came under regular attack: Corporal Brown was injured by a grenade while shopping, Corporal Giles and Trooper Mayo were wounded by grenades out on patrols, Mayo seriously when the grenade was lobbed into the turret of his armoured car. All were from 'B' Squadron. Squadron headquarters also came under mortar attack.[10, 12]

In Dhala, which was 4,000 feet up, close to the border with the Yemen, the climate was hot but dry, as opposed to the humidity of Aden. The squadron there was engaged with escort duties and patrolling. Dissidents, known by the soldiers as 'dizzies', attacked the camp one night, chipping the paintwork on one armoured car. On another night twelve tents were burnt down (only one belonging to the QDG) but the Yemen radio announced thirty killed and sixty wounded. On one occasion the Musaymir detachment found two mines outside their front gate. Early in January 1967 the dissidents attacked the camp at Habilayn one night: twenty-eight belts of Browning and forty-six rounds of 76 mm were fired, and the

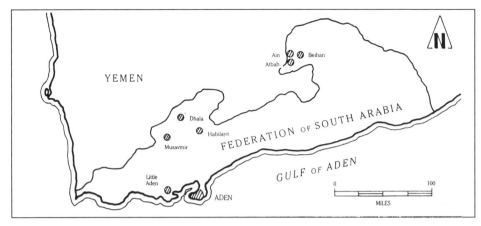

Aden

attack was beaten off. Next morning traces of blood were found. On 12 January the outposts at Musaymir and Dhala were withdrawn, and a few days later the squadron came back to Little Aden.[10, 11, 12]

In Aden there was a series of incidents from 10 to 14 February 1967 over Federation Day, during which British troops were attacked sixty-six times. Lieutenant Jenkins was caught in crossfire between the Northumberland Fusiliers and terrorists, and on the 13th a troop with the Royal Anglians in Sheikh Othman had a grenade explode under their Saracen, whereupon a section of the Irish Guards in the Saracen debussed and killing the thrower. On 28 February Sergeant Dakin was supporting a platoon of the Northumberland Fusiliers, who came under heavy rifle and automatic fire. Dakin, covered by his other vehicle, burst through a gate and arrested twelve locals, 'some of whom were caught with "warm-barrelled" pistols in their hands' ... 'We soon lost count of the number of near-misses by the grenade throwers of Sheikh Othman. '

FLOSY (Front for the Liberation of South Yemen) then called a general strike.

> Our busiest days came when strikes were called by NLF (National Libera-tion Front) or FLOSY. These were no ordinary strikes, but provided a splendid opportunity for supporters of rival gangs to take up arms and shoot at each other, or at us. We did encounter one mine which went off under Lieutenant Hulton's armoured car. Trooper Farr was extracted with a slight head wound, and Trooper Weatherlake escaped unhurt. Corporal Perks, who was in his scout car a block away, received a large part of the armoured car wing on his head. Another mine, meant for us, exploded under a minibus full of Arab children with horrifying results.[11, 12, 15]

In order to control the flow of arms and ammunition into Aden, a series of searches and checks were carried out in the desert and on the frontier under Operation Band. The regiment over this period searched 4,883 vehicles, 23,000 individuals and 449 animals; it was 'long hours with little reward', but a Land Rover containing two pistols, eight grenades and some explosives was seized by Lieutenant Stephens's troop, and nine Arabs were arrested.[11, 15]

A United Nations Mission arrived on 2 April, and during the five days that it was in Aden The Queen's Dragoon Guards were involved in more than ninety incidents, including three major street fights, each lasting for more than half an hour. 6,000 rounds of Browning machine gun and 500 rounds of other small arms ammunition was fired to cover the infantry in their searches. On 2 April a jumping jack mine exploded near a troop, luckily causing no damage. On 4 April in Sheikh Othman the 3rd Royal Anglians with Lieutenant Holmes's troop went to the rescue of the police station, which was beseiged by a mob. Holmes's Saladin was blown up on a mine and wrecked, shaking but not seriously wounding Corporal Bibey and Trooper Fordham. On 5 April 'Sergeant Robinson and Corporal Sheppard were having a furious battle with a number of terrorists, whilst trying to recover some mortar tubes from the mosque area. Terrorist machine gun fire was so severe that Sergeant Healey was directed to the mosque to help them extricate themselves. Sergeant Robinson was hit in the arm.' In Little Aden a crowd of fifty gathered to watch a scout car commander being decapitated by a wire stretched across the street. Fortunately the car's turret broke the wire. The crowd were enormously amused, and so the car commander arrested the man who was laughing the most. The frustrated United Nations Mission left Aden, but during its visit eighteen servicemen had been wounded and eight terrorists killed. At this time Lieutenant Colonel Powell took over command of the regiment from Lieutenant Colonel Muir.

Security in Sheikh Othman and Al Mansoura had been the responsibility of 3rd Royal Anglians supported by a squadron of the QDG, who in six months had coped with 459 incidents. At the end of May 1967 the 1st Parachute Regiment relieved the Royal Anglians. On 1 June a general strike in Sheikh Othman signalled the start of the battle for the area. 'This was a day which marked the beginning of a period of greatly increased sniper activity. Early in the morning a 1 Para soldier was shot dead in an O.P., and from then on throughout the day there was a constant rattle of small arms fire punctuated by grenades and explosions. Our troops were in action all day.' Two troops of armoured cars patrolled the streets, exchanging fire with the terrorists.[11, 12, 15]

The battle continued on the following days: 'Corporal Withycombe had four grenades thrown at his vehicle in ten minutes. Lieutenant Grounds's troop was involved in violent shooting incidents, causing casualties to the enemy.' Grounds had a narrow escape while pursuing a suspicious taxi at night. A rocket was fired from a building above him, skimming over one mudguard, flashing across the glacis plate and bouncing off the other mudguard to explode on the road a yard in front of the Saladin. Sergeant Aylott was also just missed by a rocket, and his troop leader, Lieutenant Gates, shot one sniper who had hit the gunner's periscope of his armoured car.[11, 15]

On 23 June Lieutenant Colonel Powell, QDG, wrote:

Tuesday 20th June was a very black day for the Army – we haven't accepted such casualties since the Korean War in the 50's. The main trouble has been that we are fighting with one hand tied behind our backs. So much of what is militarily necessary has been ruled to be politically unacceptable. This has directly affected the weapons and tactics used by us, but not those of the enemy. It is still too soon yet to be able to say what really were the causes of the shooting by the Arabian Army and Armed Police. There is no doubt that the dismissal of four of the South Arabian Army Infantry Battalion's C.O.s was a strong contributory factor – and some semblance of order was only restored when they were publicly reinstated. The trouble first started when South Arabian Army units, angered by the summary dismissal of their C.O.s, broke into their armouries and started attacking their British officers.[13]

At 10 a.m. on 20 June firing was heard from the direction of Champion Lines in Khormaksar, where the South Arabian Army was based. 'A' Squadron, led by Major Shewen, was ordered to assist the infantry. Lieutenant Jenkins's troop went to Champion Lines and came under heavy small-arms fire. A company of The King's Own Border Regiment drove up to the guardroom and, dismounting from their armoured personnel carriers and supported by fire from the armoured cars, engaged the mutineers, causing casualties. 'The troop was under continuous fire and the tyres of Jenkins' Saladin were damaged. The Border Regiment secured the armoury, losing one killed and eight wounded.

Jenkins reported the sound of unusual firing from the direction of the ranges, and Lieutenant Grandy's troop went to reconnoitre the road running between Champion Lines and Radfan Camp. They found that a three-tonner of the RCT had been shot up by the mutineers with heavy casualties. An armoured 'Pig' had punctured tyres, two Aden policemen were wounded, and there was a Land Rover with a a dead British civilian, in addition to eight dead and eight wounded RCT. The wounded were evacuated by Sergeant Dove, while Grandy gave them cover; then the bodies of the eight dead RCT soldiers and the civilian were removed.[11,15]

By 11 a.m. that day

this fighting was interpreted by certain elements within the Armed Police in Crater as the British wilfully attacking their Arab brothers. Without any warning they attacked a Recce party of the 5th Fusiliers, who were showing their successors, the Argylls, round the important spots in Crater. The Armed Police Barracks had until then been used as a Tactical Headquarters by the 5th Fusiliers in times of trouble. This time the Recce party walked head on into a carefully laid ambush in which all, but one, were killed.

At the same time in Crater town terrorists attacked the prison, releasing several hundred prisoners. Powell continued: 'Our forces, including 'A' Squadron, were forced to withdraw from the Crater area. We hold the main pass and coast road and the lip of the crater between those two roads. This allows us to look down into the bowl of the Crater and overlook the Armed Police Barracks. '[13, 15]

At midday on 20 June

Sergeant Forde — son of Sergeant Major Forde of the Bays — was asked to fly in his helicopter to retrieve four Fusiliers who were on picket duty on the heights above the Armed Police Barracks. He flew in and picked up the first pair, landing only for the briefest moment as the area of the picket was under aimed fire. The Fusiliers had jumped into the panniers on each side of the helicopter. He then went back to pick up the second pair, and when they were in the panniers, he lifted off, rose about 50 feet, then was hit in the knee, and was unable to operate the tail rotor. In spite of this he managed to land the machine, but couldn't get out. One Fusilier had a leg broken and the other severed, so couldn't move. The other, Duffey, leaped off his pannier, released Sergeant Forde, dragged him out, went back for his friend, dragged him away also, then went back for the wireless set — then, and only then, the aircraft caught fire and burned out. All three were picked up by another aircraft, and taken to safety

— Fusilier Duffey having used the rescued wireless to summon help. Duffey was awarded the DCM for his gallantry and cool-headedness, and the episode has been captured in a familiar painting by Terence Cuneo. [11, 13, 15]

In the meantime Sergeant Benford of 'A' Squadron had cleared some roadblocks on the way into Crater. When the sound of firing was heard, he moved along Queen Arwa Road into Crater towards the scene of the ambush, with some

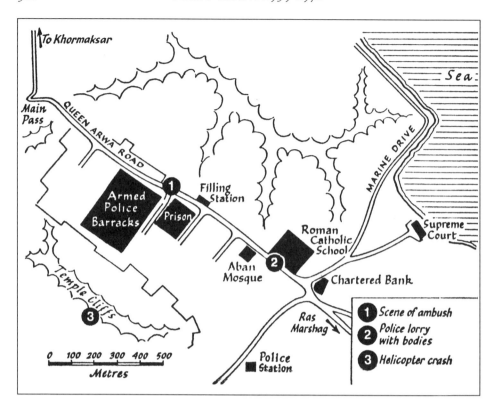

infantry following in an armoured personnel carrier. They soon came under heavy small arms fire, which was returned. The Fusilier officer in the carrier took three of his men into a house to watch the situation, sending Benfold back for reinforcements. The Fusiliers were not seen alive again. Lieutenant Stephens then took his troop back to the scene of the ambush: 'Fire was so intense my Browning machine gun was knocked out and I couldn't close the turret lid. Without cover from another Ferret we were pretty helpless. I asked permission to use my 76 mm to knock out the Bren gun position on the police roof. The answer was negative, and we were forced to retire.' At 2 p.m. an armistice was arranged with the police. Lieutenant Everitt's troop re-entered Crater, but came again under intense fire and withdrew, though not before Everitt himself and Trooper Dawes had been wounded and the armoured cars variously damaged. Everett's machine gun had been put out of action and his gunsight smashed. At 8.45 p.m. SSM Pringle, having taken over command of the wounded Everitt's troop, moved back into Crater to try to retrieve an abandoned police vehicle, which was thought to contain the bodies of those killed in the earlier ambush. As Pringle got to within fifty yards of the police vehicle he was fired at by an anti-tank rocket from the Aban

Mosque. This was the signal for heavy fire to be opened on the whole troop; a second rocket ricocheted off Pringle's car, whereupon the troop was withdrawn. The British casualties on 20 June 1967 totalled twenty-two killed and thirty-one wounded.[11, 15, 16]

On 21 June 'A' Squadron was stationed at the Main Pass and Marine Drive leading into Crater, and was engaged in sporadic exchanges of fire with snipers and machine-gunners operating from the Crater area. At long last the armoured cars were allowed to use their 76 mm guns. Crater was now sealed off and there was a period of aggressive night patrolling, with armoured car support, against diminishing opposition. Then, on the night of 3 July, a company of the Argylls, with their pipes playing, together with 'A' Squadron QDG, who flew from their aerials the red and white hackle of the Royal Northumberland Fusiliers, re-entered Crater. By 3 a.m. on 4 July it was reoccupied, the only opposition having been a few shots from the Sultan of Lahej's palace. The Chartered Bank, which commanded all the roads leading into Crater, became 'Stirling Castle', the command post of the Argylls. A signal was sent by Lieutenant Colonel Powell, QDG, to the Northumberland Fusiliers, now back in England, 'Your hackle flies again in Crater.' Major de Candole was awarded the Military Cross for his leadership of 'B' Squadron, and Major Shewen was mentioned in despatches. By 27 July 1967 1st The Queen's Dragoon Guards had left Aden, going on a well-earned block leave before reassembling at Knook Camp, Warminster.[11, 14]

The next few months at Warminster were occupied with training and in mid-November the regiment moved to Detmold to be part of the British Army on the Rhine. During 1968 the QDG was getting used to its tanks, as well as re-establishing its reputation on the field of sport. Corporals Tancock and Dakin were members of the 1968 British Olympic biathlon team, and the regiment reached the finals of the inter-regimental polo, to be beaten by the 17/21st Lancers. Lieutenant Colonel Powell and the adjutant paid a visit to Austria, renewing the old KDG associations with Vienna and establishing a close relationship with the Austrian 10th Panzer Battalion. On 19 November 1968 the regimental museum of 1st The Queen's Dragoon Guards was opened at Clive House, Shrewsbury by Major General Lord Bridgeman.[1, 14]

In 1969 The Queen's Dragoon Guards won the BAOR cross-country championship. On 12 September Lieutenant Colonel Powell handed over command to Lieutenant Colonel Lidsey. Regimental training was carried out at Soltau, and firing on the Hohne ranges. Lieutenant Mark Phillips joined the regiment and in 1970 was a member of the British team that won the World three-day event championship. In 1971 he won the Badminton Trophy. Corporals Price, Sweet, Jones and Lloyd became members of the British bobsleigh team.[17, 18]

In 1970 1st The Queen's Dragoon Guards moved from Germany to Catterick to become the Royal Armoured Corps Training Regiment, assuming their duties on 18 January 1971. A strong 'A' Squadron was detached as the Independent Armoured Squadron in Berlin, commanded by Major R. Ward, assuming its duties on 7 December 1970. The regiment's recruiting area had been widened to cover the whole of Wales, as well as Herefordshire and Shropshire. [18, 19]

SOURCES

1   E. Belfield, *The Queen's Dragoon Guards*, 1978.

2   *Regimental Journal of 1st The Queen's Dragoon Guards*, vol. 1, no. 1, 1959.

3   *Regimental Journal of 1st The Queen's Dragoon Guards*, vol. 1, no. 2, 1960.

4   *Regimental Journal of 1st The Queen's Dragoon Guards*, vol. 1, no. 3, 1961.

5   *Regimental Journal of 1st The Queen's Dragoon Guards*, vol. 1, no. 4, 1962.

6   *Regimental Journal of 1st The Queen's Dragoon Guards*, vol. 1, no. 5, 1963.

7   *Regimental Journal of 1st The Queen's Dragoon Guards*, vol. 1, no. 6, 1964.

8   *Regimental Journal of 1st The Queen's Dragoon Guards*, vol. 1, no. 7, 1965.

9   Major M. Johnston, QDG, 'The Queen's Dragoon Guards in Borneo', on loan to author.

10   *Regimental Journal of 1st The Queen's Dragoon Guards*, vol. 1, no. 8, 1966.

11   *Regimental Journal of 1st The Queen's Dragoon Guards*, vol. 2, no. 1, 1967.

12   Lieutenant Colonel T. W. Muir, QDG, account of the regimental activities from May 1966 to May 1967, on loan to author.

13   Major G. Powell, QDG, letter to his parents, June 1967, on loan to author.

14   *Regimental Journal of 1st The Queen's Dragoon Guards*, vol. 2, no. 2, 1968.

15   Julian Paget, *Last Post in Aden*, 1969.

16   S. Harper, *Last Sunset*, 1978.

17   *Regimental Journal of 1st The Queen's Dragoon Guards*, vol. 2, no. 3, 1969.

18   *Regimental Journal of 1st The Queen's Dragoon Guards*, vol. 2, no. 4, 1970.

19   *Regimental Journal of 1st The Queen's Dragoon Guards*, vol. 2, no. 5, 1971.

ABOVE: Grant tanks of The Queen's Bays at Gazala, 1942. BELOW: Captain D. C. MacCallan, The Queen's Bays, replenishing his Sherman tank, El Alamein, 1942. Photographs, Imperial War Museum.

ABOVE: A foot patrol of 1st King's Dragoon Guards, setting out from Pizzoferato, Abruzzi, Italy, 1944. Photograph, Imperial War Museum. BELOW: Lt A. R. F. Napper, 1st King's Dragoon Guards, entering Perugia in his Daimler, Italy, 1944. Photograph property of the author.

ABOVE: A Staghound armoured car and a Daimler Dingo scout car of 'D' Squadron, 1st King's Dragoon Guards, in Florence, Italy, 1944. Photograph property of the author. BELOW: A Sherman tank of The Queen's Bays moving up to Coriano, Gothic Line, Italy, 1944. Photograph property of Mr Burley, DCM, The Queen's Bays.

A Sherman tank of The Queen's Bays caught by 88mm anti-tank guns on Coriano, Gothic Line, Italy, 1944. Photograph property of I. Deighton-Gibson, The Queen's Bays. BELOW: Staghound armoured cars of 1st King's Dragoon Guards, Haifa, Palestine, 1946. Photograph property of the author.

ABOVE: 1st Armoured Division's Farewell Parade, Qassassin, Egypt, 13 September 1947. The Queen's Bays, commanded by Major G. Rich, and followed by SSM Cockroft, MM, march past the Comet tanks. Photograph property of Major J. Larminie, The Queen's Bays. BELOW: Sergeants Neil, Hood and Skillin of The Queen's Bays at Palmanova, Italy, April, 1947. Photograph property of Mr Skillin, The Queen's Bays.

ABOVE: Centurion tank of The Queen's Bays training with Jordanian infantry, Jordan, 1955. Photograph property of Major P. Gill, The Queen's Bays. BELOW: The late King Feisal of Iraq with Lt Col Manger, inspecting Centurion tanks of The Queen's Bays, Habbaniyeh, Iraq, 1955. Photograph property of Major P. Gill, The Queen's Bays.

ABOVE: Centurion tank, Mark V, of 'C' Squadron, The Queen's Bays at Sabratha, Lybia, 1957. Photograph property of Major J. Larminie, The Queen's Bays. BELOW: Sgt Stubbs in turret, Malaya, 1958. Photograph property of Major J. A. Moreton, KDG.

1st King's Dragoon Guards, Malaya 1958. ABOVE: Jungle patrol with Lt M. Thompson. BELOW: Daimler armoured car patrol. Photographs property of Major J. A. Moreton, KDG.

ABOVE: Queen Elizabeth, The Queen Mother, Colonel in Chief 1st The Queen's Dragoon Guards, escorted by Lieutenant Colonel H. Selby, MC, at the presentation of the new standard after amalgamation, Clarence House, 1959. BELOW: Regimental polo team, 1st The Queen's Dragoon Guards, Wolfenbuttel, 1960 Major J. Harman, Lt Col H. Selby, Major G. Powell, Major J. Railton. Photographs property Brigadier G. Powell, QDG.

Borneo, 1965. ABOVE: 'A' Squadron being briefed at Engkillili for a border ambush. BELOW: 'A'
Squadron 'basha', with atap roof and kajang walls, and in the foreground the ammunition bunker.
Photographs property of Lt Gen Sir M. Johnston, KBE, QDG.

Aden, 1966. ABOVE: Lt Stephens in the turret of his Saladin armoured car. Photograph property of Brigadier G. Powell, QDG. BELOW: Saladin armoured cars of 1st The Queen's Dragoon Guards in the Radfan. Photograph property of Colonel T. W. Muir, QDG.

ABOVE: Fusilier Duffey winning the DCM rescuing Sergeant Forde, QDG, from his helicopter, Aden, 1966. Painting by T. Cuneo, property of The Royal Regiment of Fusiliers. BELOW: Chieftain tanks of the 1st The Queen's Dragoon Guards, Hohne. Photograph property of the author.

Lt P. Mann, 1st The Queen's Dragoon Guards, about to go out on patrol in Northern Ireland. Photograph property of the author.

ABOVE: A patrol of 1st The Queen's Dragoon Guards in Northern Ireland. Photograph property of the author. BELOW: 1st The Queen's Dragoon Guards patrolling in Beirut, Lebanon, in Ferret armoured cars, 1983. Photograph property of Regimental Museum, 1st The Queen's Dragoon Guards.

Tercentenary Parade of 1st The Queen's Dragoon Guards, Wimbish, 1985. Photograph property of Regimental Museum, 1st The Queen's Dragoon Guards.

Gulf, 1991 ABOVE: 'A' Squadron, 1st The Queen's Dragoon Guards. BELOW: Scimitar of 'A' Squadron. Photographs property of Major H. Macdonald, QDG.

49

Catterick, Berlin, Northern Ireland, Hohne, Beirut
1971-1983

1st The Queen's Dragoon Guards became the training regiment in Catterick. Two passing off parades of recruits for the Royal Armoured Corps each month kept the regiment busy during 1971 and 1972, as well as numerous visits by senior people. A total of 2,138 recruits were trained. In September 1971 Lieutenant Colonel Johnston took over command from Lieutenant Colonel Lidsey.[1, 2]

In 1972 the colonel in chief, Her Majesty Queen Elizabeth, The Queen Mother, visited the regiment and was received on arrival by the colonel, Brigadier Llewellen Palmer. 'From that moment on the presence and character of the Colonel in Chief dominated Cambrai Barracks and its inhabitants. A feeling of confidence and pride seemed to surge from the ranks and the music from the band lent a swing to the stride.' The standard party was mounted and two squadrons were dressed in blues and carried swords, although two Chieftain tanks on the square reminded the regiment of its modern role.[3]

On the regiment's leaving Catterick in December 1972 for Germany, the inspecting general commented, 'Throughout their time in Catterick The Queen's Dragoon Guards have been an outstanding example of willing cooperation, doing their share and more of a myriad of administrative tasks which inevitably need doing in a large garrison.' In the sporting field in 1972 Lieutenant Phillips won the Badminton Trophy and was a member of the British team at the Olympic Games at Munich that won the gold medal.[3]

Meanwhile 'A' Squadron, commanded by Major R. Ward, spent two very successful years in Berlin. Its role included providing the armoured support in aged Centurions for the Berlin Brigade, patrolling the Berlin Wall and the frontier, ceremonial parades, entertaining numerous visitors, including a minister of state, one field marshal and twelve generals, and taking an impressive part in a range of sporting activities. Six weeks each year were spent training on the West German ranges at Hohne and Soltau, and the three principal ceremonial occasions – Allied Forces Day, the Queen's Birthday Parade (with two troopers mounted in full dress as outriders), and the Berlin Tattoo – exacted their share of hard work and preparation. In sport, during their two years in Berlin, the squadron reached five Army and twelve BAOR finals in a wide variety of events. Notable successes included winning the minor units' cross-country and swimming championships, the BAOR cross-country (twice), swimming (twice), and hockey, as well as the RAC hockey Jubilee Cup (twice). The pinnacle was reached when the squadron won the Army and BAOR Nordic skiing championships against all comers. In addition Sergeant Jones and Corporal Sweet were members of the British Olympic bobsleigh team at Sapporo, Japan. These sporting achievements have never been equalled by any other minor unit. Major R. Ward was awarded the MBE for his leadership of 'A' Squadron in Berlin.[2, 3, 4,5]

At the beginning of January 1973 1st The Queen's Dragoon Guards was once more reunited as a regiment in Hohne, West Germany, as one of BAOR's armoured regiments, equipped with Chieftain tanks. The regiment was ordered to form a fourth sabre squadron, as part of an organisational trial, but was allotted no extra manpower. 'D' Squadron was disbanded in September. In October 1973 Lieutenant Colonel Middleton took over command from Lieutenant Colonel Johnston.

In November 1973 Captain Phillips was married to Her Royal Highness The Princess Anne at Westminster Abbey, with all the ceremonial and panache of a royal wedding, the event being covered live by worldwide television. The bridegroom and his best man, Captain Grounds, were in the new full-dress uniform of officers of 1st The Queen's Dragoon Guards, which was being worn for the first time. The QDG provided a guard of honour with the regimental standard and band in the forecourt of Buckingham Palace, and the trumpeters and a step-lining party at Westminster Abbey itself. A message by The Queen's command to the Colonel said, 'The members of your Regiment who handed out the Order of Service Papers at Westminster Abbey carried out their important duty impeccably. May I please say how very grateful we all are to you and your Regiment.'[6]

Captain Grounds and Corporals Sweet, Price and Lloyd took part in the European and World bobsleighing championships. In February 1974 the QDG moved to Northern Ireland for a four-month tour of duty in Londonderry, with responsibility for the Shantallow area and Magiligan prison.[6, 7]

The IRA welcomed the regiment on arrival by exploding three bombs, which did little damage, 200 yards from the regiment's main gate at Fort George. On 3 March Corporal McGough was fired on while coping with a rioting crowd. Three members of a patrol under Corporal Callaghan were blown off their feet when a claymore mine was exploded behind them. Another claymore attack was made against a patrol led by Lieutenant Dickie. The commanding officer and his escort arrived at the scene a second after the mine went off, only to find Lieutenant Dickie

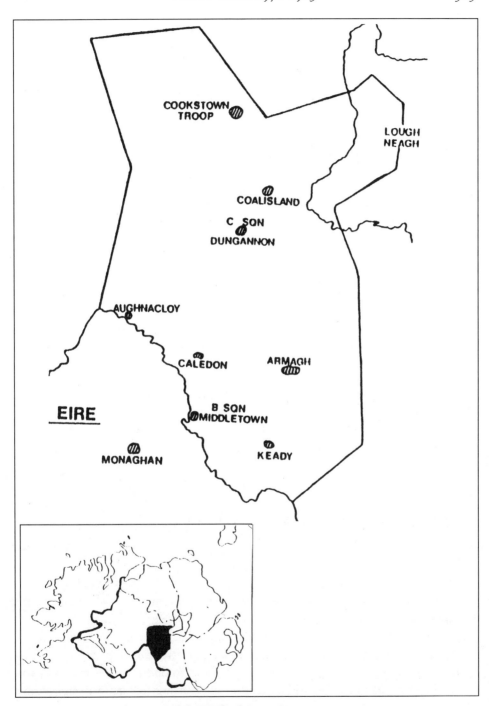

Northern Ireland Group Area, 1976

standing in the road about to give chase. Dickie greeted Lieutenant Colonel Middleton with 'Hello, Colonel, I think we have been fired at!'[7, 8, 9]

Searches and check points throughout the tour produced explosives, RPG rockets, a mortar and mortar bombs, an armalite rifle, and various rifles, pistols and ammunition. A subaltern noted in March, 'Simon Ward's troop in 'B' Squadron keep being shot at. 'A' Squadron has had shooting and aggression all day in Shantallow. We lifted three suspects at the Vehicle Check Point, and another later in the evening, but they were all allowed to go. The IRA can only grow stronger at the cost of men's lives with the limitations imposed.' On 2 June 1974 the QDG returned to Hohne in Germany. Lieutenant Colonel Middleton and Captain Grounds were mentioned in despatches for their services in Northern Ireland. Captain Phillips, stationed away from the regiment in 1974 at Sandhurst, again won Badminton.[7, 8, 9].

The regiment spent 1975 at Hohne in Western Germany as an armoured regiment, training in the autumn for another tour in Northern Ireland. In April 1975 the first QDG officer to be killed in action died, while on secondment to the Sultan of Oman's forces. Captain Mann was second in command of a company of Baluchis in the Sultan's service in Dhofar, and was posthumously awarded The Sultan's Bravery Medal. The citation read:

> The Company was subjected to repeated and well-coordinated 82 mm attacks, RPG 7, machine gun and small arms fire from three sides. Ra'ees Mann moved about the Company without regard for his own safety, encouraging the soldiers by his presence. He was himself hit by a splinter from an RPG 7 round, but continued on his way round the positions. Characteristically he did not report for treatment to his wound until much later. His actions were typical of his brave conduct during the three days during which his Company was in contact with the enemy. Ra'ees Mann showed a degree of bravery and leadership and example under fire which undoubtedly inspired the soldiers to capture their final objective, and resist enemy attempts to dislodge them subsequently. His actions were indispensable to the successful outcome of the Operation.

Mann remained in action until a month later, while out on patrol, he was killed in a brush with the enemy.[9, 10, 11]

In 1975, out of a total regimental strength of 420 all ranks, 247 were married, and the welfare of the families had become an increasingly important aspect of regimental life. During 1975 a QDG battlegroup went to Suffield in Alberta, Canada, for six weeks' training at the largest and only live firing training area in the then Western World. From 1974 to 1976 polo thrived as a sport in the regiment, with participation in every tournament in Germany. On 5 October 1975 Lieutenant Colonel R. Ward took over from Lieutenant Colonel Middleton, who was awarded the OBE for his leadership during his period of command. In January 1976 the QDG went to Northern Ireland for a second tour of duty, leaving 'A' Squadron at Hohne to look after the families, the tanks and the vehicles.[9, 10, 12]

From January to May 1976 the QDG was responsible for security in County Armagh and East Tyrone, which included the towns of Dungannon, Armagh and Cookstown, part of Lough Neagh and the border with Eire around the Monaghan salient. RHQ was in Gough Barracks, Armagh, with 'B' Squadron looking after the southern area and 'C' the northern. It proved to be an active four months, 'B' Squadron at Keady and 'C' Squadron's advance party at Dungannon being greeted within hours of arrival with exploding bombs that luckily hurt no one.

Between 3 and 20 January 1976 the QDG found two bombs, both of which were defused, and seven caches of weapons, explosives and ammunition; on the debit side three bombs exploded, causing no casualties, there were four shooting incidents involving a total of more than 200 rounds being fired at the regiment, six buses destroyed by arson, and one 'kneecapping' at Coalisland, where 'C' Squadron had a section. From 21 January to 20 February two bombs were defused, twelve caches of arms, explosives and ammunition were found and eight youths and three wanted men were arrested. In the same period there were six explosions, three murders, thirteen shooting incidents, two 'kneecappings', five rioting crowds and one case of arson.

Although figures indicate the level of IRA activity, they do not portray the sheer hard work, lack of sleep, and vigilance required of the troops. Typical daily incidents were when on the 24th January sixty to seventy rounds were fired at a vehicle check point at Middletown, or on 3 February a hijacked car was left across a road near Middletown, suspected of being booby-trapped. While it was being dealt with, an IRA recruiting poster was seen stuck to a nearby wall, which was found to have 30 lb of explosive hidden behind it, to catch the unwary souvenir hunter. On 19 February £100,000 worth of damage was caused, by arson and a bomb, to David Brown Tractors at Dungannon. On 26 February the colonel's Land Rover was involved: 'the 400 lb culvert bomb lifted the leading Rover off the ground by some 15 feet. But as a preliminary the night before, we were ambushed.' In March 'C' Squadron was redeployed to a border check point at Aughnacloy, and was greeted by an attack on the vehicle check point.

When the regiment returned to Hohne in May, it had dealt with 34 shootings, 36 bombs, 8 murders, 4 ambushes, 6 landmines, 3 cases of arson, 6 hi-jackings and 5 robberies in four months; it had found 20 assorted rifles and 3 revolvers, 2,038 rounds of ammunition, 11 magazines, an incendiary device, 2 walkie-talkie radios,

1,970 lb of explosive, and 25 detonators. On the brighter side, the band in full dress entertained the people of Armagh and played at church services, and the soldiers donated money to a local hospital, raised by organising their own sponsored run, and ran a sports day for the local children. 'The onus of operations was placed on the section commander and his six or eight men. The routine of guards, waiting, watching, manning roadblocks, and searches, all became the bread and butter of our daily lives.' Throughout this period the wives and families left behind at Hohne waited; 'The Wives Club' 'arranged many meetings, parties, shopping trips, visits and expeditions, and the highly successful coffee mornings held three times a week.'[10, 12, 13]

> As to our tour in Northern Ireland there is no doubt that we benefited enormously from our experiences. There were times of excitement, tension, danger, frustration and even boredom at the routine. We were able to claim success with large finds of arms, ammunition and explosives. Perhaps, more importantly, it was in the field of community relations that we were able to establish a healthier relationship with the local people, the majority of whom were friendly and hospitable.[7]

On the regiment's return to Germany, conversion from Chieftains to Scorpion and Scimitar armoured reconnaissance vehicles proceeded apace. Bishop Mann, himself an old KDG, and Mrs Mann, the daughter and sister of KDG, presented The Mann Trophy, in memory of their late son Captain P. Mann, to be awarded annually to the trooper or lance corporal who, over each year, contributed most to the life of the regiment. During 1976 Sergeant Francis, QDG, took part in the successful mountaineering expedition to Mount Everest. The regimental swimming team won the BAOR championship for the fourth year running, and the Army championship, having been runners-up for the previous three years. The British Olympic bobsleigh team had five QDG members, Captain Grounds, Sergeants Lloyd and D. Sweet and Corporals Price and G. Sweet. The British Olympic three-day event team had Her Royal Highness The Princess Anne and Captain Phillips. Between 1968 and 1976 the QDG had provided nine members of British Olympic teams.[10, 12]

The annual report on the QDG for 1976 commented:

> I visited The Queen's Dragoon Guards on operations in Northern Ireland and their Rear Party at Hohne, I have seen them in barracks and on the sports field, visited the Officers' and Sergeants' Messes, and taken the salute at a ceremonial, regimental parade. I have inspected the Regiment in the field, seen them on the ranges and worked with them on exercises.

This has been more than enough to convince me of their high morale, good discipline and standards. Perhaps most striking of all is the feeling of a strong family that the Regiment gives to one looking in from the outside.[10]

Early in 1977 The Queen's Dragoon Guards went to the training area at Suffield in Canada where they exercised as an armoured regiment for the final time before converting to armoured reconnaissance. The last Chieftain tanks left the regiment in October 1977 and the QDG assumed their new role on 1 January 1978. During 1977 the regiment reached the BAOR finals in nine sports, winning the angling, novices' judo, triathlon and swimming, where again the regiment won the Army championship. In December 1977 Lieutenant Colonel R. Ward handed over command to Lieutenant Colonel Pocock.[10, 14]

During 1978 the QDG settled down once again as an armoured reconnaissance regiment. The colonel in chief, Her Majesty Queen Elizabeth, The Queen Mother, presented a new standard at Hohne, nineteen years after she had presented the first QDG standard on the amalgamation of the KDG with the Bays. She was greeted on arrival by the colonel, General Sir Jack Harman.

As the old standard was marched off parade for the last time, the guard of honour, the vehicle crews, the Old Comrades and the spectators paid tribute to the symbol of a tradition which has been created over nearly 300 years by the efforts and sacrifices of our predecessors − a tradition of which we are justly proud and from which we will continue to gain inspiration and strength.[15]

Starting in 1978 and throughout 1979, The Queen's Dragoon Guards supplied one troop, and sometimes more, in a close reconnaissance role in Northern Ireland in both County Tyrone and Belfast. On 30 June 1979 Regimental Home Head-quarters moved from Shrewsbury to Maindy Barracks, Cardiff, thus recognising the predominantly Welsh character of the QDG − the Cavalry Regiment of Wales. 1979 saw the regimental swimming team regain the Army championship, while winning in BAOR for the seventh consecutive year. The 1979 Fastnet Race took place during exceptionally heavy storms and a yacht skippered by Major Moreton and crewed by Captain Hick and Lieutenants Coleman and Roberts was so badly damaged that they were forced to abandon her for a life raft, from which they were rescued by a Dutch destroyer.[16]

On 16 June 1980 Lieutenant Colonel Bond took over command of the regiment from Lieutenant Colonel Pocock. In the spring of 1980 six members of the regiment took part in a Himalayan expedition which successfully conquered KR3, a previously unclimbed 20,000-foot peak in the Himachal Pradesh region of India.[17]

In November 1980 The Queen's Dragoon Guards left Hohne, after eight years, for Lisanelly Barracks, Omagh, in Northern Ireland. The regiment's area of responsibility covered 784 square miles, which included fifty-two miles along the border with Eire, with thirty-five vehicle crossing points, only one of which was approved. The IRA tended to base themselves in the Republic, crossing into Ulster only for planned operations. 'The IRA display those qualities of Irishmen – colossal energy, daring, flair, suspicion, illogicality, a tendency to disunity and split, lack of organisation, courage, naivety, and sometimes a high degree of common-sense, coupled with extraordinary tenacity.' The colonel observed, 'The two obvious ways in which we can assist the Royal Ulster Constabulary are to kill the terrorist, or to arrest him "red-handed" with sufficient evidence to bring him successfully to justice.' These comments illustrated the frustrating, dangerous and tedious task faced by the soldiers.[17]

In February 1981 'B' Squadron uncovered a cache of arms near Carrickmore, consisting of four rifles and a Thompson submachine- gun with ammunition. 'A' Squadron was involved in two large search operations in the Castlederg and Greencastle areas of Fermanagh. A huge culvert bomb, with the explosives packed into milk churns, was discovered at Greencastle. But the main and wearying task was with incessant foot patrols in all sorts of weather. Lieutenant Colonel Bond summed it up: 'It is a frustrating task – the more we succeed in deterrence, the less likely we are to achieve tangible success; the less we deter, the higher the risk becomes for off duty UDR, RUC men and commercial targets.' A 'B' squadron troop leader described the end of a patrol:

> It had been a very wet 24 hour patrol (morale was low), we moved to the picking up point, and heard the Lynx [helicopter] approaching (morale was high). The chopper landed and out jumped the loadmaster, takes off her ear muffs, did you say HER? – and her long blonde hair fell down her back and fluttered around with the down-draught of the rotor blades (morale was bursting).[18]

'C' Squadron spent the year in Clogher. 'We put in a lot of time on the ground, the emphasis being mainly on observation posts on the border. We soon learnt how to live fairly comfortably in the field in all weathers.' A car bomb in Clogher exploded, causing some damage. In April 1982 the regiment was visited by Her Majesty Queen Elizabeth, The Queen Mother, to be welcomed by the colonel, Major General Rice. The commanding officer commented on 'the very close interest which our Colonel in Chief showed in all aspects of regimental life in Northern

Ireland'. Towards the end of the time in Northern Ireland 'A' Squadron, operating in Strabane, found a loaded revolver after an attempted murder had been foiled. 'B' Squadron observed, 'We still seem to be sitting in the same bushes, gazing at the same Irishmen, stopping the same cars, and even the weather remained true to form.' At the end of November 1982 The Queen's Dragoon Guards left Omagh for Wimbish near Saffron Walden, having in their history completed, as KDG, Bays or QDG, a total of seventy-one years and seven months' service in Ireland. Lieutenant Colonel Bond was awarded the OBE, and Major Holmes, Captains Macdonald and Daniell, and Staff Sergeants Miles and Brace were mentioned in despatches.[19]

In December 1982 Lieutenant Colonel Bond handed over command to Lieutenant Colonel Ferguson. In January 1983 'C' Squadron, commanded by Major Stewart, was sent to Cyprus to join the United Nations Peacekeeping Force, where part of the squadron relieved The Blues and Royals at Nicosia.

> From the camp we spent the next few days getting to know the buffer zone, which divides the Greeks from the Turks, in a line running east-west the whole length of the island. In Nicosia there are bullet holes in every building, and the buffer zone narrows to about twenty yards, so all you can see are thousands of Greek and Turkish soldiers glaring at each other, as you drive between them. The abandoned shops still have unfinished glasses of beer and milk sitting on the tables, untouched for the last decade since the troubles began.[19, 20]

The remaining seventy-five members of 'C' Squadron took their Ferrets by sea to the Lebanon, as the British contingent to the Multinational Peacekeeping Force in Beirut, which was attempting to maintain a fragile peace between the various warring Muslim, Druze, Christian and Israeli factions. The squadron was based in Hadath in a deplorable building, with the Israelis establishing a company position only 150 yards away, so that the QDG began to receive most of the near misses.

> There are so many different factions fighting each other that it is impossible to understand what is going on. Near here we have the Christians, and there are seven different factions of them, some fighting each other, but most fighting the Moslems. Both of those sides are supplied with weapons and ammunition by the Israelis. In return, when not fighting each other, the Christians and the Moslems fight for the Israelis. About 500 yards away there is an increasing number of PLO, who are sneaking back in, and they ambush the Israelis about once a week.[19, 20]

Within forty-eight hours of arrival two QDG troops were patrolling Greater

Beirut in their Ferret scout cars – two three-hour patrols which were undertaken daily. Twelve of the seventy-five soldiers guarded the building, two men on the roof overlooking the main Israeli supply route and two on guard at the main gate. A fatigue troop prepared the meals, kept the building clean, and maintained vehicles.

> Some recollections! Patrolling Beirut and avoiding multiple pile-ups with Lebanese who drove as if there is no tomorrow. Witnessing the devastation of Martyr Square, the opulence of Baabdah, and the squalor and stink of Sabra refugee camp. Being distracted by untouchable dusky Lebanese maidens in tight jeans and scarlet blouses, diverting our attention for potential grenade throwers. Passing grim-looking US Marines guarding the rubble remains of their Embassy on the Corniche. Being bitten by numerous plagues of mosquitoes. Avoiding the all-pervading dust thrown up by the tank tracks, and the stench of the open sewer running past the front gate of the base.

'C' Squadron was relieved by 'A', commanded by Major Boissard, at the end of July 1983. Major Stewart was awarded the MBE for his leadership of 'C' Squadron in Beirut. The summer of 1983 saw a deterioration in the situation. On 3 September the Israelis withdrew: 'the move was being shelled and our base was bracketed by medium artillery twice in ten minutes, as militias struggled to secure the old Israeli positions.' In September Lieutenant Colonel Ferguson temporarily left command of the regiment to command the British Force. On 23 October both the American and French bases were bombed and the QDG helped to free the casualties from the rubble. Throughout this period the squadron provided a cordon to protect the meetings of the Ceasefire Committee. In December the QDG were relieved by the 16/5th Queen's Royal Lancers, and the night before they left by helicopter the base was hit by four RPG rockets and some 200 rounds of .5 heavy machine gun fire, which slightly wounded a trooper, burnt out a Land Rover and destroyed the water supply. Field Marshal Lord Bramall, Chief of Defence Staff, wrote:

> Now that your squadrons have departed from Beirut, I wish to convey to you my personal admiration of their exemplary performance. Their contribution to stability in the Lebanon was outstanding and out of all proportion to what might have reasonably been expected from such a numerically small force. The professionalism, skill, hard work, diplomacy and dedication of The Queen's Dragoon Guards has been in the best tradition of the British Army and an example to us all.[19]

SOURCES

1 E. Belfield, *The Queen's Dragoon Guards*, 1978.

2 *Regimental Journal of 1st The Queen's Dragoon Guards*, vol. 2, no. 5, 1971.

3 *Regimental Journal of 1st The Queen's Dragoon Guards*, vol. 2, no. 6, 1972.

4 '"A" Squadron, 1st The Queen's Dragoon Guards', scrap book of Major R. Ward, QDG.

5 '"A" Squadron, 1st The Queen's Dragoon Guards', scrap book of Lieutenant P. Mann, QDG.

6 *Regimental Journal of 1st The Queen's Dragoon Guards*, vol. 3, no. 1, 1973.

7 *Regimental Journal of 1st The Queen's Dragoon Guards*, vol. 3, no. 2, 1974.

8 Letters of Lieutenant P. Mann, QDG, in possession of the author.

9 *Regimental Journal of 1st The Queen's Dragoon Guards*, vol. 3, no. 3, 1975.

10 '1st The Queen's Dragoon Guards', scrap book of Lieutenant Colonel R. Ward, QDG.

11 'Citation of the late Ra'ees P. Mann, QDG', in possession of the author.

12 *Regimental Journal of 1st The Queen's Dragoon Guards*, vol. 3, no. 4, 1976.

13 'Country Strife', vol. 2, Regimental Museum, 1st The Queen's Dragoon Guards, Cardiff Castle.

14 *Regimental Journal of 1st The Queen's Dragoon Guards*, vol. 3, no. 5, 1977.

15 *Regimental Journal of 1st The Queen's Dragoon Guards*, vol. 3, no. 6, 1978.

16 *Regimental Journal of 1st The Queen's Dragoon Guards*, vol. 3, no. 7, 1979.

17 *Regimental Journal of 1st The Queen's Dragoon Guards*, vol. 4, no. 1, 1980.

18 *Regimental Journal of 1st The Queen's Dragoon Guards*, vol. 4, no. 2, 1981.

19 *Regimental Journal of 1st The Queen's Dragoon Guards*, vol. 4, no. 3, 1982.

20 *Regimental Journal of 1st The Queen's Dragoon Guards*, vol. 4, no. 4, 1983.

21 Letters of Lieutenant T. Wilson, QDG, in possession of the author.

50

Wimbish, Cyprus, Tercentenary, Wolfenbuttel, Northern Ireland, Gulf War
1983-1992

Whilst 'C' Squadron soldiered in Cyprus and Beirut, the rest of The Queen's Dragoon Guards at Wimbish, near Saffron Walden, were involved with endless exercises on Salisbury Plain and at Stanford and Sennybridge, as well as security duties at Heathrow and Gatwick Airports in support of the police in countering the terrorist threat. 'B' Squadron sent a troop to an exercise near Seattle in America. During November 1983 a new guided weapons squadron, 'D' Squadron, was formed, and equipped with Striker vehicles armed with Swingfire anti-tank missiles. Captain Hick, QDG, on attachment to the Sultan of Oman's Forces, was involved in an accident which left him paralysed from the waist down.[1]

The year 1984 marked the Silver Jubilee of 1st The Queen's Dragoon Guards, and it was a year when virtually the whole regiment was able to be together at Carver Barracks, Wimbish. The only deployments were one troop sent to Belize as the armoured reconnaissance element in the British Defence Force, another on an exercise to Kenya, and 'B' Squadron had an exchange exercise with the Cavallegere di Lodi in Northern Italy in November. During the year the regiment was awarded the Wilkinson Sword of Peace in recognition of the contribution made by 'A' and 'C' Squadrons in Beirut, which was the first time such an award had been made to a cavalry regiment. The sword was presented at a special parade of the Cavalry Regiment of Wales and the Border Counties, held in the grounds of Cardiff Castle. Lieutenant Colonel Ferguson was awarded the OBE for his leadership and international diplomacy in Beirut, and Major Boissard and Lieutenant Baldwin were mentioned in despatches for the part they had played in the Lebanon.[2]

The Tercentenary Year, 1985, celebrated the 300 years of distinguished service since the raising of both The 1st King's Dragoon Guards and The Queen's Bays, the 2nd Dragoon Guards, in 1685. Lieutenant Colonel O'Brien took over command of the regiment from Lieutenant Colonel Ferguson. On 1 March, St David's Day, the Wilkinson Sword Company presented the regiment with a presentation sword, on which were inscribed all the regimental battle honours.

The Tercentenary itself was marked by celebrations over the weekend of 28/29 June 1985.

> The highlight of these two days was unquestionably the presence of our greatly loved Colonel in Chief, Her Majesty Queen Elizabeth, The Queen Mother. Few will forget the magnificent spectacle of the regiment on parade, the precision timing, the advance in review order and the dramatic charge down the full length of the runway. It was immensely heartening to see so many Old Comrades, and for us all to share the privilege of being a member of a family celebrating a very special birthday.

A delegation from the Canadian Allied Regiment, The Governor General's Horse Guards, was present, as later was a delegation from the 11th Cavalry (Frontier Force) from Pakistan.

The colonel in chief said in her address to the regiment,

> Our regiment is a family, and like all families we draw courage from the example of the past, for this gives us the strength to face the future, whatever it may hold. We remember the past because it has helped to form the present and because we are a family where son follows father and brother serves with brother. It is as a family that I would say a special word to all the wives and relatives, for to be a soldier's wife has always demanded special gifts of love and fortitude, and the part you play in sustaining your husbands is of infinite value. As I watch this Tercentenary Parade I note with pride your professionalism as armoured troops, the smartness of your mounted drill and turnout, that splash of magnificent colour from the full dress of the band, and I know that the coming years will add more honour to the pages that have already been written.[3]

On 29 July 1st The Queen's Dragoon Guards received the freedom of the City of Cardiff, an event which was marked by a parade in the city, with the regiment exercising its freedom by driving through Cardiff in sixty armoured vehicles. The QDG presented Cardiff with a cavalry sword to mark the close links between the regiment and the Principality.[3]

'B' Squadron had flown to Cyprus before the Cardiff freedom parade to become a part of the United Nations Peacekeeping Force. There they patrolled the buffer zone established to keep apart the bitterly opposed Greek and Turkish communities. A troop leader commented, 'It is one of the few chances that any soldier gets to work and befriend people from other countries, and we have all made good and long-lasting friends.' In January 1986 'B' Squadron returned from their six months in Cyprus, and in July a reinforced 'D' Squadron took up their duties with the Peacekeeping Force. During a visit by the brigadier commanding Land Forces Cyprus, while being driven by the squadron leader in a scout car commanded by a corporal, the brigadier asked the corporal what he did, and received the reply, 'Well, before we start, Brigadier, I would like to tell you that I'm a -- good bloke.' The squadron leader is reported to have died a quiet death in the driver's seat.[3, 4]

In August 1986 Sergeant Francis, QDG, was awarded the Military Medal for his bravery while on special duties in Northern Ireland, where he was wounded. 'C' Squadron carried out exercises in the USA and with BAOR in Germany. In July a QDG team climbed Chogolungma, a previously unclimbed 19,200-foot peak in the Karakorum range of Northern Pakistan, putting eight members on the summit. The expedition was made possible by the help of the QDG's allied regiment, the 11th Cavalry (Frontier Force), and so the peak was named 'Cavalry Peak' to symbolise that partnership. Captain Joyce led an adventure training expedition to Africa, during which Corporal Greatorex was drowned.[4]

In 1987 Her Majesty Queen Elizabeth, The Queen Mother, completed her fiftieth year as colonel in chief of the regiment. That Golden Anniversary was marked by 1st The Queen's Dragoon Guards parading under their colonel, Lieutenant General Sir Maurice Johnston, before their colonel in chief at her home at Clarence House on 19 February. At the subsequent reception Her Majesty was presented with a model in silver of a 2-pounder Crusader tank, with which the Bays were equipped at the Battle of Gazala. Her Majesty, in graciously accepting this gift, asked that it be held in perpetuity in the regiment, to be known as the Queen Elizabeth Trophy; it is presented annually to the sergeant in the regiment who in the opinion of the commanding officer has been both notable and selfless, and therefore earned recognition and praise.[5]

On 1 March Her Majesty again honoured her regiment by opening the new regimental museum within the walls of Cardiff Castle. In April 1987 the QDG left Wimbish after four and a half years and moved to Germany, where they were once again stationed at Northampton Barracks at Wolfenbuttel, which they had left in 1964. In October 1987 Lieutenant Colonel Boissard took over command of the regiment from Lieutenant Colonel O'Brien and laid down the regiment's priorities: first, 'to achieve a standard of excellence in all aspects of medium reconnaissance work; secondly, to maintain the regiment's quality of life; and thirdly, to maintain that happy and relaxed atmosphere that has been the hallmark of the regiment for so many years.'[5]

During their time at Wolfenbuttel one of the regiment's primary tasks was to patrol the border between West and East Germany, but there were other commitments. From February to August 1989 'A' and 'B' Squadrons were amalgamated to provide a prison guard force for the Maze Prison in Northern Ireland. The prison force had three tasks: first, to occupy the watchtowers and patrol the immediate vicinity of the prison; secondly, to maintain the security of the force's camp and the local area; and thirdly, to operate night patrols in the Lisburn area. Towards the end of their tour troops were sent to help with the security of the prison in the Crumlin Road in Belfast. During 1988 and 1989 regimental polo had been re-started, largely due to the enthusiasm of Lieutenant Colonel Boissard. In March 1990 Lieutenant Colonel MacKenzie-Beevor took over command of the regiment from Lieutenant Colonel Boissard. During April 1990 the band flew to Canada, spending time in Toronto with the regiment's allied regiment, The Governor General's Horse Guards, further cementing the close bonds between the QDG and the GGHG.[5, 6, 7, 8]

In November 1989 the first signs of the collapse of the Soviet Empire became apparent, with the breaching of the Berlin Wall and the political upheaval throughout the countries of Eastern Europe. For the regiment, patrolling the border between the two Germanies, the changes in Eastern Europe soon became more than noticeable as the inner German border disappeared and the reunification of the two Germanies started to become a reality. There were signs that the presence of a large British contingent in Germany was not as popular and accepted as it had been: farmers and local people became less willing to see major exercises carried out across their land. The financial pressures exerted by the Treasury in London reduced the amount of spares, ammunition and fuel that was available for training. With the collapse of the Warsaw Pact the threat from the huge Soviet alliance in Eastern Europe seemed to disappear and bring into question the need for so many British troops in Germany.

This early euphoria soon wore thin as the comparative stability of the Soviet empire disintegrated, to be replaced by the uncertainty and strain of many factions seeking their own forms of independence. But other dangers loomed on the horizon: in September 1990 the commanding officer was warned that the QDG featured in contingency plans to deal with Iraq's invasion of Kuwait. Shortly afterwards 'A' Squadron, under the command of Major Macdonald, became the medium reconnaissance element of 7th Armoured Brigade – the Desert Rats – the initial grouping deployed in support of the Multinational Force. There followed a period of intense preparation – the regiment worked together as a team to put 'A' Squadron into a fighting state and provided extra personnel for its war establishment. The fact that the vehicles and equipment worked so well over a long period in adverse conditions was due in main to the hard work of the other squadrons in the preparation at Wolfenbuttel prior to deployment.[8, 9]

Over the following months more QDG left for the Gulf; some to reinforce the

Royal Scots Dragoon Guards, others to the Queen's Royal Irish Hussars and the 14/20th Hussars. Groups totalling some sixty-two men went as battle casualty replacements, while other individuals deployed to a variety of other units and headquarters. Once it had been decided that the British contribution would be increased to the whole of 1st Armoured Division, the regiment hoped that the QDG would deploy in its normal role as 1st Armoured Division's reconnaissance regiment. This was not to be, and a further forty-four men went to reinforce the 16/5th Lancers, who now took over that role.

'A' Squadron's vehicles and equipment were loaded on to the *Mercandian Queen* at Bremerhaven on 27 September 1990. This vessel was described as 'a real rust bucket, which was to break down on many occasions, and thus arrive late in Saudi Arabia'. On 16 October the main body of 'A' Squadron flew out to the Gulf and was initially billeted with the US Marine Corps Reconnaissance Forces in a camp to the north of the port of Al Jubail. For the next ten days the men acclimatised to desert conditions. The vehicles finally arrived on 26 October and 'A' Squadron deployed into the desert on the 29th.[8, 9]

'A' Squadron formed a close friendship with the American Marines, cemented when there was a challenge to a rugger match, with the Marines fielding a Corps team. 'There were comments on the Queen's Dragoon Guards team's lack of size. It was a hard bruising affair, with the Squadron coming out winners four tries to nil, which led to immediate adoption by the Marines.' There followed an initial three-week training period with 7 Brigade. On 11 November 'the Squadron assembled in a tight group, at 6 a.m. with dawn breaking, surrounded by dunes, for a short service and silence — a poignant moment as one surveyed uplifted faces and prayed for the impossible.' In late November 'A' Squadron moved to train independently with the US Marines at Manifa Bay — 130 kilometres north of Al Jubail.[8, 9, 10, 11, 12]

On 6 January 1991 the squadron came under command of the 16/5th Lancers and spent the next ten days fitting and testing the new 'speech secure system'. The air war started on 17 January and preparations for the Land Offensive began in earnest. 'A' Squadron moved 300 kilometres west to the area of the Wadi Al Batin on 21 January, where 1st Armoured Division was concentrating. By now the weather had turned foul, with incessant rain, thunder and lightning.[8, 9, 10]

From 24 January to 4 February 'A' Squadron provided a reconnaissance screen in front of 4th Armoured Brigade. On 5 February 1st Armoured Division concentrated in mass formation to practise the breach crossing through the Iraqi defensive line in close cooperation with the 1st (U.S.) Infantry Division (Mechanised). Preparations for the Land Offensive were in their final phase and the squadron was detached to operate independently with the artillery group. Artillery raids took place between 14 and 23 February, with the squadron securing and screening gun positions. On 24 February 'A' Squadron regrouped with the 16/5th Lancers in the staging area south of the breach. The squadron led the reconnaissance group through the breach-head in the early morning of 25 February and refuelled in the forming- up point, Valley Forge. The 16/5th Lancers' initial mission was to attack Objective Zinc, using air and long-range artillery. However, the 7th Armoured Brigade had broken out so fast that they reached Zinc prior to the 16/5th Lancers – the latter being ordered to move to the north east of that objective. It was 1 a.m. on 26 February. The weather was dank, with the visibility reduced by the smoke from the burning Kuwaiti oil wells.

At 2.15 a.m. on 26 February orders came to attack an Iraqi position of battalion strength, codenamed Objective Lead, forty kilometres to the east. 'A' Squadron advanced in total darkness with three troops up – the guided weapons and support troops close behind. As the leading troops moved forward, they encountered wire fencing. Alert to the threat of mines, they tried to find a way through. One Scimitar hit a mine, damaging its track, but the support troop, commanded by Lieutenant Fenton, found a gap to the south, and the squadron switched its line of attack. At first light, in poor visibility, dug-in Iraqi armour was identified to the squadron's front. At 6.45 a.m. the attack started with the bombardment of an Iraqi mechanised company and tank company position with the new multi-launch rocket system (MLRS). This was followed by an air-strike which knocked out three T62 tanks, an armoured personnel carrier and motor transport. There followed a further bombardment of another Iraqi tank company's position. By 8.30 a.m. the leading troops were engaging the enemy with both 30 mm cannon and anti-tank missiles, and the Iraqis were responding. Squadron headquarters was engaged from the rear.

There were reports from 2 Troop, and the Advanced Alternative Headquarters, of Scimitars withdrawing as Iraqi armour pushed north and north east. Major Macdonald, thinking that 'A' Squadron had exposed flanks, ordered a withdrawal, but was told by the commanding officer of the 16/5th Lancers to remain in position. In the meantime Lieutenant MacLennan's troop engaged and destroyed an Iraqi T55 tank and two personnel carriers; Lieutenant Carter's troop destroyed another personnel carrier. While this was going on visibility had further decreased. Support Troop deployed its radar and detected movement within the mechanised battalion position – this it engaged with four MLRS attacks. The squadron's guided weapons were used to destroy a further four personnel carriers as they tried to escape the barrage. In the rear of the squadron position Lieutenant Carter's troop had destroyed two more personnel carriers and taken forty prisoners. At 11.45 a.m. 'A' Squadron was ordered to withdraw after five hours of continuous action, the withdrawal taking place in a growing sandstorm. By 2 p.m. replenishment was complete and all vehicles were again battleworthy.

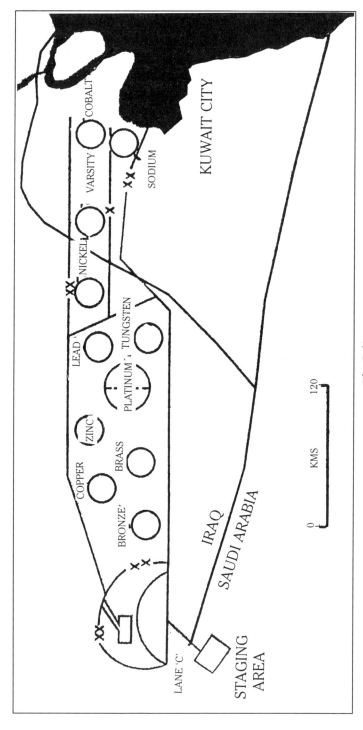

The Gulf

Late on the evening of 26 February the 16/5th Lancers were ordered to move thirty-five kilometres east in preparation for the pursuit of the Iraqi army to the Euphrates. As 'A' Squadron arrived at the forming-up point in the early hours of 27 February, it was ordered to return to 7th Armoured Brigade to provide their reconnaissance screen for the advance on Kuwait. Starting at 7.30 a.m. on 27 February the squadron, covered by attack helicopters and Challenger tanks, passed through the 16/5th Lancers in the Wadi Al Batin and led the rapid advance for sixty-five kilometres to Objective Varsity, situated in the centre of Kuwait. The final objective, codenamed Cobalt, lay astride the main road from Kuwait City to Basra, about fifteen kilometres from the coast. At 5.15 a.m. on 28 February 'A' Squadron, 1st The Queen's Dragoon Guards, closely followed by the tanks of The Queen's Royal Irish Hussars, set out for Cobalt; forty-two kilometres were covered in eighty minutes. 'A' Squadron was astride the main road just as President Bush declared the ceasefire at 8 a.m. 'The scene of devastation remains poignant with Iraqi dead lying beside the road, which was strewn with wrecked military equipment. The sky was black and overcast with smoke from the burning wells.'

The first of March was St David's Day, and the Welsh Dragon flew from the QDG armoured vehicles. For the next three days 'more prisoners were found, fed and taken to the POW cage – the wounded were treated by the doctor. We also buried their dead.' Among the trophies captured from the Iraqis was an intact T59 tank. 'A' Squadron returned to Germany with 7th Armoured Brigade, having handed over its vehicles to the RAOC at Al Jubail. The QDG were flown into Hanover aboard a British Airways Jumbo Jet on 15 March 1991 to be reunited with their families at Wolfenbuttel. The regiment had not suffered a single casualty. Major Macdonald summed up: 'The soldier was the real battle-winner – his strength to endure whatever hardship, his ability to adjust to whatever circumstance, his tenacity in combat and his humour, which makes life bearable in times of danger.'[8, 9]

A composite squadron, 'Y' Squadron, of 100 men under command of Major Andrews, went to Northern Ireland with the 1st Royal Regiment of Fusiliers, thus renewing the old links which the regiment forged with the Royal Northumberland Fusiliers in Aden in 1966. The regimental polo team won the inter-regimental cup in Germany, the first time for twenty-eight years, and reached the finals of the United Services Cup at Windsor, being beaten by the Welsh Guards. The regimental rugger side won the BAOR Cavalry Cup. In May 1991 the QDG moved from Germany to Assaye Barracks, Tidworth, as the medium reconnaissance regiment for the United Kingdom Mobile Force. On 15 June the regiment exercised its privilege of the freedom of Cardiff by marching through the city, having first attended a thanksgiving service in Llandaff Cathedral for their safe return, and for the regiment to thank the people of Wales for all the support which they had given to the QDG during the Gulf War. On 5 July 1991 the colonel, Major General R. Ward, welcomed the colonel in chief when she visited her regiment at Tidworth to welcome them home and to congratulate them on their performance in the Gulf. The regimental polo team again lost to the Welsh Guards in the final of the 1991 United Services Cup at Windsor.[8]

Following the collapse of the Warsaw Pact and the changes in Eastern Europe, British politicians felt that they could gain some advantage by carrying out a drastic reduction of the Armed Forces. A study, entitled *Options for Change*, resulted in proposals to reduce the Army and to amalgamate further many regiments that had already endured the trauma of previous amalgamations. In the Royal Armoured Corps nineteen regiments were to be reduced to eleven. The only three cavalry regiments to escape further amalgamation were 1st The Queen's Dragoon Guards, The Royal Scots Dragoon Guards and the 9/12th Royal Lancers. Whilst all members of the QDG were relieved at their survival, they took no pleasure in the disappearance of so many famous sister regiments, with whom they enjoyed such close ties. For the QDG the price they paid was conversion from armoured reconnaissance to armour, and a return to Germany after only sixteen months at Tidworth. The reduction in the Armed Forces also meant enforced redundancy for individuals, and the regiment was sad to have to bid farewell to those QDG affected in this way.

The Right Revd Michael Mann, a former KDG, preaching at the regiment's tercentenary service in 1985, had this to say:

> We live in a changing world, but the QDG have always had to cope with change – from regiments of horse to dragoon guards; an arranged marriage which, contrary to fashionable philosophy, has been outstandingly successful; the traumatic change from horsed cavalry to tanks and armoured cars, and even on occasion to infantry. We take change in our stride, pleasant or unpleasant, and we have always used it to strengthen and enhance the traditions we have inherited. And we have been able to do this because those traditions, founded as they are on the Christian values and virtues, have given us that sure foundation for the easy discipline and flair, which is the hallmark of a QDG, and which holds us together in a bond of loyalty and affection to our comrades, to the regiment, to our colonel in chief, to the Queen and to God.

1st The Queen's Dragoon Guards has always, and will always, live up to its reputation to be 'First and Foremost'.

SOURCES

1 *Regimental Journal of 1st The Queen's Dragoon Guards*, vol. 4, no. 4, 1983.

2 *Regimental Journal of 1st The Queen's Dragoon Guards*, vol. 4, no. 5, 1984.

3 *Regimental Journal of 1st The Queen's Dragoon Guards*, vol. 4, no. 6, 1985.

4 *Regimental Journal of 1st The Queen's Dragoon Guards*, vol. 4, no. 7, 1986.

5 *Regimental Journal of 1st The Queen's Dragoon Guards*, vol. 4, no. 8, 1987.

6 *Regimental Journal of 1st The Queen's Dragoon Guards*, vol. 5, no. 1, 1988.

7 *Regimental Journal of 1st The Queen's Dragoon Guards*, vol. 5, no. 2, 1989.

8 *Regimental Journal of 1st The Queen's Dragoon Guards*, vol. 5, no. 3, 1990.

9 Major H. Macdonald, QDG, 'A' Squadron, 1st The Queen's Dragoon Guards Tour Report, Gulf Theatre of Operations, October 1990 – March 1991.

10 Letter from Major H. Macdonald, QDG.

11 Letter from Captain T. R. Wilson, QDG.

12 Letter from Sergeant H. L. Barnes, QDG.

Succession of Colonels and Commanding Officers

1st King's Dragoon Guards

COLONELS	COMMANDING OFFICERS
1685 Lt Gen Sir John Lanier	William Legge
1688	Hon. Henry Lumley
1692 Gen Hon Henry Lumley	William Palmer
1704	Thomas Crowther
1707	John Deane
1713	Richard Lord Lumley
1715	Thomas Panton
1717 Col Richard Ingram, Lord Irvine	
1718	Humphrey Bland
1721 Field Marshal Sir Richard Temple, Viscount Cobham	
1733 Lt Gen Henry Herbert, Earl of Pembroke	
1737	John Brown
1742	Martin Madan
1743 Gen Sir Philip Honeywood	
1746	Timothy Carr
1752 Lt Gen Humphrey Bland	
1757	William Thompson
1759	Robert Sloper
1763 Gen John Mostyn	
1778	Anthony Lovibond
1779 Field Marshal Sir George Howard	
1784	Richard Vyse
1796 Gen Sir William Augustus Pitt	
1797	William Ann Willets (or Villettes)
1799	John Elliott
1804	Henry Fane
1805	William Fuller
1810 Gen Francis Augustus Eliott, Lord Heathfield	
1813 Gen Sir David Dundas	
1815	George Teesdale
1820 Gen Francis Edward Gwyn	

1821	Gen William Cartwright	
1827	Gen Sir Henry Fane	
1838		Hon George Cathcart
1840	Gen Hon Sir William Lumley	
1844		Henry Aitchison Hankey
1852		Anthony Spottiswode
1857	Gen Charles Murray Cathcart, Earl Cathcart	
1859		Thomas Pattle
1865	Gen Sir Thomas William Brotherton	
1868	Gen Sir James Jackson	Herbert Dawson Slade
1872	Gen Henry Aitchison Hankey	
1876		Henry Alexander
1881		R. J. C. Marter
1884		W. H. Thompson
1886	Lt Gen Sir James Sayer	
1887		G. V. C. Napier
1890		H. P. Douglas-Willan
1894		R. C. B. Lawrence
1898		H. M. Mostyn Owen
1902		S. Bogle Smith
1906		W. J. S. Fergusson
1908	Maj Gen W. V. Brownlow	
1910		F. C. L. Hulton
1914		J. A. Bell Smyth
1922		A. C. Little
1925		W. F. Chappell
1926	Lt Gen Sir Charles Briggs	
1929		S. J. Howes
1933		E. W. H. Sprott
1936		J. G. E. Tiarks
1939		J. A. Paton
1940	Brig The Earl of Gowrie	D. McCorquodale
1942		R. A. Herman
1944		A. P. C. Crossley
1944		M. J. Lindsay
1945	Brig S. J. Howes	
1946		W. B. Radford
1948		R. F. Luck
1951		M. J. Lindsay
1953	Brig J. G. E. Tiarks	
1954		W. N. Cairns
1957		H. C. Selby

The Queen's Bays (2nd Dragoon Guards)

1685	Col Henry Mordaunt, Earl of Peterborough	Sir John Talbot
1687		John Chitham
1688	Brig Gen Hon Edward Villiers	
1691		James Kirke
1692		George Carpenter
1694	Brig Gen Richard Leveson	
1699	Gen Daniel Harvey	
1704		Thomas Newcomen
1707		Edward Roper
1708		William Goring
1710		John Bland
1712	Col John Bland	
1713		James Otway
1715	Col Thomas Pitt, Earl of Londonderry	
1718		Richard Whitworth
1726	Field Marshal John Campbell, Duke of Argyll	
1733	Gen William Evans	
1740	Gen John Montagu, Duke of Montagu	
1749	Field Marshal Sir John Ligonier	Thomas Brudenell
1753	Maj Gen Hon William Herbert	
1757	Lt Gen George Germain, Viscount Sackville	James Mure Campbell
1759	Gen John Waldegrave, Earl Waldegrave	
1763		Lord Russell Manners
1773	Field Marshal George Townshend, Marquess Townshend	
1779		Sir William Arnot
1779		Sir William Innes
1782		Lord George Herbert
1794		Charles Crauford
1799		John Le Marchant
1803		James Erskine
1807	Lt Gen Sir Charles Crauford	
1814		James Kearney
1821	Lt Gen William Loftus	
1830		James Hay
1831	Gen Sir James Hay	
1837	Gen Sir Thomas Montresor	Charles Kearney
1841		Henry Wilmot Charleton

1846	William Campbell
1853 Gen Hon Henry Cavendish	
1858	William Henry Seymour
1872	Henry Holden Stewart
1873 Gen Henry White	
1874 Gen Alexander Low	
1877	Thomas W. Sneyd
1881 Maj Gen Thomas Pattle	W. H. Lee
Gen Sir Charles Walker	
1886	C. A. L. A. French
1892	C. E. Beckett
1894 Gen Sir William Seymour	
1895	J. A. Lambert
1899	James E. Dewar
1903	H. D. Fanshawe
1907	William Kirk
1911	H. W. Wilberforce
1915	A. Lawson
1919	G. H. A. Ing
1920	C. S. Rome
1921 Lt Gen Sir Hugh Fanshawe	
1924	R. H. Osborne
1928	C. A. Heydeman
1930 Lt Gen Sir Wentworth Harman	
1932	R. M. Wootten
1935	E. D. Fanshawe
1939	W. R. Beddington
1939	G. H. Fanshawe
1940	G. W. Draffen
1942	A. H. Barclay
1944	D. V. H. Asquith
1946	P. T. W. Sykes
1947	K. E. Savill
1948 Brig James Kingstone	
1950	H. J. G. Weld
1952	W. B. Manger
1954 Col G. W. Draffen	
1955	C. Armitage

1st The Queen's Dragoon Guards

1959 Brig J. G. E. Tiarks	H. C. Selby
1960	J. H. Harman

1961	Col G. W. Draffen	
1962		P. R. Body
1964	Col K. E. Savill	
1965		T. W. Muir
1967		G. N. Powell
1968	Brig A. W. A. Llewellen Palmer	
1969		J. H. Lidsey
1971		M. R. Johnston
1973		R. C. Middleton
1975	Gen Sir J. H. Harman	R. W. Ward
1977		J. I. Pocock
1980	Maj Gen Sir D. G. Rice	C. H. Bond
1982		J. G. G. de P. Ferguson
1985		E. J. K. O'Brien
1986	Lt Gen Sir M. R. Johnston	
1987		M. G. Boissard
1990		C. D. Mackenzie-Beevor
1991	Maj Gen R. W. Ward	

APPENDIX B

Dates and Names of Principal Stations and Battles

Cavalry regiments changed station frequently during the eighteenth and nineteenth centuries, and due to the scarcity of barracks were often widely spread. The stations given below refer to the place where most of the year was spent by the majority of each regiment.

1ST KING'S DRAGOON GUARDS		THE QUEEN'S BAYS (2ND DRAGOON GUARDS)
1685 Queen's Regiment of Horse		3rd Regiment of Horse (Peterborough's)
1685	London	West Country
"	West Country	
1686	Hounslow	Hounslow
"	Kent	Oxford
1687	Berkshire	Oxford
1688	Berkshire	Hounslow
1689	SLEAFORD	Ireland
"	EDINBURGH & Ireland	
1690	DUNDALK	CLAREMONT
"	BOYNE	BOYNE
1691	AUGHRIM	AUGHRIM
1692	Northampton	Wiltshire
"	Flanders	London
1693	NEERLANDEN	
1694	Flanders	Flanders
1695	NAMUR	NAMUR
"	BONEFFE	
1696	Flanders	Flanders
1697	ENGHIEN	FROMELLES
1698	Wiltshire	Ireland
1699	London	Ireland
1700	London	Ireland
1701	Windsor	Ireland
1702	Flanders	Ireland
"	LIEGE	
1703	Flanders	Portugal
"	HUY	

1704	SCHELLENBERG	Portugal
"	BLENHEIM	
1705	NEER HESPEN	Spain
1706	RAMILLIES	Spain
1707	Flanders	ALMANZA
1708	OUDENARDE	Spain
"	LILLE	
1709	MALPLAQUET	Spain
"	TOURNAI	
1710	Flanders	ALMENARA
"		BRIHUEGA
1711	Flanders	England
"	BOUCHAIN	
1712	Flanders	Ireland
1713	Flanders	Ireland

King's Regiment of Horse, 2nd Horse

1714	Midlands	Ireland
1715	West Country	Staffordshire
"		PRESTON

The Princess of Wales Own Royal Regiment of Horse

1716	Hertfordshire	Windsor
1717	Windsor	Kent
1718	Windsor	Bedfordshire
1719	West Country	Midlands
1720	West Country	Bedfordshire & Buckinghamshire
1721	Hounslow	Bedfordshire & Buckinghamshire
1722	Hounslow	Andover
1723	Hounslow	Midlands
"		Windsor
1724	England	Midlands
1725	England	Devonshire
1726	England	Northamptonshire

The Queen's Own Regiment of Horse

1727	London	London
1728	London	Nottingham
1729	London	Warwick & Coventry
1730	London & Datchet	London
1731	Canterbury	Kent & Sussex

1732	London	Midlands
1733	Southern England	Midlands
1734	Southern England	Hampshire, Shropshire, Chester
1735	Southern England	Newcastle
1736	Windsor	Northamptonshire & Windsor
1737	Lutterworth, Northampton	Coventry & Warwick
1738	Newport Pagnell	Midlands
1739	Uxbridge	Essex & Kent
1740	Hereford & Worcester	Windsor
1741	Colchester	Essex & Kent
1742	Flanders	London
1743	ASCHAFFENBURG	London
″	DETTINGEN	
1744	Hanau	Southern England
1745	FONTENOY	PENRITH

1st, or King's, Regiment of Dragoon Guards	2nd, or Queen's, Regiment of Dragoon Guards
1746	Southern England	York
1747	Herefordshire	Derby & Nottingham
1748	Norfolk	Bristol & Somerset
1749	Norfolk & Essex	Gloucester & Worcester
1750	Birmingham Riots	Sussex
1751	England	Essex & Kent
1752	Scotland	Worcester
1753	Scotland	Yorkshire
1754	York	Scotland
1755	Southern England	Scotland
1756	Southern England	Scotland
1757	London & Dorchester	Scotland
1758	ST MALO (Light Troop)	Yorkshire
″	Paderborn	
1759	CHERBOURG (Light Troop)	
″	ST MALO (Light Troop)	
	MINDEN	MINDEN
1760	CORBACH	CORBACH
	WARBURG	WARBURG
1761	VELLINGHAUSEN	VELLINGHAUSEN
	IMMENHAUSEN	CAPELNHAGEN
	FURWOHLE	FURWOHLE
1762	WILHELMSTHAL	WILHELMSTAHL
1763	Sussex	Worcester & Hereford

1764	North of England	Worcester & Hereford
1765	Scotland	London
1766	Scotland	London

2nd (Queen's) Dragoon Guards
(Official Title)
The Queen's Bays (Unofficial Title)

1767	Scotland	South of England
1768	England	Edinburgh
1769	England	Manchester
1770	Musselburgh	South of England
1771	Warwick & Coventry	London
1772	Newbury	Colchester
1773	Newbury	York
1774	Wimbledon	Scotland
1775	Wimbledon & Norwich	Midlands
1776	Norwich	Worcester
1777	Ipswich	Sussex
1778	Salisbury & Dorchester	Salisbury Plain
1779	Salisbury & Dorchester	Salisbury
1780	Exeter	Norwich
1781	Bath	Yorkshire
1782	Devizes	Scotland
1783	Coventry	Manchester & Dorset
1784	Sussex, Kent & E Anglia	Dorset
1785	Lincolnshire	Dorset
1786	York, Durham & Newcastle	Ashford
1787	Scotland	South of England
1788	Manchester & Stockport	South of England
1789	Coventry & Devon	Scotland
1790	Dorchester	Manchester
1791	Winchester & Birmingham	Exeter & Taunton
1792	Ashford	Dorchester
1793	Flanders	Flanders
"		CASSEL
"		DUNKIRK
"		LEZENNES
1794	BEAUMONT	WILLEMS
"	TOURNAI	
1795	Flanders	Flanders
1796	Romford	Romford
1797	Brighton & Salisbury	Romford, Salisbury & Southampton

1798	Swinley & Ipswich	Windsor & Croydon
1799	Croydon	Windsor & Hertfordshire
1800	Windsor & Croydon	Southampton, Wellingborough
″	Midlands & Exeter	& Peterborough
1801	Midlands & Guildford	Bristol & Bath
1802	Birmingham & Bristol	Scotland
1803	Exeter	Ireland
1804	Exeter, Arundel & Norwich	Ireland
1805	Brighton	Ireland
1806	Brighton	Liverpool, Midlands, &
″		West Country
1807	Lewes	Salisbury & Sussex
1808	York & Scotland	Sussex
1809	Scotland	Canterbury
″		WALCHEREN
1810	Ireland	Canterbury
1811	Ireland	London
1812	Ireland	Lancashire, Yorkshire &
″		Midlands
1813	Ireland	Midlands
1814	Ireland	London
1815	WATERLOO	Scotland
1816	France & Hounslow	France
1817	Lancashire & Yorkshire	France
1818	Scotland	Ireland
1819	Ireland	Ireland
1820	Ireland	Ireland
1821	Ireland	Ireland
1822	Manchester	York & Leeds
1823	Scotland	Midlands
1824	Leeds, Newcastle & Carlisle	Midlands
1825	Hounslow	Hounslow
1826	Lancashire & Yorkshire	Manchester
1827	Scotland	Ireland
1828	York, Carlisle & Newcastle	Ireland
1829	Ireland	Ireland
1830	Ireland	Manchester
1831	Ireland	York & Leeds
1832	Northern England	Scotland
1833	Canterbury & Sussex	Nottingham
1834	Brighton & Dorset	Ipswich
1835	Midlands	Ireland

1836	Manchester & Derby	Ireland
1837	Ireland	Ireland
1838	CANADA	Ireland
1839	CANADA	Ireland & North of England
1840	CANADA	Edinburgh
1841	CANADA	Edinburgh
1842	Canada	Staffordshire & Cheshire
1843	Canada & Canterbury	Ireland
1844	Canterbury	Ireland
1845	Canterbury & West Country	Ireland
1846	Midlands & Wales	Ireland
1847	Midlands & Wales	Ireland
1848	Ireland	Ireland & Edinburgh
1849	Ireland	Edinburgh
1850	Ireland	Edinburgh, York & Manchester
1851	Ireland	York & Manchester
1852	Ireland	Ireland
1853	Ireland	Ireland
1854	Ireland & Scotland	Ireland
1855	Scotland & Crimea	Ireland
″	SEVASTOPOL	
1856	Crimea & Exeter	Ireland
1857	Exeter & India	Ireland & INDIA
1858	India	LUCKNOW
1859	India	INDIA
1860	CHINA	India
″	TAKU FORTS	
″	PEKIN	
1861	India	India
1862	India	India
1863	India	India
1864	India	India
1866	India & Colchester	India
1867	Aldershot	India
1868	Midlands & North	India

2nd Dragoon Guards (Queen's Bays)

1869	Manchester	India & Colchester
1870	Ireland	Colchester
1871	Ireland	Aldershot
1872	Ireland	Aldershot
1873	Ireland	Aldershot

1874	Ireland	Aldershot
1875	Edinburgh	Ireland
1876	Edinburgh & Manchester	Ireland
1877	North of England	Ireland
1878	Aldershot	Ireland
1879	SOUTH AFRICA	Ireland
1880	South Africa & India	Ireland & Manchester
1881	India & LAING'S NEK	Manchester & Liverpool
1882	India	Manchester & Liverpool
1883	India	Manchester & Liverpool
1884	India	Aldershot & SUDAN
1885	India	Aldershot & SUDAN
1886	India	India
1887	India	India
1888	India	India
1889	India & Shorncliffe	India
1890	Shorncliffe	India
1891	Shorncliffe	India
1892	Shorncliffe & Windsor	India
1893	Windsor & Norwich	India
1894	Norwich	India
1895	Norwich & Colchester	India & Egypt
1896	Colchester	Egypt
1897	Colchester	Aldershot
1898	Ireland	Aldershot
1899	Ireland	Aldershot
1900	Ireland	Aldershot
1901	SOUTH AFRICA	Aldershot & SOUTH AFRICA
1902	SOUTH AFRICA	SOUTH AFRICA
1903	South Africa & Hounslow	South Africa
1904	Aldershot	South Africa
1905	Aldershot	South Africa
1906	Aldershot	South Africa
1907	Aldershot & Hounslow	South Africa & Hounslow
1908	India	Hounslow
1909	India	Hounslow
1910	India	Hounslow & Aldershot
1911	India	Aldershot
1912	India	Aldershot
1913	India	Aldershot
1914	India & FRANCE	Aldershot & FRANCE
"		NERY

1914		MONS
"		LE CATEAU
"		RETREAT FROM MONS
"		MARNE
"		AISNE
"		MESSINES
"		ARMENTIERES
		YPRES
1915	FESTUBERT	YPRES
"	HOOGE	FREZENBERG
"		HOOGE
1916	SOMME	HOHENZOLLERN REDOUT
"	MORVAL	SOMME
1917	MORY	FLERS COURCELLETTE
"	HARGICOURT	ARRAS
"	EPEHY	SCARPE
"	LE VERGIER	FAMPOUX
"	India	YPRES
"		CAMBRAI
1918	India	VADENCOURT
"		SOMME
"		BAPAUME
"		MONTAUBAN
"		AMIENS
"		ROSIERES
"		ALBERT
"		HINDENBURG LINE
"		ST QUENTIN CANAL
"		BEAUREVOIR
"		CAMBRAI
"		PURSUIT TO MONS
1919	AFGHANISTAN	Germany & Palestine
1920	IRAQ	Palestine
1921	Edinburgh	MALABAR
1922	Edinburgh	India
1923	Edinburgh & Germany	India
1924	Germany	India
1925	Germany	India
1926	Germany	India
1927	Aldershot	India
1928	Aldershot	India & Colchester
1929	Aldershot & Tidworth	Colchester & Tidworth

1930	Tidworth	Tidworth
1931	Tidworth & Hounslow	Tidworth & Shorncliffe
1932	Hounslow & Egypt	Shorncliffe
1933	Egypt	Shorncliffe & Aldershot
1934	Egypt	Aldershot
1935	Egypt & India	Aldershot
1936	India	MECHANISATION
''		Aldershot
1937	India	Aldershot & Tidworth
1938	MECHANISATION	
''	Aldershot	Tidworth
1939	Aldershot	Tidworth
1940	Dorset	Tidworth & France
''		SOMME
''		WITHDRAWAL TO SEINE
1941	BEDA FOMM	Marlborough
''	DEFENCE OF TOBRUK	Egypt
''	TOBRUK SORTIE	
''	RELIEF OF TOBRUK	
1942	GAZALA	MSUS
''	BIR HACHEIM	GAZALA
''		BIR EL ASLAGH
''		THE CAULDRON
''		KNIGHTSBRIDGE
''	ALAMEIN LINE	VIA BALBIA
''	ALAM EL HALFA	MERSA MATRUH
''		EL ALAMEIN
''	EL AGHEILA	
1943	TRIPOLI	
''	TEBAGA GAP	TEBAGA GAP
''	POINT 201	
''	EL HAMMA	EL HAMMA
''	AKARIT	EL KOURZIA
''		DJEBEL KOURNINE
''	TUNIS	TUNIS
''	NAPLES	
''	SCAFATI BRIDGE	
''	MONTE CAMINO	
1944	GARIGLIANO CROSSING	Algiers
''	PERUGIA	
''	AREZZO	
''	GOTHIC LINE	CORIANO

1944		CARPINETA
1945	GREECE	LAMONE CROSSING
"	ATHENS	DEFENCE OF LAMONE BRIDGEHEAD
"	Egypt	RIMINI LINE
"	SYRIA	CERIANO RIDGE
"	PALESTINE	CESENA
"		ARGENTA GAP
1946	PALESTINE	Italy
1947	PALESTINE & Libya	Italy & Chester
1948	Libya & Driffield	Chester
1949	Driffield & Omagh	Chester
1950	Omagh	Fallingbostel
1951	Omagh	Fallingbostel
1952	Hamburg	Fallingbostel
1953	Neuminster	Fallingbostel
1954	Neuminster	Fallingbostel & Tidworth
1955	Neuminster	Aqaba
1956	MALAYA	Aqaba & Libya
1957	MALAYA	Libya
1958	MALAYA	Perham Down

1st The Queen's Dragoon Guards

1959	Perham Down & Wolfenbuttel
1960	Wolfenbuttel
1961	Wolfenbuttel
1962	Wolfenbuttel
1963	Wolfenbuttel
1964	Wolfenbuttel
1965	Omagh & BORNEO
1966	Omagh & BORNEO, Persian Gulf & ADEN
1967	ADEN & Persian Gulf
1968	Warminster & Detmold
1969	Detmold
1970	Detmold, Catterick & Berlin
1971	Catterick & Berlin
1972	Catterick & Berlin
1973	Hohne
1974	Hohne
"	NORTHERN IRELAND
1975	Hohne
1976	Hohne
"	NORTHERN IRELAND

1977	Hohne
1978	Hohne
"	NORTHERN IRELAND
1979	Hohne
"	NORTHERN IRELAND
1980	Hohne
"	NORTHERN IRELAND
1981	NORTHERN IRELAND
1982	NORTHERN IRELAND
1983	Wimbish
"	BEIRUT & Cyprus
1984	Wimbish & Cyprus
"	Belize
1985	Wimbish & Cyprus
1986	Wimbish & Cyprus
1987	Wimbish & Wolfenbuttel
1988	Wolfenbuttel
1989	Wolfenbuttel
"	NORTHERN IRELAND
1990	Wolfenbuttel
"	PERSIAN GULF
1991	Wolfenbuttel & Tidworth
"	PERSIAN GULF
1992	Tidworth

Nicknames

1ST KING'S DRAGOON GUARDS

'The KDGs'   From the regiment's initials.
'The Trades Union'   From the regiment's employment in quelling industrial
                     disturbance in the nineteenth century.

THE QUEEN'S BAYS (2ND DRAGOON GUARDS)

'The Bays'   From the colour of the regiment's horses, by an order of 1767
                     for long- tailed bay mounts.
'Rusty Buckles'   The regiment, having served for many years in Ireland,
                     returned to England still equipped with the old iron horse
                     furniture.

1ST THE QUEEN'S DRAGOON GUARDS

'First and Foremost' From the regimental seniority, and during the regiment's years
                     in a reconnaissance role.

Regimental Music

1ST KING'S DRAGOON GUARDS

Slow March   Mercadante
Quick March   The Radetsky March

THE QUEEN'S BAYS (2ND DRAGOON GUARDS)

Slow March   The Queen's Bays, or Rusty Buckles
Quick March   The Queen's Bays, or Rusty Buckles

1ST THE QUEEN'S DRAGOON GUARDS

Slow March   The Queen's Dragoon Guards
              (Mercadante & The Queen's Bays)
Quick March   The Queen's Dragoon Guards
              (Radetsky and The Queen's Bays)

Regimental Badge

Prior to 1896

Officer's Full Dress	The Royal Cypher within the Garter, on a metal diamond star.
Other Ranks	On a white-metal star a gilt Garter, within which a white-metal numeral 1 on a black leather ground.
1896 to 1915	The Austrian double-headed eagle, on a scroll, inscribed 'King's Dragoon Guards'.
1915 to 1937	An eight-pointed star, surmounted by an Imperial Crown, the letters 'K.D.G.' within the Garter, in the centre of the star.
1937 to 1959	The double-headed eagle of Austria.

Prior to 1896

Officer's Full Dress	The Royal Cypher within the Garter, on a metal diamond star.
Other Ranks	On a white-metal star a gilt Garter, within which a white-metal numeral 2 on a black leather ground.
1896 to 1959	'Bays' in old-English lettering within a laurel wreath surmounted by a crown.

1959 onwards

Cap Badge	The double-headed eagle of Austria.
Collar Dogs	'Bays' in old-English lettering within a laurel wreath surmounted by a crown.

Regimental Full Dress

Dragoon Guard Full Dress

1685 to 1714	Scarlet with bright yellow facings.
1714 to 1857	Scarlet with blue velvet facings.
1847 to 1959	Tunic Scarlet with blue velvet facings.
Breeches	Blue with yellow stripe.

Helmet   Dragoon Guard pattern with red plume.
                                (Band has white plume.)

THE QUEEN'S BAYS (2ND DRAGOON GUARDS).

1685 to 1742   Scarlet with scarlet facings.
1742 to 1784   Scarlet with buff facings.
1784 to 1855   Scarlet with black velvet facings.
1855 to 1959
Tunic   Scarlet with pale buff facings.
Breeches   Blue with white stripe.
Helmet   Dragoon Guard pattern with black plume
                                (Band has white plume.)

1ST THE QUEEN'S DRAGOON GUARDS
1959 onwards
Tunic   Scarlet with blue velvet facings.
Breeches   Blue with white stripe.
Helmet   Dragoon Guard pattern with red plume.
                                (Band has white plume.)

The Uniforms of the Regiment

Both the King's Dragoon Guards and The Queen's Bays were raised in 1685 as The Queen's Horse and Peterborough's Horse respectively. The first mention of uniform occurs in 1686, when the two regiments were reviewed by James II on Hounslow Heath. The Queen's Horse [KDG] were clad in long full-skirted crimson coats, without collars, which were faced and lined with yellow, which was the Stuart colour and that of the Queen. Peterborough's Horse [Bays] were dressed in scarlet, faced in buff and lined with scarlet. The coats were decorated with a large number of buttons, down the front, on the cuffs and pocket flaps, as well as along slits at the back and side (a clothing estimate for the Scots Greys gives ten dozen tin buttons for each coat). The men of both regiments wore leather jack-boots, coming half way up their thighs, known as 'gambados', and in 1687 Peterborough's Horse wore buff breeches. Both regiments had buff gauntlets, broad crossbelts, and broad-brimmed black hats, turned up at one side and worn with a white feather. The cloaks were of red cloth. The officers of both regiments had silver lace on their hats and coats, and silver fringes on gloves and scarf, the latter worn around the waist. The troopers had a plain crimson sash and wore iron cuirasses, which were 'pistol proof', and iron helmets, known as 'pots', which were worn under the hat in action and often carried on the saddle bows otherwise. This armour was withdrawn in 1689 and reissued for a short time between 1707 and 1714. The men were armed with a sword, a pair of pistols and, later, a carbine.

Both regiments had a kettledrummer and trumpeters, who were clothed in rich costumes, and a Treasury Paper of 1692 states the cost of that of The Queen's Horse as being £612.8.2½d. For fatigues the men wore a close-fitting coat of coarse grey cloth, often with black facings and buttons.

We have little evidence for the uniform worn during the Marlborough Wars, but an advertisement for a deserter from Cadogan's Horse in the *London Gazette* of 1711 gave a red coat faced with green, broad silver lace on the sleeves, sleeves and pockets bound with narrow silver lace, green waistcoat and shag breeches, silver laced hat, brown wig with a black bag. If we replace the green of Cadogan's with the yellow of The Queen's Horse and the buff of Peterborough's, by now the 3rd Horse, we can get an accurate picture.

The accession of George I in 1714 heralded changes for both regiments, The

Queen's Horse changing their title to The King's Own Regiment of Horse, and their facings from yellow to blue. A portrait of Captain William Pritchard Ashurst (who served for a short time in The King's Horse in 1722) by Peter Tilemans shows him wearing a low cocked hat edged with silver lace, a long scarlet coat with silver lace and buttons, and very long cuffs reaching above the elbow. He wears a yellow waistcoat and breeches, buff gauntlets and no sash. A detachment of the regiment [KDG] is shown in the background riding four abreast, with trumpeters in yellow coats riding greys. This would seem to indicate that the regiment may still have been wearing yellow in 1722.

The 3rd Horse [Bays] had their title changed in 1715, for their loyalty and gallant conduct in suppressing the Jacobite uprising, to The Princess of Wales' Own Royal Regiment of Horse, but there appears to have been no change of uniform. In 1727 George II came to the throne and the regiment's title was changed to The Queen's Own Royal Regiment of Horse.

In 1742 the Duke of Cumberland had produced 'A Representation of the Cloathing of His Majesty's Household and all the forces upon the Establishments of Great Britain and Ireland 1742'. Pictures of a trooper show a red coat, with blue [KDG] and buff [Bays] facings, no lapels, brass buttons, blue [KDG] and buff [Bays] waistcoat and breeches, buff crossbelts with blue [KDG] and buff [Bays] flask cord, laced hat, red horse furniture embroidered with the Royal Arms, and a red cloak lined with blue [KDG] and buff [Bays] rolled behind the saddle.

The Earl of Stair reviewed The King's Horse on 30 October 1742, issuing the following order:

> The Horse and Dragoons to take all their small accoutrements to pieces and see that they be very well cleaned and blacked, and then put together again. The bosses, bits and curbs to be as bright as hands can make them. The boots to be as black as possible, and their knee-pieces not to appear above three inches above the boot-top. All their arms to be as bright as silver. The whole buff accoutrements to be of one light buff colour, the swords to be all brightened. The hats new cocked. Three straps to each cloak. Care to be taken that the men do not ride too long. Horses to be all trimmed and made as clean as possible. No Trooper or Dragoon to appear in the streets with his cloak on.

A Royal Warrant of 19 December 1746 changed the status of both the KDG and the Bays from horse to dragoon guards, they becoming 'Our First, or King's, Regiment of Dragoon Guards', and 'Our Second, or Queen's, Regiment of Dragoon Guards'. The old distinction of velvet facings for regiments of horse was allowed to be continued, the KDG retaining their blue and the Bays their buff velvet facings. Trumpeters were replaced by drummers and hautbois. Both regiments lost the crossbelts of the horse, together with the flask cords. Carbines were removed, and muskets and bayonets issued. The officers wore gold lace and embroidery, and a crimson silk sash over the left shoulder, while quartermasters had gold lace and the silk sash around their waist, and sergeants narrow lace on the lapels, sleeves and pockets with a worsted sash around the waist.

The clothing warrant for The King's Dragoon Guards of 1751, which was the same for the Bays except for the regimental colour, facings and distinctions, describes the uniform:

> Privates: Red coats faced blue, lapelled to the waist, lined blue, buttonholes of very narrow lace (two and two), slit sleeves, long pockets, yellow aiguillette, blue waistcoat and breeches, gold laced hat. Horse furniture red, yellow worsted and mohair lace with blue stripe one third the whole breadth. Device on housings the King's Cypher within the Garter, and Crown over embroidered. Holster Caps: G.R. and Crown embroidered. Underneath 1. D.G. in yellow on the blue stripe of the lace.
>
> Corporals: Narrow gold lace on turn-up of sleeve and shoulder strap, yellow silk aiguillettes.
>
> Sergeants: Narrow gold lace on lapels, turn-up of sleeves and pockets, pouches like the men, gold aiguillette, blue worsted sash round waist.
>
> Drummers and Hautbois: Red coat, blue facings and lining, ornamented with yellow lace with a blue stripe, long hanging sleeves fastened at the wrist, blue waistcoats and breeches, blue cloth caps, embroidered with the King's Cypher within the Garter and Crown over, the little flap red, with the White Horse and motto 'Nec Aspera Terrent', back of cap red with tassel hanging behind, blue turn-up with a drum and 1.D.G. in the middle. Cloaks: Red lined blue, blue collar, buttons set on at top in same manner as on the coat upon frogs or loops of yellow and blue lace. Watering or Forage Caps: Red turned up with blue, 1. D.G. on the little flap.
>
> Quartermasters: Sash worn around the waist.
>
> Officers: Uniform made up in the same manner as the men's. Laced, lapelled to waist and turned-up with blue, and a narrow gold lace or embroidery to the binding and buttonholes, buttoned two and two, sash over left shoulder, blue waistcoat and breeches, sword knot crimson and gold in stripes. Horse furniture: Red, with one gold lace and a stripe of blue velvet in the middle.

The main differences for the Bays were their buff facings, waistcoats and breeches, their buttons set three and three, and 2. D.G. on the appointments. The

Bays' drummers and hautbois wore the same red coats faced with blue, and the same blue breeches and waistcoats as the KDG. The horse furniture was buff with royal lace, and the Queen's Cypher within the Garter.

There are paintings in the Royal Collection at Windsor Castle by David Morier of privates of the KDG and Bays of this period, showing exactly this uniform.

In 1755 a light troop was added, the men wearing the same uniform as the rest of the regiment, except for a helmet of black japanned boiled leather with the Royal Cypher, the crown and number of the regiment in brass on the front, and having a brass crest with a tuft of stiff horsehair, coloured half red and half blue for the KDG, half red and half buff for the Bays. The light troops were disbanded in 1763, but revived again for a short time in 1778, when they were finally removed from all dragoon guard regiments and formed into three separate regiments of light dragoons.

An order of 1764 discontinued the wearing of the aiguillette and replaced it with an epaulette worn on the left shoulder. Officers discontinued the embroidered edging to their coats, and the jackboot was made to a lighter pattern. In 1764 the Bays changed their facings from buff to black. The KDG gave up their blue waistcoats and breeches for the same buff as the Bays in 1767, and the Bays' drummers and hautbois also changed their blue breeches and waistcoats for buff. This only lasted for a few years, for in 1774 both regiments changed their waistcoats and breeches to white.

A clothing warrant of 1768 allowed the coats to have turndown collars and to button up to the neck: the KDG were permitted to change the '1.D.G.' on their buttons to 'K.D.G.'; the epaulette was of the facing colour with yellow tape and fringe; black gaiters replaced white for dismounted duties, and the waist sash was of crimson with a stripe of the facing colour. An inspection report of 1768 commented, 'Officers' uniforms embroidered with gold; arms and furniture handsome. Trumpeters finely mounted'; and the following year,

> Officers' uniforms very good without any new alteration; buff waistcoats and breeches; gold embroidery; epaulettes; laced hats ... The Farriers of the Dragoon Guards to have blue coats with blue lining and blue waistcoats and breeches. To wear a small black bearskin cap, with a horseshoe on the forepart, of silver-plated metal on a black ground, and to have churns and an apron.

In 1784 the facings of The Queen's Bays were changed from buff to black. In 1787 the dragoon guards were ordered to wear their swords slung over their coats; to allow for this, epaulettes were worn on both shoulders.

The next major changes came in 1794 and 1796 when the coat was shortened to allow it to clear the saddle when mounted, epaulettes were abolished, being replaced by yellow shoulder straps with wings of red and white cloth interlaced with brass plates. As the coats were worn buttoned down the front, the waistcoats could no longer be seen. The KDG had gold lace and blue facings, while the Bays had silver lace with black facings. Breeches were plush instead of leather. The wearing of cocked hats was discontinued, and a large hat of black felt was introduced for both regiments, turned up at front and back, and worn with a red and white feather. This enormous hat was augmented in 1804 by a watering cap. There is a KDG watering cap in the regimental museum: it is a tall cylindrical cap with a peak, and the regimental monogram in brass on the front. In 1890 a Bays' watering cap of this period was exhibited at the Royal Military Exhibition; it was cylindrical in shape, of black leather, like that of the KDG but without a peak, and on the front a yellow metal plate with the royal arms surmounted by a scroll with the words '2nd or Queen's Dragoon Guards'. It also had a bag of plain leather attached to the crown. A carbine and single pistol replaced the musket and bayonet and two horse pistols. The heavy cavalry sword had a broadf straight blade, thirty-five inches long, ending in a hatchet point, the guard being a steel knucklebow.

In 1803 NCOs were ordered to wear chevrons, formed of a double row of regimental gold lace: four for a sergeant major and quartermaster sergeant, three for a sergeant, and two for a corporal. In 1808 the wearing of pigtails was abolished, although the KDG continued to wear theirs until 1809.

A major change occurred in 1812 when a new helmet was issued, often called the 'Waterloo' helmet. It had a black leather skull piece with a metal crest and plate bearing the name of the regiment and a reverse G.R. cypher. To the front of the metal crest was fixed a little tuft of black horsehair, which then swept back into a long flowing tail which hung down behind. A new coatee or jacket came into use, which did away with the bars of lace across the front and substituted broad lace from the collar down to the waist, and around the skirt and cuffs. This jacket was fastened by hooks and eyes down the front, instead of buttons. The parade breeches remained white, and were worn wth jackboots; but on active service pantaloons of grey with short boots were adopted. A narrow sword belt replaced the old broad one, and short gloves replaced the gauntlets on service.

In 1820 the Prince Regent became George IV, and his interest in dress led him to take much personal pride in the elegance and turnout of the Army. In 1822 dress regulations were issued, the short coats being replaced by a single-breasted coat decorated with cross-bar lace and fastened by eight buttons. On the lower sleeves and skirts were four V-shaped loops of lace with a regimental button in the centre, the loops being arranged in pairs. The material for overalls was changed from grey to blue. The dress sabretache became very elaborate, and the helmet was 'Roman black glazed skull and peak, encircled with richly gilt laurel leaves, rich gilt dead wrought scales and lions' heads: bear skin top'. The great bearskin crest made it a most imposing but difficult ornament to wear. In 1829 the lighter blue trousers were changed to dark blue with a yellow stripe.

A picture in the Royal Collection at Windsor Castle by Dubois Drahonet, dated 1832, shows a sergeant of The King's Dragoon Guards wearing a black leather sabretache, and in the centre a brass star surmounted by a crown, and the letters 'K.D.G.' in the centre of the star. He is wearing the black dragoon guard helmet with the high bearskin comb. His jacket is of scarlet cloth with blue velvet facings, and the cuffs of the sleeves have four loops with tassels arranged in pairs. The overalls of blue with a yellow stripe were worn by both the KDG and the Bays. A KDG inspection report of 1834 remarks: 'The helmet worn by the band has a scarlet ornament above the crest in lieu of bearskin.' Officers wore gauntlets with dress uniform, and white leather gloves with frock coat and undress.

In 1834 the black helmet was changed to an all-brass helmet with a metal crest terminating in a lion's head, which was removable to enable the large bearskin crest to be fitted for full-dress occasions. Epaulettes for full dress were 'boxed' with a bullion fringe. Metal shoulder scales were worn in undress, and the KDG had the word 'Waterloo' on a silver scroll and the monogram 'K.D.G.' also in silver. At levées the officers of both regiments wore a large cocked hat with a gilt embroidered ornament and a seventeen-inch red and white swan plume. In undress a single-breasted blue frock coat with scale epaulettes was worn, with a round blue forage cap circled by a band of regimental lace, and with a black patent leather peak. Both regiments wore a scarlet cloak with a white lining.

The KDG in Canada from 1838 to 1843 were issued with special winter clothing of a full-length blue overcoat with brass shoulder scales, and held in by a white belt and a white pouch belt; knee boots, fur gloves, and a fur cap with a flap to cover the neck and ears, and a red side flap. Colonel Cathcart, commanding the KDG in Canada, reported:

> I have been able to give every man a pair of jackboots. [They] are quite waterproof. They have the art of making them so here by putting the smooth side of cowhide leather outside, and then making a particular composition of wax and boiled oil . . . The boots are made large enough to wear a long stocking over the sock, inside of the boot, and pulled over the knee.

In 1843 a new type of brass helmet was introduced with a crest and long black horsehair tail, and this was again replaced in 1847 by the 'Albert' helmet with a central spike supporting a black plume. In 1848 the coatee was shortened when, after the Crimean War, it was replaced by a full-skirted tunic, which gave more protection to the stomach, much of the elaborate and expensive decoration being removed at the same time. The new tunic was scarlet with collar, cuffs and edging in the regimental colour, fastened down the front by eight buttons. Epaulettes were replaced by shoulder cords, and officers' badges of rank were worn on the collar. The 'Albert' helmet remained, but the colour of the plume was changed for each regiment, the KDG having red and the Bays black.

The end of the Crimean War also saw the facings of the Bays being changed again from black back to buff. In 1864 the loops of lace on the tunic collar hiding the colour of the facings were abolished, as were the loops and button on the cuffs and skirt, and an 'Austrian' knot, of yellow for the KDG and white for the Bays, was substituted on each sleeve of the tunic.

There is some evidence to show that the Bays during the Indian Mutiny, and the KDG in China, wore their brass 'Albert' helmets with the plume removed and a pugaree wrapped around the helmet. Other evidence indicates that both regiments may have worn an early type of cork pith helmet from time to time.

In 1873 the 'Albert' helmet was further simplified, much of the ornamentation being removed, and the badge on the front becoming an eight-pointed star, with the Garter and motto in the centre for officers, and the numerals '1' or '2' for the other ranks. In both regiments trumpeters and bandsmen wore white plumes. The men also were stopped using shabraques, although the officers continued to have them.

In 1876 patrol jackets replaced the stable jackets for a short time, but the following year there was a reversion to the stable jacket. Officers' badges of rank were transferred to the shoulder cords, instead of being worn on the collar. In addition, for officers' tunics, the size of the 'Austrian' knot on the sleeves indicated rank, field officers having a triple knot measuring eleven inches from the bottom of the cuff, captains a double knot of nine inches, and subalterns a single knot of seven inches' depth.

Throughout the late Victorian period undress uniform remained a blue frock coat with regimental patterns of braiding on the breast and sleeves. With this was worn the 'pill box' forage cap of blue cloth, with a gold regimental pattern lace band for officers, and for the men yellow cloth for the KDG and white for the Bays.

In 1884 the Bays contingent going to the Sudan wore grey tunics, the regimental insignia being sewn onto the sleeve, with cord pantaloons, blue puttees and brown boots, a bandolier and haversack, with a pith helmet.

With minor changes in detail, full-dress uniform remained the same for both the KDG and the Bays up to amalgamation in 1959. But the Sudan and the Boer War saw the universal introduction of more serviceable clothing than full-dress red, which tended to be used only for ceremonial occasions. The appointment of the

Austrian Emperor, Franz Joseph, as colonel in chief of the KDG led to his conferring in 1898 the use of the Austrian Habsburg double-headed eagle as the regimental cap badge. From 1899 to 1900 senior NCOs, and later possibly corporals, wore the Austrian eagle as an arm badge above the rank chevrons. The Bays' cap badge was formed of the letters 'Bays' in early English lettering surrounded by a wreath of laurel leaves, and surmounted by the Imperial Crown. During the First World War, when Austria was allied to Germany, the KDG were made, in 1915, to take down their cherished eagle badge and replace it with an eight-pointed star, surmounted by the crown, with the Garter enclosing the letters 'K.D.G.' in the centre. The Austrian eagle came back into use in 1938 and has been worn since as the cap badge of both the KDG and 1st The Queen's Dragoon Guards.

During the latter decades of the nineteenth century, khaki had been replacing the old scarlet on active service conditions. India and the Sudan established its utility, and in 1898 an experimental issue of khaki was made, to be succeeded in 1900 by an general issue of khaki serge, which both the KDG and the Bays wore in South Africa. The late Victorian pattern pith helmet was often replaced on the veldt by a floppy broad-brimmed felt hat. In 1902 an Army order laid down the form of dress for occasions when full dress was not to be worn. A peaked khaki cap was worn by all ranks; the khaki tunic was cut loose with two patch pockets on the chest. Officers' service dress jackets had a khaki collar and tie, and were worn with a Sam Browne leather sword belt. Badges of rank for officers were worn on the cuff. Breeches and brown riding boots were worn by officers, and Bedford cord breeches and puttees by the other ranks. In 1903 the KDG were issued with the much-disliked Broderick cap, and the much-prized Victorian pill-box forage cap was discontinued. The Broderick cap was of blue cloth, round in shape with a projecting rim, but with no peak; it was worn with a chinstrap, and the metal regimental badge of the Austrian double-headed eagle on a red background in the centre at the front. On their return to England in 1908 the Bays were issued with a blue Broderick cap, with the regimental badge on a white patch at the front – the patch later being changed to light buff.

From 1907 to 1914 the KDG in India wore, with their full dress, a white Wolseley helmet, with a brass spike at the crown and brass chin scales, and a red tuft on the left-hand side of the helmet.

During the First World War both the KDG and the Bays wore regulation khaki service dress, with breeches and puttees, the men wearing a brown leather bandolier over the left shoulder. After the war khaki became the normal wear of the Army, and full dress was confined to great ceremonial occasions, musical rides and levées. As a result the rather sloppy issue of the war years was smartened up to provide a walking out dress. Officers' badges of rank reverted to the shoulder.

In 1938 battledress was introduced and with it the KDG won back the wearing of the Austrian double-headed eagle; the badge being now in white metal, or silver for the officers, without the earlier scroll underneath. Battledress consisted of a loose khaki blouse, modelled on a skiing suit, with two patch pockets on the chest and buttoning at the wrists. The trousers had a pocket for a field dressing on the leg and were worn with webbing gaiters. For working dress denim overalls were introduced of the same pattern. With the mechanisation of the cavalry, the black beret was introduced for the Royal Armoured Corps, but the men of both the KDG and the Bays clung to the khaki forage cap until well into the middle years of the Second World War. Officers tended to wear khaki service dress caps, although the occasional beret was seen.

The war in the Western Desert was best known for its informality of dress – corduroy trousers, desert boots, coloured scarves, sheepskin coats. The KDG officers wore a large patch of yellow cloth beneath their metal rank badges on their battledress blouses, a regimental distinction which persisted for some years after the war. The Bays wore shoulder titles in brass, with the words 'Bays' in early English script.

Towards the end of the Second World War the KDG instituted the award of a troop leader's badge for sergeants who had commanded a troop with distinction. It consisted of the Austrian eagle collar badge worn on a background of red cloth above the chevron.

In 1947 a new patrol dress was authorised, with a blue tunic, having a high collar carrying collar badges; two breast pockets with box pleats and button, side pockets without buttons, and a cloth belt with brass buckle, later replaced by a white belt. There were steel shoulder chains, on which officers wore badges of rank, and trousers cut closely; the KDG having a yellow stripe and the Bays white. The cap was blue with a band of blue velvet for the KDG and of white for the Bays. The old gold lace belt and silver pouch box, worn over the left shoulder, continued in use.

On amalgamation of The King's Dragoon Guards with The Queen's Bays in 1959, elements of the uniforms of both regiments were combined to constitute the dress for 1st The Queen's Dragoon Guards. In full dress, still worn by the band and by individual soldiers for ceremonial occasions, the helmet of the KDG has been adopted with the red plume and the figure '1' within the star on the front of the helmet. The band wear a white plume with the helmet and aiguillettes on the shoulder. The tunic is scarlet with blue velvet facings and the Austrian knot in yellow cord on the sleeves, with the Bays' collar badges. A white waist belt with metal plate buckle is worn, and the pouch belts carry the Bays' badge on a black patent leather pouch box. The officer's tunic has gold lace, and the waist and pouch belts are of gold lace, with a silver pouch box, carrying the Austrian eagle. The trousers are close cut overalls of blue, with a broad white stripe, worn over black boots with regimental pattern spurs.

Officers' mess dress consists of a scarlet monkey jacket trimmed with regimental pattern gold lace and upright collar of blue velvet edged with gold lace, twisted gold lace shoulder cords with badges of rank. A scarlet waistcoat, edged with gold lace, is worn under the jacket. Full-dress overalls and regimental pattern spurs complete the dress.

The blue No 1 patrol dress was modified to combine elements of both regiments. The tunic carries the collar badges of the Bays and has chain mail with the shoulder title 'QDG' in early English lettering. The officers' blue patrol jacket has no pockets on the skirt, and the sword is secured by two gold-laced straps from under the tunic. The blue overalls have the Bays' broad white stripe and are worn over black boots with regimental pattern spurs. The cap is the KDG design of blue cloth with a blue velvet stripe and the Austrian eagle as cap badge. The peak of the cap has a broad rim of gold lace for the officers, a thinner band of gold lace for warrant officers, and a plain black peak for all other ranks. The pouch belts for officers are as for full dress, with other ranks wearing a white belt with the Austrian eagle on the buckle.

On amalgamation the khaki service dress of the Bays was adopted for officers, with QDG shoulder titles and Bays' collar badges, and the KDG cap and cap badge. Other ranks wear the No 2 Service Dress, with the blue KDG cap and badge, the tunic having the Bays' collar badges and QDG shoulder titles. In working dress a dark blue beret with the eagle cap badge is worn. The stable belt is royal blue. In the QDG lance corporals wear two chevrons, corporals wear two chevrons surmounted by a rank badge consisting of the Bays' emblem, which is worn by all senior NCOs. Squadron quartermaster sergeants wear four chevrons with rank badge, the whole surmounted by a crown.

APPENDIX E

Battle Honours

The battle honours shown in capital letters are those emblazoned on the standard.

KING'S DRAGOON GUARDS	THE QUEEN'S BAYS
Sleaford	
Boyne	Boyne
Aughrim	Aughrim
Neerlanden	
Schellenberg	
BLENHEIM	
RAMILLIES	Almanza
OUDENARDE	Almenara
MALPLAQUET	Brihuega
DETTINGEN	
WARBURG	WARBURG
BEAUMONT	Preston
	WILLEMS
WATERLOO	
Canada	
SEVASTOPOL	
	LUCKNOW
TAKU FORTS	
PEKIN	
SOUTH AFRICA 1879	
Laing's Nek	Abu Klea
SOUTH AFRICA 1901-1902	SOUTH AFRICA 1901-1902
	MONS
	LE CATEAU
	Retreat from Mons
	MARNE 1914
	Aisne 1914
	MESSINES 1914
	Armentières 1914
	YPRES 1914-1915

Festubert
Hooge
SOMME 1916
MORVAL

Frezenberg
Bellewarde
SOMME 1916-1918
Flers Courcellette
Arras 1917
SCARPE 1917
CAMBRAI 1917-1918
Bapaume 1918
Rosières
AMIENS
Albert 1918
Hindenburg Line
St Quentin Canal
Beaurevoir
PURSUIT TO MONS

FRANCE & FLANDERS 1914-1917
AFGHANISTAN 1919
BEDA FOMM
DEFENCE OF TOBRUK
TOBRUK 1941
Tobruk Sortie
Relief of Tobruk
GAZALA
Bir Hacheim

FRANCE & FLANDERS 1914-1918

SOMME 1940
Withdrawal to Seine
Northwest Europe 1940

Msus
GAZALA
Bir el Aslagh
Cauldron
Knightsbridge
Via Balbia
Mersa Matruh

DEFENCE OF ALAMEIN LINE
Alam el Halfa

El Agheila
ADVANCE ON TRIPOLI
TEBAGA GAP
Point 201
EL HAMMA
Akarit

TUNIS

NORTH AFRICA 1941-1943
Capture of Naples
Scafati Bridge

EL ALAMEIN

TEBAGA GAP

EL HAMMA
El Kourzia
Djebel Kournine
TUNIS
Creteville Pass
NORTH AFRICA 1941-1943

MONTE CAMINO
Garigiiano Crossing
Capture of Perugia
Arezzo
GOTHIC LINE

CORIANO
Carpineta

LAMONE CROSSING
Defence of Lamone Bridgehead
RIMINI LINE
Ceriano Ridge
Cesena
ARGENTA GAP

ITALY 1943-1944
Athens
Greece
Palestine
Malaya

ITALY 1944-1945

1st THE QUEEN'S DRAGOON GUARDS

Northern Ireland
Borneo
Aden
Lebanon
Persian Gulf

A Wife's Tale: Mrs J. Tiarks, KDG

It is a tremendous ordeal arriving as a bride in a regiment. I suffered a great deal from shyness. Suddenly to meet so many strange officers and their wives was far from easy, especially as the Colonel and the Adjutant were both bachelors and disliked women. I know I am going back many years – 1924 to be exact – but when I was married there was a ruling that no officer should marry under thirty years old, and other ranks under twenty-six. As Jack was only twenty-four when we married, he was fined £100 by the regiment. If the Army didn't accept your marriage, it meant that you got no marriage allowance, which was quite considerable.

Edinburgh is really the home of the Scots Greys, so naturally all the horses were grey. What a fine sight it is to see the regiment on parade. I don't think the grey horses were very popular with the troopers, as it takes so much more grooming to keep them spick and span. The regiment had a very good musical ride, which was in great demand. I had never seen a musical ride, and asked Jack if I could go to the covered-in-riding- school, and watch them practise. I couldn't understand what the ride was meant to be doing; it all seemed to be in a terrible muddle. I was very disappointed, and told Jack so. He roared with laughter, and explained that Sergeant Page, who was taking the ride, was horrified, and daren't open his mouth in case he shocked me, so he more or less let the ride do as it pleased. I was taken to see the ride when they had their official engagement, and I was very impressed.

When a new regiment arrives at a station, it always causes a stir among the inhabitants in the neighbourhood, especially with mothers who had daughters! We were always very welcome, and a great deal of entertaining went on. Our next move was to Cologne with the Army of Occupation. I suppose it was understandable, but we never met any Germans; they were far too proud. Inflation was running rife, money had no value at all. It didn't worry us as we did all our shopping in the NAAFI. It was a very different story for the Germans, who were hungry. Inflation became so bad that one day Jack came home with a fistful of paper marks. They were in billions! I was told to go out and spend it, and what I hadn't spent to throw the remainder into the wastepaper basket, as tomorrow it would be useless.

We were having a very gay interesting time in Cologne when disaster befell us.

Suddenly the authorities awoke to the fact that Jack wasn't thirty, so not recognised by the Army as married, and therefore not entitled to an Army quarter. He
received a large bill for the rent, back-dated. We fought the Army authorities as far
as we could, but to no avail. We had to pay. How we managed, I can't remember,
but we did feel very sorry for ourselves.

By this time I was expecting my first baby, and it was due just when the regiment
was leaving Cologne to take up duty from the French at Wiesbaden. I was driven
down to the square in Wiesbaden to see the Cameron Highlanders take over from
the French, and, believe me, it was a sight to remember. On one side of the square
stood the French in their ill-fitting, not too clean old uniforms, with unpolished
boots. Into the square marched the Cameron Highlanders with their pipes playing,
their kilts swinging – a marvellous sight -, and – would you believe it – even the
Germans cheered.

In the summer there were horse shows to add interest to the Army of Occupation. We had some very good show jumpers among the horses of the regiment.
One was called Sea Count, who became quite famous, winning many prizes. Then
there was the drum horse of the musical ride, a star among the jumpers. We were
always a little nervous, when he was leading the band, that he might see an
inviting jump, and forget his drums!

There are four squadrons in the regiment, A, B, C, and Headquarters. You took a
special interest in the squadron your husband was in, getting to know all the
wives. A club was run for other ranks' wives, which met once a week, sometimes
to sew or play whist. Tea was provided, and officers' wives were expected to
attend. As well as the club, the regiment ran a concert party. It was surprising what
talent there was. As well as all these activities, there were dances – sergeants,
corporals and troopers – and we were expected to go to them all, and dance when
requested. Sometimes this was rather difficult if you were asked to dance with a
man a little the worse for wear. Usually your plight was noticed, and you were
rescued.

Jack loved his Army life. If you wanted to take your soldiering seriously, which
Jack did, and if you were reasonably intelligent, the thing to do was to try to go to
the Staff College at Camberley. By the time Jack had done his two years at
Camberley, the regiment had moved to Cairo. I shall never forget my arrival. I had
never been abroad, other than to Germany. When I stepped out of the train at
Cairo station, it was like stepping out into an heated greenhouse. I never got used
to the heat. If you asked me what I remembered about Egypt, I would say four
things. First, the dust, which was intolerable. Then the flies, horrid insects – I hated
seeing them crawling on the faces of the children, all around their eyes. Then the
heat; I hate being in a state of sweat most of the time. Last, but not least, I would

remember the golden maw and jacaranda trees, which were everywhere, often lining the streets. Sometimes they would be out together, which was incredibly beautiful.

When the regiment left Egypt for India, they went by troopship; this meant the entire regiment travelled together. It was quite the nicest sea voyage I ever had. By this time Jack was a major and second in command, so was becoming quite important. The voyage was extremely hot, especially going through the Red Sea. By this time I was used to the heat, and so arrival in Bombay was no shock to me. On the quay was Jack's old bearer to meet him. His name was Hafiz Abdul Rahman Khan, a Mohammedan; all the other servants were Hindu. A magnificent figure of a man, he had been Jack's bearer the last time he was in India. How he got to know of the arrival of the regiment, I do not know, but there he was, delighted to see Jack; he took all responsibility for us on the journey.

Our station in India was in the State of Hyderabad, and the barracks were at Secunderabad, which is, I think, about ten miles from the city of Hyderabad. We took a train from Bombay, which was in no way like a train in England, as there was no corridor. Each compartment had its own washbasin and loo. It was very hot indeed on the train and dreadfully dusty.

We had a very nice bungalow allotted to us. The garden was fairly large with a tennis court and stabling – known in India as a compound. I tried to grow a few vegetables, and keep a few hens, but without much success. We had nineteen servants to look after us, all men, except for the sweeper – and look after us they did. An Indian servant is excellent in every way. Why you have so many servants is because they will do only one job, such as wait at table. Everything ran on oiled wheels.

Regimental life went on very much as in Egypt; work in the morning, rest in the afternoon, probably polo after tea. I often wondered how we managed to do all we did on the income we had, but manage we did. I had hoped to find some other English family with children, who would let Anne, our daughter, do lessons with them. I could find no one. However I need not have worried as Corporal Thorpe volunteered to come up each afternoon for two hours to teach Anne, provided he could get permission from the Colonel, so that was fairly easy. I had no idea whether Corporal Thorpe was able to teach. It was a most happy arrangement, he was a born teacher. I heard the following conversation. 'Miss Anne, you are not trying at all. If you don't try harder, I shall have to tell the Colonel.' The answer came back short and sharp. 'Daddy won't mind. He lets me do anything.' This I repeated to Jack, hoping, perhaps, for a little more discipline.

When Jack became Colonel, I had to take on the responsibility of all the other ranks' wives. I often said it was like an unpaid curate. I think there were about sixty to seventy families. I had to go to the hospital to visit the sick, or to congratulate the parents of the new baby. I even had to sort out matrimonial affairs on several occasions. I was certainly no Solomon, but I did my best.

The time was drawing to a close when the regiment had to move again, this time under very tragic circumstances. The powers that be decided a cavalry regiment was no longer useful in modern warfare. It meant that when we left India, we had to say goodbye to our horses. This, to Jack, was a bitter blow. For years he had worked his way up to Colonel of a cavalry regiment, only to find he had to command a tank or armoured car regiment. It nearly broke his heart.

As Jack was commanding the regiment on arrival at Aldershot, we took over the Colonel's quarters. When I was settled in, I took up the responsibilities again of being Mrs Colonel. Poor Jack had to cope with trying to make first-class riders into first-class mechanics. Not an easy task. Their hearts were not in it, and, for heaven's sake, why should they have been?

By mistake a tank took on a small Austin car — quite close to barracks. The two passengers, a man and his wife, were very shaken. The Adjutant brought them to my house. I asked the lady what she would like to drink. She said 'Coffee.' That was easy. I asked the gentleman the same question and he said 'A double brandy.' I knew there was no brandy in the house. I rang the bell and our soldier servant appeared. I said I wanted a double brandy, thinking he would say there wasn't any. He didn't. He dashed round to the mess, got the brandy, and produced it without turning a hair. When the gentleman had finished his drink, I asked if he would like anything else. He said, 'Another double brandy.' I rang the bell and gave the order, hoping our soldier servant would come up trumps again, and he did!

A subaltern was taking a tank through the village of Cocking in Sussex, when he tried to turn a corner in the village. Nothing happened and it went straight through a hedge, and landed on a rose-bed in a cottage garden. The subaltern got out of the tank, covered in confusion. An old lady came out of the cottage, and when he started to apologise, the dear old lady said, 'Don't worry, sir. Nothing exciting has happened in Cocking for years!'

All through the summer of 1939 the talk of war was on everyone's lips. As usual, I took Anne to Cornwall. Jack came with us but every day he expected a telegram recalling him from leave. After a few days of trying to be cheerful the telegram did come. Jack opened it, and said, 'We leave for Aldershot in half an hour.'

The next few days were a nightmare. To begin with, as a fighting regiment we were quite useless. This, to all the officers and men was a bitter pill, having been one of the best cavalry regiments. What trained men we did have were removed to swell the ranks of another regiment. Then the tragedy of the reservists — absolutely first-class soldiers in every way — who flowed back to fight for their country – what could they do? They knew nothing about tanks, and most of them were too old to learn.

When war was finally declared, I remember I was driving home in my car when I heard the dreadful news on the radio. On my return I met Jack, who said, 'The regiment has had its marching orders to go near Wimborne, and you have been given exactly one week to get out of our house, lock, stock and barrel!' At the end of the week, I said goodbye to my staff, who, I might say, were in floods of tears. I said goodbye to Jack, who was making for Wimborne, and went to stay with my mother-in-law, until I could find somewhere to stay near Jack.

Index